GW01237790

Mass Movements in Great Britain

THE GEOLOGICAL CONSERVATION REVIEW SERIES

The comparatively small land area of Great Britain contains an unrivalled sequence of rocks, mineral and fossil deposits, and a variety of landforms that provide a geological recordof a large part of the Earth's long history. Well-documented ancient volcanic episodes, famous fossil sites, and sedimentary rock sections used internationally as comparative standards, have given these islands an importance out of all proportion to their size. The long sequences of strata and their organic and inorganic contents have been studied by generations of leading geologists, thus giving Britain a unique status in the development of the science. Many of the divisions of geological time used throughout the world are named after British sites or areas; for instance, the Cambrian, Ordovician and Devonian systems, the Ludlow Series and the Kimmeridgian and Portlandian stages.

The Geological Conservation Review (GCR) was initiated by the Nature Conservancy Council in 1977 to assess and document the most important parts of this rich heritage. The GCR records the current state of knowledge of the key Earth science sites in Great Britain and provides a firm basis upon which site conservation can be founded in years to come. Each GCR title in the series of over 45 volumes describes networks of sites of national or international importance in the context of a portion of the geological column, or a geological, palaeontological or mineralogical topic.

Within each volume, the GCR sites are described in detail in self-contained accounts, consisting of an introduction (with a concise history of previous work), a description, an interpretation (providing geological analysis of the features of interest and assessing the fundamentals of the site's scientific interest and importance), and a conclusion (written in simpler terms for the non-specialist). Each site report is a justification of the particular scientific interest in a locality, of its importance in a British or international setting, and ultimately of its worthiness for conservation.

The aim of the Geological Conservation Review Series is to provide a public record of the features of interest in sites that have been notified or are being considered for notification as Sites of Special Scientific Interest (SSSIs). The volumes are written to the highest scientific standards but in such a way that the assessment and conservation value of the site is clear. It is a public statement of the value placed on our geological and geomorphological heritage by the Earth Science community and it will be used by the Joint Nature Conservation Committee, the Countryside Council for Wales, Natural England and Scottish Natural Heritage in carrying out their conservation functions. The three country agencies are also active in helping to establish sites of local and regional importance. Regionally Important Geological/Geomorphological Sites (RIGS) augment the SSSI coverage, with local groups identifying and conserving sites that have educational, historical, research or aesthetic value, enhancing the wider Earth heritage conservation perspective.

All the sites in this volume have been proposed for notification as SSSIs; the final decision to notify sites lies with the governing councils of the appropriate country conservation agency.

Information about the GCR publication programme may be obtained from:

GCR Unit,
Joint Nature Conservation Committee,
Monkstone House,
City Road,
Peterborough PE1 1JY.

www.jncc.gov.uk

Copies of published volumes can be purchased from:

NHBS Ltd
2–3 Wills Road
Totnes
Devon TQ9 5XN.

www.nhbs.com

Published titles in the GCR Series

1. **An Introduction to the Geological Conservation Review**
   N.V. Ellis (ed.), D.Q. Bowen, S. Campbell, J.L. Knill, A.P. McKirdy, C.D. Prosser, M.A. Vincent and R.C.L. Wilson
2. **Quaternary of Wales**
   S. Campbell and D.Q. Bowen
3. **Caledonian Structures in Britain South of the Midland Valley**
   Edited by J.E. Treagus
4. **British Tertiary Volcanic Province**
   C.H. Emeleus and M.C. Gyopari
5. **Igneous Rocks of South-West England**
   P.A. Floyd, C.S. Exley and M.T. Styles
6. **Quaternary of Scotland**
   Edited by J.E. Gordon and D.G. Sutherland
7. **Quaternary of the Thames**
   D.R. Bridgland
8. **Marine Permian of England**
   D.B. Smith
9. **Palaeozoic Palaeobotany of Great Britain**
   C.J. Cleal and B.A. Thomas
10. **Fossil Reptiles of Great Britain**
    M.J. Benton and P.S. Spencer
11. **British Upper Carboniferous Stratigraphy**
    C.J. Cleal and B.A. Thomas
12. **Karst and Caves of Great Britain**
    A.C. Waltham, M.J. Simms, A.R. Farrant and H.S. Goldie
13. **Fluvial Geomorphology of Great Britain**
    Edited by K.J. Gregory
14. **Quaternary of South-West England**
    S. Campbell, C.O. Hunt, J.D. Scourse, D.H. Keen and N. Stephens
15. **British Tertiary Stratigraphy**
    B. Daley and P. Balson
16. **Fossil Fishes of Great Britain**
    D.L. Dineley and S.J. Metcalf
17. **Caledonian Igneous Rocks of Great Britain**
    D. Stephenson, R.E. Bevins, D. Millward, A.J. Highton, I. Parsons, P. Stone and W.J. Wadsworth
18. **British Cambrian to Ordovician Stratigraphy**
    A.W.A. Rushton, A.W. Owen, R.M. Owens and J.K. Prigmore
19. **British Silurian Stratigraphy**
    R.J. Aldridge, David J. Siveter, Derek J. Siveter, P.D. Lane, D. Palmer and N.H. Woodcock
20. **Precambrian Rocks of England and Wales**
    J.N. Carney, J.M. Horák, T.C. Pharaoh, W. Gibbons, D. Wilson, W.J. Barclay, R.E.Bevins, J.C.W. Cope and T.D. Ford

Published titles in the GCR Series

21. **British Upper Jurassic Stratigraphy (Oxfordian to Kimmeridgian)**
J.K. Wright and B.M. Cox

22. **Mesozoic and Tertiary Palaeobotany of Great Britain**
C.J. Cleal, B.A. Thomas, D.J. Batten and M.E. Collinson

23. **British Upper Cretaceous Stratigraphy**
R. Mortimore, C. Wood and R. Gallois

24. **Permian and Triassic Red Beds and the Penarth Group of Great Britain**
M.J. Benton, E. Cook and P. Turner

25. **Quaternary of Northern England**
D. Huddart and N.F. Glasser

26. **British Middle Jurassic Stratigraphy**
B.M. Cox and M.G. Sumbler

27. **Carboniferous and Permian Igneous Rocks of Great Britain North of the Variscan Front**
D. Stephenson, S.C. Loughlin, D. Millward, C.N. Waters and I.T. Williamson

28. **Coastal Geomorphology of Great Britain**
V.J. May and J.D Hansom

29. **British Lower Carboniferous Stratigraphy**
P.J. Cossey, A.E. Adams, M.A. Purnell, M.J. Whiteley, M.A. Whyte and V.P. Wright

30. **British Lower Jurassic Stratigraphy**
M.J. Simms, N. Chidlaw, N. Morton and K.N. Page

31. **Old Red Sandstone Rocks of Great Britain**
W.J. Barclay, M.A.E. Browne, A.A. McMillan, E.A. Pickett, P. Stone and P.R. Wilby

32. **Mesozoic and Tertiary Fossil Mammals and Birds of Great Britain**
M.J. Benton, E. Cook and J.J. Hooker

33. **Mass Movements in Great Britain**
R.G Cooper

Mass Movements in Great Britain

R.G. Cooper
(1949–2001)

Bournemouth University,
Bournemouth, England, UK

with contributions from

C.K. Ballantyne
University of St Andrews,
St Andrews, Scotland, UK

and

D. Jarman
Mountain Landform Research,
Ross-shire, Scotland, UK

GCR Editors:

D. Brunsden
King's College, London, England, UK

and

V.J. May
Bournemouth University,
Bournemouth, England, UK

Published by the Joint Nature Conservation Committee, Monkstone House, City Road, Peterborough, PE1 1JY, UK

First edition 2007

Typeset in 10/12pt Garamond ITC by JNCC
Printed in Great Britain by Hobbs The Printers, Totton.

ISBN 1 86107 481 6

A catalogue record for this book is available from the British Library.

Recommended example citations

Cooper, R.G. (2007) *Mass Movements in Great Britain*, Geological Conservation Review Series, No. 33, Joint Nature Conservation Committee, Peterborough, 348 pp.

Jarman, D. (2007) Druim Shionnach, Highland. In *Mass Movements in Great Britain* (R.G. Cooper), Geological Conservation Review Series, No. 33, Joint Nature Conservation Committee, Peterborough, pp. 70–4.

Contents

Contents

Contents

Acknowledgements

Work on the British mass movement part of the GCR project was initiated by the Nature Conservancy Council (which existed up to 1991) and has been seen through to publication by the Joint Nature Conservation Committee on behalf of the three country agencies, Countryside Council for Wales, Natural England, Scottish Natural Heritage. The late George Black, Head of the Geology and Physiography Section of the Nature Conservancy Council, initiated the GCR project in 1977, and the 'Mass Movements', GCR *Block* (site-selection category) was considered early on in the site-selection programme. In the site-assessment and GCR site-selection phase of the project, work was co-ordinated by the late Roger Cooper, with guidance from Bill Wimbledon (then Head of the GCR Unit).

Many specialists were involved in the assessment and selection of sites, and this vital work is gratefully acknowledged.

JNCC invited Roger to build on his site-selection work, and to undertake the preparation of a text for publication for JNCC in the late 1990s, and later invited Denys Brunsden to assist Roger during the writing stages of the book in an editorial capacity.

Following Roger's death, Denys kindly took on the job of bringing the book to a publication-ready state, a task which is not to be understated, considering the amount of material that had to be sifted, which Roger had accumulated to inform his writing work. At a late stage in the preparation of the volume for publication, Vincent May, with help from Rebecca Cook (who catalogued Roger's GCR papers) also became involved, helping to ensure that the illustrative material for the book was completed, and providing some editorial input.

JNCC is grateful to all of these contributors who assisted in bringing this volume to completion, and the GCR Publication Production Team – Neil Ellis, GCR Publications Manager, and Anita Carter and Emma Durham, GCR Production Editors – are especially grateful to all of those who provided guidance while resolving technical issues during their production work.

Colin Ballantyne would like to acknowledge the collaboration of Dr John Stone of the University of Washington in carrying out dating of rock slope failures in the Scottish Highlands using cosmogenic radionuclides.

David Jarman would like to thank John Gordon (SNH) and John Mendum (BGS, Edinburgh) for their considerable assistance in the site-assessment process in the Highland areas and thereafter in advising on the site reports prepared for this volume.

Acknowledgements

The diagrams were produced to a high standard by JS Publications of Newmarket by Drs Susanne White and Chris Pamplin.

Where the content of illustrations has been replicated or modified from the work of others appropriate acknowledgements are given in the captions. The National Grid is used on diagrams with the permission of the Controller of Her Majesty's Stationery Office, © Crown copyright licence no. GD 27254X/01/00. Photographs are accredited in the captions.

Access to the countryside

This volume is not intended for use as a field guide. The description or mention of any site should not be taken as an indication that access to a site is open. Most sites described are in private ownership, and their inclusion herein is solely for the purpose of justifying their conservation. Their description or appearance on a map in this work should not be construed as an invitation to visit. Prior consent for visits should always be obtained from the landowner and/or occupier.

Information on conservation matters, including site ownership, relating to Sites of Special Scientific Interest (SSSIs) or National Nature Reserves (NNRs) in particular counties or districts may be obtained from the relevant country conservation agency headquarters listed below:

Countryside Council for Wales,
Maes-y-Ffynnon,
Penrhosgarnedd,
Bangor,
Gwynedd LL57 2DW.

Natural England,
Northminster House,
Peterborough PE1 1UA.

Scottish Natural Heritage,
Great Glen House,
Leachkin Road
Inverness IV3 8NW.

Foreword

Dr Roger Cooper
1949–2001

Born in Camberley, Surrey, Roger read Geography at the University of Hull. After completing his PhD entitled '*Geomorphological Studies: the Hambleton Hills, North Yorkshire*' in 1979, he joined the Geography Group at the then Dorset Institute of Higher Education in Bournemouth. In 1981, he began the not inconsiderable task of preparing the Landslides (Mass Movements) 'Block' for the Geological Conservation Review (GCR). By the mid-1980s, strategic changes to the departmental and curriculum structure led to Roger enjoying a rare year's sabbatical at Birkbeck College studying Geographical Information Systems under David Rhind (later Director General of the Ordnance Survey). He then provided all GIS teaching in Bournemouth's new and highly regarded MSc in Coastal Zone Management. However, soon after he returned to Bournemouth, Roger's health deteriorated and he had a brain tumour removed in September 1988. He regarded this as merely an inconvenience as he embarked on his contribution to the GCR; over the next twelve years he explored, mapped, investigated and described mass-movement features throughout Great Britain. Roger visited many of the sites identified in his initial consultation with colleagues throughout the world, surveying and mapping some for the first time. Resulting from this work, he produced the original site list that became the Mass Movements GCR Block. He brought to the description and justification for selection of the sites his usual meticulous attention to detail. Neither institutional circumstances nor his health made his task easy. Very few of his colleagues even knew that he was working on it, and yet he persevered, sometimes with little encouragement (since, by the mid-1990s, it was not regarded as relevant to his teaching), apart from those close to him.

Roger had a sharp analytical mind and he used it, not least, to approach institutional decision-making in the same spirit of peer review expected in any scholarly work. This did not always make for easy relationships, but Roger's involvement at a grass-roots level in the developments that led to the rapid transformation from Dorset Institute to Bournemouth University should not be underestimated.

Roger was an enthusiast. He was insatiably curious and enthused colleagues and students alike with a sense of excitement at discovery and gaining understanding. For many years, he edited one of the main journals in cave studies,

Foreward

'*Studies in Speleology*'. He was deeply involved in the Pengelly Cave Trust and was a caver himself. He explored and described caves on the Isle of Portland that result from the gradual toppling seawards of the limestone. He researched and wrote. When he discovered that his grandfather had been caught up in the Boxer Rebellion in China, he set out to find out about the exact circumstances and published an account of it. When he found out that John Wesley had described a Yorkshire landslide, he went back to the records and worked out how well they helped to date the landslide event.

Roger was principled, caring, precise in all his work and full of sharp wit. But these were nothing without his friendship, his intellectual and physical energy, and his belief in the future: *and* his ability to share those qualities. This volume is an appropriate memorial for a man who was above all a scholar with integrity and a sense of conviction about the place of scholarship in the world.

V.J.May
February 2007

Preface

There is such a diversity of rocks, minerals, fossils and landforms packed into the piece of the Earth's crust we call 'Britain' that it is difficult to be unimpressed by the long, complex history of geological change to which they are testimony. But if we are to improve our understanding of the nature of the geological forces that have shaped our islands, further unravel their history in 'deep time' and learn more of the history of life on Earth, we must ensure that the most scientifically important Earth science sites, which offer us evidence, are conserved for future generations to study, research and enjoy. Moreover, as an educational field resource and as training grounds for new generations of geologists on which to hone their skills, it is essential that such sites continue to remain available for study. The first step in achieving this goal is to identify the key sites, which was a primary aim of the Geological Conservation Review.

The GCR, launched in 1977, is a world-first in the systematic selection and documentation of a country's best Earth science sites. No other country has attempted such a comprehensive and systematic review of its Earth science sites on anything near the same scale. After over two decades of site evaluation, consultation with the scientific community, and site documentation, we now have an inventory of over 3000 GCR sites, selected for 100 categories covering the entire range of the geological and geomorphological features of Britain.

The minimum criterion for GCR site selection was that sites should offer the finest and/or the most representative feature for illustrating a particular aspect of geology or geomorphology. The resulting GCR sites are thus, at the very least, of national scientific importance and many of these include features regarded as either 'classic' (i.e. a 'textbook example'), internationally important or simply 'unique'. Some are, in addition, visually spectacular. Others, though less spectacular, are of considerable importance in demonstrating a particular aspect of geology or geomorphology.

The present volume is the 33rd to be published in the GCR series of books, which will be completed in over 40 volumes. It represents the results of the GCR assessment and selection programme of British Mass Movement sites conducted in the early 1980s, in describing the ultimately selected sites. These localities will be conserved for their contribution to our understanding of mass-movement processes and their manifestations. This volume summarizes the considerable research that has been undertaken on the localities. The book will be invaluable as an essential reference book to those engaged in the study of these sites and will provide a stimulus for further investigation. It will also be helpful to teachers and lecturers and for those people who, in one way or another, have a vested interest in the GCR sites: owners, occupiers, planners and, those concerned with the practicalities of site conservation. The conservation value of the sites is mostly based on a specialist understanding of the Earth science features present and is, therefore, of a technical nature. The account of each site ends, however, with a brief summary of the geomorphological interest, framed in less technical language, in order to help the non-specialist. The first chapter of the volume, used in conjunction with the glossary (contained within Chapter 1), is also aimed at a less specialist audience.

This volume deals with the state of knowledge of the sites available at the time of writing, which for the material written by the late Roger Cooper was between 1996–2000, and it must be seen in this context, although some editorial work was kindly undertaken by the editors to introduce references to more-recent publications about the sites.

However, mass-movements studies, like any other science, are ever-developing, with new discoveries being made, and existing models being subject to continual testing and modification as new data comes to light. Increased or hitherto unrecognized significance may be seen in new sites. Indeed, more recent research into Highland mass movements, separate from the original GCR writing work undertaken by Roger, has provided important new information about Scottish sites, which has been translated into up-to-date text for Chapter 2, and reports in Chapters 4 and 6, by David Jarman and Colin Ballantyne. Therefore, it is possible that further sites worthy of conservation will be identified in future years for the study of mass movements in Britain, as research continues. However, it must be stressed that the GCR is intended to be a *minimalist* scheme, with the selection for the GCR of only the best, most representative, example of a geological feature, rather than the selection of a series of sites showing closely analogous features.

Nevertheless, there is still much to learn about the GCR sites documented here, many of which are as important today – in increasing our knowledge and understanding of mass-movement processes – as they were when they were first selected.

This account will clearly demonstrate the value of British sites to mass-movement studies and the importance of the sites within the wider context of Britain's outstanding scientific and natural heritage.

N.V. Ellis,
GCR Publications Manager
January 2007

Chapter 1

Introduction

R.G. Cooper

MASS MOVEMENTS IN CONTEXT

Mass movements in the British context

Jones and Lee (1994) describe 'mass movement' as 'a broad spectrum of gravity[-driven] slope movements', of which the larger discrete movements are generally described as 'landslides'.

These mass-movement phenomena are a major influence on much of the landscape of Great Britain, but vary considerably in scale. Some mass-movement processes are shallow (operating near the land surface), slow, and affect large areas. For example, 'soil creep' has been taking place on nearly all terrestrial slopes since the retreat of glaciers during the Devensian Stage (the last glacial period of the Pleistocene Epoch in Britain, which ended about 11 500 calendar years ago). Similarly, many landscapes (e.g. Dartmoor) are mantled by 'solifluction' sheets of sediment (slow downslope-moving saturated soil or rock debris), the process that created them usually being ascribed to former periglacial (tundra-like) climatic conditions.

At the other end of the scale are deep-seated 'landslides', whose occurrence under *present* climatic conditions in Britain is relatively rare both areally and temporally (except on particular stretches of the east and south coasts). Many of these have clearly taken place in the past under conditions more conducive to mass movements and are very widespread. These mass-movement features are the principal subject of the present volume.

A study undertaken in 1984–1987 for the former Department of the Environment (DoE) by Geomorphological Services Ltd (GSL; published in 1988) in association with Rendel Palmer & Tritton, that produced an inventory of 8835 landslides in Great Britain has been analysed by Jones and Lee (1994). A major conclusion drawn from the analysis is that most inland landslides in Great Britain are relict but dormant (i.e. capable of being re-activated by engineering works, building or other disruptive activities). In contrast, coastal landsliding is a present-day process, possibly associated with rising sea level and drainage.

A particular value of the inventory has been the provision of information for local and regional planners (Clark *et al.*, 1996), who have to deal with the consequences of landslides – present-day, recent and relict – in relation to land-development applications.

No other Geological Conservation Review (GCR) volume has had the benefit of such a major survey of the features with which it is concerned carried out by another organization at a critically important time. The survey took place around the time of the period of GCR fieldwork in the 1980s. The GSL survey, however, was concerned with landslides that have been mentioned or shown in documents. Therefore it does not purport to be a complete inventory of known landslides in Great Britain. The distribution of landslides identified by the survey, described as 'ancient' and 'youthful', is shown in Figure 1.1.

The present writer (RGC) was involved as a collector of data for north-east England in the GSL exercise, which led to the production of the distribution map (Figure 1.1). It was clear that when plotted on a 1:125 000 scale map, the distribution of reported landslides in north-east England alone was likely to be a very poor representation of the true distribution of landslides actually identifiable in the field. A major reason for this was that landslides were not recorded equally well in the different surveyed areas, creating apparent, but not actual, demarcations of high- and low-density areas of landslides (see below). The demarcations, as recorded in the literature, often correlated to the boundaries between the various map areas of individual British Geological Survey maps, memoirs, and Mineral Assessment Reports. Examples included a cluster of landslides in North Yorkshire immediately west of Ripon, another around Barnard Castle in County Durham, and a group around Bellingham in Northumberland. Since it is unlikely that landslide density correlates to British Geological Survey map-sheet areas, this must indicate unevenness in the documentation between the sheets, memoirs or reports for adjacent areas. There are several reasons for such unevenness:

(a) It is clear that for areas surveyed up to some time in the 1930s, landslides were simply not marked on the resulting published British Geological Survey sheets. Examination of the one-inch sheets of the North York Moors area produced from surveys made by C. Fox-Strangways between 1880 and 1910 reveals no landslides at all. Yet the six-inch maps from which they were

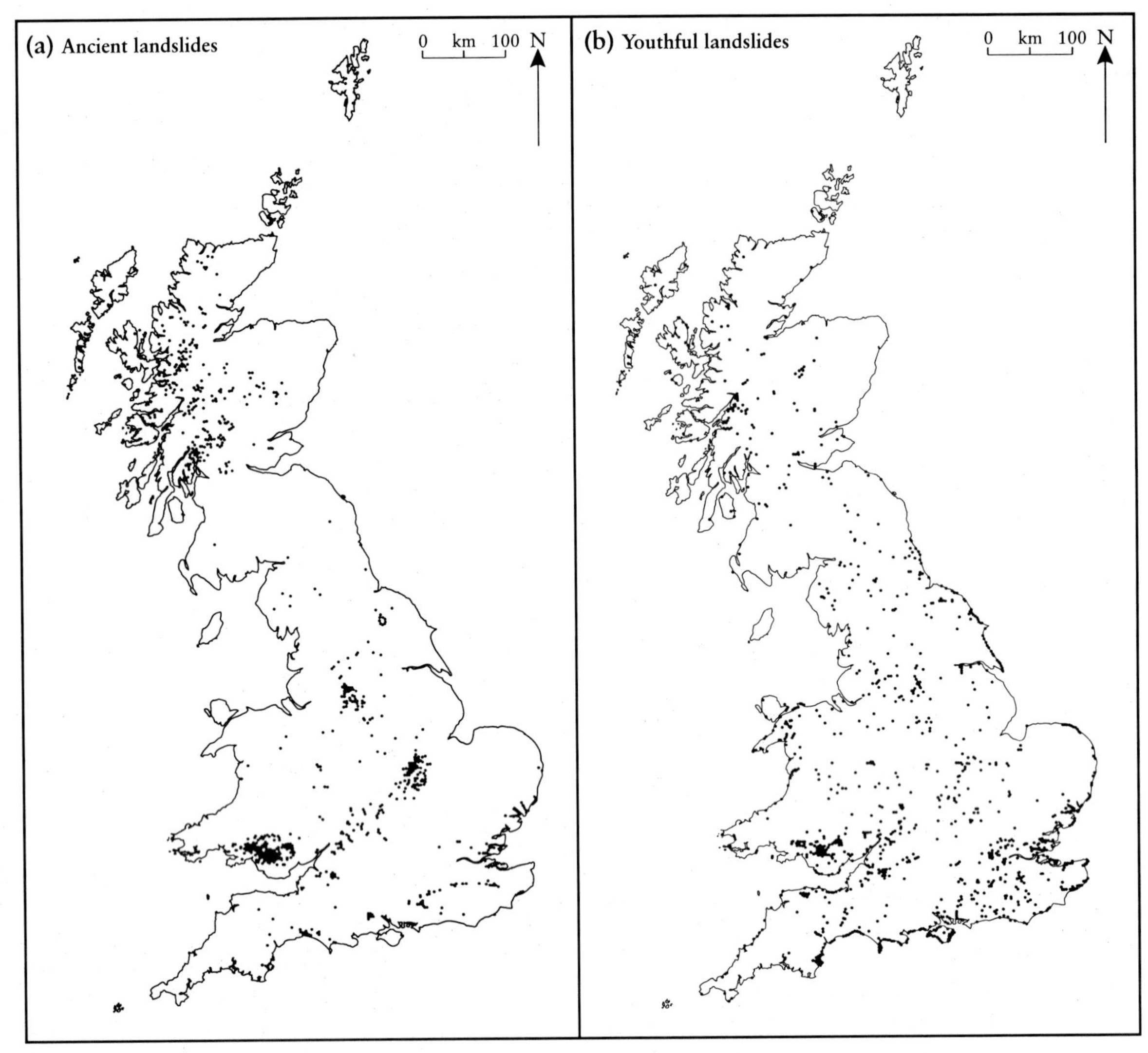

Figure 1.1 The distribution of (a) ancient and (b) youthful landslides in Great Britain recorded as a result of the survey of landslides in Great Britain commissioned by the former Department of the Environment (DoE), and completed by Geomorphological Services Ltd (GSL) between 1984 and 1987. After Jones and Lee (1994).

compiled show a very large number of slides. These do not have marked boundaries: the word 'slip' is written across the relevant slope; nevertheless, they were identified, marked and recorded by the surveyor.

(b) Even though it has been British Geological Survey practice to mark landslides on the one-inch and 1:50 000 sheets, some sheets specifically exclude them. These include one-inch sheet 50 (Hawes), published in 1971, which has a note appended to the legend: 'N.B. Landslips are not indicated on this map'.

(c) The interests and aptitudes of each surveyor have had an influence. About half of the currently available British Geological Survey memoirs have 'landslip' in their indexes. Where they do not, either there are none in the area concerned, or the surveyors were not interested in such phenomena. The latter is the present writer's (RGC's) explanation of the low number of recorded landslides in Nottinghamshire (seven in table 2.2 of Jones and Lee, 1994). The county was surveyed by geologists selected for their expertise in coal and economic geology; landslides were, therefore, not a major concern to them. The memoirs that they produced concentrate heavily on phenomena at depth; surficial geology is assigned a very minor role.

Taking a national view, even where 'landslip' can be found in the index of a British Geological Survey memoir, recorded coverage of such mass movements is very variable. The area under consideration may contain a single large and obvious slide, recorded as such and which a surveyor could hardly fail to mention, but smaller movements in the region may be overlooked or neglected. Alternatively, an area may have no *major* mass movements, but the surveyor may have a particular interest in landslides, perhaps because of Quaternary research interests, and so may record comparatively many more occurrences. Differences in the interests and aptitudes of surveyors is the only tenable explanation of why some of the recently surveyed one-inch and 1:50 000 sheets of Northumberland are replete with landslides (sheet 13 (Bellingham) has 71) while some of the adjacent sheets, with similar geology and comparable terrain – but different surveyors – show none at all.

Similar observations were made about other parts of the country by other collectors of data for the exercise, leading Jones and Lee (1994) to observe that 'the patchiness of the distribution raises questions as to the extent to which the concentrations displayed in the map [here Figure 1.1] reflect the true pattern of landslides on the ground as against spatially variable reporting'. They continue:

'It now seems certain that the pattern merely highlights those landslides which happen to have been investigated, mapped and reported, and the extent to which the total available corporate knowledge of landsliding was tapped by the survey. It is undoubtedly true that many reports of landslides published in obscure journals and old newspapers were not accessed by the survey, and the same is true of the data held in the files of numerous individual professionals, companies and even some national organizations. It must also be stressed that there must be numerous other landslides that have not yet been recorded because they exist in remote areas, are concealed by woodland, are relatively insignificant or have yet to be actually recognized as landslides. This is clearly illustrated by the results of the ... Applied Earth Science Mapping of the Torbay area (1988) which raised the total of known and reported landslides from 4 to 304. Even in the South Wales Coalfield, which has been the subject of a major landslide inventory exercise by the British Geological Survey, a detailed mapping programme in the Rhondda valleys resulted in an increase in the number of recorded landslides from 102 to 346. Clearly, in some areas, the harder you look the more examples you find. Indeed, extrapolation leads to the inevitable conclusion that the actual number of landslides in Great Britain is many times in excess of the 8835 recorded so far by this survey.'

The Torbay study referred to is described in Geomorphological Services Ltd (1988) and Doornkamp (1988). As stated by Jones and Lee (1994), the pattern of landslides displayed on the map (Figure 1.1) must be treated with caution in that it reflects under-representation of the true pattern, as an artefact of investigative interests and recording bias.

Mass movements in the European context

A large amount of research has been carried out on mass movements in Europe, particularly in relation to three broad factors: climate, topography and geology. It is worth noting, however, that this tripartite division does not create exclusive, distinct categories. Geology and topography, in particular, are intimately linked, with the geology (lithology and structure) controlling the topography in some detail. Also, few mass movements can be ascribed to a single causal factor, or even to a single type of causal factor.

(a) Climatic factors

Two initiatives by the European Commission (EC) have been concerned with Europe-wide collection and analysis of data on mass movements: the EPOCH project (Temporal Occurrence and Forecasting of Landslides in the European Community) and the TESLEC project (The Temporal Stability and Activity of Landslides in Europe with Respect to Climatic Change). The EPOCH project collected data on the past occurrence and frequency of landsliding in Europe. The UK EPOCH team then went on to extract from these data changes of geomorphological activity that may be related to climatic change in the last 20 000 years. These results are summarized in Figure 1.2 for Holocene times

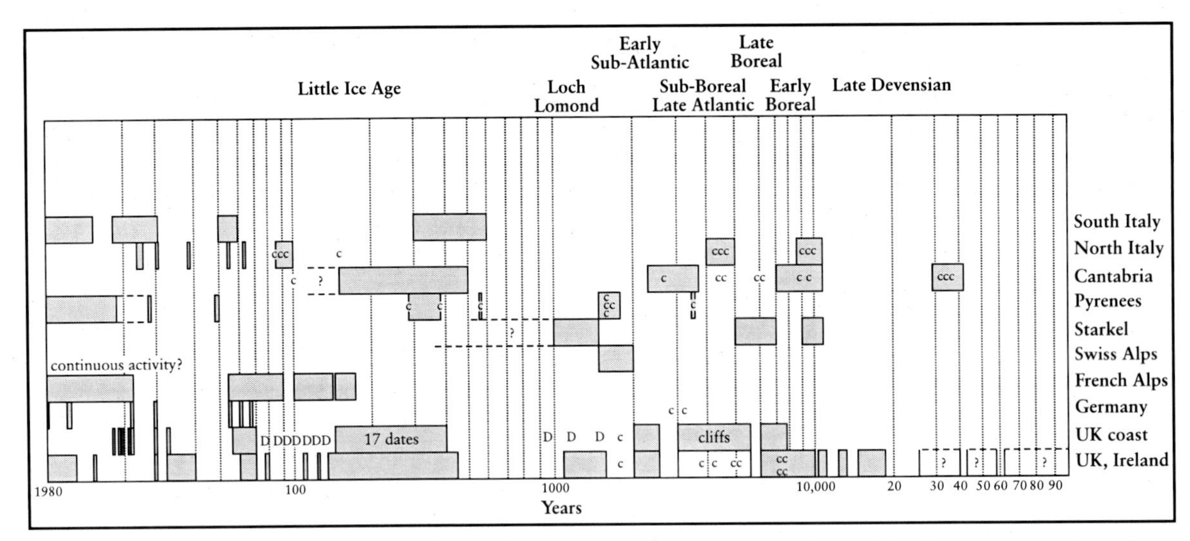

Figure 1.2 Indicative periods of major landslide activity in Europe derived from EPOCH data (c = radiocarbon (C^{14}) dates; D = important individual dates from the historical record). After Brunsden and Ibsen (1997).

based on dates of named landslides in the UK and for selected European countries, respectively (Brunsden and Ibsen, 1997; Ibsen and Brunsden, 1997). They suggest that landslide activity may be related to specific climatic periods and that the existing knowledge of this could be substantially improved.

The TESLEC project has been concerned with the effects and modelling of climate change on mass movements in Europe. This has involved continued collation of data on the past distribution of landslides. This work has shown that there are few decades when landslide events have not occurred in certain regions such as the Spanish Pyrenees and Barcelonette in the French Alps (Brunsden *et al.*, 1996a).

The 'Landslide Recognition' survey (Dikau *et al.*, 1996; the production of the survey was an initial objective of the TESLEC project), and the DoE review revealed that in Great Britain the major problem with respect to climate change scenarios is the potential for the re-activation of dormant landslide complexes, rather than the potential for first-time slides. By far the biggest potential problem is the possibility that climate change might generate widespread movement in the very large landslide complexes that lie at the foot of many of the escarpments in Mesozoic strata. These complexes are, however, rather rare on Chalk, for example the Chiltern Hills, Salisbury Plain and the Marlborough Downs, and along the North and South Downs. However, where clay is exposed at the base of a slope in Chalk strata, occasional landslide complexes are to be found, for example the Castle Hill landslide at the entrance of the Channel Tunnel, the coastal termination of the North Downs at **Folkestone Warren** (see Chapter 7), the Dorset coast, and inland at Birdsall Brow in North Yorkshire.

(b) Topographical factors

The investigations of climatic factors also required other influences on mass movements to be identified, such as unloading, sea-level rise, seasonal ground freezing, and caprock loading (Brunsden and Ibsen, 1997). However, the European chapters of a worldwide survey of the extent and economic significance of landslides (Brabb and Harrod, 1989) point to the over-riding importance of individual high-intensity weather events, rather than climatic trends, as a precipitating factor for many mass movements. As stated, this has happened in Great Britain, but only rarely. However, a landslip in Yorkshire in 1755 took place 149 days after a very high-intensity, short-duration, small area rainfall event. The reason for this time-lag is not known, but it is within the range of flow-through times from precipitation of water on the ground surface to its emergence at groundwater-fed springs in the area. An earthquake can probably be ruled out as the immediate trigger of this landslide (Cooper, 1997).

France (Flageollet, 1989) has important topographical differences from Great Britain, including relatively new mountain ranges with steep, high slopes, such as the Alps and the Pyrenees. This topography leads to a similar range of types of mass movements to Great Britain, but the proportions are different. Particularly instructive are the variations of style of failure along the Bessin Cliffs on the north coast of Normandy, at Pointe du Hoc, Raz de la Percée, le Bouffay, le Chaos and Cap Manvieux (Maquaire and Gigot, 1988).

(c) Geological factors

Generally, where areas of Europe have geological situations not found in Great Britain, they also have types of mass movement not found there. The most obvious case is the quickclay deposits of Norway and Sweden, where marine clays deposited during Holocene times have been uplifted isostatically to become part of the land surface (Gregersen and Sandersen, 1989). Since emerging from the sea, these deposits have been subject to subaerial erosion, and the salt water has leached out from the soil matrix. This leaching increases the sensitivity (s_t) of the clay from, typically, an s_t value of 3–6, to a value greater than s_t = 20. When the salt content falls below 1 g l^{-1}, the clay becomes a quickclay. An unleached marine clay remains plastic on re-moulding, but a quickclay can transform into a liquid (Bjerrum, 1954; Bjerrum *et al.*, 1969).

Norway also has a large area of hard-rock mountains, liable to rockfalls and rockslides, rather like the Scottish Highlands.

Mass movements in the global context

From the global perspective, mass movements in Great Britain are unremarkable for their size, frequency, the hazards they pose, and their overall variety. They are, perhaps, remarkable for the small proportion that are currently active, and conversely for the large proportion that are generally attributed to past climatic conditions rather than those of the present day, in particular periglacial conditions and immediate post-glacial conditions.

The reasons for this limited manifestation of mass movements here are not hard to find. The British Isles are not located close to a tectonic plate boundary, and have not been subject to volcanic activity or significant seismic activity for many millions of years, so these potential triggering mechanisms do not play a major role. Isostatic re-adjustment to the melting of the most recent Quaternary glaciers, and retreat of the last ice-sheet, may still produce minor earthquakes, but these are infrequent, and hardly of significance even in the triggering of the few active mass movements that are located in the Scottish Highlands.

The principal limiting factor for British mass movements seems to be available relief. The highest point in Great Britain, the summit of Ben Nevis, reaches only 1343 m above OD; long, steep slopes are, therefore, something of a rarity. However, Great Britain's steep slopes in upland areas have been the sites of debris flows, but comparatively only a few rockfall avalanches.

Great Britain's temperate maritime climate has seldom produced extremes of rainfall capable of giving rise to a large-scale spate of mass-movement events in an area over a timescale of a few hours. This has been known to happen, for example on Exmoor in August 1952 (Gifford, 1953; Delderfield, 1976), but such events are very uncommon, although climate change may increase their frequency in the future. Likewise, although solifluction has been an important mass-movement process across much of Great Britain, it is only in the extremes of upland Britain that conditions are sufficiently cold for periglacial processes such as solifluction to be currently taking place (Ballantyne and Harris, 1994).

On the other hand, Great Britain has an immense variety of landforms, which occur on bedrock of varied geological age and reflect differences in lithology (rock-type) and geological structure (such as faults and folds). This variety has given rise to slopes that, in valley-sides, expose rocks of greatly varying resistance to erosion, and so produce a variety of degrees of slope steepness. Many slopes of the upland areas have been steepened relatively recently by glacial erosion. Likewise many kilometres of the British coast consist of vertical or sub-vertical cliffs, of variable height. As would be expected, these features have led to the development of a great number of mass-movement sites. While the majority are unremarkable, collectively they demonstrate a variety of features associated with Quaternary erosion, scarp retreat, and landscape shaping.

There is, however, one aspect of the mass movements in Great Britain that has had a substantial, and possibly disproportionate, influence globally. This is their role in the development of knowledge of mass movements, their mechanisms and their countermeasures (Hutchinson, 1984). Thus, one of the mass-movement sites chosen for the GCR includes, arguably, the first ever large-scale landslide to be described by geologists, the Bindon landslide, part of the **Axmouth–Lyme Regis** mass-movement GCR site. In addition, a long series of studies of the behaviour of London Clay (Eocene-age deposits) has illuminated the mass-movement behaviour of all clay strata. More recently, the recognition of toppling as a separate and distinct type of slope failure has depended upon the study of British examples.

CLASSIFICATION OF MASS-MOVEMENT TYPES

The classification of mass movements into types has attracted much attention since the suggestions made in 1938 by Sharpe. Classification is dealt with in some detail in the present chapter. However, characterizing landslide type, while important scientifically, has not been the sole consideration in the selection of mass-movement GCR sites, some of which were selected on the basis of the presence at a site of an atypical or otherwise particularly interesting feature or group of features.

The classification system of mass-movement features adopted for the purposes of selecting mass-movement GCR sites in the 1980s, was originally that of Hutchinson (1968a), the overall breadth of which, including creep, frozen-ground phenomena and landsliding, indicated a convenient scope to adopt for the term 'mass movement' (Table 1.1a). Hutchinson (1968a) makes a most significant point about mass movement: 'mass movements exhibit great variety, being affected by geology, climate and topography, and their rigorous classification is hardly possible'.

Despite this general proviso, several classifications were published in the 1970s, 1980s and 1990s, including those of Zaruba and Mencl (1969), Varnes (1978), Brunsden (1979), Selby (1982), Geomorphological Services Ltd (in Jones and Lee, 1994), Hutchinson (1988; see Table 1.1b) and most notably *The Multilingual Landslide Glossary* developed by the International Geotechnical Societies' UNESCO Working Party for World Landslide Inventory (WP/WLI, 1993; see below, where the glossary is reproduced in full). Where possible the recommendations and terminology of this last-mentioned group are now used throughout this volume in order to follow international practice.

All such classifications are to some extent imperfect, in that any classification of mass movements is essentially trying to divide a continuum into classes, raising the obvious difficulty of locating the distinguishing boundaries between types. Furthermore, many sites incorporate a number of different features of a variety of mass-movement types belonging to different classes. Thus placing sites into a particular category is subject to opinion.

The introduction to Chapter 2 of the present volume addresses this difficulty in respect of the old hard rocks of the British mountains, adapting the Hutchinson (1988) schema to more specific circumstances.

Arguably, the classification used by civil engineers gives perhaps the most clearly defined and separate 'types', as it is a classification not of mass-movement types, but of *failure* types (e.g. in Hoek and Bray, 1977).

The Multilingual Landslide Glossary

The Multilingual Landslide Glossary is an international standard for the description of landslides (WP/WLI, 1993; Cruden *et al.*, 1994). Its English version is given here in full (see also Dikau *et al.*, 1996). The glossary is available in Arabic, Chinese, English, French, German, Hindi, Italian, Japanese, Persian, Russian, Spanish, Sinhala, and Tamil. While giving a comprehensive glossary of terms for the various features of a landslide (Figures 1.3 and 1.4), it also divides landslides into five types: fall, topple, slide, spread and flow (Figure 1.5). Each of these types is modified by other qualities, of which two are of particular relevance to the classification: distribution of activity (seven qualifiers of type; Table 1.1b, Figure 1.6), and style of activity (five qualifiers of type; Figure 1.7). This leaves two that are of less relevance to classification of 'type': dimensions (Figure 1.8), and state of activity (see Figure 1.5).

Therefore there are 175 ($5 \times 7 \times 5$) theoretically possible types. Of these, the editors and contributors to *Landslide Recognition* (Dikau *et al.*, 1996) choose to describe 15, which leaves one to speculate on the actual existence of field examples of the remaining 160.

Table 1.1 (a) Hutchinson's classification of mass movements on slopes (1968a); (b) Hutchinson's (1988) classification (first two levels only).

(a)

Group	Type	Subtype
CREEP	(1) Shallow, predominantly seasonal creep	
		(a) Soil creep
		(b) Talus creep
	(2) Deep-seated continuous creep; mass creep	
	(3) Progressive creep	
FROZEN GROUND PHENOMENA	(4) Freeze–thaw movements	
		(a) Solifluction
		(b) Cambering and valley-bulging
		(c) Stone streams
		(d) Rock glaciers
LANDSLIDES	(5) Translational slides	
		(a) Rock slides; block glides
		(b) Slab, or flake slides
		(c) Detritus, or debris slides
		(d) Mudflows
		(i) Climatic mudflows
		(ii) Volcanic mudflows
		(e) Bog flows; bog bursts
		(f) Flow failures
		(i) Loess flows
		(ii) Flow slides
	(6) Rotational slips	
		(a) Single rotational slips
		(b) Multiple rotational slips
		(i) in stiff, fissured clay
		(ii) in soft, extra-sensitive clays; clay flows
		(c) Successive, or stepped rotational slips
	(7) Falls	
		(a) Stone and boulder falls
		(b) Rock and soil falls
	(8) Sub-aqueous slides	
		(a) Flow slides
		(b) Under-consolidated clay slides

(b)

Class	Type	No.	Subtype
A	Rebound		
		1	Movements associated with man-made excavations
		2	Movements associated with naturally eroded valleys
B	Creep		
		1	Superficial, predominantly seasonal creep; mantle creep
		2	Deep-seated, continuous creep; mass creep
		3	Pre-failure creep; progressive creep
		4	Post-failure creep
C	Sagging of mountain slopes		
		1	Single-sided sagging associated with the initial stages of landsliding
		2	Double-sided sagging, associated with the initial stages of double landsliding, leading to ridge spreading
		3	Sagging associated with multiple toppling
D	Landslides		
		1	Confined failures
		2	Rotational slips
		3	Compound failures (markedly non-circular, with listric or bi-planar slip)
		4	Translational slides
E	Debris movements of flow-like form		
		1	Mudslides (non-periglacial)
		2	Periglacial mudslides (gelifluction of clays)
		3	Flow slides
		4	Debris flows, very to extremely rapid flows of wet debris
		5	Sturzstroms, extremely rapid flows of dry debris
F	Topples		
		1	Topples bounded by pre-existing discontinuities
		2	Topples released by tension failure at rear of mass
G	Falls		
		1	Primary, involving fresh detachment of material; rock and soil falls
		2	Secondary, involving loose material, detached earlier; stone falls
H	Complex slope movements		
		1	Cambering and valley-bulging
		2	Block-type slope movements
		3	Abandoned clay cliffs
		4	Landslides breaking down into mudslides or flows at the toe
		5	Slides caused by seepage erosion
		6	Multi-tiered slides
		7	Multi-storeyed slides

Landslide features (Figure 1.3)

(1) ***Crown***: the practically undisplaced material still in place and adjacent to the highest parts of the *main scarp* (2).

(2) ***Main scarp***: a steep surface on the undisturbed ground at the upper edge of the landslide, caused by movement of the *displaced material* (13) away from the undisturbed ground. It is the visible part of the *surface of rupture* (10).

(3) ***Top***: the highest point of contact between the *displaced material* (13) and the *main scarp* (2).

(4) ***Head***: the upper parts of the landslide along the contact between the *displaced material* (13) and the *main scarp* (2).

(5) ***Minor scarp***: a steep surface on the *displaced material* (13) of the landslide produced by differential movements within the *displaced material* (13).

(6) ***Main body***: the part of the *displaced material* (13) of the landslide that overlies the *surface of rupture* (10) between the *main scarp* (2) and the *toe of the surface of rupture* (11).

(7) ***Foot***: the portion of the landslide that has moved beyond the *toe of the surface of rupture* (11) and overlies the *original ground surface* (20).

(8) ***Tip***: the point on the *toe* (9) farthest from the *top* (3) of the landslide.

(9) ***Toe***: the lower, usually curved margin of the *displaced material* (13) of a landslide; it is the most distant margin of the landslide from the *main scarp* (2).

(10) ***Surface of rupture***: the surface that forms (or which has formed) the lower boundary of the *displaced material* (13) below the *original ground surface* (20).

(11) ***Toe of the surface of rupture***: the intersection (usually buried) between the lower part of the *surface of rupture* (10) and the *original ground surface* (20).

(12) ***Surface of separation***: the part of the *original ground surface* (20) overlain by the *foot* (7) of the landslide.

(13) ***Displaced material***: material displaced from its original position on the slope by movement in the landslide. It forms the *depleted mass* (17) and the *accumulation* (18).

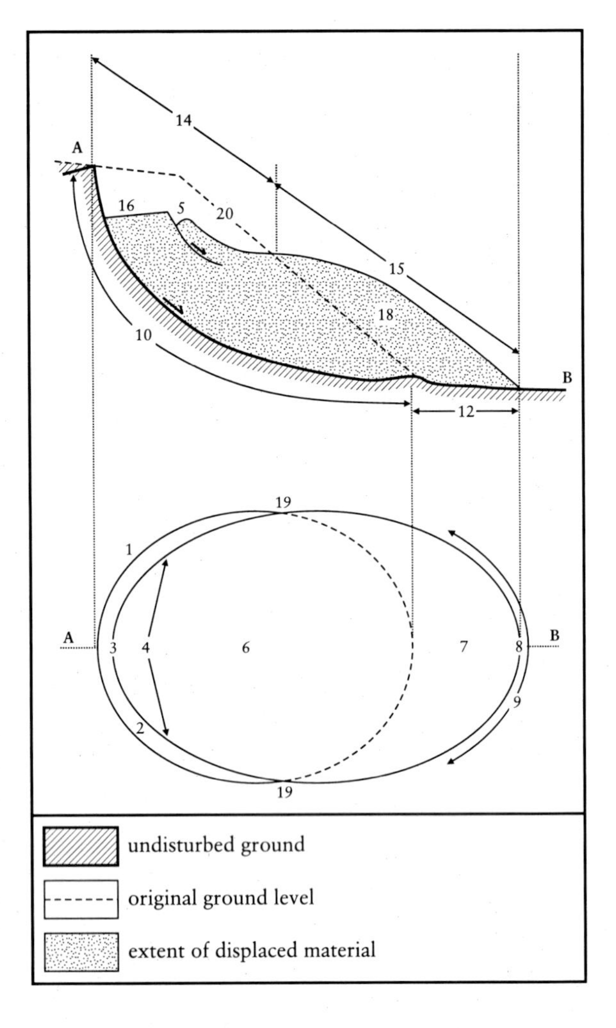

Figure 1.3 Terminology of landslides used in *The Multilingual Landslide Glossary*; profile and plan views. See text for explanation of numbers. After WP/WLI (1993).

(14) ***Zone of depletion***: the area of the landslide within which the *displaced material* (13) lies below the *original ground surface* (20).

(15) ***Zone of accumulation***: the area of the landslide within which the *displaced material* (13) lies above the *original ground surface* (20).

(16) ***Depletion***: the volume bounded by the *main scarp* (2), the *depleted mass* (17), and the *original ground surface* (20).

(17) ***Depleted mass***: the volume of the *displaced material* (13) which overlies the *surface of rupture* (10) but underlies the *original ground surface* (20).

(18) ***Accumulation***: the volume of the *displaced material* (13) which lies above the *original ground surface* (20).

(19) ***Flank***: the undisplaced material adjacent to the sides of the *surface of rupture* (10). Compass directions are preferable in describing the flanks, but if left and right are used, they refer to the flanks as viewed from the *crown* (1).

(20) ***Original ground surface***: the surface of the slope that existed before the landslide took place.

Landslide dimensions (Figure 1.4)

(1) The ***width of the displaced mass***, *Wd*, is the maximum breadth of the displaced mass perpendicular to the *length of the displaced mass, Ld* (4).

(2) The ***width of the rupture surface***, *Wr*, is the maximum width between the flanks of the landslide, perpendicular to the *length of the rupture surface, Lr* (5).

(3) The ***total length***, *L*, is the minimum from the tip of the landslide to the crown.

(4) The ***length of the displaced mass***, *Ld*, is the minimum distance from the tip to the top.

(5) The ***length of the rupture surface***, *Lr*, is the minimum distance from the toe of the surface of rupture to the crown.

(6) The ***depth of the displaced mass***, *Dd*, is the maximum depth of the displaced mass, measured perpendicular to the plane containing *Wd* (1) and *Ld* (4).

(7) The ***depth of the rupture surface***, *Dr*, is the maximum depth of the rupture surface below the original ground surface measured perpendicular to the plane containing *Wr* (2) and *Lr* (5).

Types of landslides (Figure 1.5)

(1) A ***fall*** starts with detachment of soil or rock from a steep slope along a surface on which little or no shear displacement takes place. The material then descends largely through the air by falling, saltation or rolling.

(2) A ***topple*** is the forward rotation, out of the slope, of a mass of soil or rock about a point or axis below the centre of gravity of the displaced mass.

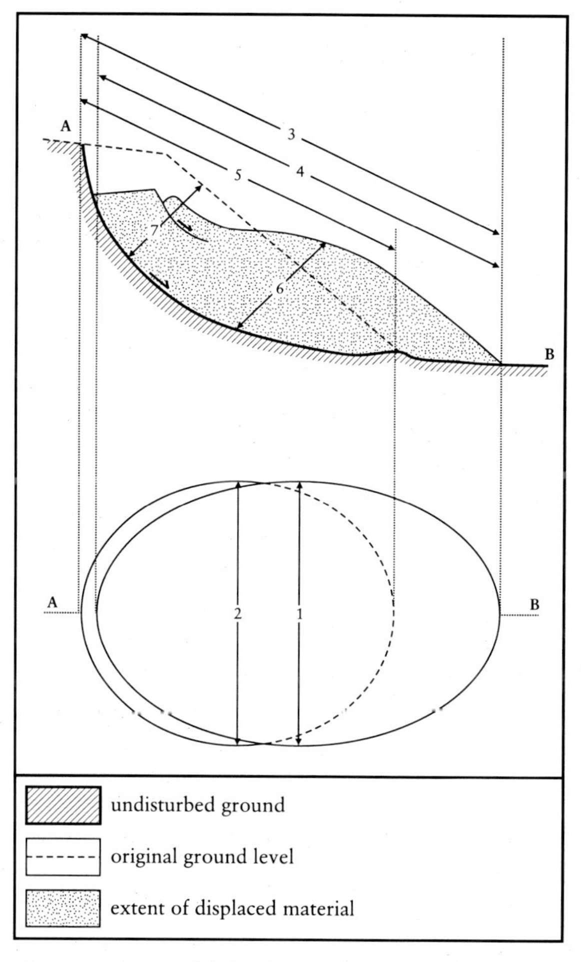

Figure 1.4 Landslide dimensions recommended in *The Multilingual Landslide Glossary*. See text for explanation of numbers. Based on WP/WLI (1993) and Cruden *et al.* (1994).

(3) A ***slide*** is the downslope movement of a soil or rock mass occurring dominantly on surfaces of rupture or relatively thin zones of intense shear strain.

(4) A ***spread*** is an extension of a cohesive soil or rock mass combined with a general subsidence of the fractured mass of cohesive material into softer underlying material. The rupture surface is not a surface of intense shear. Spreads may result from the liquefaction or flow (and extrusion) of the softer material.

(5) A ***flow*** is a spatially continuous movement in which surfaces of shear are short-lived, closely spaced and usually preserved. The distribution of velocities in the displacing mass resembles that in a viscous fluid.

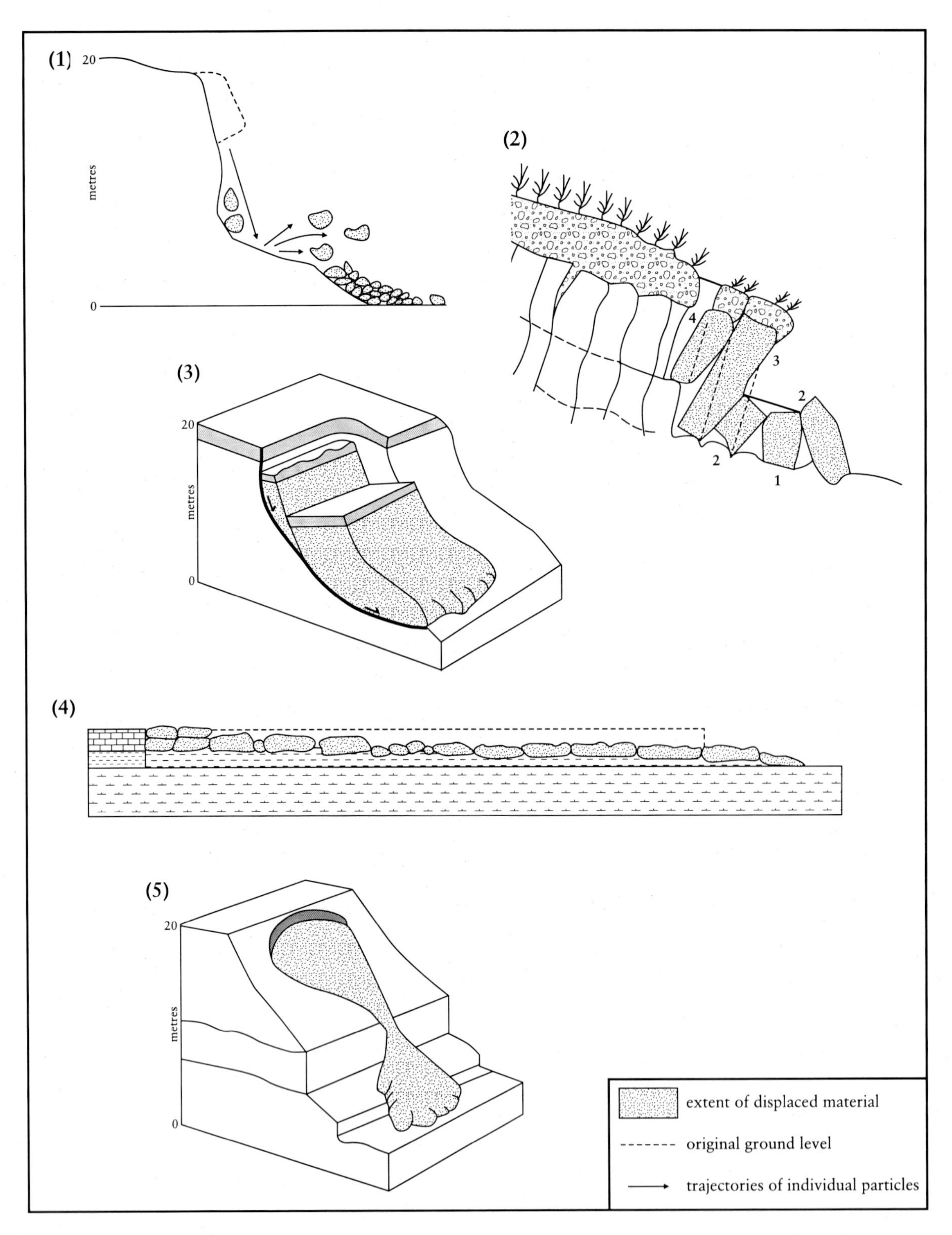

Figure 1.5 Types of landslides: (1) a fall; (2) a topple; (3) a slide; (4) a spread; (5) a flow. See text for explanation of types. After WP/WLI (1993).

States of activity of landslides (Figure 1.6)

(1) An ***active*** landslide is currently moving; the example in Figure 1.6 shows that erosion at the toe of the slope causes a block to topple.
(2) A ***suspended*** landslide has moved within the last 12 months, but is not *active* (1) at present; the example in Figure 1.6 shows local cracking in the crown of the topple.
(3) A ***re-activated*** landslide is an *active* (1) landslide which has been *inactive* (4); the example in Figure 1.6 shows that another block topples, disturbing the previously displaced material.
(4) An ***inactive*** landslide has not moved within the last 12 months and can be divided into four states: (5) *dormant*, (6) *abandoned*, (7) *stabilized*, and (8) *relict*.
(5) A ***dormant*** landslide is an *inactive* (4) landslide which can be *re-activated* (3) by its original causes or by other causes; the example in Figure 1.6 shows that the displaced mass begins to regain its tree cover, and scarps are modified by weathering.
(6) An ***abandoned*** landslide is an *inactive* (4) landslide which is no longer affected by its original causes; the example in Figure 1.6 shows that fluvial deposition has protected the toe of the slope; the scarp begins to regain its tree cover.
(7) A ***stabilized*** landslide is an *inactive* (4) landslide which has been protected from its original causes by remedial measures; the example in Figure 1.6 shows that a wall protects the toe of the slope.
(8) A ***relict*** landslide is an *inactive* (4) landslide which developed under climatic or geomorphological conditions considerably different from those at present; the example in Figure 1.6 shows that uniform tree cover has been established.

Distribution of activity in landslides (Figure 1.7)

Section 2 in each part of Figure 1.7 shows the slope after movement on the rupture surface indicated by the shear arrow in the section.

(1) In an ***advancing*** landslide the rupture surface is extending in the direction of movement.
(2) In a ***retrogressive*** landslide the rupture surface is extending in the direction opposite to the movement of the displaced material.
(3) In an ***enlarging*** landslide the rupture surface of the landslide is extending in two or more directions.
(4) In a ***diminishing*** landslide the volume of the displaced material is decreasing.
(5) In a ***confined*** landslide there is a scarp but no rupture surface visible at the foot of the displaced mass.
(6) In a ***moving*** landslide the displaced material continues to move without any visible change in the rupture surface and the volume of the displaced material.
(7) In a ***widening*** landslide the rupture surface is extending into one or both flanks of the landslide.

Styles of landslide activity (Figure 1.8)

(1) A ***complex*** landslide exhibits at least two types of movement (falling, toppling, sliding, spreading and flowing) in sequence; the example in Figure 1.8 shows gneiss and a pegmatite vein toppled with valley incision. Alluvial deposits fill the valley bottom. After weathering had weakened the toppled material, some of the displaced mass slid farther downslope.
(2) A ***composite*** landslide exhibits at least two types of movement simultaneously in different parts of the displacing mass; the example in Figure 1.8 shows that limestones have slid on the underlying shales causing toppling below the toe of the slide rupture surface.
(3) A ***successive*** landslide is the same type as a nearby, earlier landslide, but does not share displaced material or rupture surface with it; the example in Figure 1.8 shows that the latter slide, AB, is the same type as CD, but does not share displaced material or a rupture surface with it.
(4) A ***single*** landslide is a single movement of displaced material.
(5) A ***multiple*** landslide shows repeated development of the same type of movement.

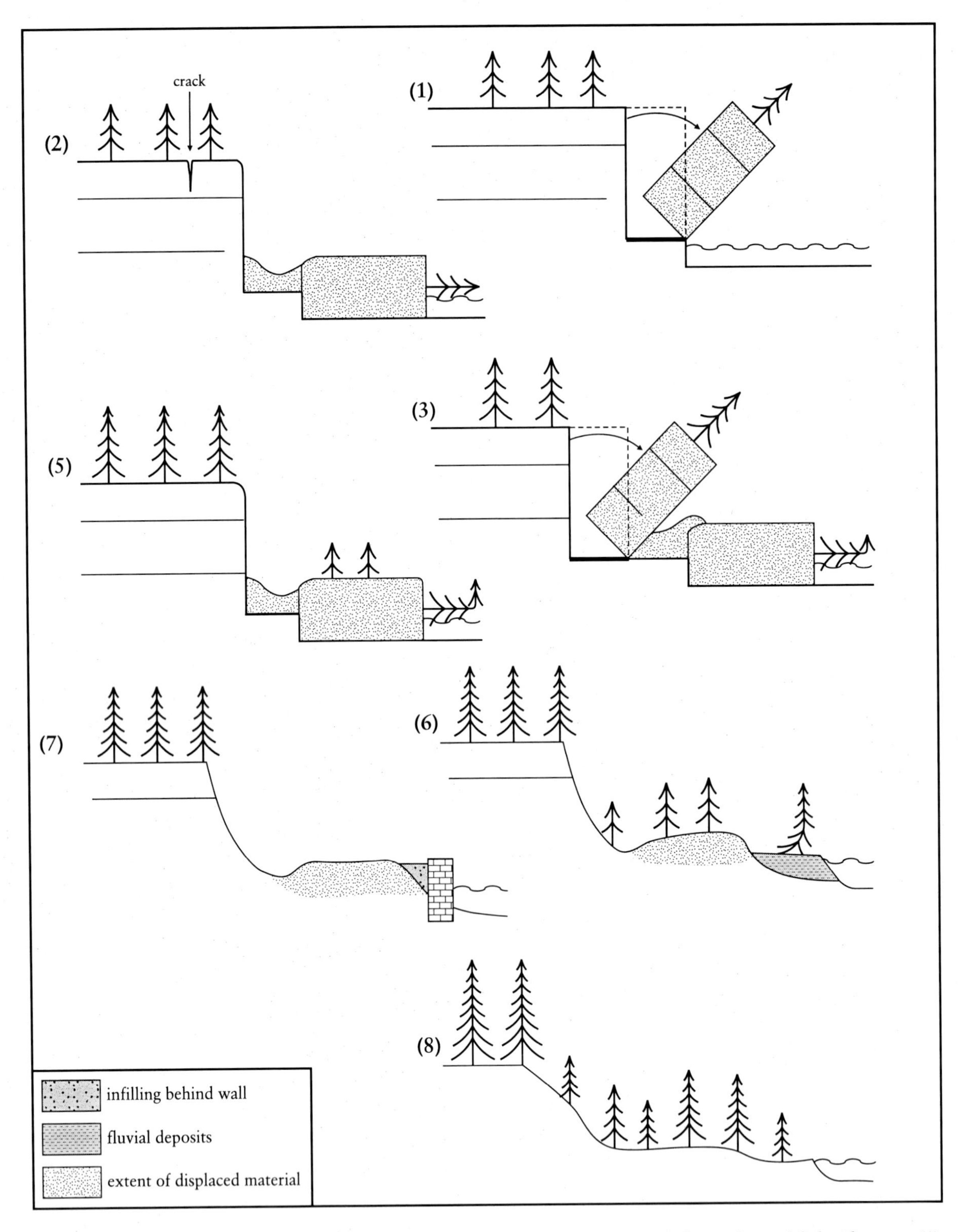

Figure 1.6 Classification of the states of activity of landslides used in the *Multilingual Landslide Glossary*: (1) active; (2) suspended; (3) re-activated; (5) dormant; (6) abandoned; (7) stabilized; (8) relict. State (4) inactive is divided into states (5)–(8). See text for explanation of states. After WP/WLI (1993).

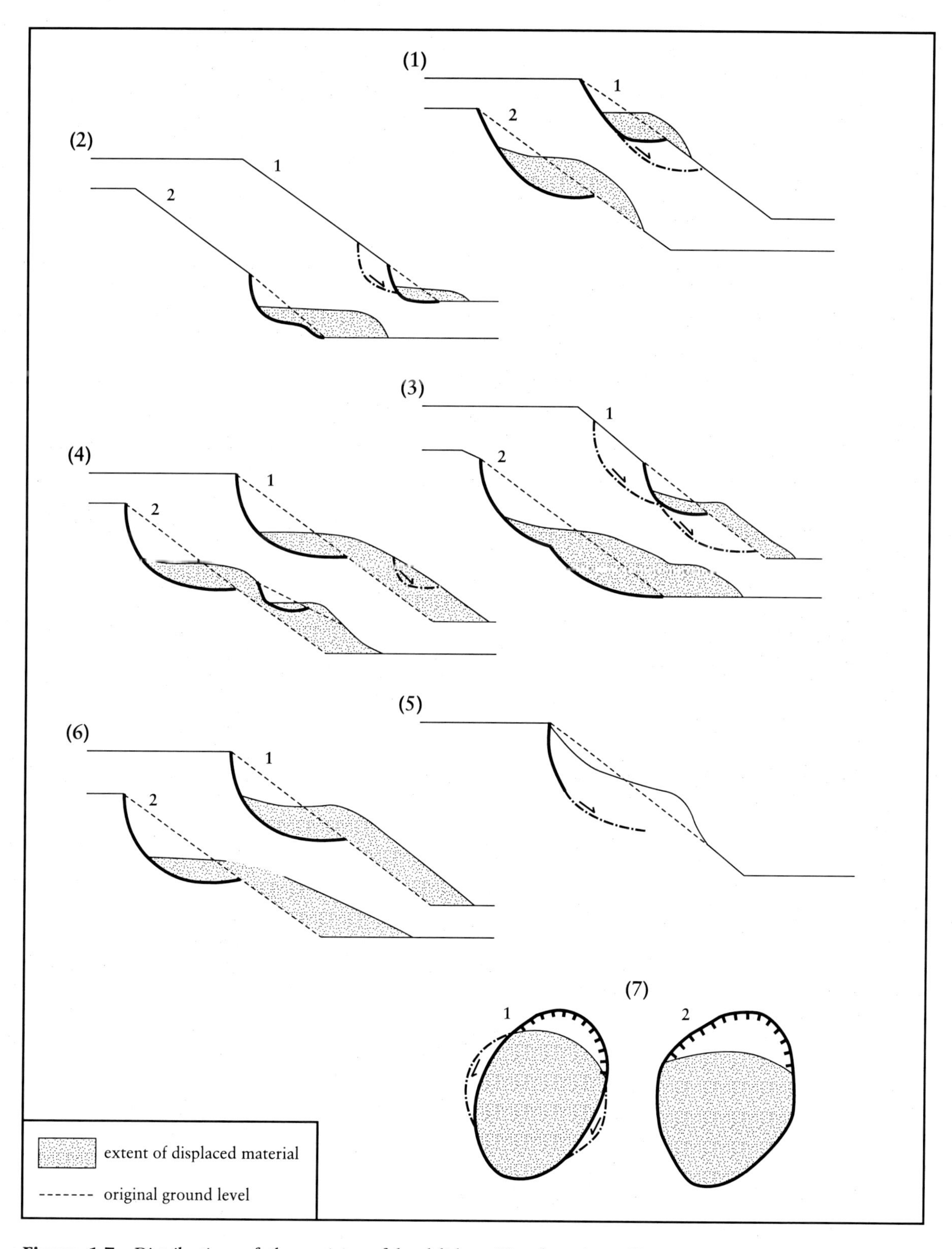

Figure 1.7 Distribution of the activity of landslides: (1) advancing; (2) retrogressive; (3) enlarging; (4) diminishing; (5) confined; (6) moving; (7) widening. See text for explanation of terms. After WP/WLI (1993).

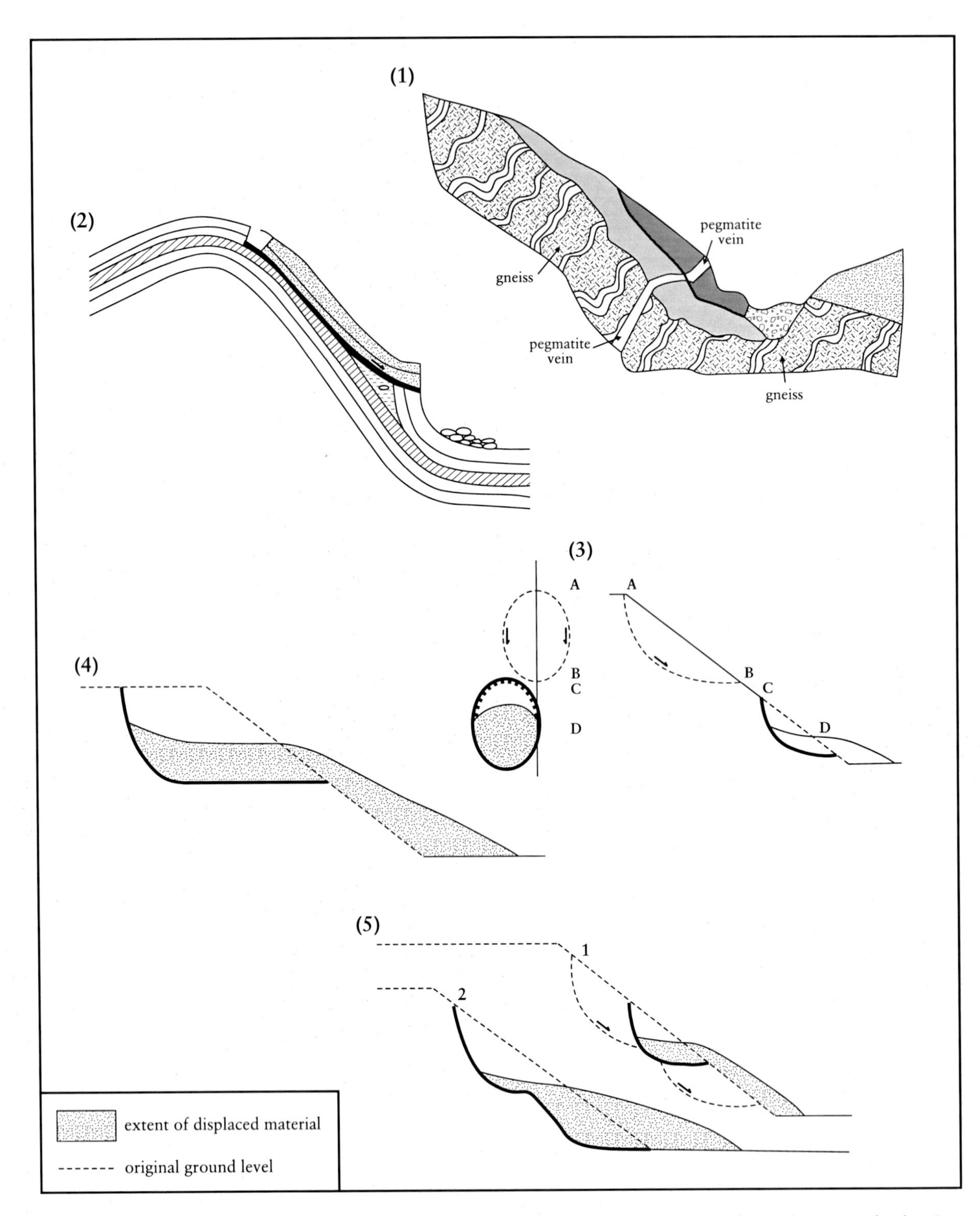

Figure 1.8 Styles of landslide activity: (1) complex; (2) composite; (3) successive; (4) single; (5) multiple. See text for explanation of terms. After WP/WLI (1993).

GCR SITE SELECTION

Methodology

The rationale, methodology and history of the selection of sites for inclusion within the Geological Conservation Review programme has been discussed in detail by Wimbledon *et al.* (1995) and in the introductory GCR volume (Ellis *et al.*, 1996). The main factors considered during the selection process can be summarized as:

(a) importance to the international Earth scientist community;
(b) presence of exceptional (classic, rare or atypical) geological/geomorphological features; and
(c) national importance for features that are representative of geological events or processes that are fundamental to understanding the geological/geomorphological history of Great Britain.

There are also the principles in GCR site selection that a chosen site should be the best available example of its kind, and that there should be a minimum of duplication of features between GCR sites.

To adapt these criteria specifically to mass movements has been particularly difficult, compared to geological (rather than geomorphological) selection categories.

Given Hutchinson's (1968a) classification, *one* particular 'type' of mass movement might be represented by *several* sites to show the different circumstances in which that 'type' typically occurs. However, during original GCR site selection it was not envisaged that, for example, the 'type' called 'rotational slips' would be represented in the ultimate GCR register by an example in strata from each of the geological periods, or in each region of the country, in which that type is found. Using the GCR ethos (Wimbledon *et al.*, 1995; Ellis *et al.*, 1996), the method of GCR site selection followed for mass movements was that set out below.

1. A first-tranche list of 23 candidate GCR sites was assembled, following literature survey and initial research. This list was circulated to relevant members of the geological, geomorphological and civil engineering communities, with the suggestion that they might delete some sites from the list, and recommend other sites not on the list.
2. The result was that 116 candidate sites were suggested for selection for the GCR by the consultees (Table 1.2) – a five-fold increase in the originally circulated list. A statistical summary was produced, which was published in February 1982 in *Earth Science Conservation* (Anon. [Cooper] in Black, 1982). The text of part of this article is reproduced here:

'...*Of these [sites suggested for consideration], 65% are located in England (13% South-East, 22% Midlands, 20% North), 24% are located in Scotland, and 11% in Wales. One third have coastal locations, and 14% are on offshore islands. Just over a quarter of the sites suggested are in Carboniferous rocks, with the Namurian of the central and southern Pennines prominent. As might be expected, the scarp-and-vale topography of the Jurassic is also a major location of recommended features (23%). Other important locations are the Precambrian (14%) and the Cretaceous (12%). Sites in the Devonian and Quaternary each make up 6% of the total, while Cambrian, Triassic and Quaternary sites each provide 3%. Permian, Silurian and Ordovician sites each provide less than 2% of the total.*

The responses to the postal survey exposed several general problems associated with site selection. Firstly, there is the problem of the transience of most medium- and small-scale mass movement phenomena. Features which yield valuable information and are educationally instructive immediately after the mass movement has taken place, may after a few years become totally obscured by the smaller-scale processes that tend to even-out irregularities on slopes. In other words, the value of such sites often resides in their freshness. There would be little point in selecting for conservation sites which are unlikely to persist, since, unlike quarry sections, mass movement sites can seldom be 'cleaned up' without destroying those features in which their academic interest resides. Secondly, the well-known mass movement classifications of Hutchinson and of Varnes include types

Table 1.2 The candidate mass-movement GCR sites suggested by the panel of experts consulted in the 1980s.

Southern England	English Midlands	Northern England	Wales	Scotland
Axminster	Alport Castles	Askrigg	Aberfan–Cilfynydd	An Teallach
Axmouth–Lyme Regis	Bredon Hill	Bilsdale	Black Mountains	Arran north coast
Bath University	Bretton Clough	Birdsall	Blorenge	Arrochar
Beachy Head	Charlesworth	Buckland's Windypit	Bodafon Mountain	Beinn a'Ghlò
Black Ven	Crowden	Canyards Hills	Cefn-y-Gader	Beinn Alligin
Blacknor Cliffs	Deer Holes	Castle Eden Dene	Craig Cerig Gleisiad	Ben Attow (Beinn Fhada)
Brighton–Saltdean	Golden Valley, Chalford	Cautley Spout	Cwmystwyth	Ben Lawers
Chale Bay cliffs	Grindesgrain Tor	Dee, Wirral	Llangollen screes	Ben Tianavaig
Charmouth foreshore	Heyden Brook	Farndale	Llyn-y-Fan Fâch	Ben Wyvis
Clovelly	Hob's House	Fremington Edge	Nant Gareg-Iwyd	Braeriach
Folkestone Warren	Jackfield	Gordale screes	Ponterwyd	Castle Ewen
Golden Cap	Lockerbrook Heights	Hilbeck Fell	Tal-y-Llyn valley	Cnoc Roll
Hadleigh	Longdendale	Holderness coast	Taren-y-Gigfran	Coire Gabhail
Herne Bay	Lud's Church	Kettleness, Staithes	Ysgyryd Fawr	Cuillin screes
High Halstow	Mam Tor	Lake District screes		Drumochter Pass
Hog's Back cliffs	Norfolk Coast	Malham		Eigg
Hooken Cliff, Beer	Northants ironstone field	Marske		Gleann an Dubh-Lochain
Hythe–Lympne–Aldington	Peter's Rock	Peak Scar		Glen Pean
Kent and Sussex Chalk cliffs	Postlip Warren	Rosedale		Glen Tilt
Keynsham railway cutting	Rockingham	Rowlee Pasture		Gribun, Mull
Maidstone cambering	Stow-on-the-Wold	Runswick Bay		Jura
Oaken Wood, Medway	The Wonder	Scarborough		Lairig Ghru
Osmington	Westend Valley	Speeton Bay		Loch Teachuis
Sevenoaks bypass	Wytham Hill	Teesdale		Lochnagar
Spot Lane, Maidstone		Wakerley		Quiraing
Stonebarrow		Whitestone Cliff		Hallaig, Raasay
Ventnor Undercliff				Rudha Gorbhaig
Warden Point				St Kilda
Winterford Heath				Storr
				Streap
				Tinto Hills

which are either of little importance or absent altogether in Great Britain. Some correspondents expressed the view that coverage should include as wide a range of examples of mass movement types as possible, while others suggested that only sites with a pronounced morphological expression, either on the surface or in section, should be considered. A third problem is that many continuously operating, small-scale processes, which are of great importance in Great Britain, do not give rise to recognizable features either on the surface or in section. For this reason they are not readily conservable. However, such processes, for example soil creep, are so widespread and commonplace that they are deemed not to require conservation at a few, specified, 'representative' sites. A pragmatic solution has been adopted, whereby, as far as possible, Great Britain's 'best' example of each mass movement type is only to be selected if that example is also a 'good' one when viewed from a global perspective.

Of the individual sites recommended in the responses to the postal survey, the complex of rotational slips on the south coast between Axmouth and Lyme Regis was the most often mentioned, closely followed by the slumps in Quaternary deposits on the north Norfolk coast around Cromer and Trimingham. Next most frequently suggested was Warden Point on the coast of the Isle of Sheppey, followed jointly by Folkestone Warren in Kent, the solifluction lobes on the Lower Greensand escarpment near Sevenoaks, the Undercliff on the south coast of the Isle of Wight, the slips at Chale Bay on the Isle of Wight, and the massive features at Quiraing and The Storr, in the Trotternish peninsula of the Isle of Skye. Other much-mentioned sites included High Halstow in Kent, the area around Bath, screes in the Lake District and near Llangollen, Mam Tor in the Peak District, and the abandoned cliff at Hadleigh Castle in Essex. Even between these most mentioned sites, there are obvious overlaps of mechanism, of surface form, and of cross-sectional features, so that the inclusion of all of them in the Review would involve unwarranted duplication. Conversely, several sites have already been assessed as suitable for inclusion, even though each was only suggested by a single correspondent.

At the close of the 1981 field season, 62% of the recommended sites had been either examined in the field, or excluded from the exercise without a visit. The latter course has been taken either through prior personal knowledge of the site in question, or because a reading of the literature shows the site to be an inferior duplicate. Many sites have been examined in aerial view at the Cambridge University Collection of Air Photographs. This has proved a most valuable aid to site selection or elimination. It is anticipated that between twenty and thirty sites will eventually be selected for inclusion in the Review; the final choice awaits completion of the programme of field visits.'

3. After sifting the original and emended lists, the remaining sites were all visited in the field during 1981 and 1982 by the present author (RGC). Photographs were taken along with some measurements, facilitating direct comparison of what might be termed 'competitive' candidate sites (the GCR being a minimalist scheme, see Ellis *et al.*, 1996). The aim at this stage was to ensure that a complete network of sites was established to *represent the variety* of mass-movement types and forms found in Great Britain. After consultation and revision, a list of 28 sites was finally produced; this list, with short descriptions, has been published in Jones and Lee (1994, pp. 242–7). This is the list (Table 1.3) that was finally adopted, and is described in the present book, with the exception of the site at Spot Lane Quarry near Maidstone in Kent, described by Worssam (1963), which was included as an example of strata exhibiting two superficial structures: cambering and gulling. However, between 1980 when the site was visited, and a return visit in 1996, a housing estate had been extended onto the area concerned. A small exposure has been preserved there on account of the fossil fauna of a gull filling (selected for the Quaternary of South-East England GCR Block), but otherwise this remnant exposure now shows camber and gull features no better than many other sites across the country.

Table 1.3 The final list of selected mass-movement sites as drawn up in the early 1980s.

Site
Alport Castles, Derbyshire
Axmouth–Lyme Regis, Devon–Dorset
Beinn Fhada, Highland
Black Ven, Dorset
Blacknor Cliffs, Dorset
Buckland's Windypit, North Yorkshire
Canyards Hills, Sheffield
Coire Gabhail, Highland
Eglwyseg Scarp (Creigiau Eglwyseg), Clwyd
Entrance Cutting at Bath University, Avon
Cwm-du, Ceredigion
Folkestone Warren, Kent
Glen Pean, Highland*
Hallaig, Isle of Raasay, Highland
High Halstow, Kent
Hob's House, Derbyshire
Llyn-y-Fan Fâch, Carmarthenshire
Lud's Church, North Staffordshire
Mam Tor, Derbyshire
Peak Scar, North Yorkshire
Postlip Warren, Gloucestershire
Rowlee Bridge, Derbyshire
Spot Lane Quarry, Kent*
Stutfall Castle, Kent
Trimingham Cliffs, Norfolk
Trotternish Escarpment, Isle of Skye, Highland (The Storr and Quiraing)
Warden Point, Kent

* Glen Pean and Spot Lane Quarry have now been deleted from the Mass-Movements GCR 'Block' (selection category) – see text.

GCR Editor's note:

A review of the Scottish Highland mass movements carried out after Roger Cooper's death showed that there were eight sites, which, as a result of recent investigations by Colin Ballantyne and David Jarman, met GCR standards. The new sites (described in Chapter 2) are listed in Table 1.4. Had this information been available at the time of the original scoping exercise, when none of these sites were suggested, there is little doubt that they would have been included. The review also showed that applying the 'minimalist' principle, one site, Glen Pean, would not now have been included in the GCR. Revised site information also became available for several of the already selected Scottish sites (**Coire Gabhail** – Chapter 4; and the **Trotternish Escarpment** (Quiraing and The Storr) – Chapter 6). Tables 1.5 and 1.6 have been revised to recognize these changes.

Table 1.4 The supplementary sites added to the GCR following recent research in Scotland.

Site
Beinn Alligin, Highland
Ben Hee, Highland
Benvane (Beinn Bhàn), Stirling
Carn Dubh, Ben Gulabin, Perthshire
The Cobbler (Beinn Artair), Argyll and Bute
Druim Shionnach, Highland
Glen Ample, Stirling
Sgurr na Ciste Duibhe, Highland

Site classification

The style and type categories from *The Multilingual Landslide Glossary*, with the codes from Hutchinson's classifications of 1968a and 1988, are shown along-side brief descriptions for each of the mass-movement sites selected for the GCR, in Table 1.5. However, the GCR deals with *sites* (areas of land with a defined boundary), and the classifications deal with the *types* of movement involved in a displaced mass, or mass undergoing displacement. Thus, **Warden Point**, for example, is recorded as *composite* in style, involving both sliding and toppling. This could give the misleading impression that at Warden Point mass-movement events characteristically involve both toppling and sliding together. In fact, Warden Point shows the results of several mass-movement events, side by side along the coast. Of these, most are slides, but one shows toppling.

Table 1.6 shows the mass-movement GCR sites described in the present volume, classified in two ways. First, by the stratigraphical order of the major geological systems in which the mass-movement phenomena occur in Great Britain. The second classification shows the broad movement mechanisms by which material moves downslope. According to this classification less than half of the sites exhibit more than one type of mass movement, but a few exhibit more than two types. There is some correlation with the areal extent of a site and the number of types present, but this is not always the case. For example, the **Axmouth–Lyme Regis** GCR site runs along about 10 km of coastline, and exhibits six of Hutchinson's (1968a) types, while Quiraing, part of the **Trotternish Escarpment** GCR site, also a very large site, exhibits just one type. Most sites of small areal extent, however, exhibit a single type of mass movement. Rotational slips (groups 6a and 6bi) are the most common; the character of this type is discussed in further detail below.

Table 1.5 The mass-movement GCR sites cdescribed in the present volume; style and type are according to the World Landslide Inventory (WP/WLI 1993), classifications are according to Hutchinson (1968a) and (1988) – described in Table 1.1a,b.

Site	Authors' classifications	Style	Type	Hutchinson categories	
				1968a	1988
Alport Castles	Mass rock creep, retrogressive rotational, translational	Composite	Slide, flow	2, 5a, 6bi	B2, D2, D4
Axmouth–Lyme Regis	Translational, rotational, subsidence	Complex	Slide, spread	5a, 6bi, 6a, 6c, 7a, 7b	D2, D3, D4
Beinn Fhada (Ben Attow)	Large-scale slope deformation, local slides, possible sags or forward topples	Complex	Spread	2, 5a	A2, B2, C1, D4, F1
Beinn Alligin	Large rockfall with excess run-out	Single	Fall, flow	5fii, 7b	E3, G1
Ben Hee	Arrested translational slide	Multiple	Slide	5a	D4
Benvane	Slope deformation and translational slide	Multiple	Spread, slide	2, 5a	B2, D4
Black Ven	Mudslides	Complex	Slide	5di	E1
Blacknor Cliffs	Block slide, slab failure	Complex	Slide, topple	5a	D4
Buckland's Windypit	Block slides	Multiple	Slide	5a, 4b	D4
Canyards Hills	Translational with breakup into ridges, lateral extension	Multiple	Slide	5a	D4
Carn Dubh, Ben Gulabin	Translational slide to flow	Single	Slide, flow	5a, 5fii	D4, E4
Coire Gabhail	Rockfalls, landslide dam, run-up opposite	Multiple	Fall	7b	G1
Cwm-du	Sub-snow solifluction sheets OR 'landslides'	Multiple	Slide	4a, 5c	E2
Druim Shionnach	In-situ slope deformation progressing to toppling	Composite	Spread	2	B2, C3
Eglwyseg Scarp (Creigiau Eglwyseg)	Active screes and relict clitter slopes	Multiple	Fall	7a, 7b	G
Entrance Cutting at Bath University	Gulls, cambers, dip-and-fault structure	Composite	Spread	4b	H
Folkestone Warren	Rockfalls, clay extrusion, rotational	Complex	Fall, slide	6bi	D2, G, H
Hallaig	Rotational slide, possibly seismically triggered	Single	Slide	6a	D2
Glen Ample					
Beinn Each	Compressional slope deformation, local rockfall	Multiple	Spread	2, 7b	A2, G1
Ben Our	Extensional slope deformation, slides, topples	Complex	Spread	2, 5a	B2, C1, D4,F1
High Halstow	Shallow successive rotational slips, hillwash, soil creep	Successive	Slide	1a, 6bi, 6c	B, D2
Hob's House	Rotational slip	Single	Slide	6a	D2
Llyn-y-Fan Fâch	Debris flow	Multiple	Flow	5c	E3
Lud's Church	Bed-on-bed translational sliding within a rotational mass	Single	Slide	5a, 6a	D2, D4
Mam Tor	Slump-earthflow	Multiple	Slide, flow	6c, 5c	D3, E3, H4
Peak Scar	Block slide, topples	Complex	Slide, topple	5a	D4, F
Postlip Warren	Large-scale gravitational slips, 'founders'	Successive	Spread	5a, 4b	H
Rowlee Bridge	Valley-bulge	Complex	Spread	4b	H
Sgurr na Ciste Duibhe	Extensional slope deformations and slides	Complex	Spread, slide	2, 5a	B2, D4
Stutfall Castle	Soil creep, earthflow, translational	Complex	Flow, slide	1a, 5c, 6b	B, D4, E1
The Cobbler	Short-travel arrested translational slide; also sub-cataclasmic	Single	Slide, fall	5a, 7b	D4, E3
Trimingham Cliffs	Blockfall, seepage failure, mudslides, rotational slip	Composite	Fall, slide	5di, 6a	D2, E1, G, H
Trotternish Escarpment					
Quiraing	Retrogressive translational slide, rockfall	Multiple	Slide, fall	5a, 7b	D4, G1
The Storr	Retrogressive translational slide, topples	Multiple	Slide, topple	5a, 7b	D4, F2
Warden Point	Rotational, topples	Composite	Slide, topple	6b	D2, F

Table 1.6 The sites described in the present volume classified by geological age and by WLI mass-movement type: (PC = Precambrian–Cambrian; Si = Silurian; De = Devonian; Ca = Carboniferous; Ju = Jurassic; Cr = Cretaceous; Eo = London Clay; Pl = Pleistocene; fa = fall; to = Topple; sl = slide; sp = spread; fl = flow; * = sites which display cambering and valley-bulging).

Site name	Geological age								Mass-movement type				
	PC	Si	De	Ca	Ju	Cr	Eo	Pl	fa	to	sl	sp	fl
Alport Castles				X							X		X
Axmouth–Lyme Regis					X	X			X		X	X	
Beinn Alligin	X								X				X
Beinn Fhada *	X									X	X	X	
Ben Hee	X										X		
Benvane	X										X	X	
Black Ven					X	X					X		X
Blacknor Cliffs					X					X	X		
Buckland's Windypit					X						X	X	
Canyards Hills				X							X		
Carn Dubh, Ben Gulabin	X										X		X
Coire Gabhail			X						X				
Cwm-du		X									X		
Druim Shionnach *	X											X	
Eglwyseg Scarp (Creigiau Eglwyseg)				X					X				
Entrance Cutting at Bath University *					X							X	
Folkestone Warren						X			X		X		
Glen Ample													
Beinn Each *	X								X				X
Ben Our	X									X	X	X	
Hallaig					X						X		
High Halstow							X				X		
Hob's House				X							X	X	
Llyn-y-Fan Fâch			X										X
Lud's Church				X							X		
Mam Tor				X							X		X
Peak Scar					X					X	X		
Postlip Warren					X							X	X
Rowlee Bridge *				X								X	
Sgurr na Ciste Duibhe	X										X	X	
Stutfall Castle						X					X		X
The Cobbler	X								X		X		
Trimingham Cliffs								X	X		X		
Trotternish Escarpment													
Quiraing					X		X		X		X		
The Storr					X		X			X	X		
Warden Point							X			X	X		

Representativeness

Since 1980 a focusing of GCR objectives has taken place, whereby 'representativeness' is a term now used to encapsulate many of the 18 selection criteria recommended in 1992 (Gordon, 1992). At the time of the original selection process (1980s), GCR sites were not selected on the basis of their ability to represent mass movements in different geological formations or areas of the country, but rather to create an inventory of the most important mass-movement sites in Great Britain by mass-movement type. In reconsidering the Mass-Movements GCR Block in the light of the more focused objectives in the late 1990s (when the present volume was commissioned), sites were reconsidered against a scheme of stratigraphical and, thereby, areal representativeness (compare with the US system of geological site selection for conservation; Cooper, 1985). This re-focusing has brought about a change to the approach to the present mass-movement GCR volume, such that the text is divided into chapters on the basis of stratigraphy (age of the geological strata in which the mass movements occur; Figure 1.9).

'Representativeness' involves the notion of what is typical, or 'archetypal', but it is important to note that 'atypical' or 'exceptional' sites may provide insights into the nature of 'type' examples, and this is also a criterion for the GCR.

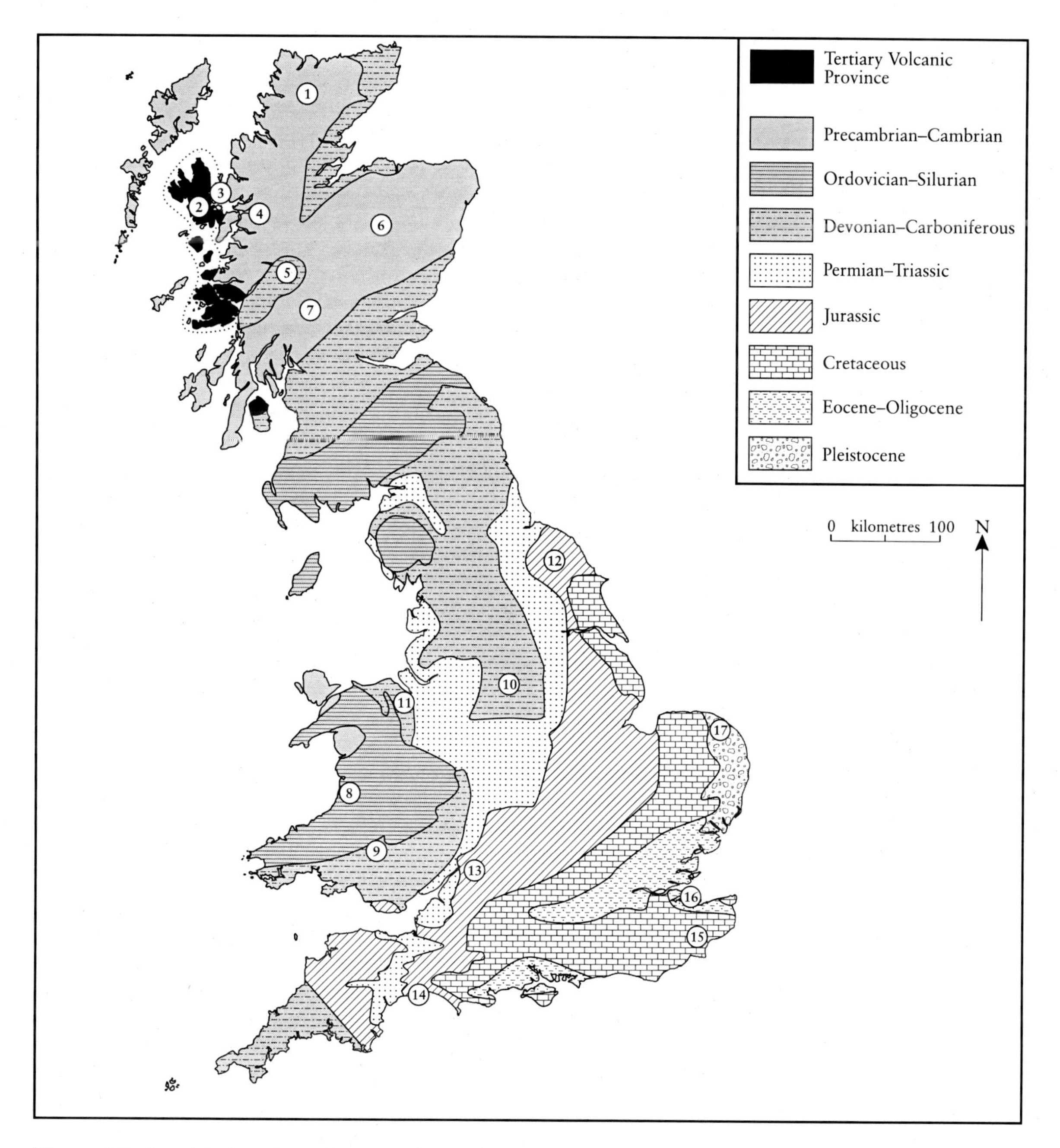

Figure 1.9 Simplified geological map of Great Britain, with the general locations of the mass-movement GCR sites numbered: (1 – Ben Hee: 2 – Trotternish Escarpment: 3 – Hallaig; Beinn Alligin: 4 – Beinn Fhada; Sgurr na Ciste Duibhe; Druim Shionnach: 5 – Coire Gabhail: 6 – Carn Dubh: 7 – The Cobbler; Benvane; Glen Ample: 8 – Cwm-du: 9 – Llyn-y-Fan Fâch: 10 – Hob's House; Alport Castles; Canyards Hills; Lud's Church; Mam Tor; Rowlee Bridge: 11 – Eglwyseg Scarp: 12 – Peak Scar; Buckland's Windypit: 13 – Postlip Warren; Entrance Cutting at Bath University: 14 – Axmouth–Lyme Regis; Black Ven; Blacknor Cliffs: 15 – Folkestone Warren; Stutfall Castle: 16 – High Halstow: Warden Point: 17 – Trimingham Cliffs).

Revision of the GCR in the future

Mass-movements studies, like any other science, are ever-developing, with new discoveries being made, and existing models being subject to continual testing and modification as new data come to light. Increased or hitherto unrecognized significance may be seen in new sites. Therefore, it is possible that further sites worthy of conservation will be identified in future years for the study of mass movements in Britain, as research continues. However, it must be stressed that the GCR is intended to be a *minimalist* scheme, with the selection for the GCR of only the best and most representative example of a geological feature, rather than the selection of a series of sites showing closely analogous features.

Legal protection of GCR sites

V.J. May and N.V. Ellis

The list of GCR sites has been used as a basis for establishing Earth science Sites of Special Scientific Interest (SSSIs), protected under the Wildlife and Countryside Act 1981 (as amended) by the statutory nature conservation agencies (the Countryside Council for Wales, Natural England and Scottish Natural Heritage).

The SSSI designation is the main protection measure in the UK for sites of importance to conservation because of the wildlife they support, or because of the geological and geomorphological features that are found there. About 8% of the total land area of Britain is designated as SSSIs. Well over half of the SSSIs, by area, are internationally important for a particular conservation interest and are additionally protected through international designations and agreements.

About one third of the SSSIs have a geological/ geomorphological component that constitutes at least part of the 'special interest'. Although some SSSIs are designated solely because of the importance to wildlife conservation, there are many others that have both such features and geological/geomorphological features of 'special interest'. Furthermore, there are localities that, regardless of their importance to wildlife conservation, are conserved as SSSIs solely on account of their importance to geological or geomorphological studies.

Therefore, many SSSIs are composite, with site boundaries drawn from a 'mosaic' of one or more GCR sites and wildlife 'special interest' areas; such sites may be heterogeneous in character, in that different constituent parts may be important for different features.

Many of the SSSIs that are designated solely because of their Earth science features have interesting wildlife and habitat features, underlining the inextricable links between habitat, biodiversity and the underlying geology and geomorphology.

It is evident from some of the individual site reports in this volume, describing sites in coastal locations, that the conservation interest of the geomorphological features is likely to be affected by shoreline management activities outside of the site itself, especially where the GCR sites lie within large sediment-transport cells. A number of the sites have landslide toes which extend below low-water mark of spring tides. However, since SSSI notification of GCR sites presently extends to mean low-water mark in England and Wales and low-water mark of spring tides in Scotland, there is no statutory protection of these landslide toes below low-water mark, unless they are co-incidentally part of some other conservation designation (e.g. Special Protection Areas or Special Areas of Conservation – see below).

International measures

Presently, there is no formal international conservation convention or designation for geological/geomorphological sites below the level of the 'World Heritage Convention' (the 'Convention concerning the Protection of the World Cultural and Natural Heritage'). World Heritage Sites are declared by the United Nations Educational, Scientific and Cultural Organisation (UNESCO). The objective of the World Heritage Convention is the protection of natural and cultural sites of global significance. Many of the British World Heritage Sites are 'cultural' in aspect, but the Giant's Causeway in Northern Ireland and the Dorset and East Devon Coast ('the Jurassic Coast') are inscribed because of their importance to the Earth sciences as part of the 'natural heritage' – the Dorset and East Devon Coast World Heritage Site is of particular relevance here insofar as it was the outstanding geology and coastal geomorphology (including sites described in this volume and other sites described in the *Coastal Geomorphology of Great Britain* GCR volume (May and Hansom, 2003) that include mass-movement phenomena).

In contrast to the Earth sciences, there are many other formal international conventions – particularly at a European level – concerning the conservation of wildlife and habitat. Of course, many sites that are formally recognized internationally for their contribution to wildlife conservation are underpinned by their geological/geomorphological character, but this fact is only implicit in such designations. Nevertheless, some of the sites described in the present volume are not only geomorphological SSSIs, but also *habitat* sites recognized as being internationally important. These areas are thus afforded further protection by international designations above the provisions of the SSSI system.

Special Areas of Conservation (SACs)
Of special relevance to the present volume are those coastal and mountain habitats that are dependent upon coastal or mountain geomorphology and are conserved as Special Areas of Conservation (SACs). In 1992 the European Community adopted Council Directive 92/43/EEC on the conservation of natural habitats and of wild fauna and flora, commonly known as the 'Habitats Directive'. This is an important piece of supranational legislation for wildlife conservation under which a European network of sites is selected, designated and protected. The aim is to help conserve the 169 habitat types and 623 species identified in Annexes I and II of the Directive.

Special Protection Areas (SPAs)
Special Protection Areas are strictly protected sites classified in accordance with Article 4 of the EC Directive on the conservation of wild birds (79/409/EEC), also known as the 'Birds Directive', which came into force in April 1979. They are classified for rare and vulnerable birds, listed in Annex I to the Birds Directive, and for regularly occurring migratory species.

Although SACs and SPAs are identified for the conservation importance of their biological features, individually or collectively, many also include scientifically important geomorphological features.

GCR site selection in conclusion

It is clear from the foregoing that many factors have been involved in selecting and protecting the sites described in this volume. Sites rarely fall neatly into one category or another; normally they have attributes and characteristics that satisfy a range of the GCR guidelines and preferential weightings (Ellis *et al.*, 1996). A full appreciation of the reasons for the selection of individual sites cannot be gained from these few paragraphs. The full justification and arguments behind the selection of particular sites are only explained satisfactorily by the site accounts given in the subsequent chapters of the present volume.

ORGANIZATION OF THE MASS-MOVEMENTS GCR VOLUME

The original plan for this volume was to divide it into chapters on the basis of mass-movement type. Thus, there would be a chapter on rotational slide sites, another on bedding-plane controlled slide sites, and so on. It was quickly realized that this would fail to represent adequately the network of GCR sites actually selected. In particular it separated some sites, which, when placed together, illustrated very well the variety of mass movements found in particular areas of the country, for example the southern Pennines. Since most of the sites illustrate complex landslides involving several types of failure, rather than single mechanisms, classification would be difficult.

However, a succession of chapters, some of which were based on mass-movement type, while others were based on regional considerations, gave a disorganized impression. Accordingly the present stratigraphical arrangement was adopted. This is still less than ideal. While it works well in highlighting the main mass-movement producing systems in Great Britain: Carboniferous, Jurassic and Cretaceous strata (which together account for 75% of the landslides identified in the DoE survey; Jones and Lee, 1994), it is less successful for sites in other geological systems. Since all of the mass movements in Great Britain represented by the sites described in the present volume have taken place in Quaternary times, the relevance of the age of the rocks in which they have taken place is indirect. More significant factors include the attitude of bedding, the frequency of jointing, and above all the succession of lithological types cropping out down a slope or a coastal cliff. In particular, and this is the key to the prolific numbers of landslides in the Carboniferous, Jurassic and Cretaceous age rocks, is the presence of soft, 'incompetent' strata cropping out downslope or 'down cliff' from hard, jointed, 'competent' strata. An attempt to develop an ad-hoc order of presentation for the present volume was based on characteristics of physiography and geological succession at the selected sites. However, the arrangement of chapters by geological system for the purposes of publishing the accounts has been retained (Figure 1.9).

COMMENTS ON SOME GENERAL ASPECTS OF THE SITES SELECTED

In addition to this introductory chapter in which general matters of relevance to the whole book are discussed, each of the following chapters has an introductory section in which geomorphological principles pertinent to the sites described in that chapter are discussed. However, some issues are described in the following text, which are of relevance to more than one of the chapters of site descriptions.

Movement

Mass-movement sites have in common that they represent the *results* of mass movements, i.e. movements that have already taken place; in other words, in only some of the selected GCR sites has the actual occurrence of movement been detected and recorded as it occurred. Movement may be detected in two main ways: measurement and eyewitness accounts. Measurement is generally carried out by identifying a fixed point (or points) on the ground surface and marking it/them with wooden or metal pegs. Its precise position is then surveyed, generally by triangulation from two locations whose positions are already known, by marking on a recent aerial photograph of the site, or using GPS techniques. After a period of time, perhaps one year later, the process is repeated using the same survey points (and taking new aerial photographs, GPS or laser-measurement data). A difference in the position of the marked point will indicate that mass movement has indeed taken place, and data can be recorded about the distance and direction of the movement. The problem with measurement of this type is that it is only worthwhile at a location where movement may be expected to take place over the surveying period, for example a location where movement is believed to have taken place in the previous year. However, the technique has been successfully used at East Pentwyn, Bourneville, Ironbridge, **Mam Tor**, St Mary's Bay, **Black Ven**, Stonebarrow, **Folkestone Warren**, St Catherine's Point, and the north coast of the Isle of Sheppey.

Eyewitness accounts exist for the 1839 movement of part of the **Axmouth–Lyme Regis** coast, now a National Nature Reserve. Eyewitness accounts were also collected by John Wesley, the Methodist preacher, of the collapse of Whitestone Cliff, North Yorkshire, in 1755 (reproduced in Jones and Lee, 1994; see also Cooper, 1997). Such accounts also exist for movements at, for example, Black Ven in Dorset, Robin Hood's Bay in North Yorkshire, and on the north coast of the Isle of Sheppey in Kent. Black Ven in Dorset is an important site where one can rely on seeing mudslides in motion, if visiting at the right time of year and after a suitable spell of wet weather. Black Ven has been intensively studied, with a complete record of movements for over 50 years. Stonebarrow, the next cliff to the east of Black Ven, has displacement and pore-pressure records for three years in the late 1960s, and the slides at Lyme Regis are currently heavily monitored. Many other records for short periods are associated with sites which require engineering stabilization works (e.g. Mam Tor). The fact remains, however, that many of the mass-movement GCR sites are only known in terms of simple morphological or geological descriptions.

Mudflows, mudslides and earthflows

Mudflows are generally taken to be rheological flows of material that consist predominantly of clay-sized particles, under the influence of gravity, and sufficiently wetted for the moisture content to be above the 'Plastic Limit'.

Mudslides are taken to be similar to mudflows, except that they experience shearing at the contact with adjacent solid material. This zone of shearing is usually as sharp as a knife cut, with a 'scraped off' soft layer immediately above. The shear surface will be polished and striated. Deep-seated slides and extrusion layers may have a thicker zone of displacement. Mudslides can form within mudslides as they dry out, but still they are bounded by separate, clear shears (Brunsden, 1984).

This distinction largely became acknowledged with the publication of an important paper by Hutchinson and Bhandari (1971), in which it was explicitly recognized that many of the mass movements previously described as mudflows actually advance by sliding on discrete boundary shear surfaces, and that such mass movements are better termed 'mudslides', a term used by Fleming (1978) and by Cailleux and Tricart (1950). It was demonstrated that very often the surging forward of a 'mudflow' was caused not by flowage, but by undrained loading of its rearward parts, the whole mass moving downslope by sliding. However, the term 'mudflow' is still valid for very fine-grained flows, but it is also an old term for mudslides. Hutchinson's 1968a 'climatic mudflows' (see Table 1.1a above) are now mudslides (Brunsden, 1984).

In the World Landslide Inventory (WP/WLI 1993) classification, the American usage 'Earthflow' is preferred. Buma and van Asch state in *Landslide Recognition* (Dikau *et al.*, 1996) that 'the American usage 'earthflow' is replaced in European literature by 'mudslide''. However, 'earthflow' is used by Skempton *et al.* (1989) in describing part of the landslide at **Mam Tor** (see Chapter 5). Varnes (1978), the principal American source on such matters, does not endorse this one-to-one correspondence in terminology. Stating that earthflows range in water content from above saturation to essentially dry, he places mudflows at the wet end of the scale, as 'soupy end members of the family of predominantly fine-grained earthflows'. This neglects the important observation that the 'stiffer' forms slide on discrete surfaces.

Undrained loading

Hutchinson and Bhandari (1971) provided an expanded account of a suggestion made by Hutchinson (1970) which applies to many mudslides and also to a variety of other types of mass movement. They observed that many 'mudflows' were advancing downslope by shearing on slopes that were of considerably lower angle than the slope of limiting equilibrium for residual strength on the sliding surface and groundwater co-incident with and flowing parallel to the slope surface. For example, with slopes at Bouldnor, Isle of Wight that have residual strength of $c_r' = 0$, $\phi_r' = 13.5°$ (where c_r' is residual cohesion and ϕ_r' is angle of internal friction), it was shown using infinite slope analysis (Skempton and Delory, 1957) that the lowest slope angle at which sliding could occur is 6.1°. Measurement of these slopes showed that they stand at angles as low as 3.9° (Hutchinson and Bhandari, 1971). They suggested that the sliding is brought about by the virtually undrained loading of the headward parts of the mudslides by debris discharged from steeper slopes to the rear. This undrained loading develops a forward thrust in the rear part of the mudslide, where the basal slip-surface is inclined fairly steeply downwards, giving rise to shearing movements on very low angle slopes (Figure 1.10), even at slopes of zero or negative inclination for short distances (Hutchinson and Bhandari, 1971).

Collapse of caprocks

There is a group of mass movements, generally characterized by a hard but possibly jointed caprock, which does not have to be thin and can be several tens of metres in thickness, overlying a stratum or series of strata characterized by 'incompetence', the inability to support the overlying 'competent' caprock at locations where erosion has cut down to expose the incompetent strata. This can lead, according to local circumstances, to one or more of a variety of recognized mass-movement types, in the terms of Hutchinson (1988) rebound associated with naturally eroded valleys, post-failure creep, and complex failures of types (1) cambering and valley bulging, and (2) block-type slope movements.

This phenomenon has been more widely accepted in continental Europe and other parts of the world than in Great Britain (Brunsden, 1996a). Possibly, British workers, who naturally are those who have been most closely concerned with British mass movements, have been too circumspect. Why invoke a thick mobile stratum when a thin one will do? This shows confusion between theory and verification. In a highly empirical subject like geology such theorizing must give way to evidence that shows nature to be more complex than expected.

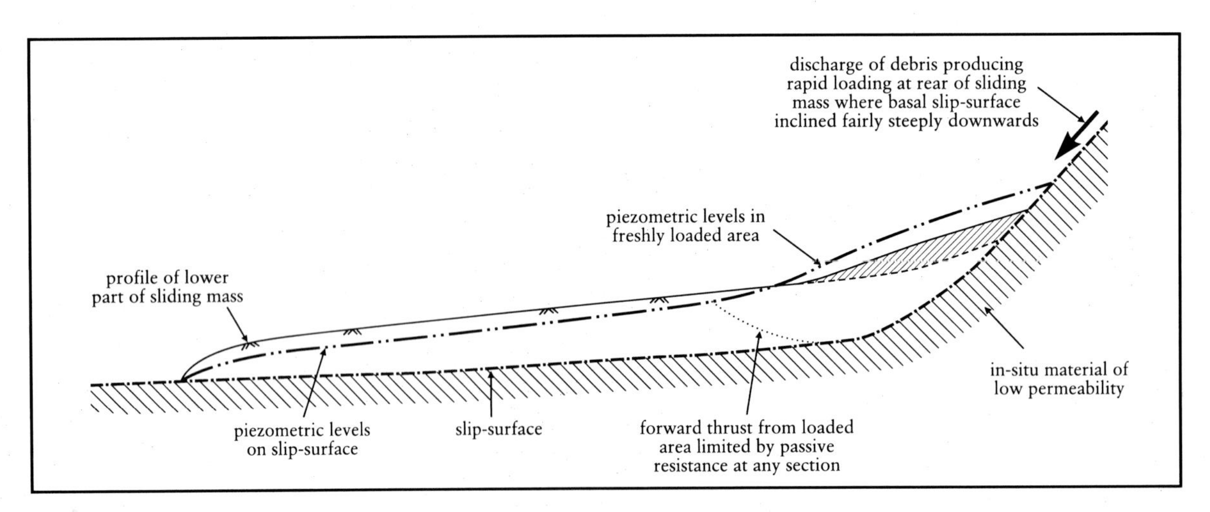

Figure 1.10 The example model of undrained loading suggested by Hutchinson and Bhandari (1971).

Non-circular failure surfaces

Some failures take place over a surface which, when seen in section, has the form of an arc of a circle. A rotational slip over such a surface results in the slipped mass tilting backwards, and the form of the surface enables slipping to happen without the slipped mass breaking up. This observation has been used by geotechnical engineers to provide a simple method of analysing such 'circular failure', using 'circular failure charts' (see, for example, Hoek and Bray, 1977). This, in turn, has led to the expectation that many failure surfaces will have the form of a circular arc. Thus, many slipped masses which have rotated backwards are assumed to have rotated on a circular arc.

That this perception has been recognized as over-simple is illustrated by one of the differences between Hutchinson's 1968a and 1988 mass-movement classifications (Tables 1.1a and 1.1b). The term *listric* ('spoon-shaped'), used in Table 1.1b, refers to a surface that is at all points concave upwards, but of which the radius of curvature decreases downslope. This naturally causes the slipped mass to crack and break up. A further point tending to make circular failures rather unusual is that in sedimentary rocks, at least, the sedimentary sequence is rarely massive enough to be effectively anisotropic with respect to physical properties. As a result, whenever a failure surface meets a pre-existing plane of weakness, it tends to follow it, whether it be a fault, a joint or a bedding plane. An important result of this is that, in many cases, the failure plane may be a non-circular concave-upwards curve beneath the upper parts of a landslip, but follows a sub-horizontal planar bedding beneath the downslope parts (this argument is from Varnes (1978), although Barton (1984) traces it to Taylor (1948); see Figure 1.11).

Rib and Liang (1978) point out that downslope decrease in the curvature of the failure plane produces tension and ultimate failure in the slump block owing to lack of support on its uphill side. This can lead to the formation of a graben in the rear of the slope (Figure 1.11). However, Barton (1984) has observed, from the opportunities that exist for the examination of 'rotational' slips in cross-section, that often the only concave-upwards segment of a slip-surface is of small radius of curvature, at the foot of a straight, steeply dipping segment, and grading into the angle of the bedding on its downslope side (Figures 1.12 and 1.13). He goes so far as

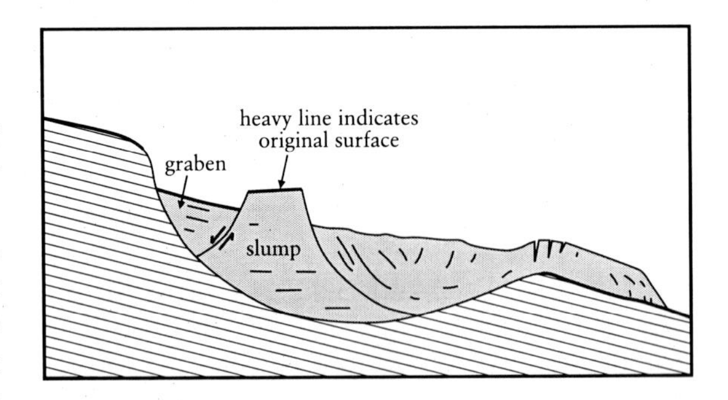

Figure 1.11 Illustration of a 'circular' failure in which the slump block rotates uphill and the graben rotates downhill (after Taylor, 1948). In more recent literature the 'graben' morphology is generally interpreted as being diagnostic of planar failure surfaces (non-circular) often related to the dip of the bedding.

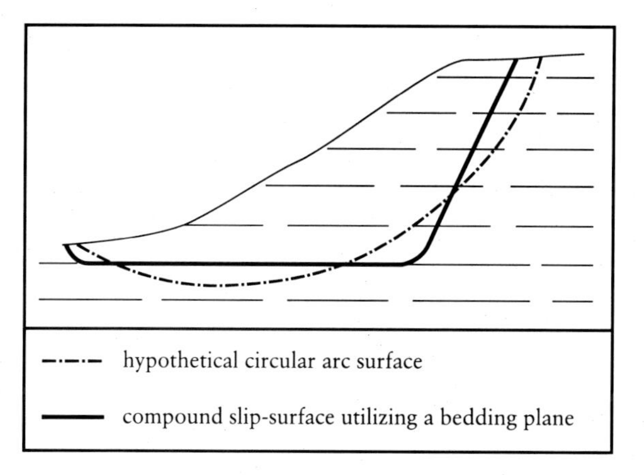

Figure 1.12 The shape of a landslide shear surface in stratified soil with horizontal bedding compared with a hypothetical circular arc surface. After Taylor (1948).

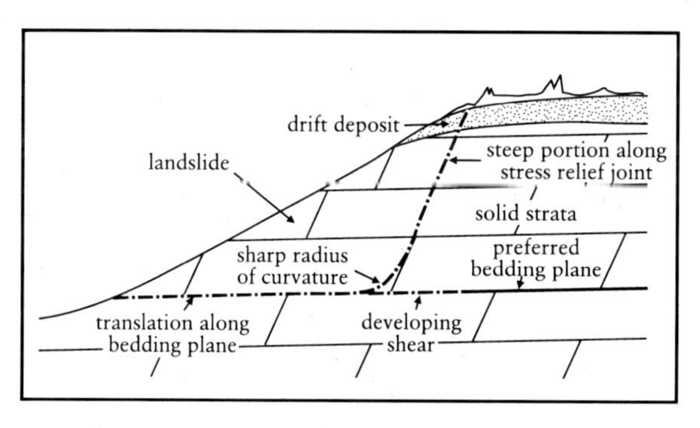

Figure 1.13 The main characteristics of compound landslides with flat-lying bedding. After Barton (1984).

to suggest that this is such a common observation worldwide that it should be 'regarded as the norm and such a surface should be assumed until, and unless, definite evidence to the contrary is obtained'. This conclusion is amply borne out by the mass-movement sites selected for the GCR. Further, although this is not mentioned in its accompanying text, the diagram illustrating a 'single' landslide in *The Multilingual Landslide Glossary* (Figure 1.8) shows a failure surface of this type.

GENERAL CHARACTERISTICS OF GCR SITE DESCRIPTIONS

The length and detail of each site description herein has been determined by the volume of research that has been published on the site. Generally, the most significant sites in terms of the development of understanding of mass movements in Great Britain are those that have received the most detailed study, often over a long period of time, and over which contending views may have developed. On the other hand, some sites have been selected about which very little has been written, but which nonetheless exhibit features of special interest. In these cases the text concentrates on general description rather than detailed scientific explanation. It is hoped that this will provide an incentive, justification and/or rationale for further research.

Overall, the site descriptions vary considerably in length, detail, and degree of illustration. To have imposed a rigid uniformity on the descriptions would have failed to give an accurate impression of the variety of mass-movement sites to be found in Great Britain, and would have failed to do justice to the most intensively studied sites, those with innovative and/or enterprising methods of study, and those with the longest history of study.

Consideration was given to providing each site description with a stereopair of aerial photographs, so that the physiographical expression of mass movement at each site may be illustrated. However, there is a risk that at some sites this could lead to an inappropriate concentration on the physiographical aspects of the site and not their scientific causes/importance *per se*. Also, woodland or forest vegetation tends to obscure or smooth over such features as viewed from above.

Many of the site descriptions are provided with cross-sections. Where they are not, this is because no reasonably accurate cross-section has been published. However, some of the sites are illustrated by slope profiles measured by the present author (RGC). In all cases these were measured in successive 1.5 m ground lengths using a slope pantometer (Pitty, 1966). They run directly downslope, and are orthogonal to the contours. In order to avoid this orthogonal line appearing as a curve in plan, locations for measurement were selected where the contours were roughly parallel to each other. All profiles were originally plotted at a scale of 1:400, and are drawn without vertical exaggeration.

Chapter 2

Mass-movement GCR sites in Precambrian and Cambrian rocks

INTRODUCTION TO THE MASS MOVEMENTS IN THE OLDER MOUNTAIN AREAS OF GREAT BRITAIN

D. Jarman

The older mountain ranges of Britain – the Scottish Highlands, the Southern Uplands, the Lake District, and the northern half of Wales (Figure 2.1) – have long been prized for both their exceptional landscape value and their scientific interest. They were fashioned during the Caledonian Orogeny 480–390 million years ago, mainly in metamorphic rocks of Precambrian, Cambrian, Ordovician, and Silurian age, but with some contemporaneous igneous intrusions. Mass movements in these ranges differ considerably in character, cause, mechanism, and geomorphological effect from those on the Devonian, Carboniferous, Mesozoic and Cenozoic cliffs, escarpments and valley

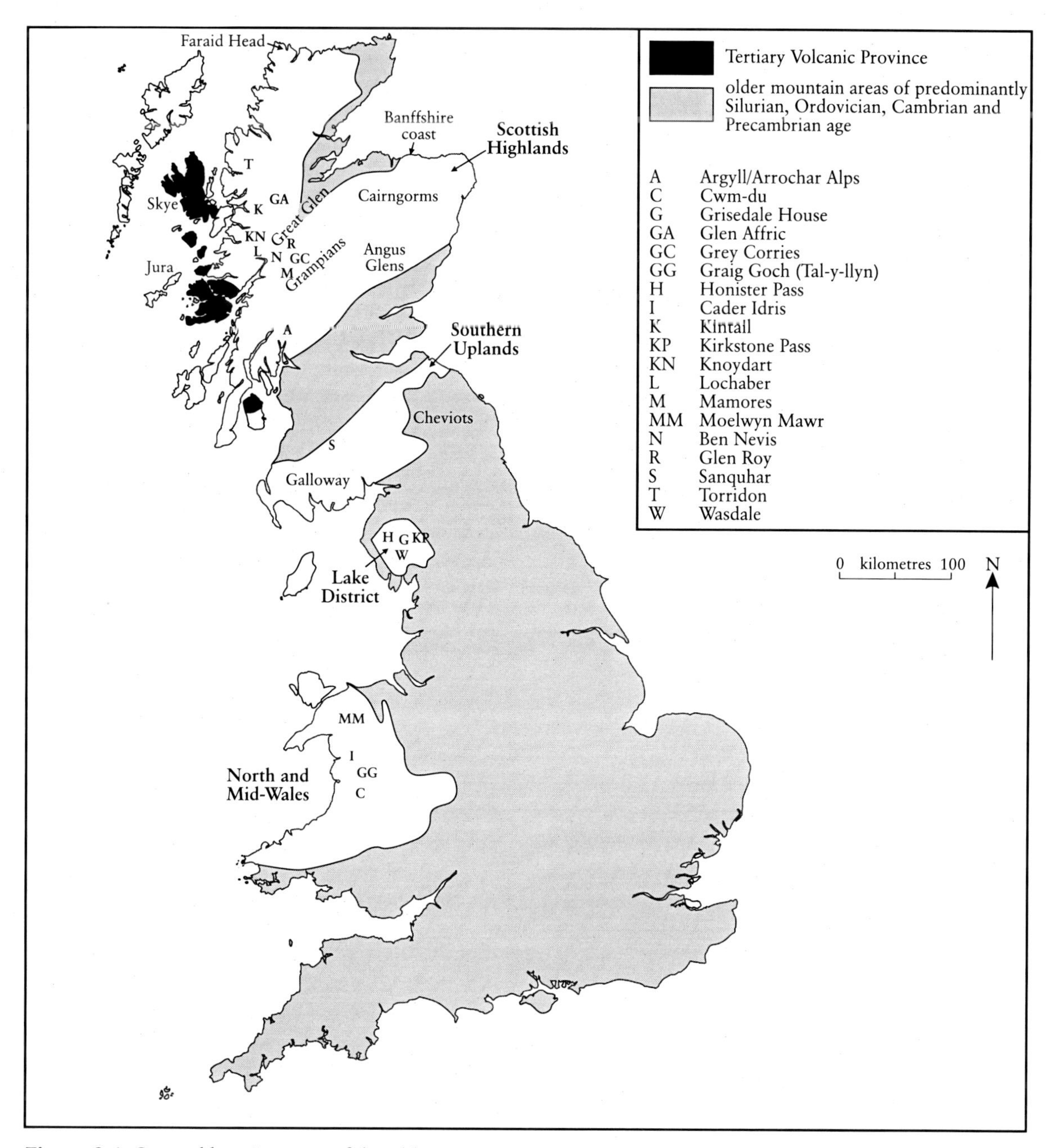

Figure 2.1 General location map of the older mountain ranges in Britain. Locations of GCR sites in this chapter are shown in Figure 2.13. Other sites within the older mountain areas are **Cwm-du** (see GCR site report, Chapter 3) and **Coire Gabhail** (see GCR site report, Chapter 4).

slopes described in following chapters of the present volume.

This chapter focuses on deep-seated mass movements in bedrock ('rock slope failure' – RSF). Mass movement in superficial deposits and mass wasting of rockfaces, both active and relict, is of course prevalent in many parts of these mountains, and may become more significant again as climate change favours more extreme events. However manifestations such as screes, debris cones, solifluction terraces, and rock glaciers are well represented by sites in the Quaternary volumes of the GCR series, *Quaternary of Scotland* (Gordon and Sutherland, 1993), *Quaternary of Wales* (Campbell and Bowen, 1989) and *Quaternary of Northern England* (Huddart and Glasser, 2002).

Paraglacial rock slope failure

Rock slope failure is principally a 'paraglacial' phenomenon in the British mountains. In other words it is not a direct product of glacial or glacio-fluvial activity, but is a geomorphic response to glaciation and deglaciation, a process of stress-release and re-equilibration (Ballantyne, 2002a). Nearly all of the recorded rock slope failures are on the flanks of glacial troughs (including those submerged as fjords), or in the corries/cwms which feed them. Non-paraglacial failures can also be found on sea cliffs (e.g. Faraid Head, Durness; the Banffshire coast), and with instructive rarity above fluvial gorges (e.g. Arnisdale by Loch Hourn (Figure 2.2), Craig Maskeldie in the Angus glens (NO 386 795)).

A special case arises with the Tertiary [Palaeogene] Volcanic Province of the Western Highlands and Inner Hebrides (Figure 2.1). Here numerous extensive mass movements occur where massive basalts overlie less competent strata. They are more comparable in behaviour and in expression with the coastal sites of England, and following the scheme of this volume are dealt with in Chapter 6. Many of them are still active or could easily be re-activated, whereas nearly all of the rock slope failures in the old hard rocks of the mountains are long dormant or 'metastable'. The failure sites selected at the **Trotternish Escarpment** in Skye (see Chapter 6) also represent the extensive plateau-rim failures formed on Devonian and Carboniferous lavas in the Midland Valley of Scotland (Evans and Hansom, 1998; see also Craig Rossie GCR site report in Stephenson *et al.* (1999)).

Figure 2.2 Beinn Bhuidhe rock slope failure, Arnisdale, Loch Hourn, Western Highlands (NG 860 113). A typical armchair slide with slope toe exceptionally undercut by deep fluvial (rather than glacial) incision, and thus not directly a paraglacial rock slope failure. (Photo: D. Jarman.)

Extent of rock slope failure and state of knowledge

By contrast with mass movement in the sedimentary slopes of Britain, there has been remarkably little research into rock slope failures in the mountain areas. This is partly because of their isolation and low geohazard status, and partly because they are enigmatic features, difficult to address within the canons of either geomorphology or engineering geology.

Scotland

Most is known about the Scottish Highlands. Here, a systematic but unpublished study of aerial photographs yielded 364 extant rock slope failures (Holmes, 1984), and together with incomplete British Geological Survey mapping and other sources gave Ballantyne (1986a) a total population of 495 in the mainland Highlands. Detailed field survey indicates that in some areas this might be augmented by 20% (Jarman, 2003a). Three unpublished theses provide information on collections of sites including some valuable geotechnical analyses (Watters, 1972 – 20 cases; Holmes, 1984 – 27 cases; Fenton, 1991 – 44 cases). A few individual sites have been recorded in the literature, but these have generally not been of seminal status (e.g. Beinn nan Cnaimhseag; Sellier and Lawson, 1998).

In the Southern Uplands, rock slope failure is sparse and low-key, with isolated cases, for example in the Galloway mountains (Cornish, 1981), north-east of Sanqhuar, and in the Cheviots (W. Mitchell, University of Durham, pers. comm.). Ballantyne (1986a) recorded 24 sites.

England

Systematic investigation of rock slope failure in the Lake District is beginning to emerge, with around 50 sites affecting at least 5.5 km^2 in total already identified (Wilson *et al.*, 2004), . Some are quite substantial, with one at Robinson–Hindscarth (Buttermere) being large (1.7 km^2) and significant in UK terms (Wilson and Smith, 2006). Others display bold features, notably the antiscarps on Kirkfell, Wasdale (Wilson, 2005) (Figure 2.3) and the Fairfield complex (see Figure 2.10).

Figure 2.3 Kirk Fell rock slope failure, Wasdale, Lake District. A classic virtually in-situ slope deformation, with an antiscarp 600 m long crossing the summit plateau and others on the south-west flanks. (Photo: P. Wilson.)

Wales

In northern and mid Wales, there is no systematic survey of paraglacial mountain rock slope failure, although active failure in coastal old hard-rock exposures is of continuing interest (cf. Nichol, 2002). A few individual sites have been reported in Snowdonia (e.g. Curry *et al.*, 2001; Rose, 2001), and some in the Berwyns such as that damming Llyn Moelfre (SJ 180 285) (Hutchinson, unpublished data). One substantial site selected here at **Cwm-du** (see Chapter 3) represents behaviour in weakly indurated Silurian metasediments, but is most noted for the uncertainty surrounding its origins. Another site at Tal-y-llyn near Cader Idris has been thoroughly investigated (Hutchinson and Millar, 2001, fig. 48) and is notable as the largest landslide dam in Britain. A kilometre of glacial trough wall cut in Ordovician metasediments collapsed, with 50×10^6 m^3 of debris impounding the lake of Tal-y-llyn, once 2.5 km and now 1.6 km long. A substantial extension to the failure scar has become arrested after short travel. This site is of international significance (Nichol, 2002).

GCR site selection

When the original shortlisting of mass-movement sites was made in 1982, only three sites were put forward in the mainland Highlands, reflecting the dearth of published investigations. Of these, two were major discoveries arising out of unpublished PhD theses – Glen Pean (de Freitas and Watters, 1973) and **Beinn Fhada** (Holmes and Jarvis, 1985). The third Scottish site at **Coire Gabhail** (see Chapter 4) was famous as the landslide-blocked Lost Valley of Glencoe; it is the only known rock slope failure on high-strength Devonian lavas in the Highlands.

Three GCR sites selected for the Quaternary of Scotland GCR Block (Gordon and Sutherland, 1993) are also relevant to the subject of the present volume. One is in Dalradian quartzite and two are in Precambrian Torridonian sandstone: Beinn Shiantaidh on Jura (Gordon and Mactaggart, 1997) and Baosbheinn in Torridon are now regarded as more probably rock slope failures rather than rock glacier and protalus rampart cases respectively, while **Beinn Alligin** has been described in the present volume (as well as in Gordon and Sutherland, 1993), as cosmogenic dating (Ballantyne and Stone, 2004) has largely resolved the controversy over its mode of emplacement (although it still does not fully merit the designation of 'sturzstrom').

Systematic characterization of rock slope failures in several parts of the Highlands (Jarman, 2003a,b; Hall and Jarman, 2004), and of all 140 larger failures in the Highlands (Jarman, 2006) has demonstrated their great diversity. It has also underlined the previously overlooked importance of their contribution to erosion and landscape shaping over Quaternary times. It became clear that rock slope failure in the older mountain areas could not adequately be represented by just six sites, of which three were in lithologies where failure is exceptional and relatively small-scale. Of the three in metasedimentary rocks, **Cwm-du** (Chapter 3) has been studied mainly as a quasi-glacial deposit; and while Glen Pean and **Beinn Fhada** are two of the largest and most impressive failures in Britain, they are of rather similar character and setting, and both occur within similar geological contexts in the North-west Highlands.

In reviewing the Highland rock slope failure sites in 2003, to ensure that the GCR encompassed the full spectrum of characteristics (including geological context, type of failure, landshaping effects), eight additional GCR sites were proposed and are described in the present chapter – **Beinn Alligin**, **Ben Hee**, **Benvane**, **Carn Dubh**, **The Cobbler**, **Druim Shionnach**, **Glen Ample** and **Sgurr na Ciste Duibhe**. The Glen Pean site was recommended for deletion (where the original interpretation is now found to be implausible, cf. Jarman and Ballantyne, 2002); this type of rock slope failure is better represented by **Beinn Fhada**.

This introduction sets out the general context for understanding the diversity and significance of rock slope failure in the older mountain areas, starting with the main characteristics that the selected sites seek to represent.

Representing the diversity of geology and structure

Lithological controls

The vast majority of rock slope failures occur in the metamorphic lithologies (Ballantyne, 1986a). This is unsurprising given that most of the older mountain ranges are constructed from them.

But significant failure can occur in every lithology, including Torridonian sandstone (e.g. **Beinn Alligin**), volcanic lavas (e.g. **Coire Gabhail**, Chapter 4), and granite (e.g. Lundie, NH 164 114). A notable complex on granite affects 3 km on both sides of Strath Nethy, beside Cairn Gorm (Hall, 2003). This complex falls within the Cairngorms GCR site, selected for the Quaternary of Scotland GCR Block (Gordon and Sutherland, 1993). Current investigations at this site by the British Geological Survey suggest an unusual combination of glacial, periglacial and paraglacial activity.

Metamorphic rocks in the British mountains range in age from dominantly Precambrian in the Scottish Highlands and Islands, to mainly Ordovician and Silurian in the Southern Uplands, Lake District and North Wales. In the Highlands, the Moine and Dalradian Supergroup rocks are mainly composed of metamorphosed and deformed sandstones, siltstones and mudstones. Here the term 'schist' has been applied commonly to the more indurated Highland rocks, with the dominant psammitic (i.e. sandy) variants being termed 'quartz schists'. In the Palaeozoic ranges the metasedimentary rocks are often less indurated, with the generic term 'slate' including friable greywackes (see **Cwm-du** GCR site report, Chapter 3). But rock slope failure occurs across all metamorphic types and grades, including those of igneous origin such as the ancient Lewisian gneisses and the Borrowdale volcanic rocks. It tends to be more widespread in slaty and interbedded strata, where mica-rich cleavage and foliation surfaces facilitate sliding, but can equally well operate in blocky to massive and relatively uniform psammite terrains such as the central Grampian Highlands (Hall and Jarman, 2004).

Structural controls

Structurally, the metamorphic rocks are more prone to develop deep-seated failure planes, by virtue of profound tectonic activity during the Caledonian Orogeny and (in the North-west Highlands) earlier orogenies. As well as the foliation (schistosity) surface, three or more joint-sets are commonly present (Watters, 1972) (Figure 2.4), so that potential sliding surfaces, depending on their pervasiveness, can be available on most slope aspects. By contrast, granite is generally more sparsely jointed, and it 'springs' on shallow fracture surfaces that develop parallel to the present or original slope. Torridonian sandstone also tends to fail at joint-block scale, although slices of cliff have collapsed on near-vertical joints, and mass creep

Figure 2.4 A typical small crag in Moine psammites displays four distinct discontinuities (the foliation or schistosity surface and three joint-sets), which have released a miniature wedge failure. (Photo: D. Jarman.)

has occurred on gently dipping bedding planes (e.g. Beinn Bhàn, Applecross (NG 800 450)).

Mountain-building processes have left the metasedimentary rocks inclined at all angles from sub-horizontal to sub-vertical, and in all scales and intensities of folding. The textbook ideal for large-scale sliding is a smooth surface inclined close to the peak or residual friction angle (Hoek and Bray, 1981), typically 20°–40° for schists. Any gentler, and friction prevents sliding, any steeper, and the surface becomes less likely to have been undercut by glacial trough steepening; ultimately it becomes a self-supporting wall. However, rock slope failure occurs freely in rocks inclined at all angles and showing every degree of folding and contortion, if not on the foliation or bedding surface then on joint-sets that cut through the contortions, and if not by sliding alone then by creep, sag, buckle or topple, or any combination.

The sites selected here show that while geological controls can be direct and obvious (e.g. **Ben Hee**), rock slope failure can develop in a wide range of contexts, and in some cases without an obvious relationship to any observable structures (e.g. **Beinn Fhada**, **The Cobbler**). The relatively straightforward analyses and predictions that can be made for failures in the regular sedimentary strata of Britain become more elusive in the older mountain areas.

Representing the diversity of rock slope failure types

The general introduction to this volume follows the classification of Hutchinson (1988), but observes that any attempt to classify mass movements is unsatisfactory because firstly they are on a continuum, and secondly most are complex, embodying several modes of failure. This is especially true of the older mountain areas. Characterization of rock slope failures in the older mountain areas like the Scottish Highlands (Table 2.1; Figure 2.5) has adapted the Hutchinson schema to reflect prevailing modes there, with five broad categories spanning the continuum (Jarman, 2006):

- compressional deformation
- extensional deformation
- arrested translational sliding
- sub-cataclasmic slide/collapse
- cataclasmic slide/collapse

Slope deformation

The first two categories (compressional and extensional deformation) cover slope deformations where the lateral margins are diffuse, and downslope movement is limited. They tend to be extensive, and account for 64% of the larger (> 0.25 km^2) Highland rock slope failures

Table 2.1 Characteristic types of large rock slope failures (RSFs) in the Scottish Highlands and Lake District. Adapted from Jarman (2006) and Wilson *et al.* (2004). See Figure 2.5 for explanation of terms.

		Scottish Highlands	**Lake District**
RSF size	0.25–0.49 km^2	67	5
	0.5–0.99 km^2	61	1
	1.0–1.99 km^2	16	1
	2.0–3.0 km^2	3	–
total RSFs	0.25–3.0 km^2	**147**	7
RSF predominant mode			
rockslides (all degrees of arrestment/disintegration)		**54**	4
of which	cataclasmic	3	–
	sub-cataclasmic	14	–
	arrested short–medium travel	37	4
slope deformations		**92**	3
of which	extensional (sag and creep)	68	3
	compressional (rebound) including Cluanie hybrids	24	–
Association with glacial breaches (including tributary troughs)			
	main watersheds	55	1
	secondary watersheds	56	1
	no close association	36	5

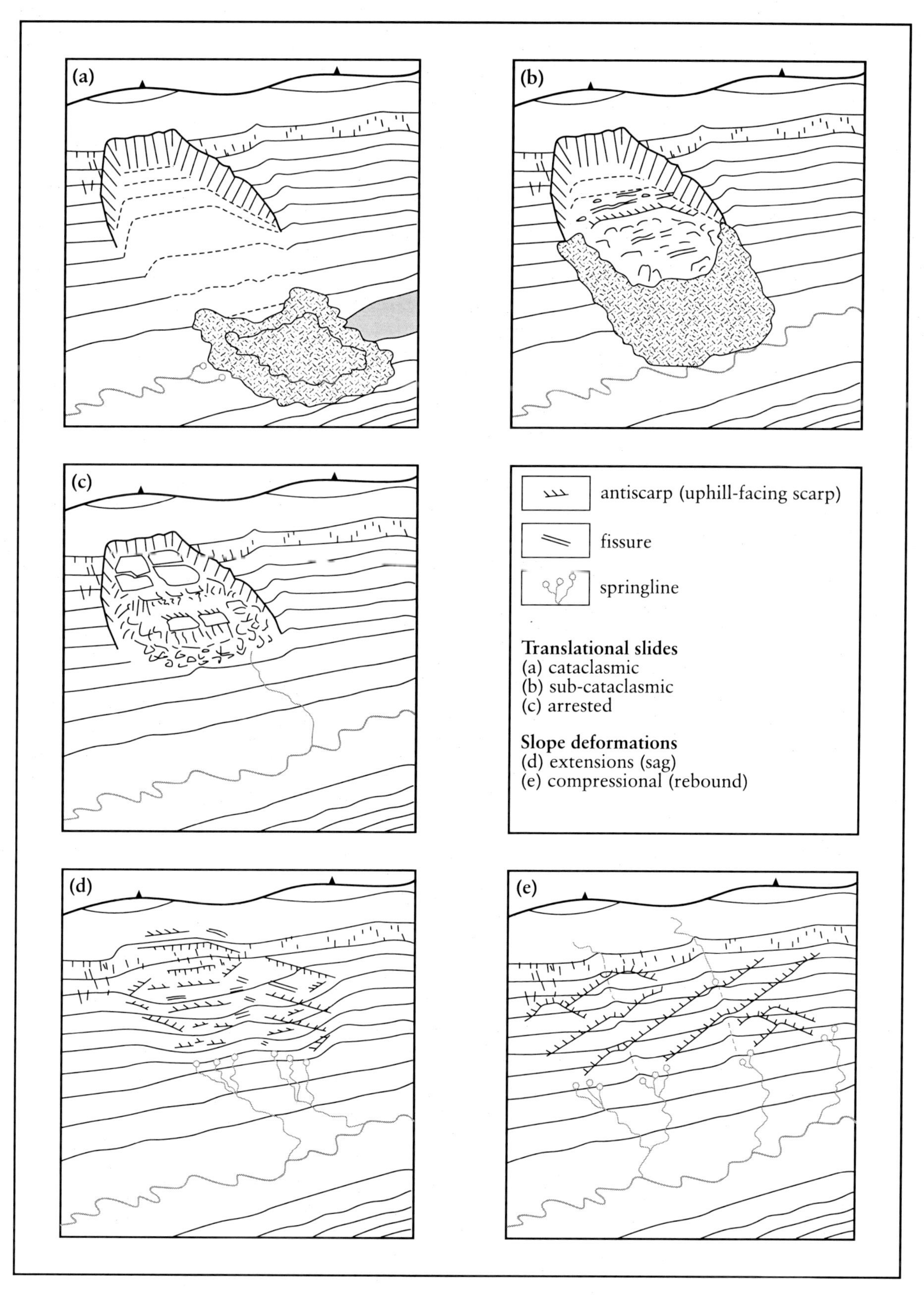

Figure 2.5 Characteristic types of larger-scale rock slope failure identified in the Scottish Highlands. The plateau rim location is typical of less intensely dissected terrain. In more acute relief, headscarps may daylight below or behind the crest, or split the ridge. After Jarman (2006).

(Jarman, 2006). Extensional deformations display creep, sagging, or bulging usually with some headscarp development and tension features such as furrows, fissures and trenches. They include the 'sackungen' first described in the Alps. Compressional deformations are less well attested, but correspond to Hutchinson's 'rebound' category. They are virtually *in situ*, lacking in tension features, and evidenced mainly by disrupted drainage (dry slope with basal springline). The distinction between these types is subtle, and often best attested by the character of their antiscarp arrays (see below). Many slope deformations combine both compressional and extensional indications.

Varying degrees of deformation are represented at **Beinn Fhada**, **Benvane**, **Sgurr na Ciste Duibhe**, and **Glen Ample**. These are very large sites, affecting at least a square kilometre of valley-side; Beinn Fhada and Ben Our (Glen Ample) are two of the largest rock slope failures in the older mountain areas, and are of exceptional significance for their clarity of surface expression and scope for research into slope deformation processes.

Arrested translational sliding

Arrested translational sliding is the prevalent mode of failure in the mountain areas, accounting for 60% of all rock slope failures in the South-west Highlands (Table 2.2), although only 25% of the larger sites across the Highlands. Here a well-defined mass has failed and travelled some distance downslope before becoming arrested in a more-or-less coherent state. There is usually a well-defined source cavity, which may be wedge or armchair-shaped. The failed mass may have antiscarps, open fissures, detached megablocks, and basal debris-lobes or block piles. The reasons for arrestment are poorly understood, but probably reflect inadequate lubrication, rough or corrugated failure planes, and lack of vertical relief; such arrested slides seem less apparent in higher mountain ranges of the world.

Arrested slides are represented at **The Cobbler** (very short travel from a headscarp but much dislocated), **Benvane** (long travel from a classic source bowl, low-key lobe) and **Ben Hee** (short travel from a corrie headwall). Good examples of armchair-source slides are at An Sornach in Glen Affric (Holmes, 1984; Jarman, 2003c), Glen Fintaig in Lochaber (Watters, 1972), and Beinn Tulaichean at Balquhidder (Jarman, 2003a) (Figure 2.6), while Tullich Hill above Glen Douglas (Luss Hill) is an instructive wedge-source complex (Jarman, 2003d) (Figure 2.7).

A hybrid category between the in-situ compressional deformation and the translational slide is common in the Cluanie area of the Western Highlands (Jarman, 2003b), and elsewhere on steeply inclined high-competence lithologies, such as the Mamores and Grey Corries. It is represented by **Druim Shionnach**.

Cataclasmic and sub-cataclasmic failure

Cataclasmic rock slope failure is so termed because it is not literally 'catastrophic', but describes where the failed mass has disintegrated and travelled to the slope foot or beyond (Jarman, 2002, 2006). Fully cataclasmic events are very uncommon in Britain, and are here represented by **Beinn Alligin**, and by **Coire Gabhail** (Chapter 4). The term 'sturzstrom' has occasionally been borrowed from high mountain areas to indicate rockslides with excessive runout, but even Beinn Alligin scarcely warrants this designation, lacking the vertical relief and the jump-off bench normally required.

Sub-cataclasmic failure is commoner, where although substantially disintegrated, the debris lobe has barely reached the slope foot, and remains only conditionally stable. **Carn Dubh** (Ben Gulabin) represents this mode with two small debris-tongues. Other tongues analysed include Mam na Cloiche Airde (Watters, 1972) and Gleann na Guiserein (Bennett and Langridge, 1990; Ballantyne, 1992), both in Knoydart, Mullach Fraoch-choire in Glen Affric (Holmes, 1984) (Figure 2.8), and Burtness Comb in the Lake District (Clark and Wilson, 2004). Moelwyn Mawr in Snowdonia (Rose, 2001) may well be similar.

A variant of sub-cataclasmic failure is the crag collapse leaving a coarse blocky debris-pile on the slope. In the South-west Highlands, 50% of sites display an element of (sub-)cataclasmic behaviour, the massive schistose grit lithology being more conducive to it than in areas of slabby metasediments. A fine example is on Beinn an Lochain North (Watters, 1972; Holmes, 1984) (Figure 2.9).

Complex failures

Many rock slope failures demonstrate several modes of failure, as at **Sgurr na Ciste Duibhe**.

Table 2.2 Rock slope failure (RSF) incidence, character, landshaping effect, and association with breaching in the Southern Highlands and Kintail area (including clusters 1, 5 and 7 in Figure 2.13). Updated from Jarman (2003a,b). Note: sites may be in more than one character or landshaping category. See Figures 2.15 and 2.18.

	Southern Highlands				**S. Affric/Kintail/Glen Shiel**
	1W	1E	2	**Total**	7N/8S
Number of RSFs	**119**	**40**	**13**	**172**	**54**
< 0.25 km^2	86	33	8	**127**	33
0.25–0.99 km^2	31	6	4	**41**	17
1.00–3.00 km^2	2	1	1	**4**	4
Extent of RSF (km^2)	**27.9**	**7.0**	**5.2**	**40.1**	**18.6**
average size (km^2)	0.23	0.17	0.40		0.35
% of densest core area affected by RSF	7.7	7.2	16.7		6.0
extent of core area (km^2)	112	40	26		41
RSF character (number of)					
arrested translational slides	48	20	11	**79**	25
sub-cataclasmic failures	35	21	6	**62**	6
slope deformations	6	10	5	**21**	23
incipient failures	28	10	5	**43**	5
not ascertained	26	8	1	**35**	–
Landshaping contribution					
glen and trough widening	89	17	9	**115**	38
corrie enlargement	13	13	1	**27**	11
corrie initiation	11	2	1	**14**	–
spur truncation	39	11	6	**56**	9
crest sharpening, arêtes and horns	39	16	7	**62**	19
ridge reduction	8	5	0	**13**	23
potential watershed breaching/ dissection	3	2	2	7	6
elimination of mountain blocks	12	4	1	**17**	2
Association with evolving glacial breaches					
at a 'recent' or enlarging breach	20	5	5	**30**	27
near a breach (< 2 km downflow)	24	15	4	**43**	
in a side trough rejuvenated by a breach below					11

Hell's Glen (Holmes, 1984) dramatically illustrates a progression from extensional deformation, producing a 15 m-deep antiscarp trench slanting across the midslope, with large translated masses breaking away into detached megablocks up to 60 m high, and some cataclasmic collapse debris reaching the glen floor.

Representing the landshaping effects of rock slope failure

Paraglacial rock slope failure has played a significant, if generally unremarked, role in shaping the present mountain topography. Its geomorphological impacts are numerous (Jarman, 2003a,b; Table 2.2), and may appear as isolated incidents (e.g. **Carn Dubh** (Ben Gulabin)) or, where its occurrence is dense, as a more generic contribution to Quaternary landscape evolution.

Arêtes and horns

The most striking effects of rock slope failure can be seen where summit ridges have been narrowed and incidented. This has been recognized in attributing arêtes in the ranges around Ben Nevis to rock slope failure (Bailey and

Figure 2.6 Beinn Tulaichean, Balquhidder, Southern Highlands (NN 420 196). A classic short-travel arrested translational slide from a splayed armchair source which splits the summit ridge. Despite the blocky veneer and spray fan, the failed mass is substantially intact, with double-decker-bus-sized fissures in the upper area. (Photo: D. Jarman.)

Figure 2.7 Tullich Hill rock slope failures, Luss Hills, South-west Highlands (NS 292 998). A translational slide complex from a multiple wedge source; the inner cavity in the west (left) rock slope failure is 30 m deep. (Photo: D. Jarman.)

◂**Figure 2.8** Mullach Fraoch-choire, Glen Affric (NH 102 187). A sub-cataclasmic debris-lobe 20 m thick with fine levées almost reaches the stream. It emanates from a narrow source pocket, and descends 380 m in 800 m. (Photo: D. Jarman.)

Maufe, 1916), and in interpreting the horn of Streap near Glenfinnan as a summit sliced by a slide scar (Watters, 1972). **The Cobbler** has been selected to display these summit ridge effects, while at **Sgurr na Ciste Duibhe** the summit mass has been lowered bodily by about 10 m, leaving a fretted arête. Good examples in the Lake District are the horns of Helm Crag (Grasmere (NY 325 090), and the Cofa Pike arête on Fairfield (NY 355 129; Figure 2.10).

Corrie development

It has been speculated that rock slope failure may play a part in 'seeding' corrie development (Clough, 1897; Peacock *et al.*, 1992; Turnbull and Davies, 2006; see **Cwm-du** GCR site report, Chapter 3). Certainly in the minority of cases where failure occurs within a corrie, it is contributing to corrie enlargement and ultimate

Figure 2.9 Beinn an Lochain North rock slope failure, Arrochar Alps (NN 217 083). A sub-cataclasmic rock-slide, partly fallen from the cliffs (behind which deep fissures indicate incipient increments), but with a sliding component that has sharpened the summit ridge (top left) to an arête. (Photo: D. Jarman.)

Figure 2.10 Fairfield rock slope failure, Lake District (NY 355 120). An extensive translational slide, with long-travel masses arrested above the floor of Grisedale, which breaches through to Grasmere and Dunmail Raise just to the west (right). The headscarp encroaches into the north-east ridge (left) to shape the arête of Cofa Pike. (Photo: D. Jarman.)

destruction, whether by reduction of the enclosing arms (e.g. **The Cobbler**) or breaking through the headwall (e.g. **Ben Hee**).

Valley widening and ridge reduction

Most rock slope failures contribute to the general processes of valley widening and spur truncation (Table 2.2), both directly and indirectly, by providing weakened slopes and debris masses ready for erosion and evacuation by glaciers in the next cycle (Bentley and Dugmore, 1998). By the same token, rock slope failures encroach into ridges and pre-glacial plateau remnants, with source scarps and fractures often daylighting by as much as 50–100 m behind the crest or rim (e.g. A' Chaoirnich in the Central Grampians) (Jarman, 2004a). Such effects are seen particularly well at **Beinn Fhada**, and in lower-relief contexts in **Glen Ample** and at **Benvane**.

Where mountain ranges are dissected by breaching, the isolated mountains become more vulnerable to concerted attrition by rock slope failure, until their crests are lowered sufficiently to permit glacial over-riding and reduction to subdued relief. Examples of mountains at various stages of isolation by breaching and where rock slope failure is encroaching on several fronts include Beinn an Lochain in Argyll (Watters, 1972), Ciste Dhubh in Kintail (Jarman, 2003b), An Dùn in the Gaick Pass (Jarman, 2004a), and Na Gruagaichean in the Mamores (Figure 2.11).

Antiscarps

Antiscarps are one of the most conspicuous indicators of rock slope failure in the mountain landscape. These uphill-facing scarplets occur both in translational slides, as the moving mass begins to disaggregate (**The Cobbler**), and in slope deformations, where they may extend laterally for hundreds of metres (e.g. the exceptionally fine array on **Beinn Fhada**). They may develop intricate lattices (**Benvane**) and platy structures (**Glen Ample**). In the Lake District, a notable case occurs on Kirk Fell (Wilson, 2005; Figure 2.3). Antiscarps typically reach up to a few metres high, and where only decimetric may be hard to see on the ground. In Glen Shiel and Kintail, and a few other isolated cases (including Kirkstone Pass in the Lake District – Wilson *et*

Figure 2.11 Na Gruagaichean rock slope failure complex, Mamores, Lochaber (NN 195 650). The twin summits (centre and right) are divided by a 140 m-deep gash, the source of a very large wedge slide that has been substantially evacuated leaving a SW-facing bowl that is not a corrie in origin or by adaptation, the floor of which is extensively ruptured with antiscarps up to 3 m high. Another large rock slope failure encroaches onto the south ridge (right), and a third slide lobe sharpens the north-west ridge (left-centre). (Photo: J. Digney.)

al., 2004), they can attain 5–10 m, with **Beinn Fhada** having the highest classic midslope antiscarps. **Druim Shionnach** is exceptional, with a 14 m antiscarp, but this may have developed as part of a graben structure opposite the source scarp.

Research is needed to understand why and how antiscarps develop. Their incidence and character may illuminate the extent to which deformation is either extensional in unsupported steep slopes, or compressional, driven by differential glacio-isostatic rebound stresses between valley floor and summit ridge. The factors governing antiscarp height are especially unclear. It could be an indicator of rebound stress intensity, but will also depend partly on rock-type and strength, and on length of exposure to weathering. Extant heights may be much reduced from their original levels, partly by crest degradation but mainly by trench infilling. It is rare to find datable bedrock exposures on an antiscarp, or a fault-like fracture in the few exposed cross-sections. Some antiscarp arrays may have been subdued by the Loch Lomond Stadial (LLS) glaciers, or re-emerged after them (**Beinn Fhada**; Beinn Odhar Bheag, Glenfinnan, NM 850 775).

A special case of apparent antiscarp occurs where the failed mass includes the summit ridge, and the fracture plane 'daylights' behind the crest. This creates the phenomenon known as a 'split ridge', common in the Alps as a 'doppelgrat' (Crosta, 1996). **Sgurr na Ciste Duibhe** displays aspects of this tendency, but a remarkably clear example extending intermittently for over 1.5 km and attaining a source scarp height of 10–15 m has recently been recognized on Aonach Sgoilte in Knoydart (Figure 2.12).

The spatial distribution and root causes of rock slope failure

Perhaps the most puzzling aspect of the rock slope failure phenomenon is its irregular spatial distribution. Even on mountain ranges most conducive to it, failure is far from endemic: it may be abundant, sparse, or absent on valley-sides of similar scale, steepness and geological character. Where rock slope failure is common, it is seldom obvious why one section of valley slope should have failed rather than adjacent sections; local factors are clearly important. Given that most failures appear to date from early or mid-Holocene times, this suggests that at periods of maximum rebound and climatic stress, extensive areas of valley-side were close to the limits of stability.

Figure 2.12 Aonach Sgoilte rock slope failure, Knoydart (NG 836 020). The ridge is split over 1.5 km by a source fracture daylighting behind the crest, often 30 m downslope on the north (right). The summit mass seen here has slipped south by 10–15 m. The gentler slope below is extensively antiscarped, one being 500 m long and reaching 7.5 m in height. A wedge slip has left the shadowed notch. The rounded headscarp crest (contrasting with **Sgurr na Ciste Duibhe**) could indicate relatively ancient inception. (Photo: D. Jarman.)

Association with ice limits

Two studies have addressed the wider distribution of rock slope failure, though regrettably neither has been published to expose the debate. Holmes (1984) found that in nearly all cases, some force, augmenting gravity, was required to mobilize translational sliding. He sought to demonstrate a close correlation between failure locations and the upper limits of the Loch Lomond Stadial (LLS) glaciers, as identified by Sissons and his school. He envisaged that excessive meltwater pressures at deglaciation would trigger a spate of failures. Unfortunately, where the LLS was a valley-full glaciation with nunatak ridges exposed, there is an inevitable co-incidence between steep valley-sides where rock slope failure is most likely to occur, and the upper limits of the glaciers. Where the LLS achieved near-icecap coverage at its centres, as is now thought likely (Golledge and Hubbard, 2005), such a correlation obviously cannot occur. In fact, failure can be found at all levels within any glaciated valley system, both within and well beyond the LLS outer limits. Furthermore, this model does not explain adequately the absence of rock slope failure from apparently suitable terrain well within the LLS outer limits such as Ardgour (see Figure 2.13). Nor does an engineering model developed for discrete, compact translational block-sliding account for the large proportion of diffuse and laterally extensive slope deformations. Finally, the model cannot account for the rock slope failures that have been dated (directly or inferentially) to several thousand years after deglaciation.

Figure 2.13▸ Spatial distribution and size of 140 larger rock slope failures (RSFs) (> 0.25 km^2) in the mainland Scottish Highlands (distribution of all rock slope failures is similar). Rock slope failure is clustered on main watersheds that have been breached and displaced during Pleistocene times. It is scarce in ranges away from the watersheds, in the far north where ice cover was thinner, and in the eastern Grampians where glacial dissection is less intense. Sites reported in this chapter are shown. After Jarman (2006).

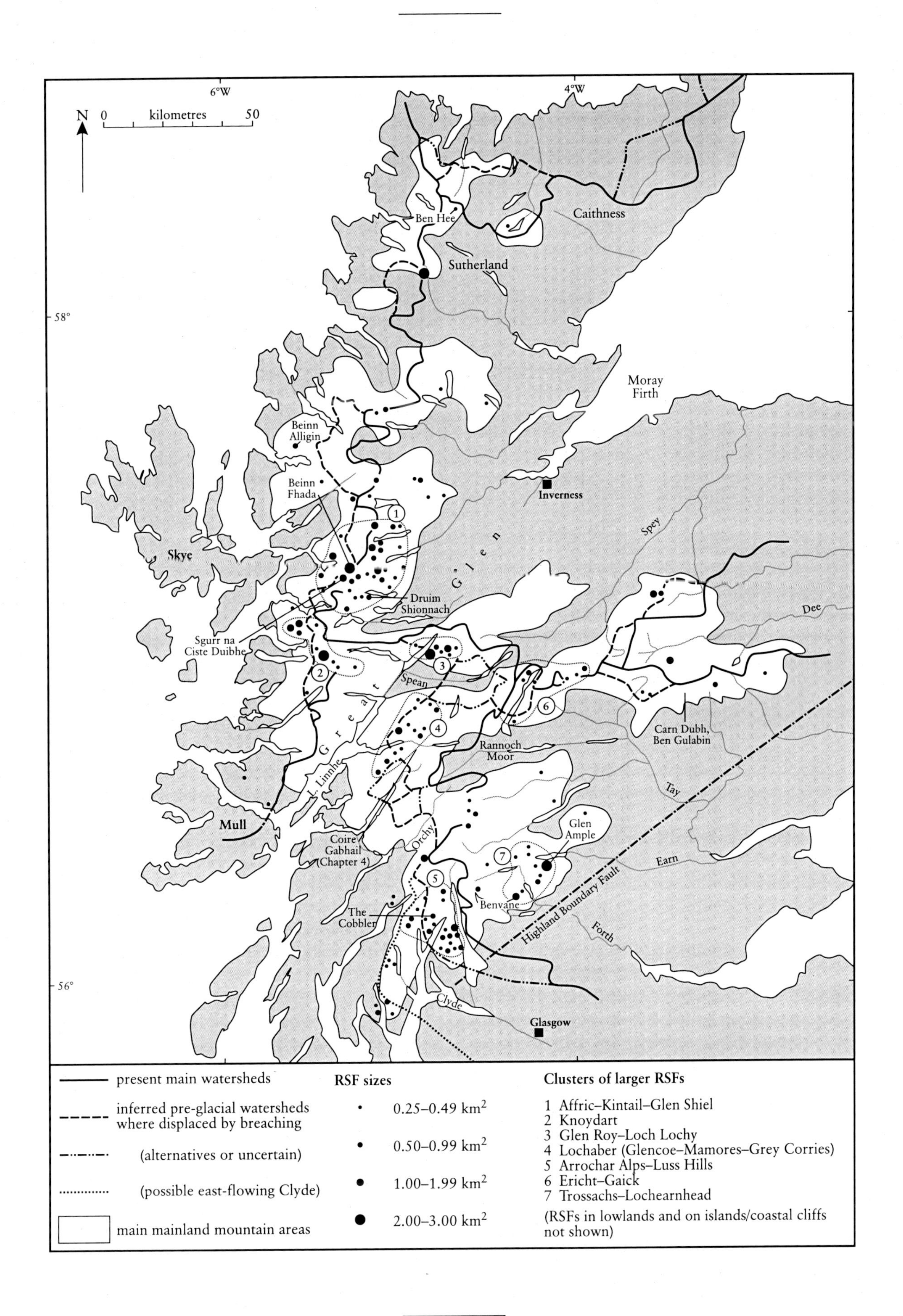

6°W
4°W
N
0 kilometres 50
58°
56°
Ben Hee
Caithness
Sutherland
Moray Firth
Beinn Alligin
Beinn Fhada
Inverness
Skye
Druim Shionnach
Great Glen
Spey
Dee
Sgurr na Ciste Duibhe
Spean
Carn Dubh, Ben Gulabin
Rannoch Moor
L. Linnhe
Tay
Mull
Glen Ample
Coire Gabhail (Chapter 4)
Orchy
Earn
Highland Boundary Fault
Benvane
The Cobbler
Forth
Clyde
Glasgow
1
2
3
4
5
6
7
present main watersheds
inferred pre-glacial watersheds where displaced by breaching
(alternatives or uncertain)
(possible east-flowing Clyde)
main mainland mountain areas
RSF sizes
0.25–0.49 km²
0.50–0.99 km²
1.00–1.99 km²
2.00–3.00 km²
Clusters of larger RSFs
1 Affric–Kintail–Glen Shiel
2 Knoydart
3 Glen Roy–Loch Lochy
4 Lochaber (Glencoe–Mamores–Grey Corries)
5 Arrochar Alps–Luss Hills
6 Ericht–Gaick
7 Trossachs–Lochearnhead
(RSFs in lowlands and on islands/coastal cliffs not shown)

Association with neotectonic activity

Fenton (1991, 1992) took the spatial incidence of rock slope failure in the North-west Highlands as an indicator of significant seismic movements after deglaciation. This built upon a model that assumed that plate-tectonic stresses suppressed by the weight of the icecap would undergo a period of accelerated 'catching-up', triggering earthquakes along ancient faultlines (Davenport *et al.*, 1989). Unfortunately, while many failures are located on faults, that is mostly because valleys have been eroded along them; conversely, many failures are as distant from main faults as is possible in the Highlands (Jarman, 2003a). Furthermore, the derivation of earthquake magnitudes from rock slope failure size and proximity was based on evidence from contemporary earthquakes in tectonically active mountain ranges, and therefore subject to interpretation in its wider application.

It is rather more likely that a phase of modestly elevated seismic activity followed deglaciation, as rebound stresses varied in relation to the local thickness of ice cover and intensity of erosion. The concept of blocky isostatic recovery was first developed at Glen Roy (Sissons and Cornish, 1982) and may retain some validity (Stewart *et al.*, 2000), although firm evidence for the 'neotectonic fault scarp' displacements found by Fenton is still lacking (cf. Firth and Stewart, 2000). Work in alpine ranges suggests that 'topographic amplification' of seismic shocks can enhance their shaking effects at crests and peaks (Ashford and Sitar, 1995). This might be explored in cases such as **Sgurr na Ciste Duibhe** and **The Cobbler**.

Association with glacial breaching

A possible explanation currently being explored views rock slope failure as a response to locally exceptional slope stresses, on top of those generically induced by glaciation/deglaciation. A spatial association can be observed in every failure cluster with glacial breaches of main and secondary watersheds (Jarman, 2003a,b; 2006; Tables 2.1 and 2.2). The Central Grampian cluster demonstrates a near-100% association between rock slope failure and the breaches of Loch Ericht, Garry, and Gaick (Hall and Jarman, 2004) (see Figure 2.14). Conversely, failure is sparse or absent in mature troughs and glens that have long adapted to efficient ice discharge, and on plateau rims away from breaches (see Figure 2.15). The inferred cause of rock slope failure is that breaching has involved concentrated glacial erosion of bedrock either in the breach or within a few kilometres downstream (by augmentation of ice catchment area), or in rejuvenated tributary valleys. Bulk erosion by glaciers streaming through such breaches could be an effective way of over-steepening slopes, setting up rock-mass stresses, and daylighting fallible discontinuities (see Figure 2.16). On deglaciation, the rebound stresses are significantly greater than those from generic glacio-isostatic recovery alone, provoking failure in the most susceptible locations.

The extent of glacial breaching was recognized by Linton (1949), and its importance in the dissection and reduction of the mountain areas by Haynes (1977a). It can only occur when the iceshed becomes offset from the watershed. The main watersheds of the Highlands have shifted east and north by 5–30 km over the Quaternary Period along much of their length as a result of glacial breaching (Jarman, 2006; see Figure 2.13). The spatial association of paraglacial rock slope failure with this dense incidence of breaching is therefore of considerable interest for future research, although it must be emphasized that many failures are not in close proximity to obvious breaches, and that there is an inevitable degree of self-correlation between rock slope failure and steep, narrow passes (Figure 2.17).

Rock slope failure clusters

Most rock slope failures in the Highlands fall within the main clusters identified for the larger sites (Jarman, 2006; Figure 2.13):

Location	*number of sites > 0.25 km²*
Affric–Kintail–Glen Shiel	29
Knoydart	8
Glen Roy–Loch Lochy	7
Glencoe–Mamores–Grey Corries	13
Arrochar Alps–Cowal–Luss Hills	20
Loch Ericht–Gaick	7
Trossachs–Lochearnhead	9
TOTAL (7 clusters with 5 sites or more) > 0.25 km²	*93*
Total (all larger RSFs)	140

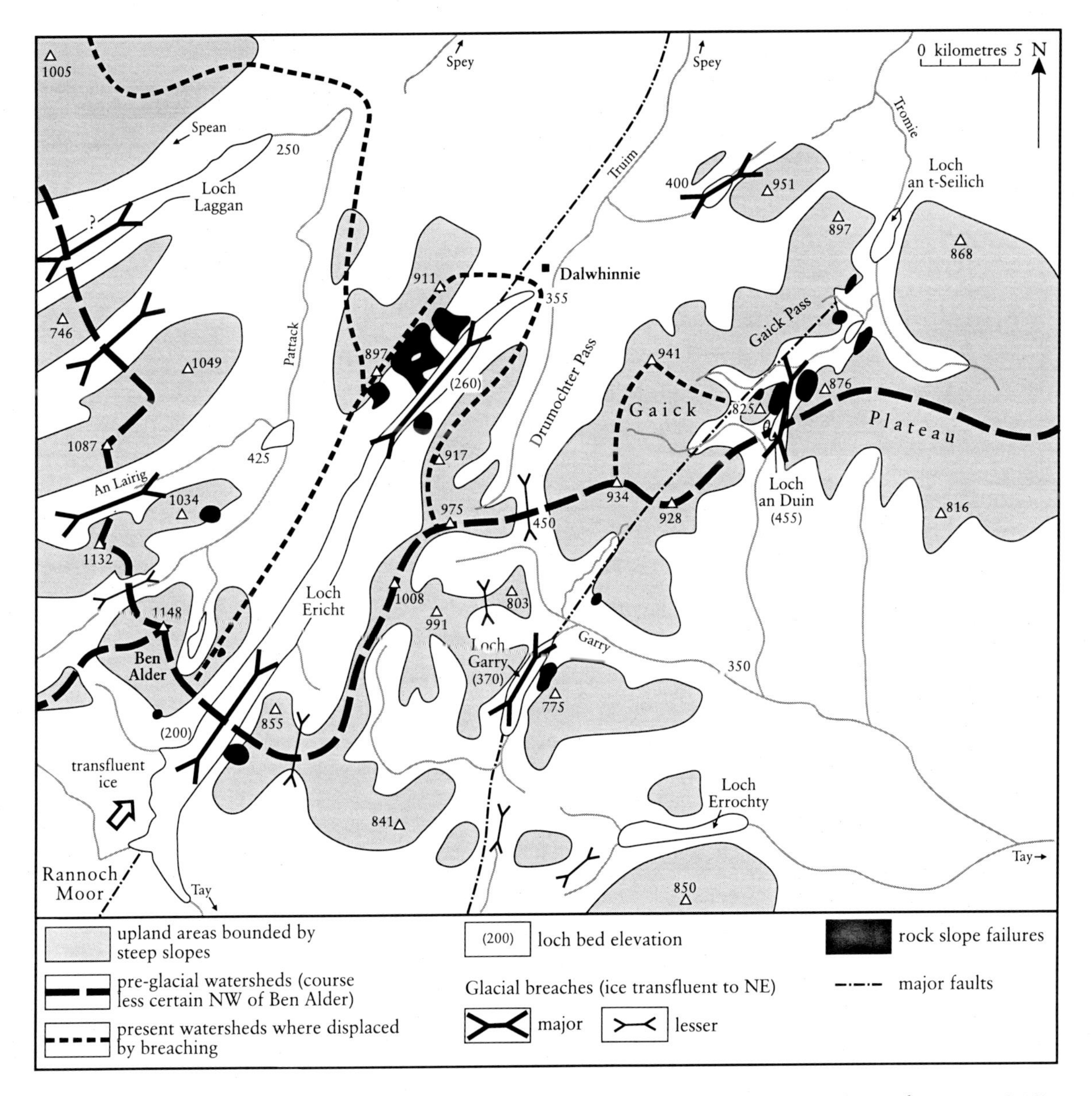

Figure 2.14 The Central Grampian Highlands, including the Loch Ericht–Gaick cluster (cluster 6 in Figure 2.13). All rock slope failures shown true to scale. Failure only occurs in or downvalley from breaches, and is entirely absent from other valley-sides and plateau rims. Failure concentrations in the Loch Ericht and Gaick Pass breaches may point to their recent origin or enlargement; rock slope failure absence from Drumochter Pass (main road/rail corridor) may indicate its earlier development. Adapted and revised from Hall and Jarman (2004).

It may be that these clusters indicate where glacial breaching and erosion have been most active in Devensian times, and thus where shifts in ice centres and/or dispersal patterns have been most marked. Work is required to scope and calibrate deglaciation slope stresses of all kinds, and to refine icecap models in light of rock slope failure information. The intensity of dissection by breaching increases dramatically from east to west across the Highlands, as in microcosm in the Lake District and North Wales (Clayton, 1974, after notes by D. Linton; Haynes, 1977a, 1995). Failure clusters may indicate where dissection is further intensifying, and where the 'Clayton Zones' of dissection intensity are migrating eastwards (see Gordon and Sutherland, 1993, fig. 2.1).

The sites selected here mostly fall within the two densest rock slope failure concentrations first recognized by Holmes (1984) – the

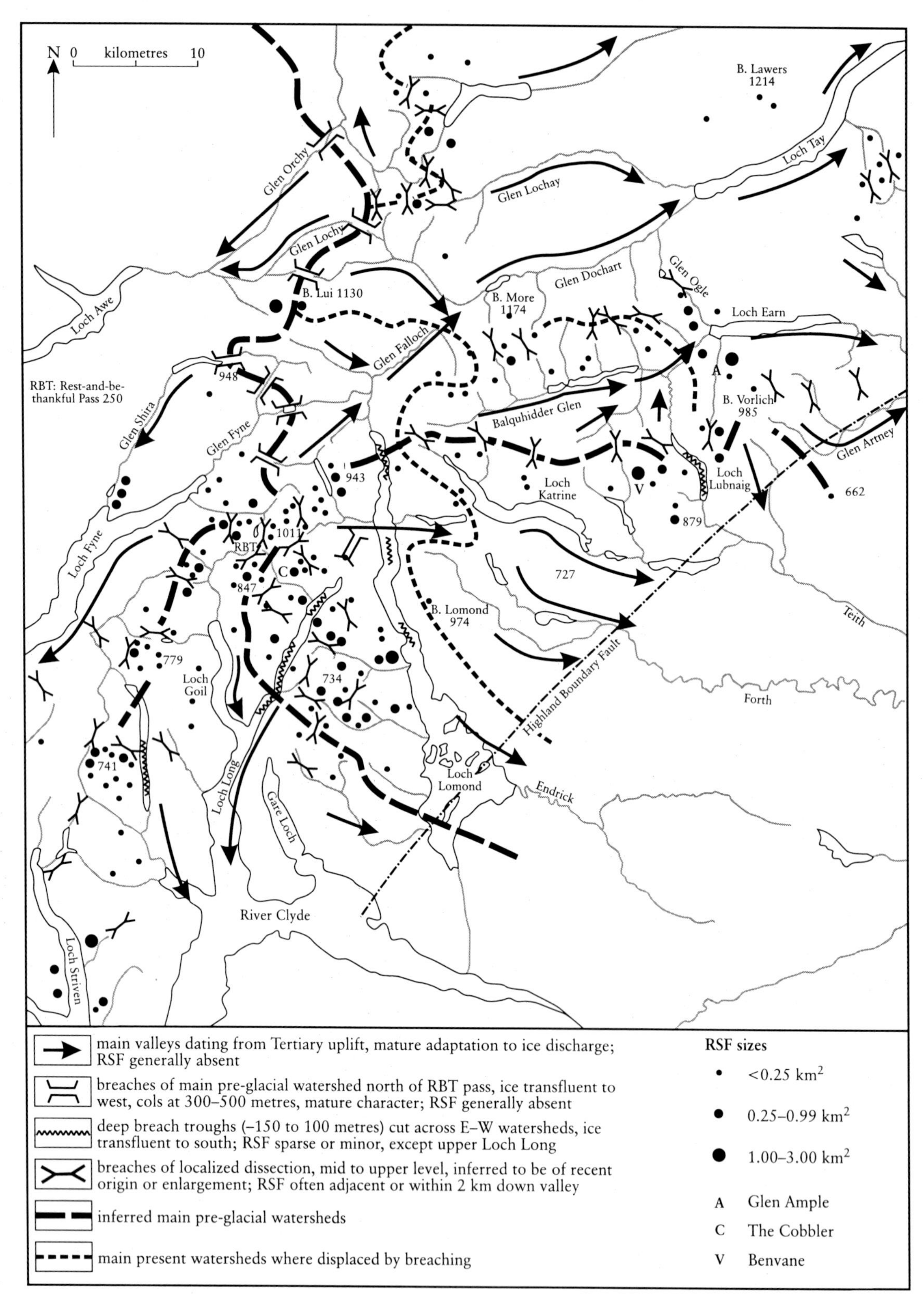

Figure 2.15 The Southern Highlands, an area of intense rock slope failure (RSF) activity, including the Arrochar–Cowal–Luss and Trossachs–Lochearnhead clusters (clusters 5 and 7 in Figure 2.13). Failure is scarce or absent in main pre-glacial valleys and some breaches of the main watershed, despite their slopes and geology being susceptible to it. Its paucity along the deep breach trench of Loch Lomond is surprising. Note mini-clusters top-centre and top-right, where locally intense breaching occurs across main and secondary watersheds. The locations of three sites (**Glen Ample**, **The Cobbler**, **Benvane**) are shown. Adapted and revised from Jarman (2003a).

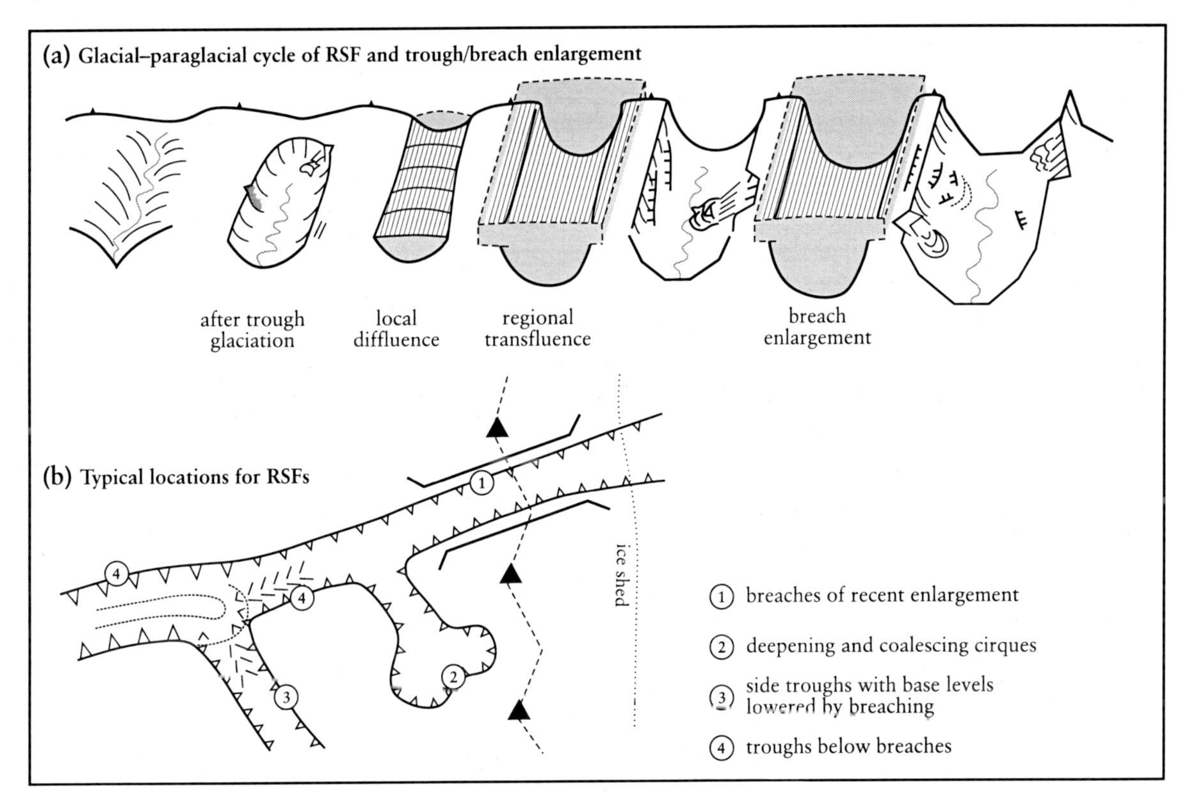

Figure 2.16 (a) The inferred association of rock slope failure (RSF) with breach incision and enlargement over repeated glacial/paraglacial cycles. (b) Typical locations for rock slope failure responding to exceptional deglaciation stresses directly and indirectly associated with glacial breaching. After Jarman (2006).

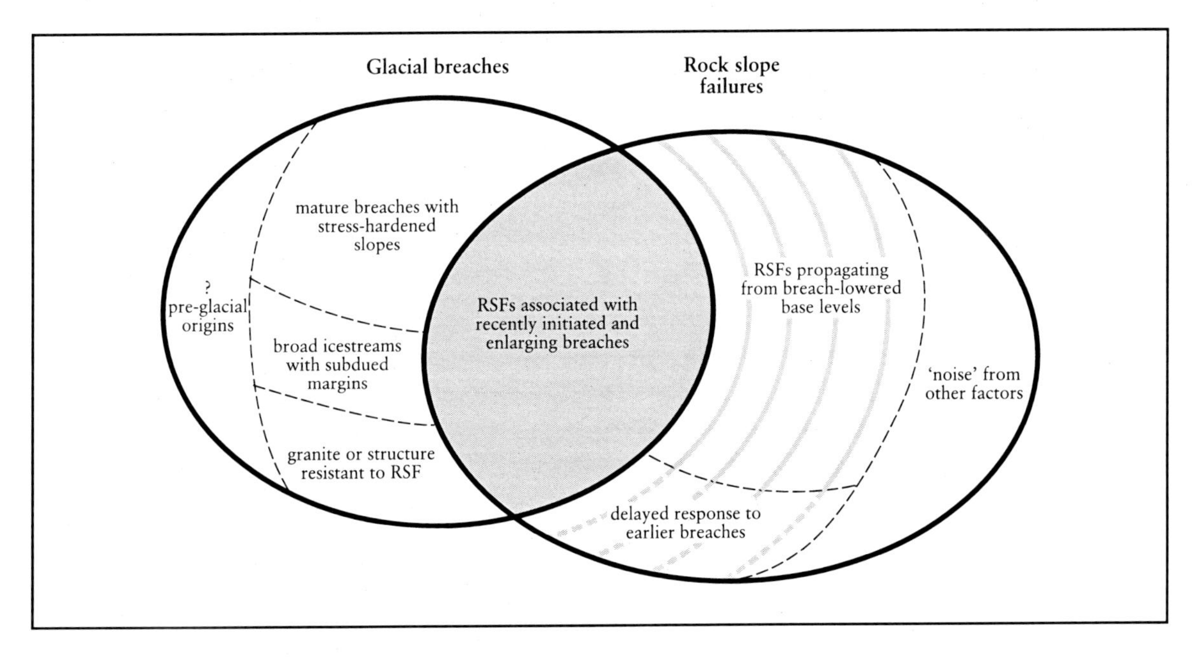

Figure 2.17 The nature of the association between paraglacial rock slope failure and glacial breaching. This suggests why many breaches may lack rock slope failure, and acknowledges that not all failures can be attributed to breaching. After Jarman (2006).

Southern Highlands, and Affric–Kintail (Figures 2.15 and 2.18). A cluster around Glen Roy is touched upon under that GCR site report in the *Quaternary of Scotland* GCR volume (Gordon and Sutherland, 1993), an important site, not least because the sequence of Parallel Roads enables close dating of the different rock slope failures (Fenton, 1991; Peacock and May, 1993). Two sites in Glen Shiel are notable in interpreting the major breach west from Glen Cluanie: **Sgurr na Ciste Duibhe** is immediately above the 'canyon of adjustment' on one of the highest steep slopes in Britain, while **Druim Shionnach** is in one of the side troughs rejuvenated by it. **Beinn Fhada** is just downstream of the next breach north, and on the inferred pre-glacial watershed. By contrast, **The Cobbler** is in the heart of an area intensely dissected by primary and secondary breaching, with failure ramifying into the rejuvenated side valleys. The **Ben Hee** rock slope failures may be a response to local northwards breaching. **Benvane** and **Glen Ample** are close to immature breaches of a major secondary divide, in lower relief near the Loch Lomond Stadial outer margin.

Some of the major Lake District rock slope failures are adjacent to breaches, such as Honister, Grisedale (Figure 2.10), and Kirkstone, or above troughs possibly augmented by transfluent ice such as at Wasdale. Conversely, failure is almost absent from more mature valleys such as Langdale, Borrowdale and Ullswater. In North Wales, the Tal-y-llyn trough (into which the Graig Goch landslide descended) appears to have been augmented by ice breaching across the main divide.

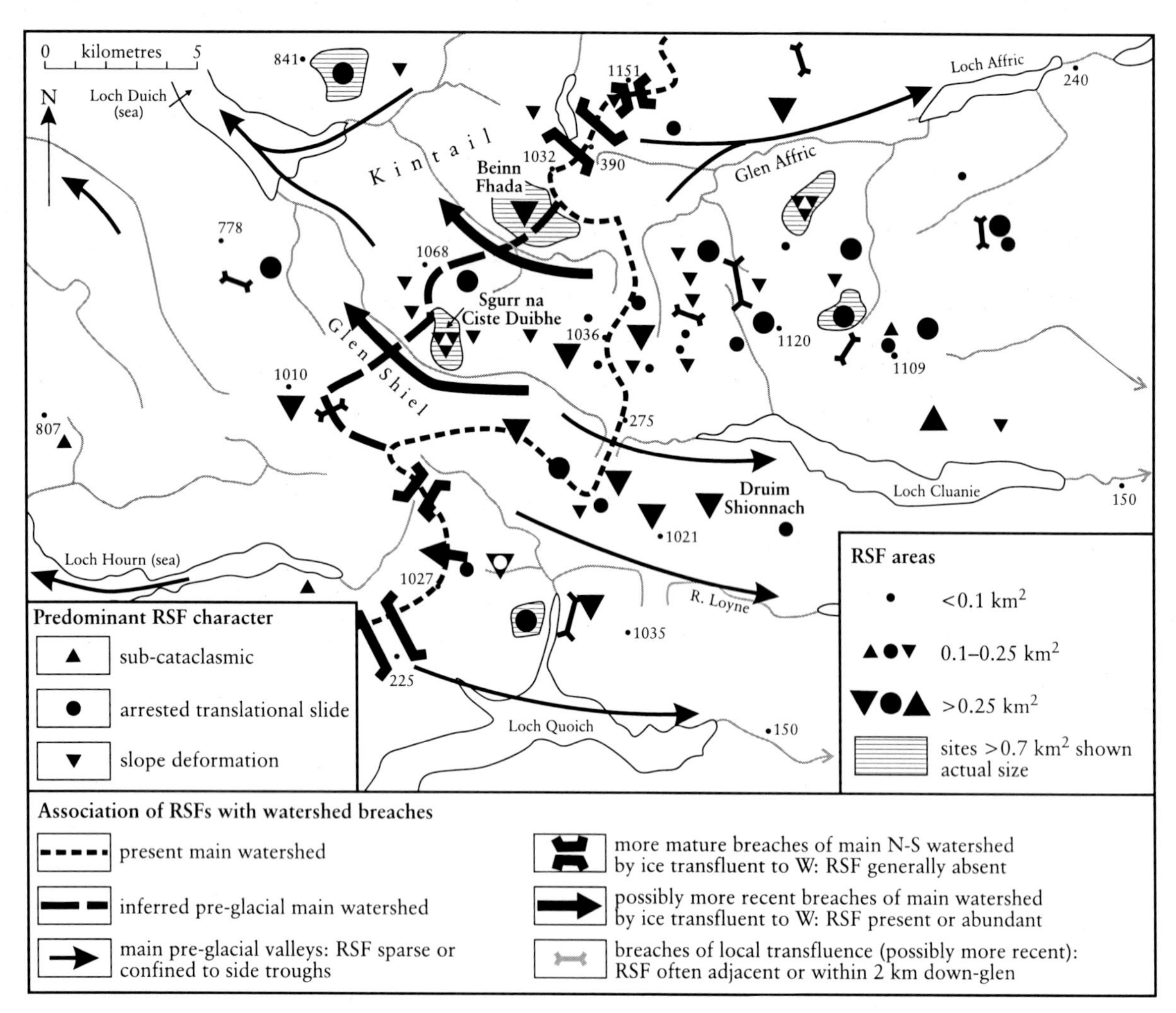

Figure 2.18 Part of the Affric–Kintail–Glen Shiel rock slope failure cluster (cluster 1 in Figure 2.13). Rock slope failures tend to be located near breaches of the inferred pre-glacial watershed, and in side troughs rejuvenated by breaching. Note the locations of three GCR sites – **Beinn Fhada**, **Druim Shionnach** and **Sgurr na Ciste Duibhe**. Adapted from Jarman (2003b).

It is probable that the actual triggers for individual rock slope failures were a combination of factors such as elevated water pressure, seismic shaking, freeze-thaw, and progressive failure (cf. Ballantyne, 2002a). These factors have probably all applied in varying intensities and combinations. The sites selected here represent locations where each factor has been invoked, and where further research is merited. Some of the sites described here are simple, others complex in mode and probable triggering mechanism. Some of the mass-movement features are relatively recent, dated as mid–late Holocene (**Beinn Alligin**; **Coire Gabhail** (Chaper 4)); others probably occurred around final deglaciation or even earlier, or have evolved in stages (**Sgurr na Ciste Duibhe**).

Scale and international comparions

Rock slope failure sizes display a typically skewed size distribution (Tables 2.1 and 2.2), with many small cases of purely local impact. The depth to which failure extends is seldom known, except in rare cases where a cavity has been largely evacuated, but is generally assumed to be in the order of tens of metres for smaller rock slope failures, rising to 100–150 m for larger cases (Holmes, 1984; Fenton, 1991). Inferred volumes reach 10×10^6 m^3 for translational slides, and may exceed 100×10^6 m^3 for the very largest deformations (Table 2.3).

The scale of rock slope failures in the mountain areas of Britain is of course limited by virtue of the available relief being less than in ranges such as the Scandes and the Alps. Even so, the largest cataclasmic rock slope failure in Britain (Graig Goch in North Wales) is of the same order of magnitude as the largest yet reported in Scandinavia at Kärkevagge (Jarman, 2002); albeit at approximately 50×10^6 m^3 these are an order of magnitude smaller than the ten largest in the Alps, let alone the extraordinary Flims failure which reaches $12\,000 \times 10^6$ m^3 (Abele, 1974).

With slope deformations, the typical scale in the Alps involves relief of 1000 m, a depth of 100 m, and a volume of 100×10^6 m^3 (Brückl and Parotidis, 2005). It is interesting that such dimensions are approached by the largest or highest rock slope failures in the Highlands (Table 2.3).

In a European context rock slope failures such as **Beinn Fhada** and Glen Pean (de Freitas and Watters, 1973) are of significant scale and bear instructive comparison as steep mountain deformations. **Sgurr na Ciste Duibhe** is the highest and steepest failed slope in Britain, and since it displays some progression from in-situ deformation to sliding, offers a valuable benchmark for international studies of critical stability thresholds. By contrast, Ben Our (**Glen Ample**) is exceptionally extensive in very low relief, and no international comparators have yet been found for it. In terms of exhibiting the direct impact rock slope failure can have in shaping mountain scenery, **The Cobbler** can hold its own with any international comparison.

Incidence and impact of rock slope failure during the Quaternary Period

Clearly this paraglacial contribution to erosion in mountain areas can be very substantial locally. Where rock slope failures encroach significantly into corries, ridges or pre-glacial land surfaces, the local rate of scarp retreat can be vastly greater than by all other means of erosion over a glacial cycle.

With many such glacial–paraglacial cycles over the Quaternary Period, the cumulative impact of rock slope failure will have been very substantial, especially in the most geologically susceptible areas (Evans, 1997; Hall and Jarman, 2004; cf. **Trotternish Escarpment** GCR site report, Chapter 6). However, the total contribution of paraglacial rock slope failure to the huge volumes of erosion experienced by the higher mountain ranges of Britain is as yet difficult to assess. Extant landslides yielding debris ready for evacuation by the next glaciers have relatively small volumes (typically less than 1×10^6 m^3), whereas almost intact slope deformations such as Ben Our (**Glen Ample**) with much larger volumes may not be much more vulnerable to erosion than the surrounding terrain.

The incidence of rock slope failure may have changed as Quaternary landscape adaptation to ice progressed, and as glaciations increased in severity. It seems likely that paraglacial failure would have been fairly endemic in the relatively weak early Quaternary glacial–interglacial cycles, as fluvial valleys with interlocking spurs became adapted to efficient ice discharge as straightened and deepened troughs. In the great ice-sheet glaciations of mid-Quaternary times, transfluence across watersheds would have initiated the process of glacial breaching, provoking a fresh round of intensive but less widespread failure.

Table 2.3 Large rock slope failures (RSFs) in the Scottish Highlands for which data are available. After Jarman (2006). Sites are listed from the north, with the Great Glen separating the North-west Highlands from the Grampians. Note the disproportionate number of large RSFs studied north-west of the Great Glen, where foliation (F) is rarely as conducive to sliding as in the Southern Highlands. Most studies are of (sub-)cataclasmic RSFs or slope deformations, rather than conventional arrested slides. *Continued opposite.*

RSF	**Ref.**	**Mode**	**Area** km^2	**Vol.** $\times 10^6$m^3	**Depth** m	**H/S** m	**A/S** m	**Slide plane**	**Comments**
Loch Vaich, Ross-shire	2	ext def	0.5		>50?		2	J low angle	Short-travel (50 m) slip, forward toppling on 60° F
Sgurr Bhreac, Fannich	2,3	ext def	0.82	36?	?	30	Sm	?	Sackung with lattice of fissures
Beinn Alligin, Torridon	1,2	cata	0.52	3.5	200 20 #1	60	–	42°	Acute faulted wedge in sub-horizontal sandstone
Glenuaig, Strathcarron	2	ext def	0.7		?	<5	<2	F 15°+	Short-travel sliding slump, incipient fissuring
Sgurr na Conbhaire, Monar	2,6	sub-cata	0.35		150		2	F 30–40°	Long-travel (150 m) slump onto lower slope
Sgurr na Feartaig, Strathcarron	2	ext def	0.9		100?	2	yes	F	Short-travel block slide
An Socach, Monar	2	comp def	1.0	–	20?	nil	sm	–	500 m long, linear A/S diffuse margins – rebound
Carn na Con Dhu, Mullardoch	2,3	slide	1.46	61?	120?	12 35?	2	not on F/J	Short-travel slide/slump, A/S <200 m long on strike of J1+J2
An Sornach, Affric	2,3,4	ext def	0.75	13?	30?	42	3	not F/J	Slip with A/S lattice > bulge > collapse. Rebound 5 m A/S
Mullach Fraoch-choire, Affric	3	sub-cata	0.2	0.73	20	10	–	J2,3 29°	Slide tongue within 1.1 km^2 slope deformation
Sgurr na Lapaich, Affric	2,3	comp def	0.3	7?	100?	10	–	–	Ridge crest failure, possibly seismic/rebound faulting
Beinn Fhada, Kintail	2,3, 4,6	comp def	3.0	112 #2	100?	none	10	–	~8 sub-horizontal A/S < 700 m long, main ones are 5–8 m high
Sgurr na Ciste Duibhe, Glenshiel	5	ext def	1.25	5–10 #3	80	15	(11) 5	not F/J	Summit lowered ~10 m > long-travel slide in deformation
Sgurr a' Bhealaich Dheirg, Glenshiel	2	comp def	0.7		100?		6	–	Bulging slide, rebound A/S < 200 m long

Footnotes:
'Ref.': reference sources are (1) Ballantyne, 2003; (2) Fenton, 1991; (3) Holmes, 1984; (4) Jarman, 2003c,d,e, 2004a, and present volume; (5) Jarman, 2003b; (6) Watters, 1972.
'Mode': cata = cataclasmic; sub-cata = sub-cataclasmic; ext def = extensional deformation (sag, creep); comp def = compressional deformation (rebound).
'Area': RSF size is here taken as the gross area including source cavity, since most cases are incompletely evacuated. British Geological Survey mapping of RSF is variable and incomplete, but recent sheets only map as 'landslips' disturbed ground, thus excluding both source areas and semi-intact slope deformations. The gross area best indicates the geomorphological impact of the RSF, but clearly requires adjustment when volumetric calculations are made.
'Vol.'(-ume) and maximum '**Depth**' should be seen as broad estimates, especially sites marked '?' where the depth cannot readily be assessed.
#1 depth figures are for cavity (ref. 2) and debris tongue (ref. 1);
#2 volume (ref. 3) assumes there is a failed mass with a boundary at ~100 m, no volume can be calculated if the failure partly dissipates at depth;
#3 volume and depth are for main cavity within larger deformation.
'**H/S**' = headscarp (rear scarp, source scarp) maximum height.
'**A/S**' = antiscarp (obsequent scarp, counterscarp, uphill-facing scarp) maximum height – figures in brackets are graben trenches or uphill faces of large slipped masses.
'**Slide plane**': F = foliation or schistosity surface;
J = joint-sets (in order of significance);
RFA = residual friction angle.

Table 2.3 – *continued.* For key see the footnotes on the previous page.

RSF	Ref.	Mode	Area km²	Vol. ×10⁶m³	Depth m	H/S m	A/S m	Slide plane	Comments
Ciste Dhubh, Affric	5	sub-cata	0.46	7	80	30	2	?	Corrie-floor source, toe reaches river in breach glen
Druim Shionnach, Cluanie	5	comp def	0.55	–	150?	25	(14) 3	–	Top A/S is outer half of graben > Cluanie bulge
Meall Buidhe, Knoydart	6	ext def	0.5		40?	30	7?	J1,2 < 44°	Broad slump zone
Mam na Cloiche Airde, Knoydart	6	sub-cata	0.26		40	20	–	35°	Semi-intact masses > slope debris > 5° flow slide
Glen Pean, Knoydart	6	comp def	2.5	–	60?		–	J1 28–48°	A/S array on strike of F/J2, cataclasmic slide to west
GREAT GLEN									
Streap, Glenfinnan	6	Slide	0.25		25	75	(10)	J1 36°	Long-travel arrested sub-cataclasmic summit has lost top ~15 m, seismic trigger?
Beinn an Lochain W, Arrochar	6	ext def	0.34		15?	<5	<2	F 20–30°	Thin, undeveloped slide, upper tier beside
The Cobbler SW, Arrochar	4	slide	0.62	8–10	30	28	(10) 6	not F	Four-panel, short-travel, disintegrated slip
Hell's Glen, Cowal	3,6	ext def	0.52	1.75 (+)	60	15	(15) 5	J2 40–50°	Topple block slips and collapses in broad slump
Mullach Coire a' Chuir, Cowal	3	slide	0.57	9.6?	20	50	(12) 2	F + J2	Part-collapsed sliding topple on stepped surface
Meallan Sidhein, Loch Striven	6	slide	0.75		70	40	–	F 25–32°	Slip in phyllite, effective F dip 20°, equals RFA
Tullich Hill West and East	4	slide	1.25 in total			40	8	not F	Short-travel, multi-phase, slump complex
Benvane, Trossachs	4	def/ slide	1.25	25	20–30	26	3	not F	Deformation progresses laterally to slide
Ben Our (**Glen Ample**), Lochearnhead	4	def	2.90	100–200?	150?	4	4	–	Platy deformation with basal slumps

Many of these breaches are now relatively mature. It is therefore likely that the incidence of failure has diminished into the last (Devensian) glaciation. The extant population of rock slope failures may thus be relatively modest, and its spatial distribution probably indicates where glacial erosion and other destabilizing factors were concentrated latterly.

Rock slope failure commonly occurs with a delayed reaction time of hundreds or even several thousands of years after deglaciation. Thus although many failures are within the boundaries of the Loch Lomond Stadial (LLS), this was short-lived (*c.* 12 900–11 500 years BP) and carried out little fresh erosion. These rock slope failures are probably responding to stresses induced during the Last Glacial Maximum, which peaked approximately 22 000 years BP and deglaciated approximately 15 000 years BP.

However, paraglacial responses as delayed as at **Beinn Alligin** (7000 years after deglaciation) and **Coire Gabhail** (9000 years after deglaciation) may be exceptional. No large rock slope failure movements are known within the mountain areas during recorded history, with even conspicuous cases such as Glen Kinglass (NN 190 096) of *c.* 1700 AD (Clough, 1897) only amounting to 70 000 m³ (Holmes, 1984). Progressive failure in one area of Norway is still leading to catastrophic collapses (Bjerrum and Jørstad, 1968) but this is unknown in Britain, the creeping failure in fjord-type cliffs affecting road and rail at Attadale, Loch Carron (NG 914 377; Watters, 1972) perhaps being the nearest approximation.

Significance of rock slope failure in the older mountain areas

Understanding mass movements in lowland Britain is of critical importance for civil engineering, geohazard awareness, and active geomorphological processes such as coastal retreat, but there are different reasons for studying paraglacial rock slope failure in the mountain areas, and for conserving key sites:

- It is a significant if overlooked contributor to glaciated landscape development, especially as an agent of selective linear erosion (notably breaching), where rates of valley incision and widening can be orders of magnitude more rapid than normal.
- It has played a remarkable role in shaping many mountain summits and ridges, and is a potential key component in Earth science and landscape interpretation and geotourism development (cf. Brown, 2003).
- It links with other nature conservation and environmental history interests, in that failed and deformed slopes greatly increase habitat niche availability, ameliorate microclimate, and thus enhance biodiversity. Because these slopes are typically drier and warmer, and more fertile and sheltered, they have played an important part in the colonization of the mountains by early man and in the survival of subsistence farming and contemporary land utilization in otherwise inhospitable terrain.
- Finally, the spatial distribution of rock slope failure has potential to inform and calibrate efforts to inform the analysis of shifting ice centres and dispersal patterns over the Devensian glacial period, with possible benefits for palaeoclimate change studies.

The sites selected here represent all of these aspects of rock slope failure to varying degrees, but by comparison with well-studied fields such as coastal landslips and glacial deposits, there is considerable scope to augment the site coverage as further failure research proceeds. In particular, sites in Lochaber, the Lake District, and North Wales would reflect local diversity of expression, while no sites have yet been subjected to state-of-the-art geotechnical investigation, or to slope-stress modelling in relation to underlying causes and trigger events.

BEINN FHADA (BEN ATTOW), HIGHLAND (NH 000 185–NH 021 185)

C.K. Ballantyne and D. Jarman

Introduction

The majority of large (> 0.25 km^2) rock slope failures in Scotland take the form of rock slope deformations that lack runout of debris. The largest area of rock slope deformation occurs on the south-west side of Beinn Fhada (Ben Attow), a 1032 m-high peak between the head of Loch Duich and upper Glen Affric. Slope deformation at this site involves the entire mountainside over an area of about 3 km^2, and was estimated by Holmes (1984) to involve the displacement of 112×10^6 m^3 (roughly 300 million tonnes) of rock. Although this estimate must be regarded as approximate in view of the uncertainty concerning the depth of deformation, it implies that the Beinn Fhada feature is probably the largest rock slope failure on the Scottish mainland. It is also significant as the site of the most impressive suite of antiscarps (uphill-facing scarps) in Britain.

The Beinn Fhada rock slope deformation was first interpreted by Watters (1972) as a translational landslide over a deep failure plane. Holmes (1984) and Holmes and Jarvis (1985) re-interpreted the site in terms of deep internal rock deformation, expressed at the surface by antiscarps produced by joint-guided block-flexural toppling. Fenton (1992) proposed that deformation reflected sliding failure triggered by a high-magnitude earthquake. Jarman and Ballantyne (2002) provided a more comprehensive description of the site, attributed rock slope deformation to paraglacial stress-release following deglacial unloading, and proposed two further possible models to explain the assemblage of associated landforms. Jarman (2003e) further reviewed the characteristics of the site, noting inconsistencies between the field evidence and a translational sliding model, and considered possible reasons for exceptionally large-scale rock slope deformation at this location.

Description

Setting

The southern slopes of Beinn Fhada rise 750–900 m above the adjacent valley floor of Gleann Lichd at average gradients of 30°–35°, forming one of the most extensive uninterrupted bedrock slopes developed in metamorphic rocks in the Scottish Highlands (Figures 2.19–2.21). Above the slope crest a remnant of pre-glacial surface forms an undulating plateau up to about 1 km wide. The mountain is underlain by psammites and semipelites of the Moine Supergroup of Neoproterozoic age (May *et al.*, 1993). The rocks have been deformed, folded and metamorphosed and the bedding now generally dips slightly eastwards. Much of the psammite is coarse-grained and gneissose. In the area of rock slope deformation, Holmes and Jarvis (1985) detected four main discontinuities: one parallel to foliation; one joint-set striking parallel to, and dipping steeply into, the slope (J1); and two sets that dip gently westwards (J2 and J3).

Glacially moulded rock outcrops occur up to 950 m on Beinn Fhada and 1050 m on the neighbouring summit of Sgurr nan Ceathreamhnan (Ballantyne *et al.*, 1998a) suggesting that during the last (Late Devensian) glacial maximum the summit plateau of Beinn Fhada lay under a thin cover of glacier ice, and that a westwards-moving icestream *c.* 1000 m thick occupied Gleann Lichd. During the subsequent Loch Lomond Stadial of *c.* 12.9–11.5 cal. ka BP a valley glacier moved westwards down Gleann Lichd. The surface of this glacier descended from about 600 m near the present-day watershed to about 300 m in lower Gleann Lichd (Tate, 1995).

The area of rock slope deformation

The area of rock slope deformation has been described by Jarman and Ballantyne (2002). It lacks distinct lateral margins and is defined by a remarkable array of antiscarps (Figures 2.19–2.22). On the upper half of the slope these achieve lengths of up to 800 m and heights of up to 10 m, but those above the slope crest and on the lower slope are 1–5 m high. All are composed of intact bedrock, with steep upslope faces and sharp crests, implying formation after deglaciation (Figure 2.22). The antiscarps are aligned parallel to the slope contours, though some descend gently westwards or south-eastwards across the slope and some converge (Figure 2.20). Small antiscarps occur within the area of renewed glacial occupation of Gleann Lichd during the Loch Lomond Stadial, descending to within 50 m of the valley floor.

The slope exhibits three large-scale bulges or convexities separated by intervening depres-

Figure 2.19 The Beinn Fhada rock slope failure seen from the west across Gleann Lichd. Unbroken antiscarps extend up to 800 m across the 30°–35° glacial trough side. Deformation extends for 3 km along the valley and onto the pre-glacial upland surface, reaching the south top (1000 m) in the background, and affecting 3.0 km^2. (Photo: D. Jarman.)

Figure 2.20 Vertical aerial photograph of the Beinn Fhada rock slope deformation. The three major slope convexities are highlighted by antiscarp arrays, as is the extent of the pre-glacial land surface with its structural lineaments. (Photo: Crown Copyright: RCAHMS (All Scotland Survey Collection).)

sions and highlighted by the convex-outward planform of associated antiscarps (Figure 2.21). Antiscarps also extend on to the adjacent plateau and across the depression between the western and central bulges, indicating that slope deformation was more extensive than the bulges alone suggest. According to Fenton (1991), the area of rock slope deformation appears to be generally under compression, though a minor area of shallow translational sliding failure occurs near the eastern margin of the major slope deformation (Figure 2.21). The toe of the slope is apparently intact, with no evidence for displacement towards the valley axis. Springs emerging near the foot of the area of deformation appear to be fed by groundwater movement through a sub-surface zone of fractured rock upslope.

The slope crest above the main area of deformation is marked by a degraded rock scarp up to 90 m high. Weathered rock crops out locally in the scarp face, which diminishes in height northwestwards above an oblique rock ramp (Figure 2.21) and merges eastwards with prominent

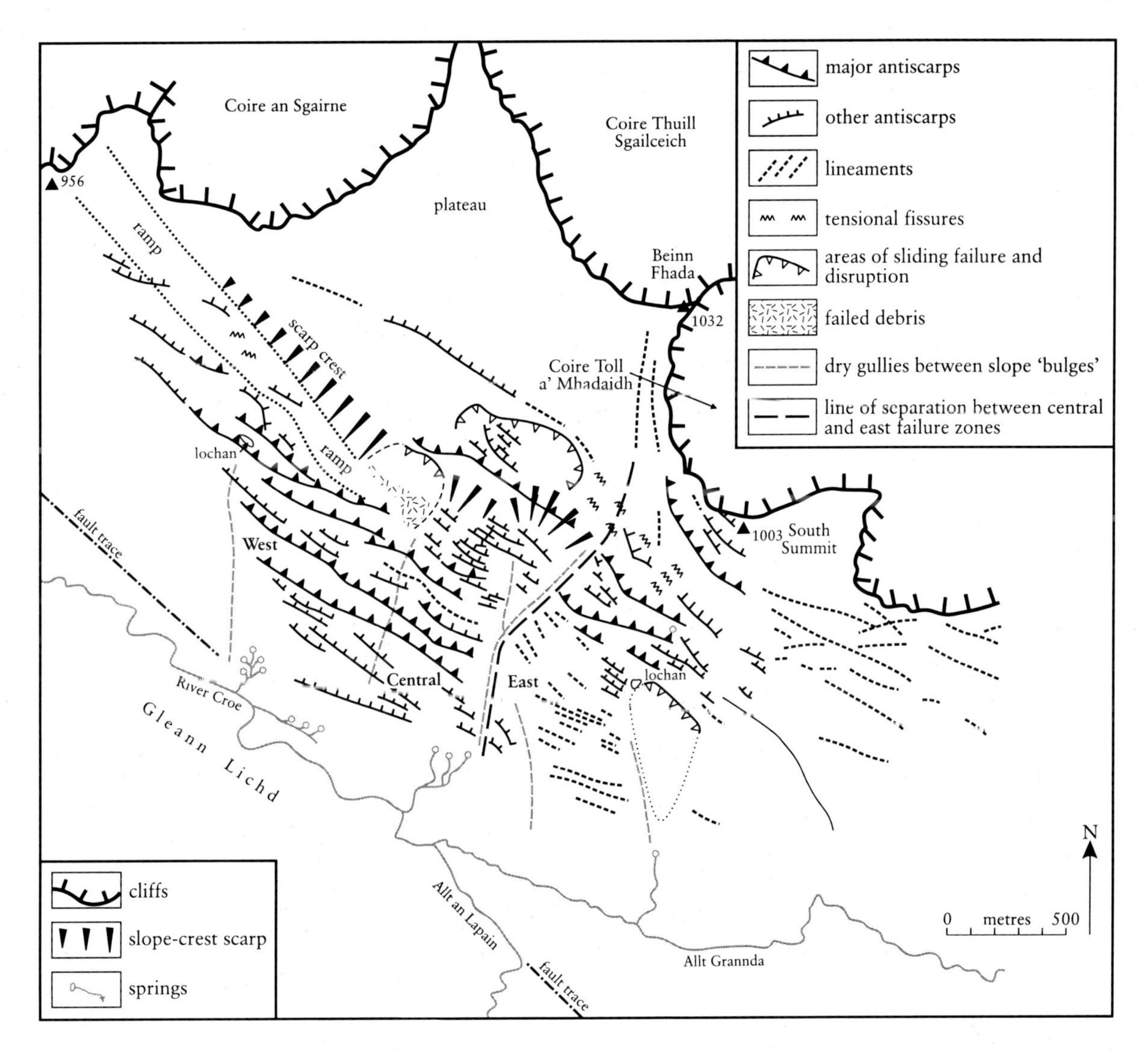

Figure 2.21 Geomorphological map of the Beinn Fhada rock slope deformation, showing the location of prominent antiscarps, lineaments, areas of localized sliding failure and the three slope convexities or bulges (labelled 'West', 'Central' and 'East'). The map is based on the aerial photograph in Figure 2.20, with an average scale of 1:22 500. After Jarman and Ballantyne (2002).

antiscarps. Above the central part of the scarp, the slope crest has failed and a spread of coarse debris extends a short distance downslope. The western and central convexities terminate upslope at the scarp foot, suggesting that the scarp represents a failure plane (Holmes and Jarvis, 1985). Above the slope crest, the plateau surface is crossed by structural lineaments, some of which have been re-activated by rock-mass deformation. Above the central slope bulge a shallow depression with a 10 m-high headscarp and internal antiscarps marks a site of incipient sliding failure (Figure 2.21). Above the eastern slope bulge, antiscarps up to 4 m high extend upslope to the summit ridge. Offset lineaments and tensile fractures separate the central and eastern parts of the plateau, implying differential movement and deformation of two adjacent rock masses.

Interpretation

Mechanics of deformation

The morphological characteristics of the Beinn Fhada failure zone are characteristic of those of deep-seated gravitational slope deformation (Chigira, 1992; Soldati, 2004), and in particular of *sackung*-type ('sagging') slope deformation (Hutchinson, 1988). The latter has been widely

Figure 2.22 Antiscarps near the top of the western slope bulge. The farthest of the three antiscarps rises 6–8 m (locally 10 m) out of the slope. (Photo: C.K. Ballantyne.)

identified in areas of high, steep mountain slopes and is thought to reflect deformation at depth (Bisci *et al.*, 1996; Rizzo and Leggeri, 2004). It is particularly common on mountain slopes steepened by glacial erosion (Radbruch-Hall, 1978; Bovis and Evans, 1996). At such sites deformation has been initiated by stress-release as overlying and adjacent ice downwastes, unloading and debuttressing the rock mass and altering the orientation of the principal stress field. Two main displacement models have been proposed. Most researchers assume that high confining pressures in the central part of the slope permit deformation at depth, without development of a continuous failure plane, though shearing surfaces may be present at the top and/or base of the slope where confining pressures are lower (e.g. Mahr, 1977). Others have proposed that the zone of deformation is seated on a continuous shear surface (e.g. Savage and Varnes, 1987; Bovis and Evans, 1996).

Both models have been invoked for rock slope deformation at Beinn Fhada. Watters (1972) suggested translational sliding along a continuous deep failure-plane, with outward movement of the slope foot initiating sequential upslope separation of large slices of rock; the antiscarps he interpreted as debris-filled tension cracks between individual slipped blocks. This solution appears flawed. The geometry of the slope does not permit reconstruction of a single planar shear-plane, there is no evidence for outward movement of the slope foot, and the antiscarps suggest compression, not extension, of the slope surface. Conversely, Holmes and Jarvis (1985) proposed that displacement of the slope had been accommodated internally within the rock mass. They suggested that translational failure may have occurred along J2 joints at depth, and interpreted the antiscarps in terms of block-flexural toppling along inward-dipping (J1) joints, initiated by a minor topple along the lower slope. There is, however, no evidence for the latter. Jarman and Ballantyne (2002) suggested two further explanations. The first involves movement along a deep failure-plane or shear zone, with associated antiscarp formation by compressional block-flexural toppling. The second involves glacio-isostatic rebound of the valley floor and lower slope with the development of compressional (reverse) fault-swarms across the slope, forming antiscarps.

Jarman (2003e) has outlined arguments against development of a continuous failure-plane under the zone of slope deformation. The main points are:

(1) the apparent absence of suitably aligned joint-sets;
(2) the slope configuration precludes development of a single planar surface connecting the headscarp and toe slope;
(3) the absence of a headscarp above the eastern part of the area of deformation;
(4) the absence of lateral flank scarps or rupture zones;
(5) the absence of evidence for toe slope displacement, or of a failure plane at the slide toe.

These considerations appear to favour an alternative visco-plastic explanation of slope deformation. In terms of this explanation, the antiscarps represent fracture and toppling of near-surface rock masses over a region of deep deformation caused by micro-fracturing of rock under high pressure (Radbruch-Hall, 1978). The rock scarp at the head of the western and central parts of the slope deformation may be a failure plane representing deep translational sliding of the uppermost part of the slope deformation; its absence above the eastern slope convexity is consistent with evidence for differential displacement of the central and eastern parts of the area of slope deformation. Essentially, this explanation resembles that of Holmes and Jarvis (1985), but without invoking extensional toppling initiated near the foot of the slope.

Timing and mode of displacement

The sharpness of antiscarp crests implies that slope deformation post-dates ice-sheet deglaciation. The presence of antiscarps on lower slopes demonstrates that deformation, though possibly initiated during or after ice-sheet downwastage (*c.* 17–15 cal. ka BP), continued after deglaciation at the end of the Loch Lomond Stadial (*c.* 12.9–11.5 cal. ka BP). Both Watters (1972) and Holmes and Jarvis (1985) suggested that high water-pressures during deglaciation may have aided slope deformation by decreasing the interlayer strength of the rock mass. Although elevated joint-water pressures have been identified elsewhere as instrumental in gravitational deformation of slopes (Bovis and Evans, 1996), the presence of antiscarps within the area re-occupied by ice during the Loch Lomond Stadial suggests that deformation continued after final deglaciation. Fenton (1992) has suggested that slope deformation was triggered by a high-magnitude seismic shock due to fault re-activation by differential glacio-isostatic unloading, but this explanation appears to rest solely on the proximity of the Gleann Lichd and Strathconon faults. Deep-seated gravitational slope deformation is widely considered to be a gradual phenomenon (Bisci *et al.*, 1996; Bovis and Evans, 1996) extending over centuries or millennia until post-deglaciation strength-equilibrium conditions are regained. At Beinn Fhada, the lack of runout debris, the essentially intact nature of displaced blocks and the continuity of antiscarp crests are consistent with gradual rather than abrupt displacement.

Wider significance

The Beinn Fhada rock slope deformation represents a type of paraglacial (glacially conditioned) slope failure that is widespread in the Scottish Highlands, particularly on metasediments. Such deformations are characterized by the absence of evidence for continuous failure planes or lateral flank scarps, and by the development of antiscarps, tensional crevices at or near the slope crest, slope bulging and headscarps, though the last three features may be absent or poorly developed. Although present in several parts of the Scottish Highlands, such deformations are particularly well-represented in the mountains around Glen Shiel, 6 km south of Beinn Fhada (Jarman, 2003b). Examples in this area include the north-west slope of **Druim Shionnach** (NN 070 090) and the south slope of Sgurr a' Bhealaich Dheirg (NH 023 140), both of which exhibit slope bulging, antiscarp development, headscarps and tensional fissures at the slope crest.

Jarman (2003e) has suggested that the large extent of rock slope deformation at Beinn Fhada may reflect the exceptional width of the plateau at this site, as stresses may be higher within slopes below broad mountain ridges (Beck, 1968; Gerber and Scheidegger, 1969). Alternatively, the exceptional size of the Beinn Fhada failure may simply reflect the occurrence of two or three contiguous areas of slope deformation on an uninterrupted slope of unusually high and steep relief. Jarman (2003c,e) also suggested an association between the location of rock slope failures and breaches mainly attributable to glacial erosion (Linton, 1949). In the case of the Beinn Fhada rock slope deformation, enhanced slope steepening may have resulted from accelerated glacial sliding velocities as the icestream occupying Gleann Lichd at the last glacial maximum descended westwards from or across the watershed to the valley floor (see Figure 2.18).

Conclusions

The Beinn Fhada rock mass deformation is the best documented example of *sackung*-type deep gravitational slope deformation in Great Britain. It is important for several reasons. The site exemplifies all of the principal characteristics of such slope deformations, namely the development of slope convexities (bulges), the formation of widespread antiscarp arrays, a possible exposed failure plane at the crest of part of the deformed rock mass, diffuse lateral margins without flank scarps, and the absence of foot-slope deformation or evidence for a continuous failure plane. In terms of surface area and estimated volume it is not only the largest such slope deformation in Great Britain, but also the most extensive rock slope failure on the Scottish mainland. The height (6–10 m) and length (600–800 m) of the most conspicuous antiscarps are almost without parallel in Scotland, and the overall area of deformation is unusual in exhibiting evidence for complex deformation in the form of three distinct large-scale slope bulges, one of which has moved differentially relative to the other two.

Like many major rock slope failures in the Scottish Highlands, the development of the Beinn Fhada rock slope deformation can be attributed to slope steepening by glacial erosion and rock-mass weakening associated with deglacial unloading and paraglacial stress-release. Displacement was probably gradual and aided by high joint-water pressures. In common with similar slope deformations in Scotland and elsewhere, the absence of a continuous failure plane makes analysis of failure geometry and progression speculative. Though there is widespread acceptance that *sackung*-type gravitational slope deformation reflects a combination of deep-seated sliding, formation of antiscarps by joint-guided toppling and deep-seated deformation under high confining pressures, the validity of this explanation in the context of the Beinn Fhada rock slope deformation remains conjectural.

SGURR NA CISTE DUIBHE, HIGHLAND (NG 988 143)

D. Jarman

Introduction

Sgurr na Ciste Duibhe (1027 m) is one of the Five Sisters of Kintail in the Western Highlands. It is the clearest case of rock slope failure controlled by a basement fault. It has been advanced – with insufficient justification – as evidence for high-magnitude seismic events following the last deglaciation (Fenton, 1992). Its summit area has been lowered by about 10 m as a result of creep or slippage, and is the most striking example in Britain of this kind of paraglacial landscape modification. It also demonstrates classic rock slope failure features including ridge-top depressions and arêtes. The deformation progresses downslope into a sliding failure of unusual geometry, which probably reaches the floor of Glen Shiel; if so, this is the greatest vertical extent of rock slope failure in Britain at almost 1000 m, approaching alpine scale.

This is one of the 20 largest rock slope failures in the Highlands, and is within one of the densest clusters situated in the Glen Shiel–Affric area (Figure 2.18). While this cluster may have neotectonic associations, it is also within an area of anomalously low valley interconnectivity (Haynes, 1977a) and may point to relatively recent breaching of the main watershed by transfluent ice (Jarman, 2003b). Sgurr na Ciste Duibhe stands directly above the gorge of the River Shiel at the probable locus of the breach (Linton, 1949).

Description

The mountain ridge on the north side of Glen Shiel is continuous above 725 m OD for 13 km, one of the longest in Scotland, and matched by that on the south side. For 1 km, the ridge is dislocated by the intersection of the Glen Shiel Fault-swarm with the crest of Sgurr na Ciste Duibhe (Figures 2.23–2.25). The main fault passes 100 m behind the summit, where a 10 m red rock step on the north-east spur is prominent on the skyline (Point 985). The fault then crosses the ridge 300 m east of the summit, interrupting it with a 15 m crag (Point 935). Between these incidents, the fault is expressed as an arête across a small corrie head, partly

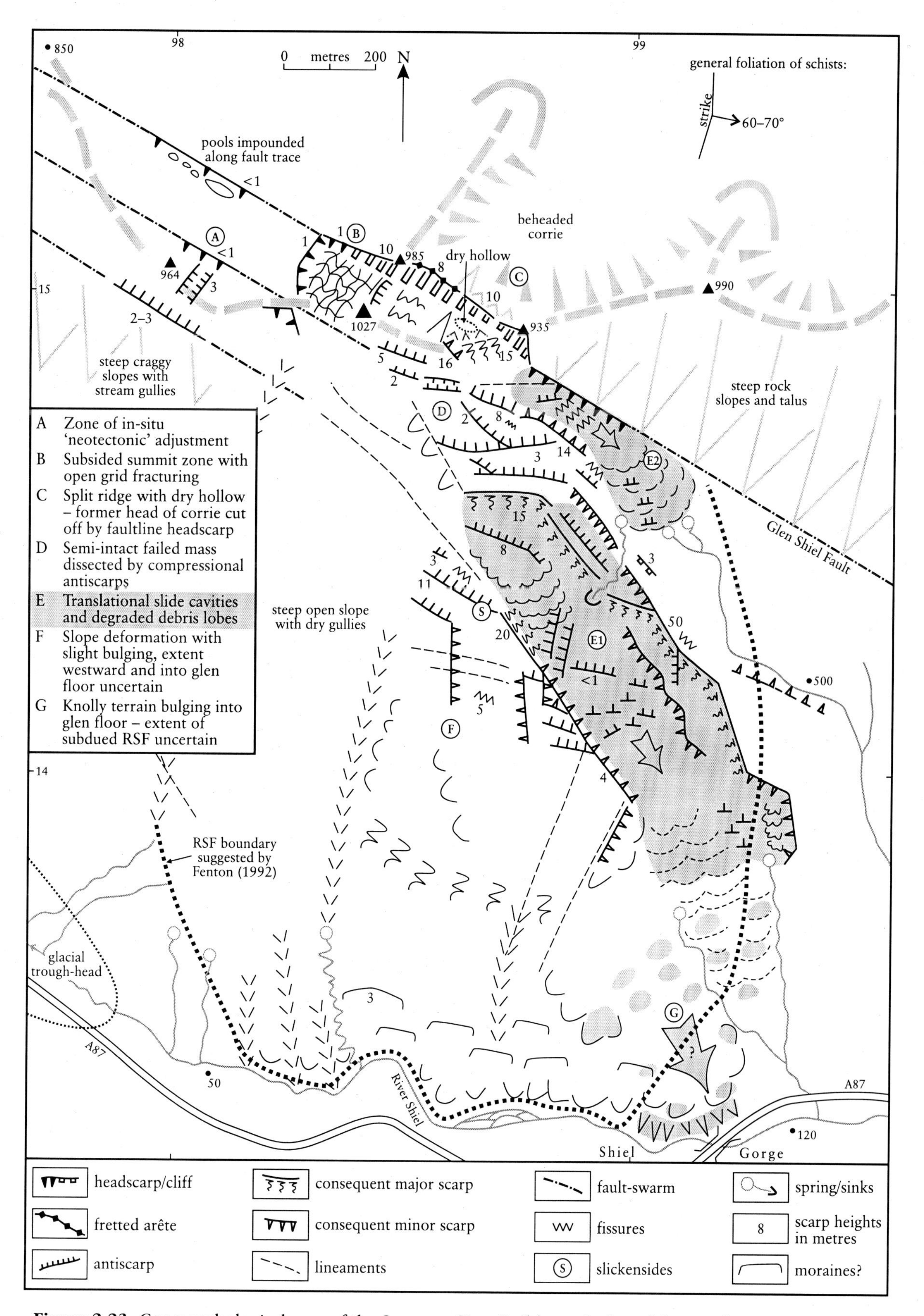

Figure 2.23 Geomorphological map of the Sgurr na Ciste Duibhe rock slope failure. After Jarman (2003b).

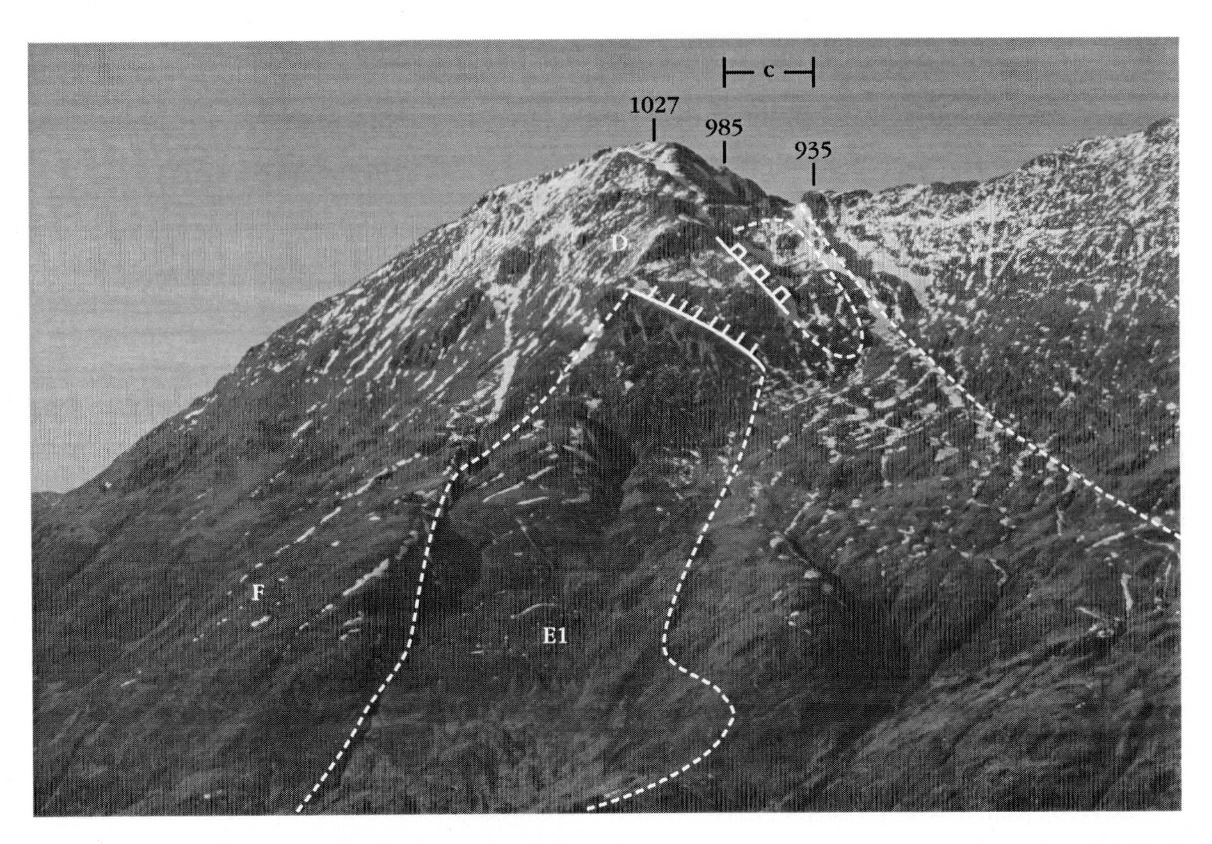

Figure 2.24 Failed slope rising 950 m above Glen Shiel breach, viewed from the south, showing the Glen Shiel Fault (diagonal pecked line), skyline dislocation, lowered summit and some of the failure zones marked on Figure 2.23 (Zone C is shown on Figure 2.26). (Photo: D. Jarman.)

smooth but partly fretted, and trapping two dry depressions (Figure 2.26). West of the red rock step, the scarp diminishes to 1 m before returning at right angles to the summit ridge. The fault trace, however, continues WNW, impounding small ribbon pools before recrossing the ridge; to the ESE the trace descends gradually into Glen Shiel. In all it is distinct for 5 km (Fenton, 1992).

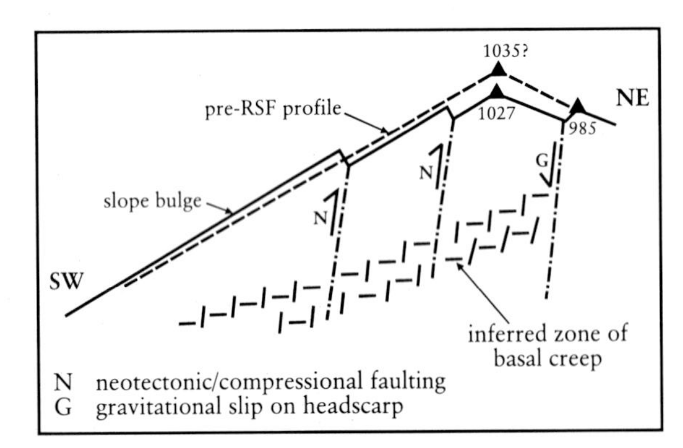

Figure 2.25 Sgurr na Ciste Duibhe rock slope failure (RSF); notional cross-section of the summit area. After Jarman (2003b).

The summit area is a coarse blockfield (Zone B, Figure 2.23), more disrupted than is usual for such periglaciated terrain; a 2 m-deep hole beside the trig point indicates tension, as does a 3 m × 3 m trench across the ridge to the west. Antiscarps up to 5 m high have developed along parallel members of the fault-swarm on both sides of the crest.

To the south-east of the summit, an area 250 m by 150 m is dislocated by a striking series of antiscarps up to 3 m high on two steep diagonals, with large tension fissures in between (Zone D, Figure 2.27). The uphill face of this mass presents a ragged cliff 8–14 m high to a graben-like depression, out of which decreasingly coherent debris has moved to the south-east, with a springline at its foot (Zone E2). The downhill face of this dislocated area is a crag reaching 15 m high, so close below the lowest antiscarp that in places there remains a rock parapet only 4 m wide. This crag is the top part of an approximately 60 m-high source scarp to the main sliding failure. The slide cavity (Coire Caol – Zone E1) is contained by prominent

Figure 2.26 Five Sisters summit ridge dislocated by subsidence along the Glen Shiel Fault. View looking south-eastwards from Point 985. Ridge breaks are 10–15 m high; note the dry hollows, lower-right (Zone C). (Photo: D. Jarman.)

Figure 2.27 Antiscarps and graben wall below the summit (Zone D). The bulging toe (hiding the road) creates a local gorge. The slide cavity with fissured debris-mass can be seen in the foreground. The long parapet/antiscarp and uphill-facing crag are marked on Figure 2.24. (Photo: D. Jarman.)

sub-parallel scarps 300 m apart, orientated NNW–SSE at 45° to the fall-line. The cavity reaches 50–80 m deep by contour extrapolation, with the scarp on its east being considerably higher than that on the west, and stepped and embayed with signs of incipient retreat above. In the head of the cavity a large monolithic failed mass remains, presenting an antiscarp up to 8 m high to its source; its surface is convex and subdued, with distinct decimetric antiscarplets and furrows. Below this, the main cavity is not fully evacuated: its floor also has antiscarplets, and a substantial stream sinks beneath it. Beyond the confines of the cavity, a very degraded debris-lobe descends at least to springs at 300 m OD, and probably to 200 m OD.

On the west side of this slide cavity, a broad midslope convexity has distinct indicators of tensional failure in its upper parts, including a 5 m-deep fissure, and some minor antiscarps (zone F). At its head at 650 m OD, the cavity scarp breaks back as an 11 m-high rockface with slickensiding on its smooth surfaces (Gordon, pers. comm.). The lowest part of this convexity forms a bluff protruding into Glen Shiel to form a local gorge with anomalous local topography; it has an absence of rock outcrops or surface wetness, but displays no positive indicators of failure. In its lee to the west, a broad embayment beneath the convexity has a series of debris ridges with pronounced uphill faces, reaching 3 m high, which however are not typical antiscarps. To the west again, a steep open slope extends almost unbroken from summit to valley floor, lacking surface-water drainage and the broken crags typical of the lower glen. Several springs rise near the slope foot, but when observed with the River Shiel in spate, their flows were modest and adjacent gullies remained dry.

Fenton (1992) includes all of these areas within the rock slope failure boundary, giving a possible maximum area of 1.9 km^2. Holmes (1984) identified only a limited area of 0.14 km^2 near the summit, as does the 1:50 000 geological map. The failure certainly extends over at least 0.95 km^2 including the in-situ summit ridge deformation, and probably over at least 1.25 km^2 (Jarman, 2003b). The volume of the main cavity is of the order of 5–10 × 10^6 m^3. Efforts to assess the overall volume of the rock slope failure would be very difficult; suffice it to say that it appears deep-seated, with Fenton (1991) estimating a depth in excess of 50 m.

The bedrock consists of Neoproterozoic age Moine psammites and subsidiary semipelites of the Morar Group; the bedding and main foliation in the rocks dips at 50°–75° to the east or south-east. It is probably co-incidental that this rock slope failure is developed in stratigraphically equivalent rocks to those underlying the **Beinn Fhada** failure. No detailed structural analysis has been made, but Fenton (1991) suggests fractures dipping south more steeply than the slope at about 45°. Fluid alteration has occurred along the near-vertical Glen Shiel Fault, giving rise to a weak band of oxidized and degraded rock: this is visible as an eroded trench and may have aided mass slippage.

Interpretation

This complex rock slope failure demonstrates a clear progression from in-situ deformation along the ridge west of the summit, through spreading, creep, and embryonic sliding failure in the summit area and upper slopes, to sub-cataclasmic failure of a large segment of the midslope.

The configuration of the whole complex is unusual in being orientated up to 90° away from the fall-line trend of the glen wall. This suggests, as at **Beinn Fhada** nearby, that joint-sets conducive to translational sliding are not well developed here. Indeed, analyses of 12 sites north-west of the Great Glen by Watters (1972) and Holmes (1984) suggest that both schistosity surface and principal joint planes are commonly inclined above 60°, by contrast with the South-west Highlands (e.g. **The Cobbler**).

Here four structural components can be identified. The Glen Shiel Fault is clearly the backing scarp for the whole rock slope failure; Fenton (1992) describes it as the 'headscarp', but movement is as much along it as down it. Secondly, the schistosity strikes orthogonally to the fault, and provides the return scarp and internal step-down scarps to the failed summit mass. Thirdly, the parallel members of the fault-swarm have allowed graben-type spreading and subsidence of the summit mass and short-travel sliding to the south-east. Some elements of these parallel faults appear to have been displaced south as part of a broader slope spreading. Fourthly, the main slide cavity is controlled by a north-west–south-east joint-set that may be cognate with the main joint-set identified at **Beinn Fhada**.

It is inferred from the 10 m height of the red rock step on the north-east spur that the summit of Sgurr na Ciste Duibhe has been lowered by 8–10 m (Figure 2.25). Although the return scarp to the north-west is only 1 m high, the step scarps within the disrupted blockfield make up the difference. Such lowering of a major summit by paraglacial rock slope failure would be unique in Scotland: close parallels occur at Carn na Con Dhu near Glen Affric (NH 07 24) and Beinn Bhreac by Loch Lomond (NN 32 00) where large sections of summit plateau have subsided. The area affected by lowering here extends 100 m north-east and north-west of the summit, and 200 m to the south-east, one of the most substantial encroachments into a mountain ridge by rock slope failure in Britain.

Equally remarkable is the failed mass south-east of the summit, which is semi-intact rather than disrupted, but is carved up by curving antiscarps on alignments discordant with the general structure (Figure 2.27). These suggest that the mass has begun to move by basal creep plus forward toppling, with both compressional antiscarps (Fenton, 1991) and tension fissures. Poised above the main rock slope failure cavity and with only limited lateral restraint, this mass may be only marginally stable. Analogous failed masses occur at Ben Lawers (NN 63 41) and Ben Vorlich by Loch Lomond (NN 29 12). Their survival may indicate a long-term lack of significant seismic shaking, for which precariously balanced rocks are a recognized indicator.

Although the main rock slope failure cavity is unusual in its slantwise orientation, the residual mass in the neck is a common feature of such slides (e.g. Meall Cala, Figure 2.39), as is the low-key disturbance of the floor (e.g. Beinn an Lochain west, Argyll, NN 215 076). The latter feature may indicate post-slide decompressive recovery, though the underground watercourse suggests incomplete removal (compare with Beinn Each, **Glen Ample**). The considerable volume of the cavity is not readily accounted for by debris below, implying partial evacuation by the last glacier. Very few rock slope failures within the Loch Lomond Stadial limits have been identified as pre-dating the Loch Lomond deglaciation (cf. **Beinn Fhada**). Sgurr na Ciste Duibhe provides a good opportunity to test this issue, by investigating the integrity of the knolls in Zone G (which could be a glacially over-ridden slipped mass), and the provenance of the atypically substantial lateral-moraine type mounds to its west.

To the west of the slide cavity, the broad midslope convexity has definite indicators of failure in its upper and eastern margins, but minimal indications of downhill displacement. As a hybrid between in-situ deformation and translational sliding, it fits the model identified in this area of the Cluanie-hybrid type failure exemplified at **Druim Shionnach**. In a context of high-angle schistosity planes and joint-sets in steep relief, decompressive forces are envisaged as conducive to slope bulging, with formation of grabens and antiscarp arrays, but no development of a sliding surface or debris lobe. Indeed, the Sgurr na Ciste Duibhe rock slope failure could have evolved initially as a Cluanie-hybrid failure affecting much of the glen slopes, with the more disrupted upper half then slicing off diagonally by progressive creep and sub-cataclasmic sliding. Lack of surface drainage and the powerful basal springs confirm deep-seated failure.

It is unclear whether the translational slide has at some stage reached the narrow glen floor and temporarily blocked it. Major landslide dams are rare in the British mountains, although common in other ranges, the dams impounding Llyn Tal-y-llyn being a good exemplar (Hutchinson and Millar, 2001). The rock slope failure toe below the knolls is exceptionally steep without being a rock gorge, and the River Shiel immediately downstream is atypically braided through coarse angular debris. This could be a result of fluvial breaching of a small dam, without re-activating the slide lobe.

One of the most widely recognized indicators of rock slope failure in mountain ranges is the split ridge or *doppelgrat* (Radbruch-Hall *et al.*, 1976; Crosta, 1996). This occurs where the source scarp daylights behind the crest, creating an uphill-facing scarp often confused with anti-scarps. Sgurr na Ciste Duibhe exemplifies the ridge-top depression in Britain, other instances being found on Ben Challum (NN 38 31) and Helm Crag, Lake District (NY 325 090) and most strikingly Aonach Sgoilte (Figure 2.12). The sharpening of a crest into an arête by rock slope failure is also seen locally here, but is better-developed farther east along the ridge between Saileag and Sgurr a' Bhealaich Dheirg (Figure 2.28), and on Aonach Meadhoin.

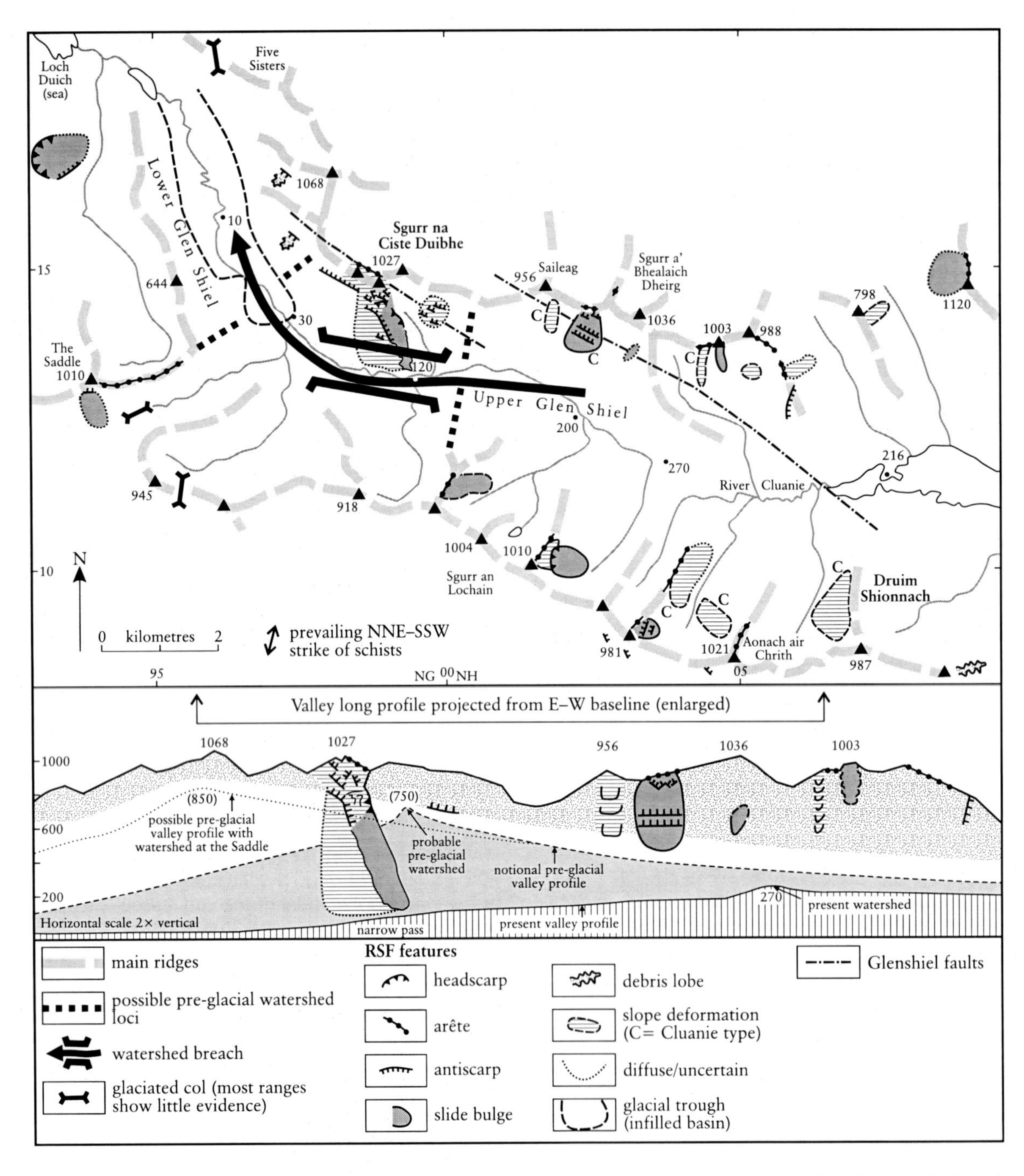

Figure 2.28 Map and long-section of the Glen Shiel breach with reconstructed pre-glacial watershed and associated rock slope failures on the north side of the main valley and in the trough corries of the south Cluanie ridge (e.g. **Druim Shionnach**). After Jarman (2003b).

Neotectonic activity

Fenton (1991, 1992) identified Sgurr na Ciste Duibhe as a prime example of a rock slope failure controlled by basement faulting, and triggered by neotectonic activity along it in response to post-glacial stress-release. From fault dimensions, he estimated a paleoseismic event here of magnitude 6.0–6.6 M_S, and from the incidence of rock slope failure and sediment liquefaction, magnitudes of 4.5–5.9 M_S. The implausibility of this model has been noted in the 'Introduction' to the present chapter, and the polyphase nature of the complex rock slope failure here further militates against it. In particular, the midslope slide cavity is markedly more subdued than either the lowest side slip into it or the sharply etched antiscarps and arête above. Fenton (1992) further proposed that the ridge crest has been offset by approximately 15 m in a sinistral sense, as a result of post-glacial intra-plate shearing, and attributed the ridge-crest hollow to this movement. The model for this was the proposed 160 m lateral displacement along the Kinlochhourn Fault nearby (Ringrose, 1989), which has been refuted by Firth and Stewart (2000). Here, there is no lateral offset to the north-east spur or the adjacent corrie-head gullies: an apparent offset of this scale where the fault cuts the east ridge is attributable to sliding out of the 'graben' section of the rock slope failure (Zone E2).

However, it remains possible that this and other rock slope failures have been partly triggered by seismic shocks of local origin, as the terrain recovered from differential ice loadings of 700–1000 m between ridge and glen, and from concentrated slope-foot erosion. Kintail is currently a focus of minor seismicity, and there is a marked cluster of rock slope failures here (Jarman, 2006). The high available relief, the existence of a sizeable basement fault crossing a narrow ridge, and the evidence of slickensiding all make Sgurr na Ciste Duibhe one of the prime sites in Britain for detailed examination of paleoseismicity.

Glacial breaching

Since the precise trigger mechanisms for the rock slope failure may prove difficult to determine, it is more fruitful to consider why this major failure complex has occurred here, within a dense cluster of such failures. Glen Shiel is a short west-flowing valley typical of the Western Highlands where the main watershed lies close to the coast, with a major glacial breach at its head (270 m OD) through to Glen Cluanie (Figures 2.18 and 2.28). Sgurr na Ciste Duibhe stands above the foot of its steep descent at the point where it becomes a glacial trough only 50 m above OD. However, Linton (1949) places the pre-glacial watershed at this point, implying that possibly 500–750 m of erosion has occurred here over the Quaternary Period, by a combination of vigorous trough-head excavation by the west-flowing valley glacier, and glacial breaching at times when the iceshed lay to the east of the watershed. Haynes (1977a) observes anomalously low valley interconnectivity in Kintail, and the breaches of the main watershed are at their highest elevations here. Haynes attributes this to a subtle combination of resistant geology, a high mountain axis aligned north-east–south-west, and an ice dome centred on the range deflecting other icestreams around it. In this context, the Glen Shiel breach may be relatively young, and bulk erosion may have been concentrated at the foot of Sgurr na Ciste Duibhe. This will have provoked rock slope failure partly by exposing potential failure planes, but primarily by generating rock mass stresses sufficient to propagate deep fracturing and compensatory outward movement.

In identifying a strong association between possible recent breaches and the incidence of rock slope failure in this area, in contrast to an absence of rock slope failure along mature glen sides, Jarman (2003b) suggests that this may indicate a shift or intensification of ice dispersal patterns in Devensian times. In this analysis, Sgurr na Ciste Duibhe is the most significant rock slope failure in this area.

Conclusions

Sgurr na Ciste Duibhe is a key site for showing structural controls on rock slope failure, and paraglacial mountain summit shaping. The failure complex demonstrates vertical progression from in-situ deformation to long-runout sliding over almost 1000 m of valley-side. It is clearly associated with a basement fault zone, and is the best site on which to test theories of elevated seismic activity around deglaciation. The mountain top owes much of its present shape to rock slope failure, with fine examples of arêtes, ridge-top depressions, and antiscarps; its fractured and lowered summit may be unique in Britain. The failure stands above one of the most deeply excavated glacial breaches in Britain, and is the most significant member of the Kintail cluster in exploring the association between recent bulk erosion, generation of deep-seated rock-mass stresses, and their release in various modes of slope failure.

DRUIM SHIONNACH, HIGHLAND (NH 070 090)

D. Jarman

Introduction

Druim Shionnach is the type example of the 'Cluanie-hybrid' mode of rock slope failure identified by Jarman (2003b) and associated with steeply inclined and highly indurated metasediments in the Cluanie–Glen Shiel area and elsewhere. It has the essential characteristics of the compressional semi-intact slope deformation, with no signs of disintegration or debris mass, but it also has the pronounced headscarp and extensional spreading features found in translational slides. Such transitional character is valuable in understanding the mechanics of rock slope failure initiation and development. This failure also shows unusually clear-cut geological controls, and has produced a remarkably steep, smooth slope bulge (Figures 2.29–2.31.

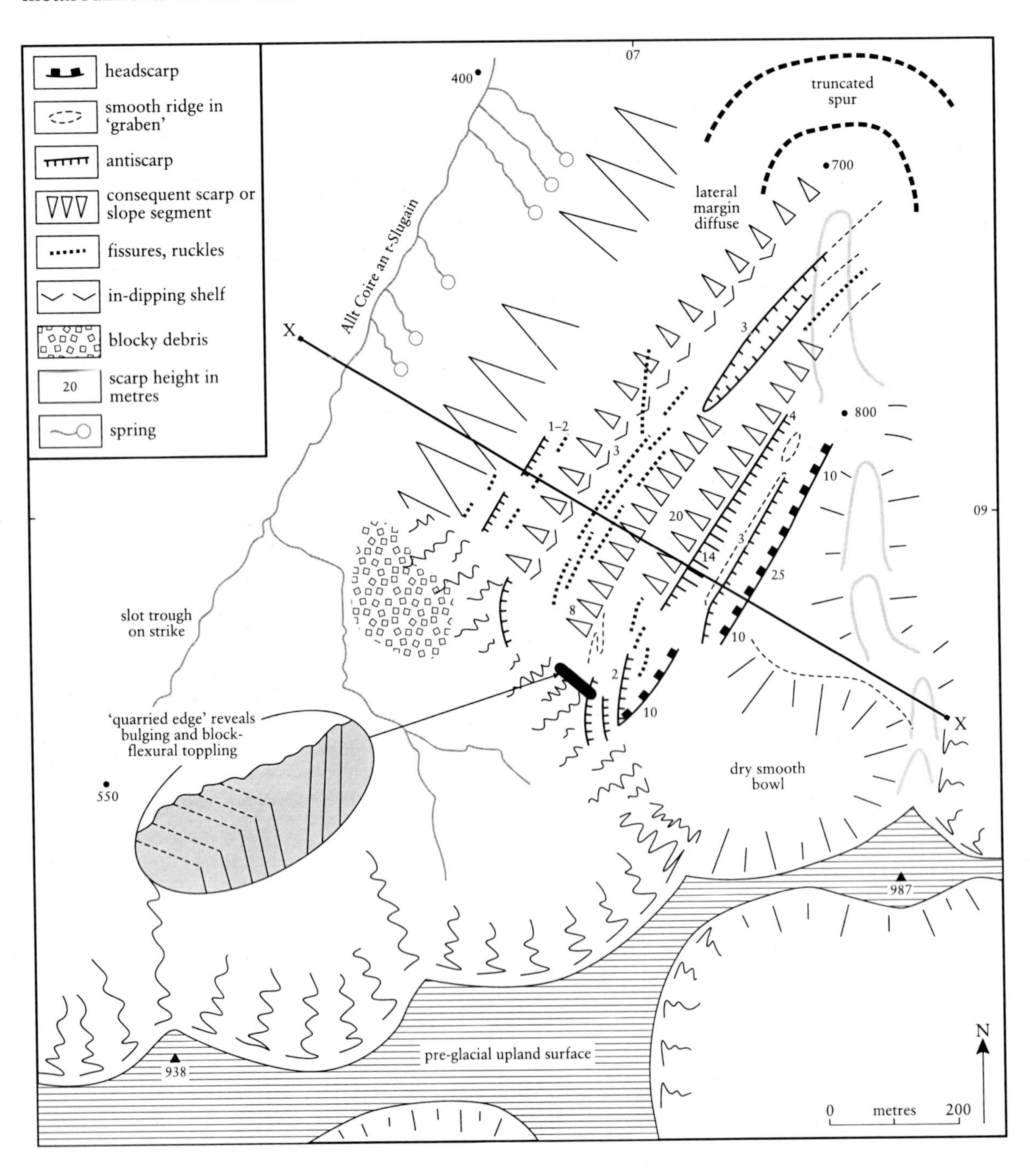

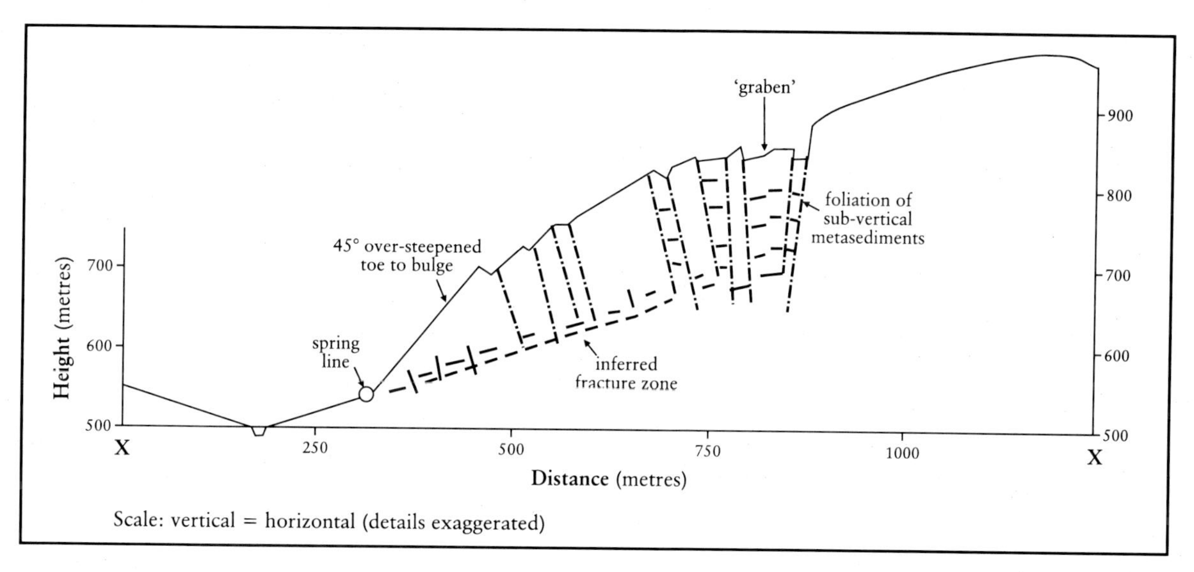

Figure 2.30 Section X–X on Figure 2.29, showing over-steepened bulge and graben progressively tilting failed slices away from source. After Jarman (2003b).

Figure 2.31 The Druim Shionnach rock slope failure is on the flank of a side trough off the main Cluanie pre-glacial valley. The smooth and bulging failed slope is wider than the apparent source scarp and notably lacks surface drainage and a clear lateral margin. (Photo: D. Jarman.)

The ridge on the south side of the Cluanie–Glen Shiel breach valley is notable for its succession of compound corries. Most of these have rock slope failures on their flanks (Figure 2.28), but not on their headwalls (as at **Ben Hee**); generally these failures are of Cluanie-hybrid type. The role of failure in lateral enlargement of corries is evident at Druim Shionnach, and westwards along the ridge the process has progressed further, with interfluves first being narrowed to arêtes and then reduced to rounded rumps.

◂**Figure 2.29** Geomorphological interpretation of the Druim Shionnach rock slope failure based on OS 1:25 000 mapping. After Jarman (2003b).

Description

The failure is located on the west side of the north shoulder of Druim Shionnach, a 987 m-high peak above the head of Loch Cluanie (Figure 2.32). Its rockwall headscarp is 25 m high in the centre, quickly reducing to 10 m on either side. It confronts an impressive 'antiscarp' across a broad (50 m) trench within which lies a smooth linear island with subsidiary antiscarp. The main 'antiscarp' reaches 14 m in height, and is steep grass with a bedrock rim (Figure 2.30). This considerably exceeds the normal height range for antiscarps, which are generally below 5 m, and even on **Beinn Fhada** only locally approach 10 m.

Behind the 'antiscarp' the open valley-side descends from 850 m to 700 m OD in a series of smooth slope facets and broad benches. These are level or slightly in-dipping, and incidented with abundant minor ruckles, fissures, and antiscarplets less than 2 m high (Figure 2.29). Only towards the north-east end does one of these contour-parallel lineaments open out into a smooth broad trench with 3 m outer antiscarp. These features fade out onto the open north shoulder, but can be traced slightly beyond its brow. Below 700 m OD, the main valley-side steepens to an angle of approximately 45° down to a springline at 500–550 m OD (Figure 2.30). This is exceptionally steep for a slope developed in foliated bedrock with no exposed crags, and is above the the typical peak friction angle for sliding (Watters, 1972). The northern extent of this over-steepened slope is diffuse.

By contrast, the southern flank of the failure is abruptly truncated by a broken slope about 100 m high which faces the compound head of Coire an t-Slugain (Figure 2.29). It is most unusual for almost the whole of a failure from headscarp to toe to be revealed in deep cross-section in this way. Some minor antiscarps and blockfalls have developed along it, but it is generally intact. Interestingly, the upper part displays block-flexural toppling, which has been widely identified as a mode of rock slope failure (Zischinsky, 1966; de Freitas and Watters, 1973; Holmes, 1984; Hutchinson, 1988), but is here only a few metres deep, and merely a process of superficial creep within an already failed mass (Figures 2.29 (inset) and 2.32).

The total area affected by rock slope failure is given as 0.33 km^2 by Holmes (1984). The 1:50 000 geological map indicates an extent of approximately 0.4 km^2, and including in-situ deformation of the north shoulder the area may reach 0.55 km^2. The failure has formed within

Figure 2.32 The Druim Shionnach rock slope failure top surface, seen in close-up from the summit ridge to the south-west. The prominent peak is the 14 m-high antiscarp facing the source scarp across a half-graben. Note the block-flexural toppling in the near-vertical metasediments, revealed in section in the foreground (see inset, Figure 2.29). (Photo: D. Jarman.)

massive gneissose psammites of the Glenfinnan Group. These Neoproterozoic-age rocks have been deformed by several orogenic events and metamorphosed to a high grade. Here they are tightly folded, steeply dipping and strongly indurated. The bedding and main foliation strike NNE, and dips range about the vertical from 75° eastwards and westwards (May *et al.*, 1993). Locally, pelite bands create micaceous planer surfaces that are more prone to parting.

Interpretation

Geological control is strongly evident at Druim Shionnach, as in other 'Cluanie-hybrid' type rock slope failures. Here, the headscarp is co-incident with the strike of the foliation, which is sub-vertical at the crest and dips 80° glenwards near the toe. Although no geotechnical analysis has been conducted, it may be inferred from the steep, intact bulge and in particular its lack of lateral restraint that joint-sets inclined valley-wards at angles conducive to sliding are not present; the main structural controls here are orientated north-west–south-east.

The major feature at Druim Shionnach is therefore interpreted as a kind of 'graben' (Jarman, 2003b), where the failed mass has moved outwards in response to decompression. The 'anti-scarp' is thus not a typical adjustment feature within a deforming mass, as at **Beinn Fhada**, but the face of an unusually broad tension trench, hence its exceptional height. A comparable case occurs on the Arrochar Ben Vorlich above Loch Sloy (NN 29 11). The subsidiary feature within the 'graben' has probably not literally subsided as in a rift valley, but reflects the stepping or doubling of the trench.

Within the failed mass, a progression can be seen from tensional features in the upper half to compression in the bulge. There is no evidence of downslope sliding movement, unless the toe has been glacially trimmed, and the scale of displacement evident in the source zone has presumably been accommodated by internal deformation and creep. A schematic cross-section (Figure 2.30) suggests that such deformation may be at least 100 m deep, consistent with the assumptions of Holmes (1984) and Fenton (1991).

A peculiarity of Druim Shionnach is its discordance with the pre-failure topography. Not only do the tension features transgress onto the north shoulder, they also continue south-west across the mouth of a smooth open bowl that contrasts markedly with the typical scalloped compound corrie head. The steep south-west flank to the rock slope failure might suggest that a late (Loch Lomond Stadial?) corrie glacier has quarried the edge of a formerly more extensive failure (Jarman, 2003b), but this is seen as glaciologically less likely. The glacial/paraglacial evolution of Coire an t-Slugain is complex, and the timing and contribution of rock slope failure to corrie development may fruitfully be explored here.

Wider landscape evolution

The south Cluanie ridge affords an instructive overview of rock slope failure as an agent of mountain landscape evolution (Figure 2.28). This is one of the longest high-level ridges in Scotland, continuously above 700 m for 15 km. The main ridge has become asymmetrical as a result of glacial trough–corrie development exploiting the strike of the schists, with its centre-line offset by up to 2 km to the south from an original median between Glen Quoich and Glen Shiel, and by up to 500 m between the extant summits, which typically lie off the main ridge out on the projecting spurs. However, failure has probably contributed less to this headward corrie erosion (by contrast with **Ben Hee**) than to their lateral expansion. All the extant failures are on the corrie flanks, or in their headwall angles, and display Cluanie-hybrid character, notably north-west of Aonach air Chrith. Those north-east of Sgurr Beag and Sgurr an Doire Leathain have shaped the spur crests into arêtes, the latter having progressed into a large sliding slump. Rock slope failure has therefore contributed to corrie amalgamation, creating the compound corries typical of the Moine Supergroup (Gordon, 1977), and to the reduction of the intervening ridges to a level where they have become subject to glacial scour. By contrast, there is minimal failure on the long steep south flank of the main ridge above Glen Quoich, implying that that valley has long since adjusted to ice discharge; unlike Glen Shiel its head has not become breached.

Other nearby examples of the Cluanie-hybrid type rock slope failure type occur in similar geological and topographical contexts on the northern arêtes of Sgurr a' Bhealaich Dheirg, while on its south-west flank a notable failure in the breached valley of Glen Shiel has affinities despite being orthogonal to the strike of the schists. This failure has 7 m antiscarps noted by Fenton (1991) as 'pop-ups' co-inciding with the Glen Shiel Fault-swarm. 'Cluanie-hybrid' type character can also be observed in some of the failures in the Mamores–Grey Corries cluster, again on steeply dipping schists and quartzites.

Conclusions

Druim Shionnach is a well-defined rock slope failure that illustrates transitional character between the slope deformation and the sliding mass, with both compressional and extensional elements. It has several exceptional features, notably a 14 m-high 'antiscarp' and graben structure, and a rare exposure in cross-section of block-flexural toppling. While in itself its mountain-shaping role is limited to trough–corrie widening, other rock slope failures of this 'Cluanie-hybrid' type in the vicinity clearly contribute to the evolution of the whole ridge and to arête development on its spurs.

BENVANE (BEINN BHÀN), STIRLING (NN 533 122)

D. Jarman

Introduction

Benvane is a major reference site for the diversity of rock slope failure modes and features in the old hard rocks of Britain (Figures 2.33–2.35). It displays a lateral progression from in-situ slope deformation to translational sliding. It is located on some of the lowest relief to give rise to extensive failure in the Scottish Highlands. Its boundaries are notably distinct, and it displays one of the finest and most extensive lattice antiscarp arrays in Britain, here, unusually, exhibiting three distinct orientations. The failure extends up to 120 m behind the brow of the broad ridge over a distance of 1 km, and is a telling indicator of the effect failure can have in wholesale reduction of relief in mountainous

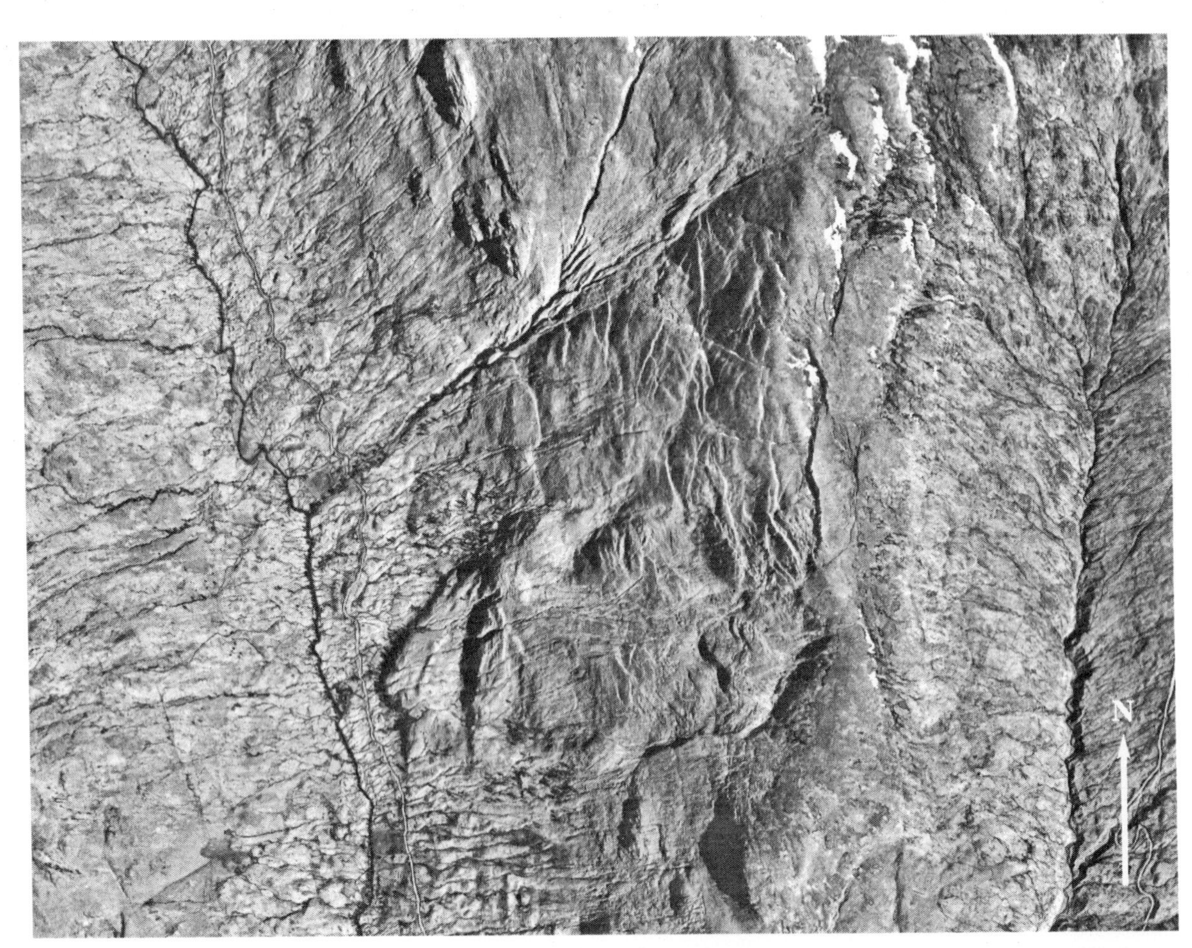

Figure 2.33 Vertical aerial photograph (1989) of Benvane, with sun from the east accentuating the array of antiscarps, scarplets and benches. (Photo: Crown Copyright: RCAHMS (All Scotland Survey Collection).)

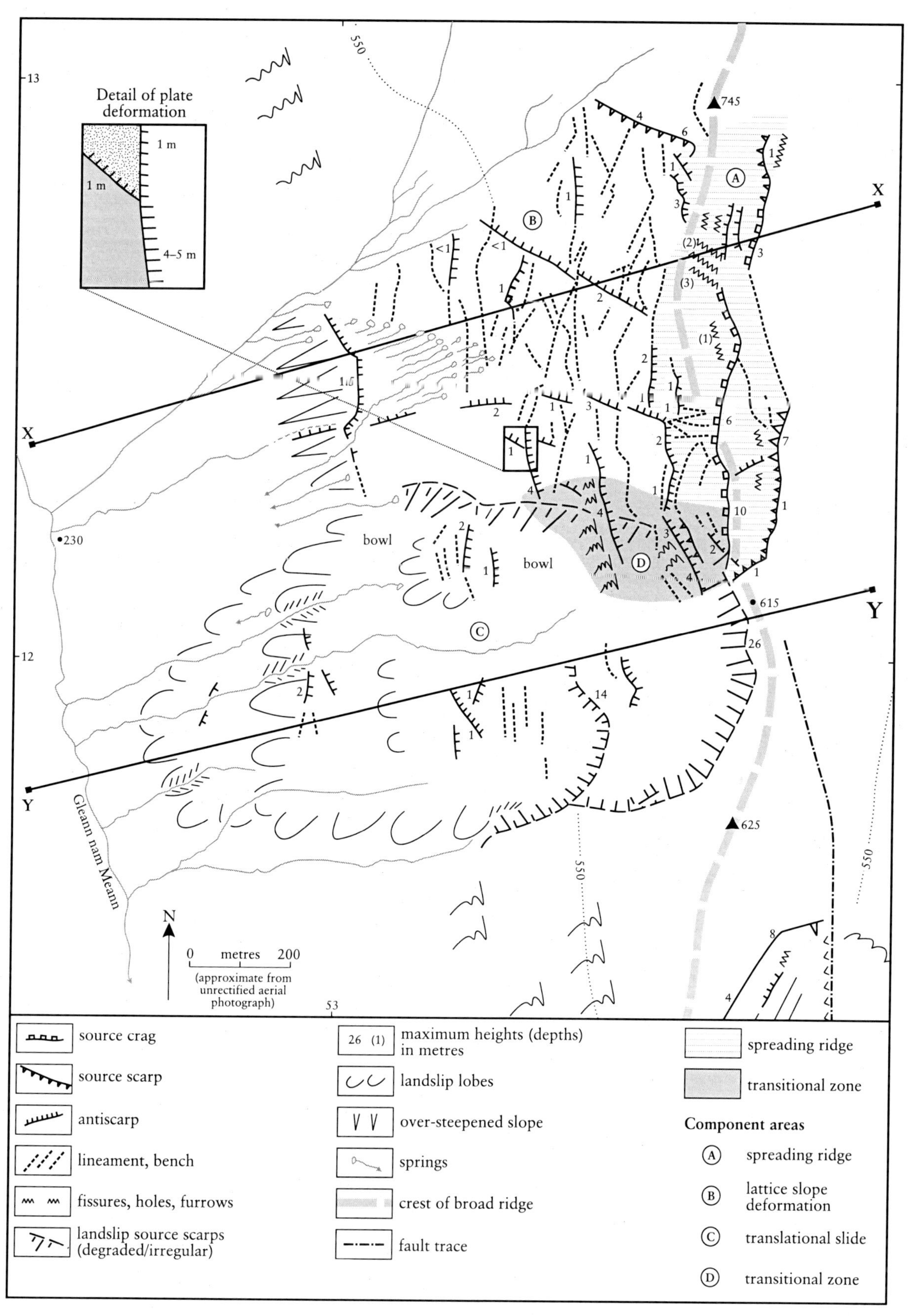

Figure 2.34 Geomorphological interpretation of the Benvane rock slope failure complex. Based on unrectified aerial photograph with field verification.

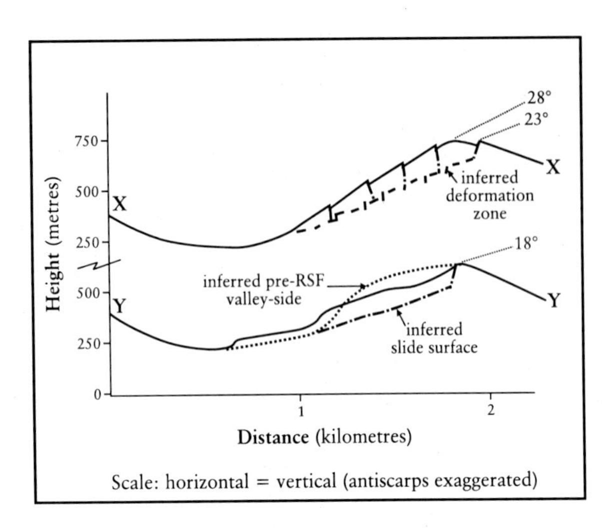

Figure 2.35 Sections X–X through the deformation zone and Y–Y through the translational slide. For locations see Figure 2.34.

areas. With an area of 1.25 km^2, Benvane is one of the ten largest rock slope failures in the Highlands.

Description

The rock slope failure affects the west side of the broad, gently undulating south ridge of Benvane in the Trossachs area of the Southern Highlands. It has four distinct components: spreading ridge, lattice slope deformation, translational slide, and transitional zone (Figures 2.34 and 2.35). The ridge itself has failed for over 1 km south of the 745 m south top (NN 537 129), with a discontinuous zig-zag headwall locally expressed as a 5 m rockwall and 7 m grass scarp. This is set back 120 m behind the crest of the ridge at the north end and 80 m behind at the south end. At the north end, an array of 14 decimetre-scale furrows and antiscarp steps between crest and source scarp illustrates the pervasive fracturing of the bedrock.

The northern half of the rock slope failure is a slope deformation which presents a remarkable network of antiscarps well seen on aerial photographs (Figure 2.33). They follow three alignments; contour parallel (NNW–SSE), gently diagonal (NNE–SSW), and a few more widely spaced on a steeper diagonal (WNW–ESE) which nears vertical lower down. By contrast with the **Beinn Fhada** antiscarp array which attains heights of 5–10 m, this array generally attains less than 1 m in height, with many elements merely being benches; the WNW–ESE set locally reaches 3 m. Some of the intersections are finely detailed, and suggest platy tilt-outs (Figure 2.34 (inset); cf. An Sornach, Jarman 2003c). This area of virtually in-situ deformation extends down to a more pronounced (1.5 m) lowest antiscarp at 420 m OD, which steps across the slope within a broad area where springs issue. Below it, a smooth, steep slope split by a vertical dry rupture, and a subdued slump zone, may extend the failure down to 350 m OD. The northern margin of this deformation is defined by a deep gully complex that cuts off at least eight antiscarps or benches.

The southern half of the rock slope failure has developed into a long-travel translational slide (Figure 2.36). It has a degraded grassy headscarp 26 m high at the apex of an irregular bowl (Coire Dubh) almost biting into the crest of the ridge at 610 m OD. The slide mass has descended almost to the slope foot at 230 m OD, protruding as a broad lobe with a 20–25 m-high basal rampart. Despite its viscous appearance, there is no evidence of disintegration to blocky or slumped debris in the failed mass. It displays a staircase of four widely spaced risers 4–30 m high; each tread has occasional 1–2 m antiscarps indicating some retention of coherence. The boldest riser could match intact slope steps to the south, representing an approximately 300 m translation downslope; it is cut by two streams with gorges in apparently intact bedrock, suggesting rafting of large failed slices. One of these streams emanates from a spring, but the other sources at 500 m OD within the bowl, indicating that the rock slope failure is well-consolidated. The southern flank of the failure is well defined and displays a classic transition, with the source scarp rapidly reducing in height until the slip protrudes from the slope to form a flank rampart that reaches 20 m high.

The most striking rock slope failure features are developed in a fourth component, transitional between the slope deformation and the slide bowl (Figure 2.37). The north flank of the bowl is highly irregular but attains 20 m in depth in places. It is incidented by four prominent antiscarps. The second highest is a wafer of quasi-intact slope with a 4.5 m-high rock antiscarp as its uphill face that extends out into the floor of Coire Dubh; its downhill face is a crag above a zone of slippage with tensional antiscarps. Adjacent to the headscarp of Coire

Figure 2.36 Benvane slide bowl and lobe, viewed from the west from Meall Cala, with lattice slope deformation upper-left; extensive springs can be seen above the lowest antiscarp centre-left. (Photo: D. Jarman.)

Figure 2.37 View north across the degraded headscarp of the slide bowl (Coire Dubh) to the fresher crags of the transitional zone. The extent of incipient encroachment into the broad ridge is indicated with a broken line. (Photo: D. Jarman.)

Dubh, another wafer with a fretted crest has slipped down approximately 10 m, opening a 4 m-deep tension trench. At the neck where it still attaches to the deformed slope, it transmutes into trifurcating antiscarps 0.5–2 m high. At the head of this transitional sector, a short 10 m rock crag is the most pronounced feature of the rock slope failure. It breaks back to become an obtuse wedge source scarp to a subsided section of the spreading ridge, which is some 400 m long and extends 80 m in from the ridge crest (Figure 2.38). A series of tension trenches up to 2 m wide and deep occur above the north side of this subsided wedge, and in its floor, on the WNW–ESE orientation, with the sense of movement being southwards down-ridge as much as valley-wards (cf. **Sgurr na Ciste Duibhe**).

Deranged drainage is a standard indicator of rock slope failure (Holmes, 1984) and here the pattern may provide pointers to the structure and sequence. It is unusual to have substantial streams flowing over a large slide lobe and incising its risers. It is also unusual to have extensive springs above the lowest antiscarp in the slope deformation zone, with the main source on a remarkable 50 m-wide front, although more conventional springs also occur at the north end of this antiscarp. The deformation otherwise lacks any surface drainage, even dry fluvial gullies as seen on **Beinn Fhada** (Jarman and Ballantyne, 2002).

Geologically, the underlying bedrock consists of Dalradian (late Cambrian–Precambrian) arenite and semipelite of the Ardnandave Sandstone Formation, which dip at 30°–55° south-east into the slope. The rock slope failure lies just north-west of a broad, large-scale monoformal fold hinge, termed the 'Downbend', which separates rocks to the south-east that dip steeply south-east from those to the north-west that dip at shallower angles. The 1:50 000 geological map shows a minor NE-trending fault across the north part of the site. It only marks the Coire Dubh area (0.55 km^2) as a 'landslip', whereas Holmes (1984) identified 0.91 km^2 from aerial photographs, and the full extent is 1.5 km^2. It is difficult to assess the volume of the translated debris in the southern half, since the cavities are only partially evacuated, and the slide plane is unknown. At a conservative depth of 20 m as expressed in the bounding features, the order of magnitude is approximately 10×10^6 m^3. There is even less evidence for the volume affected by deformation in the northern half. A plane exposed at source scarp and slope foot would slice 50–60 m off the ridge, and at an average depth of 25–30 m the failed, but still quasi-in-situ, mass could amount to an additional 15–18 $\times 10^6$ m^3. The total failure volume is thus very substantial in Scottish Highland and north European terms (cf. Table 2.3).

Benvane lies within a structurally related cluster of rock slope failures (Figure 2.39). The aerial photographs show its NNW–SSE ridge-splitting component continuing as an erosional lineament, which then controls a shallow, arrested 0.24 km^2 landslip on the east side of the ridge in Gleann Casaig. To the north of the deformation, the geological map shows similar features recurring sporadically for 2 km along the midslopes almost to the col above Glen Buckie. On the opposite side of Gleann nam Meann, the south ridge of Meall Cala is nicked by a small but striking failure which well represents the classic acute wedge form of rock slope failure. It has a 20–30 m-high source plane on the dip of the schists, and a 12 m detachment scarp crag on the south flank.

Interpretation

In the absence of any geotechnical studies of Benvane, it can only be observed that the large translational slide has developed contrary to the dip of the schists into the slope, and has a generalized surface gradient of less than 20°, implying an exceptionally low-angle failure surface near the lower limit for sliding (Figure 2.35). The slope deformation component may have failed to progress to actual sliding for the same reasons: its steepest gradients of 28°–35° are well above the residual friction angle for schists, but the plane exposed at source scarp and slope foot is approximately 23° (Figure 2.35). This deformation is of a platy character, with the strong but widely spaced WNW–ESE and north–south joint-sets guiding the rupturing of the main plates, and the contour benches and scarplets relaxing the stresses. The whole slope is in compression, with no open fissures. By contrast, the broad ridge above is in tension, with numerous furrows and trenches; their comparatively innocuous state suggests long inactivity.

No dates are available for Benvane, and it is not obvious whether the two main components

Figure 2.38 (a) View south down the spreading ridge to the slide bowl. The 6 m-high source scarp to the subsidence graben is not the limit of encroachment in to the ridge, the true headscarp being just visible on the left edge. (b) A close-up view across the slide bowl of the transitional zone extending into the spreading ridge. (Photos: D. Jarman.)

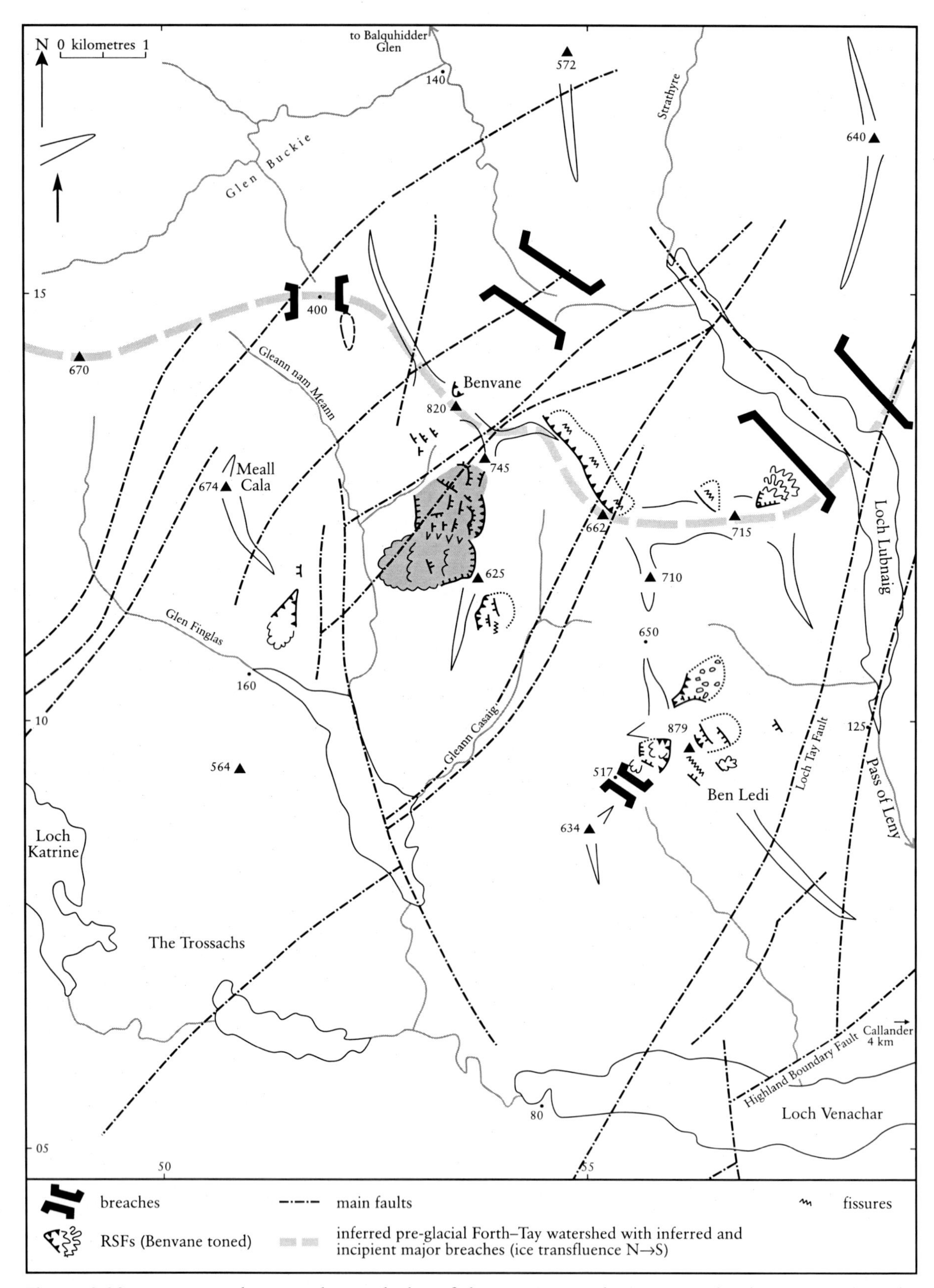

Figure 2.39 Benvane and surrounding rock slope failures (RSFs) in their topographical context. This sub-cluster may be associated with glacial transfluence south-east across local watersheds, the breaches being at varying stages of development. Unlike the **Glen Ample** sub-cluster immediately to the north-east, there is no specific association with main faults.

have evolved simultaneously, or by later sliding within the original deformation. The slide bowl and lobes are comparatively degraded, whereas the deformation features are fairly sharp and well preserved given their small scale. However, the transitional area between them has much larger and bolder features, and represents progressive encroachment of the slide into the deformed slope and ridge, with two substantial secondary slips into the bowl at midslope, and incipient encroachment into the ridge above. Indeed the trench fissures around the obtuse wedge bite suggest that the zone of actual translational failure has migrated up-ridge northwards as well as headwards.

It is not obvious why extensive and deep-seated rock slope failure should occur in such an area of relatively low relief, well removed from the main mountain cores, and from channels of intensive selective glacial erosion such as Loch Lubnaig (Linton, 1940). Yet there are 10 failure complexes affecting 2.9 km^2 of the 43 km^2 around Benvane (Figure 2.39), a density of 7% which compares with some of the highest in Scotland (Jarman, 2003a,b). The cognate rock slope failure cluster at **Glen Ample** 8 km to the north-east records an exceptional density of over 16%.

Benvane therefore provides an excellent locus for research into the mechanics of rock slope failure, and the reasons for its occurrence. Of the two triggers commonly invoked, elevated water pressures at deglaciation seem irrelevant to a compressional slope deformation and gently spreading ridge. Benvane and Ben Ledi appear to have been nunataks during the Loch Lomond Stadial (Holmes, 1984), but most of the site was probably ice covered, and hence the delicate deformation antiscarps and low-level slide lobe must post-date it. A high-magnitude seismic trigger is at odds with the presence of large erratic boulders on the steep sides of some antiscarp trenches, and the slide is far from cataclasmic. There is no major fault crossing the site, although a branch of the Loch Tay Fault passes down Gleann Casaig. However this cluster of rock slope failures, and the Glen Ample cluster nearby, are close to the outer limits of the Loch Lomond ice, and probably at a point where the Devensian icecap gradient was in steep transition from highland to lowland terrain. Differential glacio-isostatic recovery may have been most acute here, generating slope stresses sufficient to provoke deformation and sliding in vulnerable situations.

Rock slope failure is sparse or absent along the W–E-trending main valleys of pre- or early-glacial origin, such as Loch Katrine/ Venachar and Balquhidder, a pattern found throughout the Southern Highlands (Figure 2.15). The concentration of rock slope failure in N–S-orientated side-valleys may indicate an early stage in their enlargement by transfluent ice from the north. Glen Finglas and its two tributaries are a relict of the pre-glacial dendritic Forth drainage system (Linton and Moisley, 1960). Their heads show signs of glacial over-riding, and the 400 m OD col west of Benvane is an incipient glacial breach of the pre-glacial Forth–Tay divide (Linton, 1940). It is possible that glacial downcutting even in this lower-relief area has been sufficient during the Devensian glacial to destabilize the slopes of Benvane and its neighbours. Similar lattice anti-scarp arrays occur close to developing breaches in Glen Luss (NS 28 95) and near Tyndrum on Beinn Chaorach (NN 35 32).

The extant mountain-shaping effects of the Benvane rock slope failure are modest, but the scale of incipient encroachment is so great as to render the whole south ridge vulnerable to reduction and eventual 'divide elimination' (Linton, 1967).

Conclusions

The Benvane rock slope failure complex is an outstanding example of quasi in-situ slope deformation, with one of the finest and most extensive lattice antiscarp arrays in Britain. It displays lateral progression to a deep but degraded, long-travelled but coherent translational slide. The interface between these two zones is made conspicuous by rock slope failure features of much fresher character. Deformation encroaches into the broad summit ridge scale, and demonstrates the past and potential contribution of failure to large-scale erosion. Benvane also affords instructive comparison with the **Glen Ample** failure cluster as a platy deformation on steeper valley slopes, as compared to more gently sloping upland. It provides an excellent basis for research into the mechanisms, triggers, and underlying causes of rock slope failure in mountain areas of relatively lower relief.

GLEN AMPLE, STIRLING (NN 596 160–NN 610 215)

D. Jarman

Introduction

The area south of Lochearnhead, Perthshire is of exceptional significance as having the highest known density of rock slope failure in the Highlands, with seven failures affecting 16% of the 26 km^2 Glen Ample area (Figure 2.40). Core areas of other failure clusters studied do not exceed 8% (Table 2.2). It is also one of the most pronounced concentrations of significant failures along the line of a major basement fault, the Loch Tay Fault. The two principal failures, at Ben Our and Beinn Each, are essentially in-situ slope deformations with marked structural expression. Their extensive, but often delicate, ground rupture features are possible indicators of neotectonic activity: high-magnitude seismic shocks following deglaciation have been proposed but not yet confirmed in the Highlands (Stewart *et al.*, 2000). These structures provide unusually clear evidence of how deep-seated deformation can develop in some of the gentlest relief known to be affected by paraglacial rock slope failure. Ben Our is of exceptional significance for its extent and its unique platy structure.

Description

Glen Ample is a short (8 km) side-valley off of Loch Earn, with rather open slopes and less than 500 m relief to its immediate rims. It narrows at its head into a minor glacial breach of the main Forth–Tay divide south to the major breach of Loch Lubnaig (Figure 2.39). There are several lesser rock slumps and slides on its flanks, and just south of the pass is a more extensive anti-scarped zone. Directly above the pass stands the impressive Beinn Each rock slope failure. At the foot of the glen, one of the three largest rock slope failures in the Highlands occupies most of the low rounded hill of Ben Our. Being located at the junction of Glen Ample with the trough of Loch Earn, it is unusual in responding to slope stresses in directions almost 90° apart.

On the opposite side of Loch Earn, a small but striking rock slope failure has a deep, narrow wedge cavity and a slide lobe exhibiting creep in the last century. The west side of Glen Ogle has a chain of crag collapses across which a railway was engineered without re-activating them. Its rounded nose has signs of deformation, progressing to sliding slumps, complementary to Ben Our. Together with some rock slope failures on the south-west side of Loch Lubnaig, this dense cluster stands apart from the general concentration along the main Highlands watershed close to the west coast (Figure 2.13), and it occurs in some of the lowest relief in the Highlands to support rock slope failure.

Geologically, Glen Ample comprises Dalradian metasedimentary rocks (late Precambrian–early Cambrian in age). These are mainly arenites, semipelites, and pelites, with a distinctive intercalation of 'Green Beds' (which include reworked volcanic detritus) on the south-west slope of Ben Our. The rocks of Ben Our contain a higher proportion of schistose pelites and semipelites. The rocks are of greenschist to lower amphibolite metamorphic grade, typified by biotite and garnet growth. Structurally, Glen Ample lies close to the Highland Boundary Fault, where the relatively flat-lying Dalradian rocks become downfolded into a large monoformal structure called the 'Downbend', whose axis trends north-east. Thus while the beds dip gently on Ben Our, they are steeply inclined on Beinn Each. Glen Ample has formed by enhanced erosion along the line of the Loch Tay Fault, a sub-vertical NNE-trending structure. There is no recorded information on joint-sets, but there appear to be no obvious bedding or cleavage dips that can easily account for the incidence of rock slope failure in this area (J. Mendum, British Geological Survey, pers. comm.)

Ben Our (Beinn Odhar)

When seen on an aerial photograph, or in ideal snow or light conditions, the pervasive platy deformation of Ben Our is very unusual for a hill of relatively unassuming height and character (Figure 2.41). The summit area is almost flat, with tops at 730 m and 740 m OD separated by a shallow graben-like saddle. A swarm of scarplets runs behind the summit above the col to the south and below the broken crags on its east. To the west, these scarplets converge into a major scarp reaching 4 m in height, which runs for 600 m above the fluvial cleft of Coire Mheobhith. Since this scarp faces uphill (north) it appears to be an antiscarp, but together with the scarplet

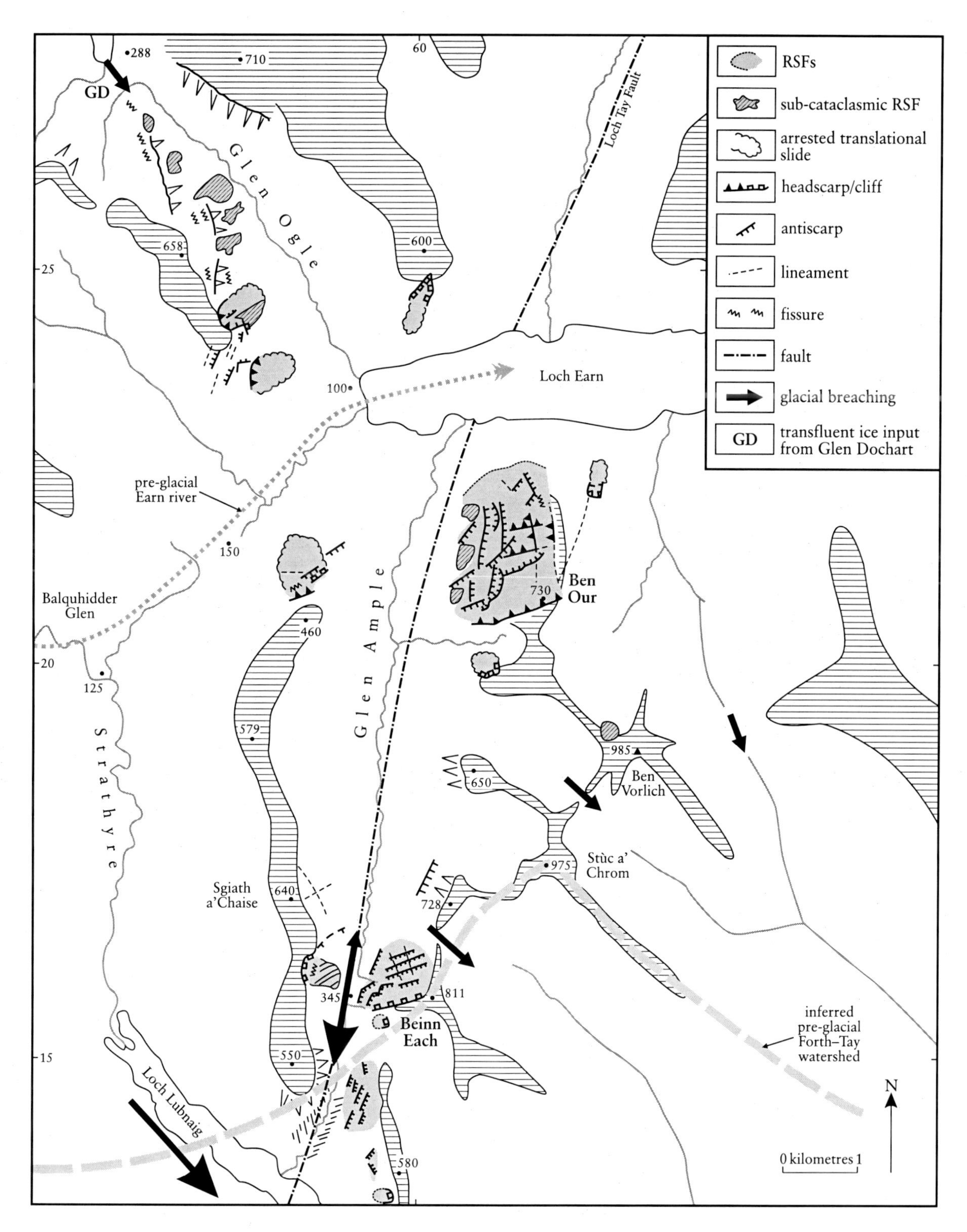

Figure 2.40 The Glen Ample rock slope failure (RSF) cluster in relation to the Loch Tay Fault and immature glacial breaches.

swarm girdling the summit, it is in fact the source fracture from which the whole mass of the hill has slipped slightly away (Figure 2.42).

The broad shoulder north from the summit is split by a fracture 1.2 km long, in places a mere furrow, but generally a sharp step typically 1–2 m

Figure 2.41 Vertical aerial photograph (1989) of the Ben Our rock slope failure. (Photo: Crown Copyright: RCAHMS (All Scotland Survey Collection).)

and locally up to 5 m high. This is another source scarp to the subsidence, with slight movement along or away from it, although tension fissures and steps also occur on the brow east of it. The manner in which the west-trending and north-trending source scarps interact in the summit area is unclear.

The boldest feature of this rock slope failure is a ragged tear scarp which scythes across the whole site in two arcuate sweeps, reaching 18 m high (Figures 2.41 and 2.42). Above this feature the gentle upper slopes are split by a tension scarp and furrow, and have minor antiscarps. Below it to the north, the central part of the failure has subsided but with no open fissures or indications of significant spreading. It has fractured into a series of rectilinear plates, framed in the east by step-down scarps up to 6 m high, but in the lower ground farther west by antiscarps reaching 4 m high. The south-west end of this ragged tear scarp propagates into a series of nested slip hollows, culminating in a conspicuous promontory that has crept out into Glen Ample. This pattern of short-travel upward-propagating movement has been described at Tullich Hill (Jarman, 2003d).

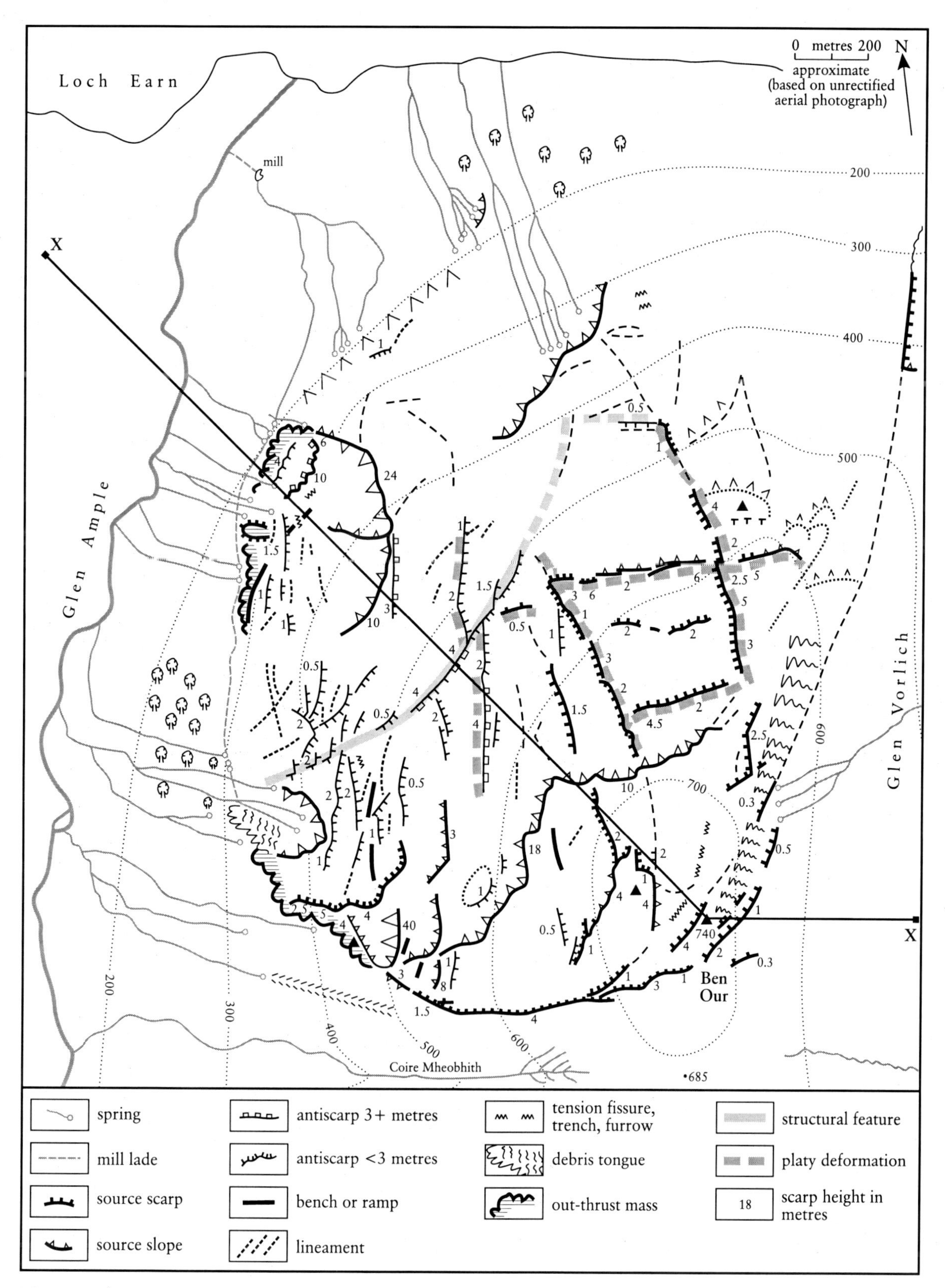

Figure 2.42 Geomorphological interpretation of the Ben Our rock slope failure, based on the unrectified aerial photograph with field verification.

All along the Glen Ample side, the glacially steepened lower slope has a dense array of antiscarps and lineaments on several orientations, generally below 2 m in scale. Locally these become tension trenches. At the south and north ends, these features are framed by nested subarcuate scarps where deformation is progressing to outward movement. These source scarps reach 24 m high in the north, and locally 40 m in the south. Both these creep masses have substantial flank ramparts and toe bulges; the northern one contains the only striking rock-mass dislocation in this rock slope failure, with a 10 m head crag and rocky, antiscarped topple masses. Between these failed masses, the middle section appears substantially intact, but is crossed by steeply inclined antiscarps associated with a major fracture crossing the site. Along the steepened slope above Loch Earn, there are break-lines but only one isolated antiscarp near the foot.

The whole west and north perimeter of the site has the most extensive and abundant effusion of springs of any large rock slope failure. The springs cluster below the creep masses and major fractures, mostly on two levels. Slump bowls in superficial deposits, some very large, are actively developing and migrating upwards. There is no surface water drainage on the site, nor any sign of dried-up former watercourses save along the southern boundary, indicating the pervasiveness of fracturing.

The extent of failure is reasonably clear except on the north shoulder and east slope, where irregular terrain continues down towards Loch Earn and into Glen Vorlich. A separate, rectilinear rock slope failure sourcing at midslope close to the angle of these valleys has a long debris-lobe; its west flank scarp is an extension of a distinct furrow lineament. The total extent is at least 2.9 km^2. Holmes (1984) only identified the most conventional slope failure areas above Glen Ample, totalling 0.17 km^2.

Possible contributory geological controls include bedding and foliation surfaces dipping NNW at approximately 45° along the southern boundary and shallowing northwards to as low as 5°: in combination with suitable joints, these could facilitate the translational sliding along the west flank, and extensional creep within the core of the deformation. Massive rock 'Green Bed' units on the lower slopes and more schistose lithologies near the summit within an interlayered structure may also have assisted mass translation. A NNE-trending fault passes just east of the summit, and may also have assisted in releasing the failed mass, whereas the main back feature to the lower failed area may be faulted as it parallels the inferred trace of the Loch Tay Fault (J. Mendum, British Geological Survey, pers. comm.).

Ben Our – interpretation

Ben Our is a unique site. The extensive platy dislocation on the gentle upper slopes is unprecedented, and may be attributable to tensional spreading in two directions on a convex valley junction. Most of the deformation occurs over only 200–300 m of vertical relief, at a gradient of only about 18°: the slope foot in Glen Ample is glacially steepened to 27°, but produces only a bluff about 100 m high (Figure 2.43). Ben Our is almost an inverse of **Benvane** geometrically, with extensional platy deformation on gentler upper slopes, as against compressional latticing on steeper slopes below a spreading ridge, an instructive contrast for further investigation.

It is difficult to invoke localized glacial erosion as the prime factor here, by contrast with the comparably large deformation at **Beinn Fhada** (Jarman, 2003e). If a failure surface is interpolated from the scarplets behind the summit to the slope foot springline in the north-west corner, then at 14° this is too low for conventional translational sliding (Figure 2.43). Such a surface is unlikely to exist as a planar throughgoing discontinuity: the component blocks are more likely to have their own 'floors', whether clear-cut or transitional, staircasing down the slope as joint-sets intersect above the NNW-inclined foliation surface. This would favour tensional spreading (as along the ragged tear) and assist translational movement on the outer slopes, but its concave nature may have restrained mass creep. The interpolated 'failure plane' would simplistically give a general depth of failure in excess of 150 m, which is comparable with the maxima proposed for major rock slope failures in high relief in the North-west Highlands (Fenton, 1991). Such a depth would give a very large total failed volume of the order of 100–200 × 10^6 m^3, comparable with **Beinn Fhada**. The actual extent to which the rock mass has lost structural integrity remains somewhat conjectural until geophysical surveys are conducted, but may well be considerably less.

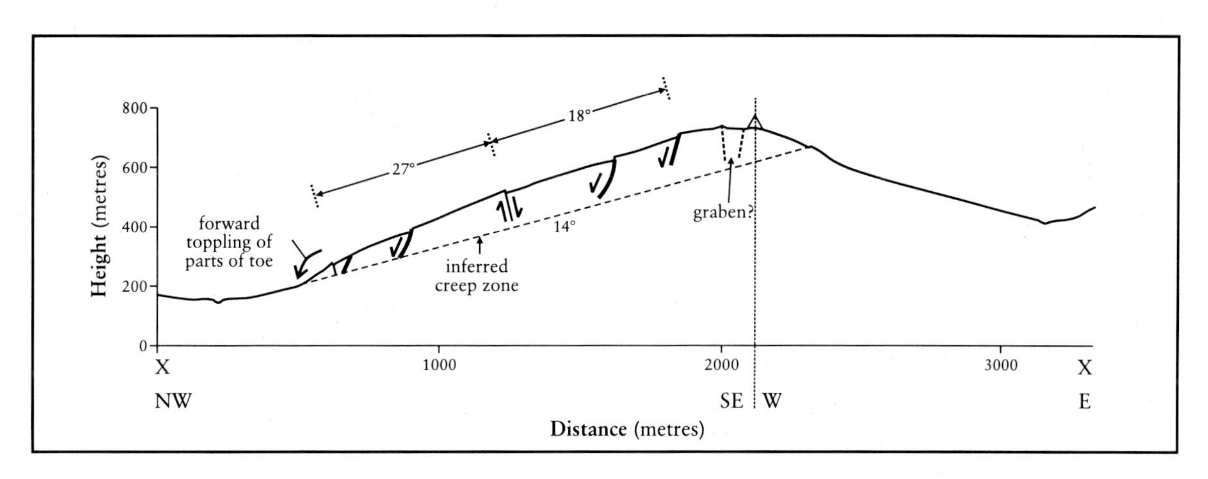

Figure 2.43 Section X–X across Figure 2.42 with inferred failure behaviour on a very low angle creep surface.

Beinn Each

Although Beinn Each is, like Ben Our, a nearly in-situ slope deformation, the mode of expression is very different (Figure 2.44). Almost all of the features are compressional (antiscarps, ramps and benches breaking upwards out of the slope), in contrast to Ben Our where the structure is extensional (downward breaks in slope), except for a band across the waist.

The most striking component at Beinn Each is a smooth dome, only about 200 m square, tilting out from the approximately 30° valley-side to an angle of approximately 40°. It is fractured by 'noughts-and-crosses' antiscarps typically 3–5 m high but reaching 9 m at one intersection, unusually high for sites south of the Great Glen. Visual impressions suggest that sinistral displacement of about 30 m may have occurred along the lower contour-parallel anti-scarp. This extrusion has been likened to 'egg-box architecture' and 'celtic knotwork' (Figure 2.45).

This intense, but localized, deformation is at the lowest corner of the rock slope failure, only 150 m above the col at the head of Glen Ample. From it a sub-horizontal lineament extends north for 800 m, the whole width of the slope failure, below the steeper part of the valley-side. It snakes in the manner of an uncoiled rope, as do the lowest antiscarps on **Beinn Fhada**, but is generally no more than a broad bench (Figure 2.44, feature A). Midway, it is intersected by two pronounced lineaments trending diagonally south-west–north-east across the deformation and emerging in places as sharp, but modest, antiscarps (Figure 2.44, features B and C). From its north end, a ramp ascends south-east and intersects B and C in a nexus of small plates emulating in miniature **Benvane** and An Sornach (Jarman, 2003c). The ramp becomes a structural weakness across gentler upper slopes (Figure 2.44, feature D) towards the headscarp, and meets the bold scarp of the west shoulder of Beinn Each at the point where it has collapsed in a pile of massive blocks. Immediately beneath this bold scarp runs a final lineament (Figure 2.44, feature E). This starts from the extruded dome, extends uphill for 600 m as a well-defined, if discontinuous, antiscarp 2–6 m high, and continues as a trace onto the skyline. In places, it is the axis of a swarm of closely spaced lesser antiscarps; it resembles similar cliff-foot locations in the Mamores (e.g. Stob Ban, NN 147 648). Other features run parallel to these main lineaments and help to confirm the extent of deformation as about 0.5 km^2.

There are few obvious geological controls here (J. Mendum, British Geological Survey, pers. comm.), with the bedding and cleavage in the schists dipping too steeply to the NNW for translational sliding. The strike of the structure is however roughly parallel to lineaments B and C. A fault trace extending north-east from near Beinn Each summit may continue south-west to form the prominent southern 'headwall' to the rock slope failure, with an east–west fault causing an additional 'break up weakness' within the failure. Note that E and B correspond to fault traces mapped on the eastern side of Beinn Each.

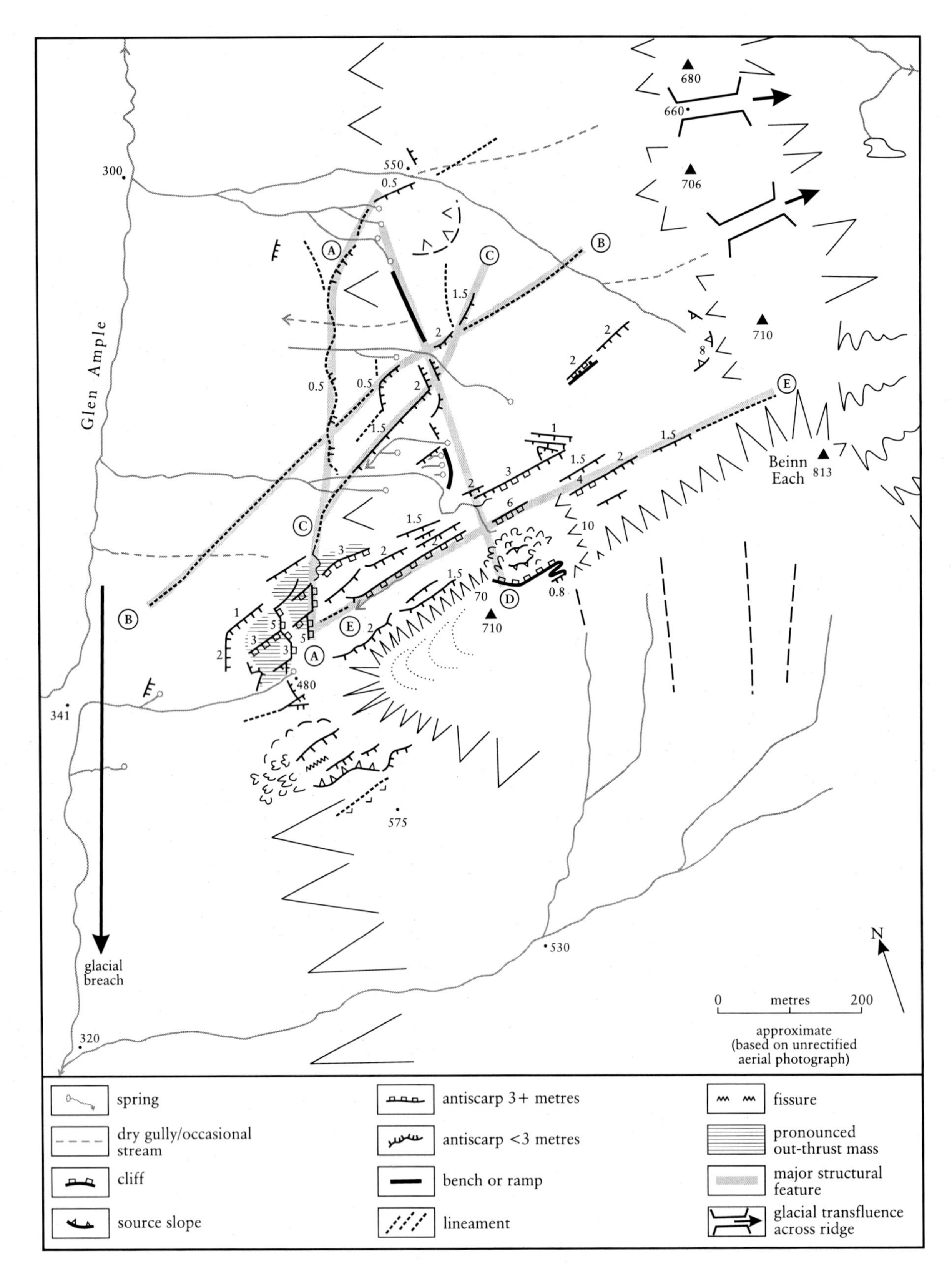

Figure 2.44 Geomorphological interpretation of the Beinn Each rock slope failure. Features A, B, C, D and E are described in the text.

Figure 2.45 Close-up of the Beinn Each rock slope failure nexus, suggesting extrusion of a component of the failed mass with fracturing along two joint-sets. (Photo: D. Jarman.)

Beinn Each – interpretation

Unlike Ben Our and **Benvane**, it is not at all clear where the Beinn Each rock slope failure originates. The bold scarp of the west shoulder cannot be interpreted as a 70 m headscarp since there is no sign that the failed mass has subsided or bulged out into the glen to any great extent. The collapsed section of this scarp, with its incipient encroachments into the ridge, is in effect a separate failure, as is that on the south-west flank of its nose (Figure 2.44). Nor is there any evidence of a source configuration along the heavily scoured north ridge. Nevertheless, this failure does occupy a broad wedge-shaped depression of subdued relief between steeper rockier bluffs.

A further sharp contrast with Ben Our, and with most deformations including Benvane, is the drainage pattern. Rather than a springline along the slope foot, there are major effusion zones at the upper and lower ends of lineament D, several springs associated with lineaments B and C, and streams following irregular courses over the rock slope failure. This could indicate that deformation is unusually shallow or partial, or alternatively that it is of such antiquity that it is reconsolidating. Examples such as Beinn an Lochain West and Beinn an Fhidleir in the Arrochar Alps (NN 215 076), and Na Gruagaichean in the Mamores (NN 227 105; Figure 2.11), suggest that removal of a previous failed layer might unload the surface sufficiently to re-activate deep-seated weaknesses. The 70 m-high scarp at Beinn Each is not the source cavity for the extant slope failure, but in this analysis it could represent the thickness of a previous failed layer now removed. The source of earlier failure could have been on the present north ridge, lowering it to enable over-riding by ice. Evacuation of such a failed layer could have been partly by landslipping *en masse*, since the present overall slope angle of approximately 26° is just feasible for translational sliding; selective quarrying of a weakened slope may also have played a part.

The present col at the head of Glen Ample may have been lowered by 100–200 m from its pre-glacial position. Although relatively modest, this localized erosion could have augmented the rebound stresses, especially if it has occurred mainly in Late Devensian times.

The most intense deformation occurs along the foot of the bold scarp, and at its nose. This is where the greatest thickness of material will have been removed in earlier glacial–paraglacial cycles, and closest above the breach where the valley-side is over-steepened along the 500 m contour. The 'eggbox' extrusions (or forward topples) must be close to the point of shearing or collapse, and their survival suggests vertical slices pinned at depth by the intersecting lineaments.

Interpretation

Remarkably, these major rock slope failures have not previously been published or commented upon. Holmes (1984) identified small sites along the Glen Ample slope foot at Ben Our, but nothing at Beinn Each. The British Geological Survey has yet to publish revised mapping of this area, although officers are aware of these sites.

The ages of these rock slope failures have not been investigated. The features are generally sharp but not unusually fresh except in the lowest collapse at Ben Our, indicating an earlier Holocene date. Their relatively low elevation has encouraged vegetation, and possibly protected them from periglaciation during the Loch Lomond Stadial. It seems more probable that such large deformations are a response primarily to the Last Glacial Maximum and its deglaciation: the main dislocations possibly occurred during the Windermere Interstadial, with the finer details emerging after final deglaciation.

Explaining the Glen Ample cluster is problematic. The association with glacial breaching and hence rapid erosion is initially attractive, since the Beinn Each rock slope failure and the sub-cataclasmic failure opposite stand directly above the Ample–Lubnaig breach. If Loch Lubnaig is accepted as a major breach of the former Forth–Tay divide (Linton, 1957), then Glen Ample is a subsidiary and possibly later-formed breach. The pass is of modest capacity, and may only be accommodating local ice displaced from flowing out north by transfluent ice down Glen Ogle (Figure 2.40). It does not appear to have been enlarging vigorously, and indeed lies almost transverse to regional ice outflow.

If glacial erosion has done no more than activate local slumping or forward toppling along the rock slope failure toes, then other causes must be sought. High-magnitude seismic shocks have been inferred from rock slope failure clustering and other indicators, especially where they co-incide with major faults (Fenton, 1991), although the evidence is weak (see 'Introduction', this chapter). The Loch Tay Fault runs along Glen Ample, and is one of the main Caledonian (NE–SW-trending) faults, with 7 km of strike-slip movement and up to 1 km of vertical displacement (Treagus, 2003); however there is no recorded present seismic activity along it. Some of the antiscarps run broadly parallel to this fault, but most interesting is the lineament which extends for over 1 km from near the summit of Ben Our, linking but essentially outwith it main and north-east failures (Figure 2.40. Any neotectonic origins for this feature, or for the rock slope failures themselves, must remain speculative at present. While Glen Ample is one of the best candidates for a rock slope failure cluster to be associated with post-glacial fault re-activation, it is nevertheless more likely to be a co-incidence of selective valley erosion along a suitable line of weakness.

This rock slope failure cluster is unusual in being located well to the east of the main watershed and former ice divide of the Highlands (Figure 2.15). Other easterly failures are associated with deep transectional breaches of the Grampian watershed (Hall and Jarman, 2004) or with vigorously enlarging trough heads such as Glen Clova. It may be that the location of this cluster in relation to the Pleistocene ice-sheets is significant. It lies close to the outer limit of the Loch Lomond Stadial (Holmes, 1984, after Sissons). It also lies close to the Highland–Lowland boundary at Callander, where the Devensian and earlier icecaps were in transition from high mountain-based domes to icestreams, possibly with relatively steep surface gradients and reductions in average thickness. On deglaciation, the regional glacio-isostatic rebound stresses may have diminished rapidly over relatively short distances, including from west to east across Glen Ample. Generally, such differentials are resolved by gradual deformation, or remain locked in. Where local rock structure is conducive, they may conceivably provoke ground rupturing, perhaps where additional local factors such as valley erosion apply.

Both Ben Our and Beinn Each have one unusually long and relatively continuous lineament running diagonally across the terrain in a north-east orientation. These are not parallel with the Loch Tay Fault, but may be with the strike of the foliation surface or a major joint-set (no geotechnical survey has yet been conducted) and do correspond to secondary fault orientation. They are expressed as antiscarps for much of their lengths. One hypothesis worthy of exploration is that initial ruptures occurred along these lineaments, which may roughly parallel the regional ice contours at their steepest. These ruptures would not necessarily involve high-magnitude seismic shocks, nor occur at the same time. They would trigger, or co-incide with, delamination of the surface to a depth of tens of metres, along another joint or quasi-bedding plane, or along a more irregular self-creating surface. The failed layer would fracture into slices or plates depending on terrain and geology, and would be freed to resettle in such a way as to minimize residual rock-mass stresses. Here, they have not developed into translational slides, but have progressed by creep and subsidence. As a result, surface water is channelled along the main fractures to emerge in springs.

A similar interpretation has been developed for the An Sornach rock slope failure in Glen Affric (Jarman, 2003c), which was previously advanced as neotectonically triggered (Fenton, 1991). Other small clusters of significant slope failures in apparently marginal, lower-erosion locations, which might be accounted for by regional glacio-isostatic rebound gradients, include Loch Striven and Glen Shira (see Jarman, 2003a), Strathfarrar (see Jarman and Reid, 2003), and west Knoydart.

The landshaping effects of these rock slope failures are relatively subtle. Despite its low profile, Ben Our is still the bulkiest of the promontories encircling the head of Loch Earn. Removal of all of the failed material would lower the whole hill by possibly as much as a hundred metres; alternatively, removal of the material below the ragged tear would widen the glen and accentuate the promontory for an interim period, as on the Sgiath a' Chaise ridge opposite. The Beinn Each rock slope failure is similarly tending to isolate the resistant summit core, which may originally have been an extended shoulder of Stuc a' Chroin rather than a separate peak; it is also helping to enlarge the Ample–Lubnaig breach.

Conclusions

Glen Ample is of considerable interest for its anomalously high density of rock slope failure, in a relatively isolated and low-relief location. Ben Our is one of the three largest rock slope failures in the Highlands, and its extensive platy deformation is unique in Britain. It is unusual and instructive in being located on a valley junction corner, exposing the hill to slope stresses in several directions. Beinn Each has a remarkably bold and intricate antiscarp array, on intersecting alignments, in marked contrast with the parallel array on **Beinn Fhada** and the filigree lattice on **Benvane**. Both Ben Our and Beinn Each clearly exhibit slope deformation with only limited progression into downslope separation. Their extraordinary expression, extent, and enigmatic origins have attracted international attention. This is a locality where it is worth investigating whether neotectonic seismic movements have acted as a trigger for rock slope failure, but even then they may be no more than ancillary. Other hypotheses such as differential (glacio-)isostatic rebound stresses are necessarily more speculative, but the sites are ripe for detailed geotechnical examination. The cluster offers great scope for exploring the fundamental causes and spatial distribution of rock slope failure in glaciated ancient mountain areas.

THE COBBLER (BEINN ARTAIR), ARGYLL AND BUTE (NN 260 058)

D. Jarman

Introduction

The Cobbler is a distinctive triple peak in the Arrochar Alps, in the South-west Highlands of Scotland. It is the most striking case in Britain of a mountain shaped by rock slope failure. Although limited failure has been recorded here its full extent and topographical impact has only recently been recognized (Jarman, 2004b). The main slope failure extends over the whole upper south flank of the mountain, and its source arête includes the Summit 'tor' and the spectacular horn of the South Peak. The North Peak is deeply fissured and overhanging, with indications of previous failure below. The well-known profile of three peaks closely grouped around a small corrie is therefore attributable to pervasive paraglacial rock slope failure (Figure 2.46).

The Cobbler is at the heart of the largest rock slope failure cluster in the Scottish Highlands (Figures 2.13 and 2.15). Several different modes of failure are evident on The Cobbler, ranging from in-situ fracturing through coherent translational slides with varying degrees of arrestment to a fully disintegrated rockslide. These kinds of rock slope failure are widely distributed in the 'old hard rock' uplands of Britain, and those on The Cobbler are particularly characteristic of the massive schists of the Arrochar Alps–Cowal area (Clough, 1897).

Description

There are four distinct zones of rock slope failure on The Cobbler, affecting 0.84 km^2 or 10% of the area of the mountain (22% of the area above 500 m OD), a remarkable intensity (Figure 2.47).

The main rock slope failure on the south face of The Cobbler is one of the 50 largest in the Scottish Highlands at 0.62 km^2, and is a translational slide complex in various stages of disintegration and arrestment. The 1901 geological map marks 'landslip' across the slope, but the current inset map only indicates two small slips. Holmes (1984) identified only the west part (0.20 km^2) from aerial photographs. The slide has four components or 'panels' over its 1200 m width, which emerge onto a broad midslope bench or 'alp' at 450–600 m OD (Figures 2.48 and 2.49).

Panel 1 (westmost) has the appearance of a chaotic pile of blocks several metres across. Its flank rampart is 5–15 m high, while its exceptionally steep (40°) toe rampart – which barely reaches the 'alp' – is at least 30 m high, suggesting the order of depth of the failed debris. However, the extent of disintegration is deceptive: the upper half is semi-intact, with only local fissuring, blocky dislocation, and short antiscarps in a grassy sheepwalk. The steeper lower half is extensively disrupted, but the aerial photograph shows organization into quasi-antiscarps, which are up to 6 m high and 150 m long. The antiscarps are much broken up, and appear to have emerged by forward rotation of contour-parallel rock slices on a convex slope. This indicates a considerable degree of coherence, given that the failed mass has spread fanwise from an 8 m-wide and 8 m-deep tension trench along the ridge west of the summit.

Panel 2 is separated from Panel 1 by an open gully on the general alignment of the NNE–SSW faults common in this area. The failed mass has descended appreciably further, despite appearing less disrupted, apart from some very large masses that protrude irregularly. One frac-

Figure 2.46 The distinctive profile of The Cobbler from the ESE, with the main rock slope failure on its left flank and a small rockslide into the breach col on the right. Both North Peak and South Peak may be the remnants of former corrie arms truncated by rock slope failure. The wide skyline nick may also result from a headwall collapse, but only small debris-lobes remain in the corrie. (Photo: D. Jarman.)

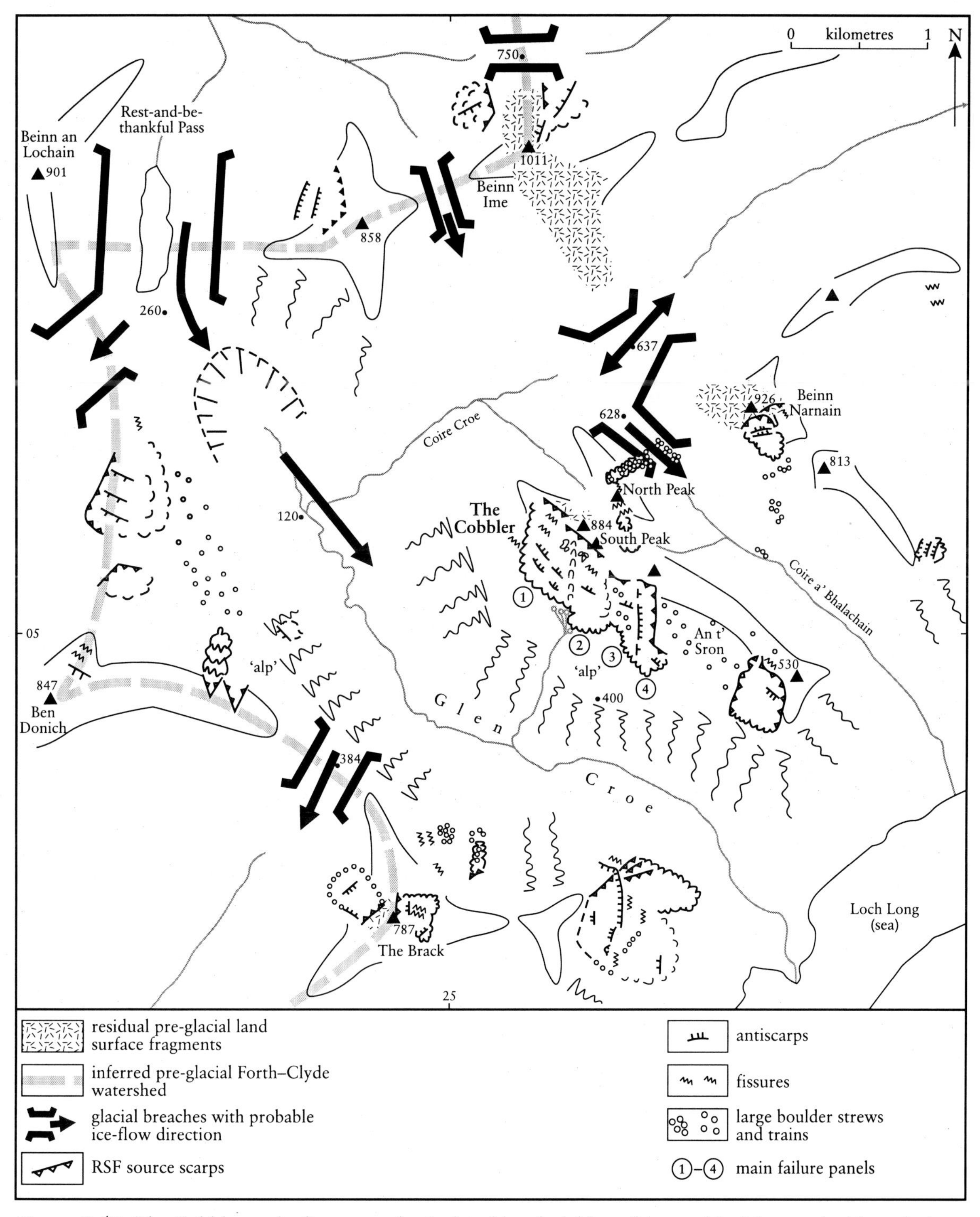

Figure 2.47 The Cobbler and adjacent peaks, isolated by glacial breaching and incision, and with rock slope failure (RSF) encroaching into the pre-glacial land surface.

tured slice on the west side is 25 m long and 8 m deep, and projects by 6–10 m. To the east, these protrusions resemble intact bedrock, but occasional deep fissures and basal springs confirm that most if not all are failed. Unlike Panel 1, a high but irregular headscarp is present, reaching 28 m west of the summit, and 12 m at the arête south-east of the summit above a tract of impenetrably jumbled megablocks, while the South Peak presents a 60 m rockface to the least

Figure 2.48 Vertical aerial photograph of the The Cobbler, showing contrasting land surface types and modes of erosion on its flanks. (Photo: Crown Copyright: RCAHMS (All Scotland Survey Collection).)

distressed part of this panel. A stack of four triangular wedges steps down the upper junction between Panels 1 and 2, so that the upper part of the dry gully acts as a flank scarp to Panel 2, reaching 14 m in height. Panel 2 has also travelled farther out onto the 'alp' than Panel 1, with slopes around 30° and a toe only 6 m high, suggesting greater fragmentation and fluidity. A series of powerful springs along the toes of Panels 1 and 2 confirms the pervasiveness and interconnectivity of deep fracturing.

Panel 3 by contrast is grassy and fully consolidated, with a probably thinner failed mass only betrayed by decimetre-scale furrow swarms and a small array of < 1 m antiscarps. It has a weak headscarp 8–10 m high rather below the summit ridge, and a presumed degraded toe rampart 8 m high. The debris of Panel 2 appears to have encroached onto it, although there is some continuity in the pattern of small antiscarps.

Panel 4 (eastmost) is a distinct cavity between contour-orthogonal flank scarps up to 12 m high. In effect a section of Panel 3 has slipped out, leaving a higher but still degraded 15 m headscarp, and with a subdued debris-tongue draping over the edge of the 'alp'.

The boundaries of the main rock slope failure are clearly defined, by comparison with many that have diffuse margins. The head and east flank scarps are almost continuous, and there is no spray fan below the toe ramparts. There is no

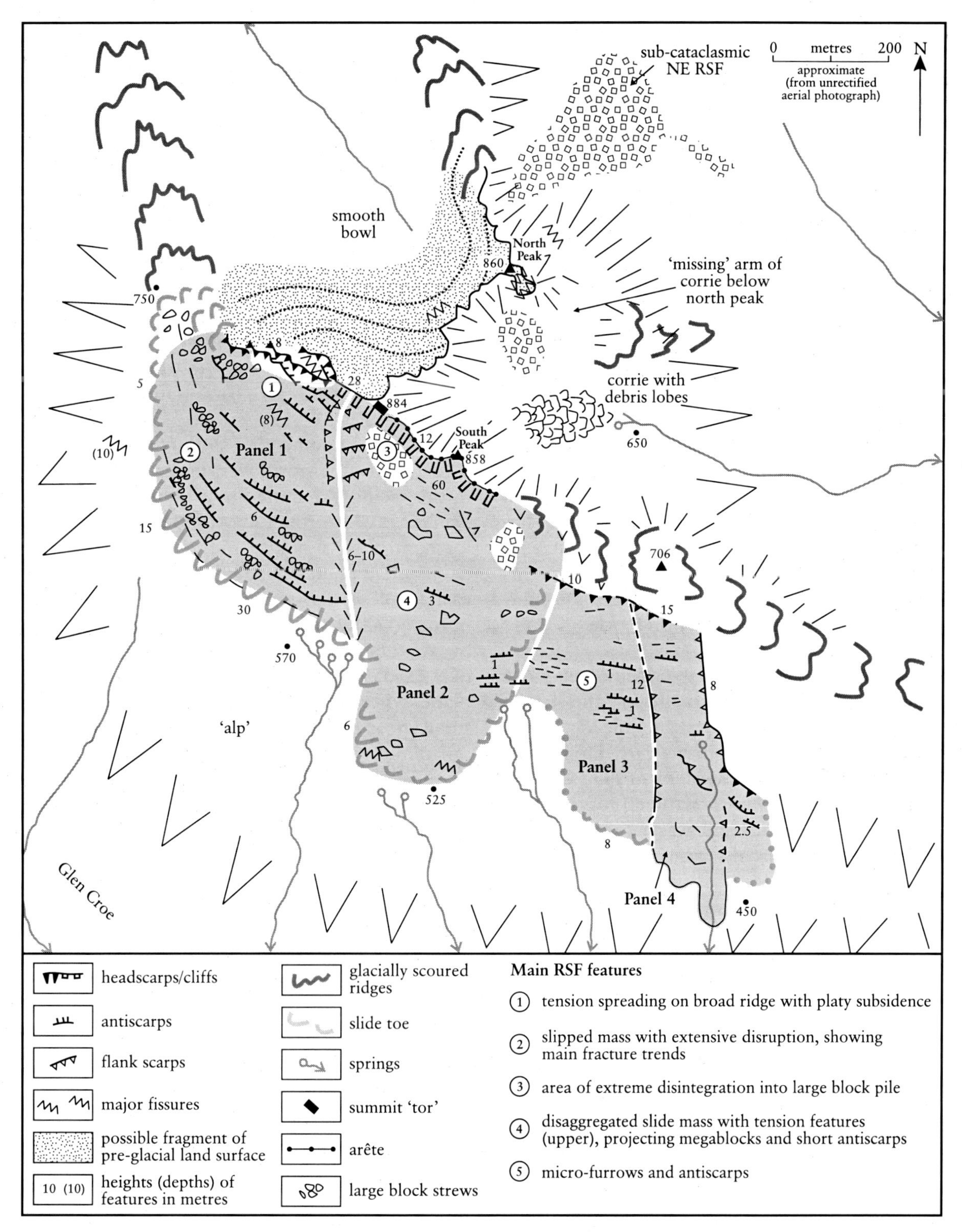

Figure 2.49 Geomorphological map of The Cobbler, based on Figure 2.48.

west flank scarp, unusually, since the outer end of the summit ridge has collapsed laterally. It is impossible to give an accurate volume for this rock slope failure, in view of its short travel and unknown failure surface, but applying conservative estimates for average depths of 20/15/5/10 m to the four panels yields an order of magnitude of 8–10 × 10^6 m^3.

Farther along the south-east ridge, the An t-Sron rock slope failure (Figure 2.47) is a rather subdued 'armchair' translational slide, similar to Panels 3–4 (Figures 2.48 and 2.49). It has distinct approximately 6 m-high source and flank scarps and minor holes and ruckles, with one open trench above indicating incipience.

In the corrie floor the 1901 geological map marks another 'landslip', unrecorded on the current map or by Holmes (1984). Two modest degraded blocky debris-lobes have probably descended as rockslides from the North and/or South Peaks. The North Peak itself is deeply fractured, with several climbing routes penetrating its caves. It might be classed as a limiting case of rock slope failure *in situ*.

On the north-east ridge a quite different type of rock slope failure is extant (Jarman, 2004b). Its source is a wedge scar biting into the crest just behind the North Peak, and the debris has run out almost to the slope foot as a sub-cataclasmic failure (see 'Introduction', this chapter). Debris from similar earlier slides may core the bouldery morainic hummocks in Coire a' Bhalachain.

The geology is typical blocky schistose arenite, semipelite and pelite of the Beinn Bheula Schist Formation within the Dalradian Supergroup (late Precambrian–early Cambrian). This unit is laterally equivalent to the Ben Ledi Grit Formation but has a greater proportion of pelite and semipelite, and was metamorphosed under upper greenschist facies conditions. The regional bedding and foliation vary from sub-horizontal to moderately north-west dipping (J. Mendum, British Geological Survey, pers. comm.).

Interpretation

The prevalent mode of rock slope failure in the Dalradian schists of the Southern Highlands is the short-travel arrested translational slide (Jarman, 2003a). On The Cobbler, the main and An t-Sron failures fall within this category. A geotechnical analysis of The Cobbler is not available, but Watters (1972) and Holmes (1984) back-analysed 14 rock slope failures in the Beinn Bheula Schist Formation in this area. They found peak friction angles ranging from 34°–50° and residual friction angles around 24°, increased by i values typically of 1°–5°, depending on coarseness of asperities and orientation of microfolds. This renders somewhat conjectural attempts to interpret the failure mechanics of large complex rock slope failures such as these, where actual sliding surfaces are not revealed, are of unknown depth and state of weathering, and may not even follow pre-existing joint planes.

Translational sliding in the metamorphic lithologies, which predominate in the Highlands, is facilitated by the multiplicity of discontinuities, both in joint-sets and especially in the foliation or schistosity surface (Figure 2.4). On The Cobbler, the schistosity surface generally dips at 25°–40° to the north-west. Since the failures are orientated from SSE through to WSW, this surface can have only contributed marginally to sliding, notably in the spreading of the north-west end. The joint-sets in this vicinity identified by Holmes (1984) and Watters (1972) are all above 50° and too steep for controlled sliding. They generally provide detachment planes, as seen here in the head and east flank scarps. Sliding may therefore have occurred along a staircase of joints, as modelled by Watters (1972) at Beinn an Lochain West (NN 215 076), or possibly by inventing a suitable plane where none pre-existed, as a series of small shear facets (Jarman, 2003c).

Holmes (1984) and Watters (1972) concluded that the translational slides in this area probably required a small additional force augmenting gravity in order to trigger shearing failure. Watters identified a toppling component (de Freitas and Watters, 1973), and both invoked elevated cleft water pressures; chemical weathering of abundant chlorite and muscovite may also have played a role in lubricating slide surfaces. Separation of large rock units by low-angle block gliding is common in this area, notably at Ben Donich (NN 22 05), Carnach Mor (NS 13 99), Hell's Glen (NN 18 05), and The Steeple (NN 20 00), and may account for the tensional spreading seen at The Cobbler (Panels 1–2, Figures 2.48 and 2.49). However, the source area is almost certainly above the Loch Lomond Stadial limit which Holmes (1984) identified as the likely locale of abundant meltwater to pressurise clefts, the Arrochar peaks having formed nunataks at this stage (cf. 'Introduction', this chapter).

The north-east rock slope failure is sub-cataclasmic in having fully disintegrated but not travelled bodily to the slope foot or beyond. It has no contained toe, and an extensive 'spray fan', indicating relatively rapid travel. Such failures are rare in the Highlands. The sub-cataclasmic rock slope failure on Beinn an Lochain North (Watters, 1972; Figure 2.9) approaches it in character, but is fully contained by a stickier matrix; it was interpreted as a rock glacier (Ballantyne and Harris, 1994) but is now seen as a conventional rockslide. The extensive near-vertical fissuring around North Peak might suggest that this north-east failure was simply a very large rockfall, but the wedge-shaped cavity and a retained cubic mass that has slipped 2–3 m from the rim could indicate initial release as a high-angle slide of more alpine character, as at **Beinn Alligin**.

A possible seismic trigger for rock slope failure has often been invoked but has yet to be established (see 'Introduction', this chapter). Against this, The Cobbler is 10 km from the nearest major basement fault and its rock slope failure components are unlikely to be of the same age, but it does lie on a current zone of seismic attenuation (Carlisle to Kyle of Lochalsh). The coherence and containment of its translational slides are suggestive of progressive creep. However, an unusually large fissure 75 m long and 10 m deep just beyond the toe of Panel 1 (NN 255 056) has affinities with tear or impact fractures beside large failures in Scandinavia (cf. Dawson *et al.*, 1986; Jarman, 2002).

No dates are available for The Cobbler rock slope failures. It is most probable that the translational failures here developed soon after final deglaciation, but since they extend to the summit, they owe their inception to the much greater stresses induced by the Late Devensian glaciation. In the main rock slope failure, Panels 3 and 4 are subdued, with degraded source and toe areas, and may be earlier than Panels 1 and 2; they may even be the rump of similar events since removed by glaciers, and here resemble the two-tier rock slope failure at Beinn an Lochain West (Watters, 1972; cf. Beinn Each (**Glen Ample**)). Erratic boulders are strewn over the upper parts of Panels 3 and 4 and across much of the upper south-east ridge (Figure 2.47) and may be ice transported following earlier rock slope failure episodes, as they occur up to 600 m OD, rather above the toe of Panels 1 and 2 (Figure 2.49). Boulder trains from probable failure sources are surprisingly uncommon: Carn Ghluasaid (NH 140 119) offers an unusually clear case

Panels 1 and 2 appear considerably 'fresher', and to encroach onto Panel 3, but no conclusions can be drawn as to their relative ages. Although Panel 2 has descended further than Panel 1 (Figure 2.50), they may have originated at the same time, with the eastern component able to travel farther owing to the slant of the basal 'alp' and suitably orientated release planes. The arrestment of Panel 1 at a high angle approaching 40° is a common feature of Scottish rock slope failure. Why such failed masses do not disintegrate cataclasmically on such steep and unconfined slopes has yet to be addressed, but possible explanations include dewatering, locking-up of large component slices, and lack of available relief by comparison with alpine ranges.

Landscape evolution

The main rock slope failure source area along the summit ridge has resulted in one of the most striking mountain landscapes in Britain, and has only recently been recognized as the product of paraglacial failure intersecting a corrie headwall (Jarman, 2004b). Prior to failure, the corrie rim would have cut into a level or gently south-west dipping smooth ridge, as it still does between the Summit and North Peak. This can be demonstrated if the flatter elements of the upper rock slope failure are re-instated to their source scarps. The exceptional rock tower of the South Peak may have been exaggerated by a small wedge slide yielding a blocky debris-pile a short way to the south-east. The North Peak may also have been reduced to a fractured stump by failure of a former north-east arm to the corrie during earlier interglacials: the prominent schistosity surfaces dip back into the corrie at angles (20°–30°) highly conducive to sliding and toppling.

The shaping of mountain summits by rock slope failure into arêtes and horns is also seen in this area at Ben Lomond and Beinn an Lochain (Figure 2.9). It is attributable to failure planes daylighting behind the crest, leaving a sharp edge that has not suffered rounding by glacial over-riding or periglacial weathering, as have most summit ridges in the schists.

If the west summit ridge of The Cobbler is re-assembled to pre-rock slope failure condition, a small fragment (approximately 0.05 km^2) of pre-

Figure 2.50 The main rock slope failure complex seen from the south-west across Glen Croe, showing the panels mapped in Figure 2.49. Panel 2 (on the right) has travelled further than Panel 1 to expose the arête culminating in the South Peak (Point 858). The level pre-glacial summit surface of Beinn Narnain (Point 926, right background) suggests the character of The Cobbler before it underwent more intense paraglacial rock slope failure. The summit tor (Point 884) stands out from the vestigial pre-glacial skyline. (Photo: D. Jarman.)

glacial upland surface can be identified. Behind the main headscarp and ridge-splitting trench, a small (< 2 m) dogleg depression and fissures indicate incipient extension of the rock slope failure by up to 25 m into the residual pre-glacial surface. The remarkable summit 'tor' may be a corestone left from stripping of a deep-weathering regolith (cf. Hall, 1991). The next peak to the east, Beinn Narnain, has a more extensive plateau surface of approximately 0.5 km^2 (Figures 2.47 and 2.50), and likewise has lost a quadrant of it to failure, leaving a projecting fractured arête. Other surrounding peaks show similar traces of an undulating pre-glacial upland surface, and encroaching rock slope failures.

In the core mountain area of 87 km^2 around The Cobbler, 7% of the total land surface is affected by rock slope failure, one of the highest densities so far identified in Scotland (Table 2.2); the figure could in fact be higher with many small failures. This rises to 10% for The Cobbler itself, to its bounding waters. This concentration lies within the largest cluster of rock slope failures in Scotland, with over 100 significant (> 0.02 km^2) cases in 500 km^2 west of Loch Lomond. No reasons were advanced for the existence of this cluster by Holmes (1984), nor following its publication by Ballantyne (1986a). Explanations related to Loch Lomond Stadial limits or high-magnitude seismic shocks are inadequate, although suitably inclined foliation and bedding, and the style of metamorphic grade, increase the incidence of rock slope failure (see 'Introduction', this chapter).

The tentative association between rock slope failure incidence and glacial breaches of possibly recent origin or active enlargement can be investigated here (Jarman, 2003a), if less straightforwardly than in the Kintail cluster (see **Sgurr na Ciste Duibhe**). The Arrochar Alps and surrounding hills are intensely dissected to such a degree that most are isolated, and few high-level linking ridges survive (Haynes, 1977a). This reflects their location at the junction of the catchments of the Clyde estuary, Loch Fyne, and the River Forth prior to its beheading by the cutting of Loch Lomond (Linton and Moisley, 1960) (Figure 2.13). These pre-glacial radiating valley systems have been over-run by ice centred at various times to the west, east, and particularly north of the triple watersheds, although probably not from as far as the Rannoch Moor ice centre, as suggested by Linton (1957) and Boulton *et al.* (1991). As a result, almost all of the peaks in this area are separated by glacially breached cols, some relatively narrow or deep.

The Cobbler is isolated by high-level breaches at 630 m OD and by the deep trough of Glen Croe, which descends from the Rest-and-be-thankful Pass (260 m OD) to sea level at Loch Long (Figures 2.15 and 2.47). Linton and Moisley (1960) show Glen Croe as a probable breach of the main pre-glacial Clyde–Forth watershed; it has certainly been incised in response to over-deepening of the Loch Long fjord. However, while the sub-cataclasmic north-east rock slope failure is directly above the breach into Coire a' Bhalachain, the main and An t-Sron failures are separated from the Glen Croe trough by an 'alp'. If recent breaching through the Rest-and-be-thankful Pass from the north has created high, local, slope stresses, then the rock slope failure incidence might suggest this has been more by widening at upper levels than by trough deepening. A similar incidence at higher levels is found on Ben Donich and The Brack opposite (Figure 2.47); this is not always the case, as rock slope failure is concentrated at lower levels along Loch Long and in Glen Kinglas, and extends from crest to foot of the deep Loch Sloy breach (Figure 2.15).

Conclusions

The Cobbler clearly demonstrates the effects of large-scale rock slope failure in creating mountain landforms such as arêtes and horns. The contribution of paraglacial slope failure to bulk erosion, valley widening, dissection of mountain massifs, and progressive elimination of ridges and summit areas has been given little consideration, but in susceptible areas its cumulative impact over the Quaternary Period is likely to be very considerable. The Cobbler and the surrounding Arrochar Alps and Cowal hills provide an important opportunity to understand this process, and to seek to quantify and date it. Containing the densest large cluster of rock slope failures in Scotland, they provide an exceptional locus for research into its causes and its possible significance as an indicator of shifting ice centres and dispersal patterns. Relicts of the pre-glacial land surface on The Cobbler and its neighbours provide scope for landscape reconstruction, in one of the upland areas of Britain most intensely dissected by glacial breaching.

The Cobbler itself has a variety of types of rock slope failure representative of the Dalradian rocks of the South-west Highlands. As one of the best-known and most idiosyncratically shaped mountains in Britain, in a conspicuous and accessible location in the Loch Lomond and The Trossachs National Park, The Cobbler is also well suited for research into geomorphological education, for geotourism, and for widening public awareness of landscape origins.

BEN HEE, HIGHLAND (NC 430 343)

D. Jarman and S. Lukas

Introduction

Ben Hee provides one of the best examples in Britain of Holocene rock slope failure activity within a corrie, and is one of the largest such cases. It also clearly represents the arrested translational slide mode of failure which predominates in the Scottish mountains. The failure complex has several components of rock slope failure arrested at different stages of development, which suggest progression of activity both downslope and laterally along the corrie headwall. Unusually extensive deposits in the corrie floor below may represent material reworked from failure in a previous interglacial period. The concept of rock slope failure as a major factor in glacial/paraglacial erosion over repeated cycles is in its infancy (Evans, 1997; Ballantyne, 2002a; Jarman, 2002), and Ben Hee offers a testbed for research into its scale and mode of operation.

The rock slope failure has encroached substantially into the headwall of the corrie, and further activity will lead to breaching into the adjacent corrie and dissection of the mountain block. Viewed in conjunction with neighbouring slope failures, Ben Hee affords an excellent illustration of the role of large-scale mass movement in initiating, enlarging, and then eliminating corrie-type landforms.

Description

Ben Hee (873 m) is a large mountain in northern Scotland separated from the Reay Forest massif by a glacial breach, and standing above the glacially modelled Caithness–Sutherland intermediate surface (Figure 2.51). An Gorm-choire is a large corrie that has operated, latterly at least, as a true glacial 'cirque', its sub-parallel sides reflecting structural controls rather than evolution as a glacial trough head. The corrie is approximately 1 km wide and 2 km long, and faces ESE. Its south flank is a 200 m-high crag, its north flank is a gullied and embayed scree-slope, and below its low headwall (2–15 m high) is a large landslip complex (Figure 2.52). This has three components laterally, unified by a continuous source trench that splits the smooth, broad ridge between the two summits. At the south corner, a small shallow wedge has dropped in several slices by 6 m, and encroaches into the broad ridge by 30 m; below it, a small disintegrated debris-mass clings to the headwall. At the north end, a similar small segment lowers the skyline by virtue of daylighting behind the crest, leaving a 5 m-deep hollow. The flanking rock buttress is crazed with tension fractures, and coarse, unstable, rockfall deposits below may have formed relatively recently.

In between, the main failed mass itself has three distinct tiers (1, 2 and 3 on Figure 2.52) below an obtuse wedge source. This fracture

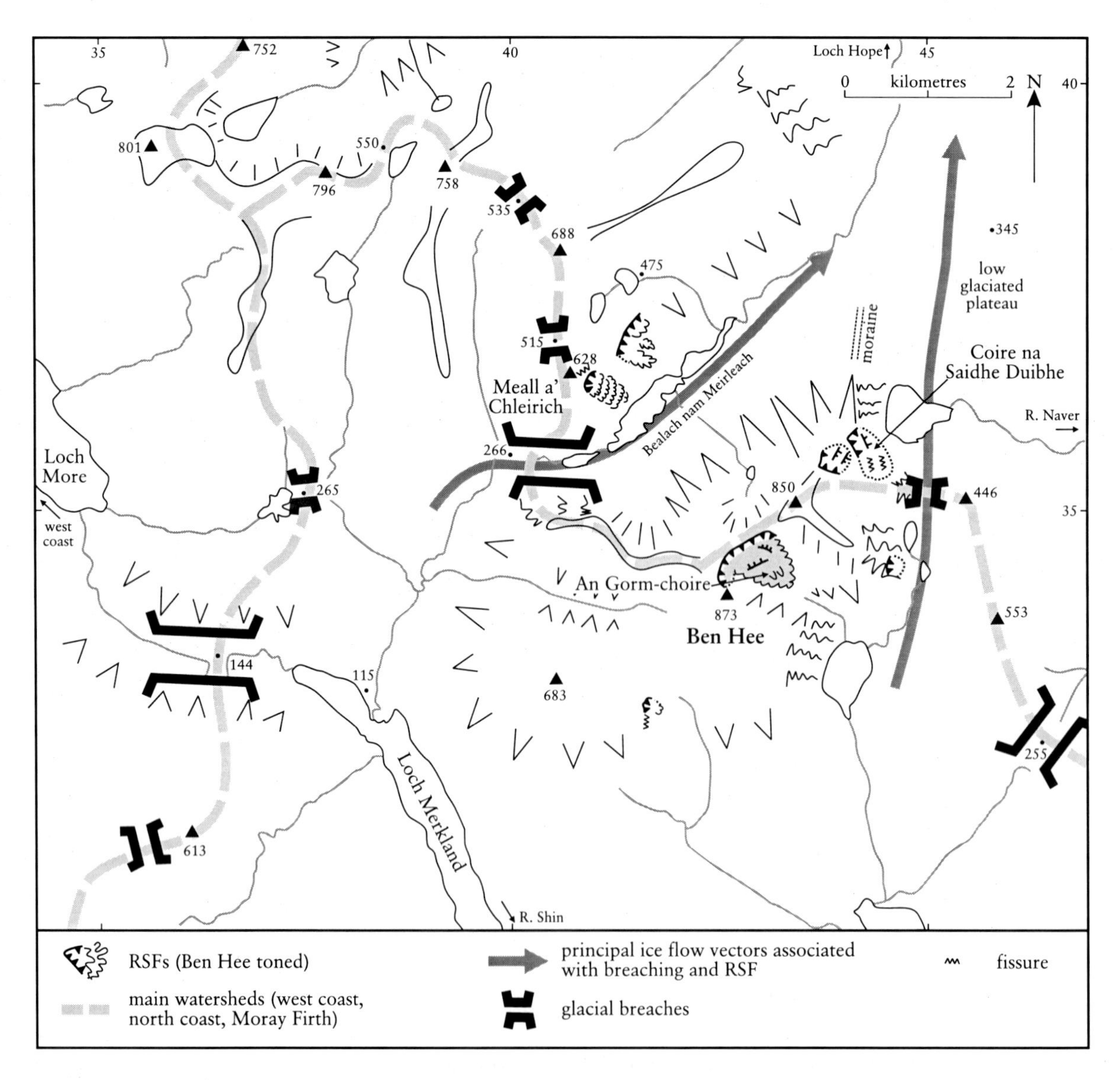

Figure 2.51 The Ben Hee rock slope failure (RSF) cluster, with glacial breaches and related ice movements.

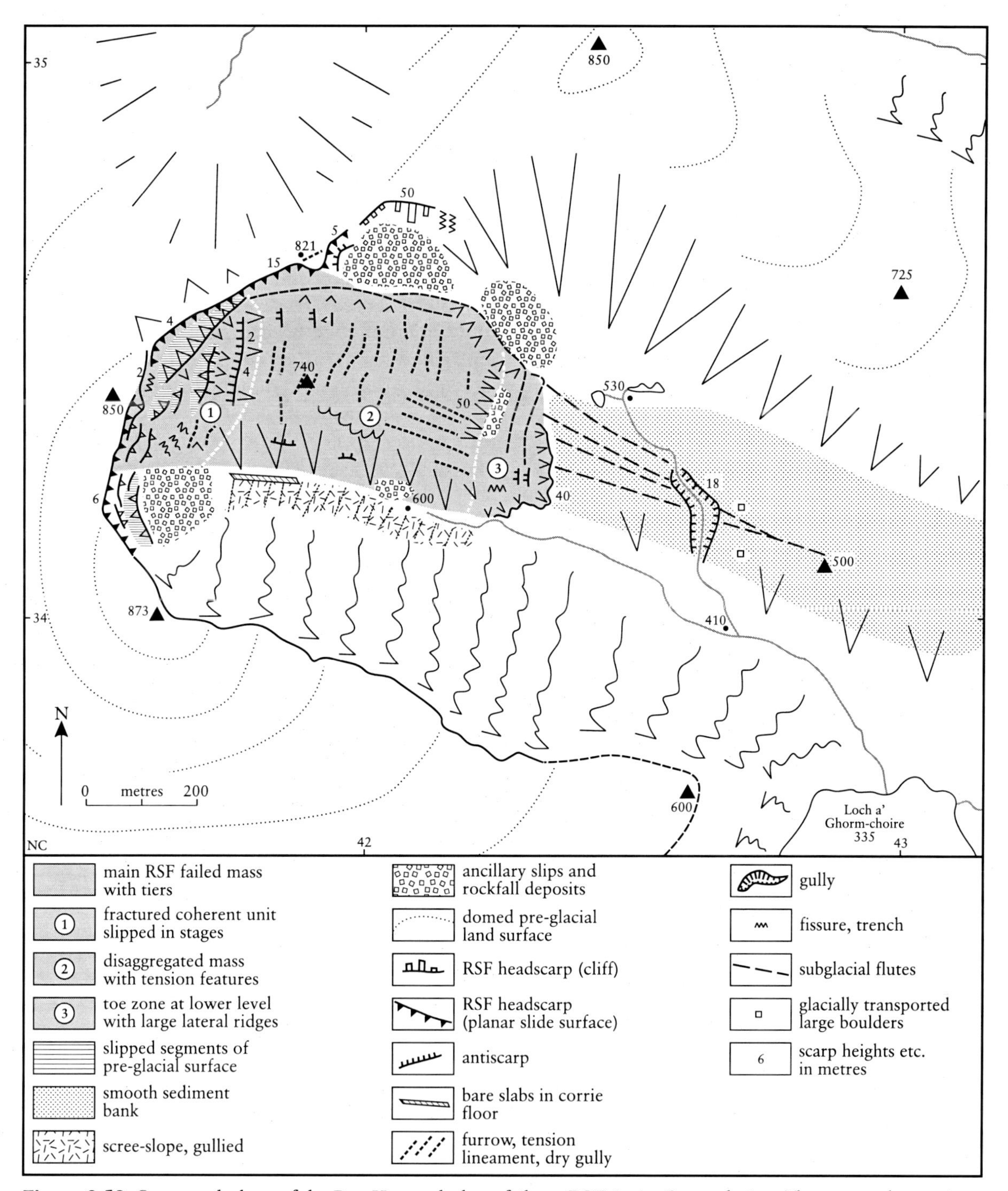

Figure 2.52 Geomorphology of the Ben Hee rock slope failure (RSF) in An Gorm-choire. The encroachment into the undulating pre-glacial surface is considerably greater than the extent of the 'slipped segments'.

begins at the upper (south) end as narrow 1–2 m-deep trenches, becoming a low angle (20°–25°) headscarp which attains 15 m in height at its north end in the col. In the obtuse angle, Tier 1 is a semi-intact slice of summit ridge 200 m long by 50 m wide, which has dropped by 2–4 m; its outer parts have broken away and slipped a little further, forming minor antiscarps. Periglacial blockfields cover its surface and are identical to those found on the summit ridge.

Below this uppermost sector of recognizable provenance, the one major antiscarp of the rock slope failure crosses the full width of the main slip, typically less than 2 m high, but locally attaining 4 m. Its crest is about 15 m lower than the semi-intact surface above. This antiscarp defines the start of Tier 2, which is much the largest, and is a disaggregated mass extending out into the corrie for about 400 m. Its gently sloping upper surface also retains the decimetre-scale blockfield characteristic of the summit ridge, but appears to have a randomized topography (Figure 2.53). However the network of shallow tension hollows and minor (< 2 m) antiscarps is organized along two main axes, down and across corrie. Tier 2 terminates in a steep rampart to the lower corrie, about 50 m high (Figure 2.54), with few signs of rock slope failure apart from minor blockfalls. However its long south flank of even greater height has extensive side-slipping with some antiscarp development. By contrast, its north flank starts as a modest linear dry gully separating the main failed mass from the north component; the dry gully then plunges into the northern bay of the corrie.

Below this rampart, Tier 3 is a terrace of different, grassier character bearing weak signs of rock slope failure with hollows, antiscarplets, and an area of tensional dissection. Its northern part has a series of bold lateral ridges up to 10 m wide and 5 m high, but these are probably tension gashes possibly exploited by fluvial erosion (Figure 2.54). Its southern part extends markedly further out into the corrie floor than the main failed mass above. The toe of this tier rises 30–40 m above the corrie floor, and is sub-arcuate.

The corrie floor is a long, rather narrow trench, with exposed scoured bedrock in its middle reach. Here several 'pods' of fractured but coherent rock have probably been glacially entrained from a rockstep just above. This scoured area at 500 m OD hangs well above Loch a' Ghorm-choire at 330 m, which is outwith the corrie proper. The north side of outer Gorm-choire is occupied by a remarkably smooth bank of sediment consisting of a massive, matrix-supported diamicton that is over-consolidated, shows numerous fissures and contains predominantly sub-angular clasts. It is cut by an 18 m-deep gully. The top surface of this sediment accumulation is traversed by four distinct smooth ridges 0.5–1.0 m high, 1–2 m wide, and up to 600 m long. They emanate from the toe of the rock slope failure, converge, are

Figure 2.53 View across Tiers 1 and 2 of the main failed mass, and both flanking increments, from the north rim to the summit of Ben Hee. The reconstructed pre-deglaciation crest follows the axis of the picture, whereas the pre-Quaternary crest probably curved more to the left between gentle domes. (Photo: D. Jarman.)

Figure 2.54 View up An Gorm-choire from the smooth sediment bank, with sub-glacial flute ridges emanating from beneath the rock slope failure toe. Above them, Tier 3 has lateral lineations which contrast with the amorphous slumping mass of Tier 2. The pecked line denotes the pre-failure skyline, reconstructed in Figure 2.56, and inferred to have been lowered by up to 35 m. Pre-glacial surface remnants survive at top-left and top-right. (Photo: D. Jarman.)

cut by the gully, and fade out eastwards. A few large glacially transported boulders lie on the sediment bank. The gully contains a small stream emanating from lochans ponded up by this bank in a broad side bay of Gorm-choire.

To the north of Gorm-choire, a broad shoulder of Ben Hee throws out three spurs, which, with their intervening bays, are abruptly truncated on the east by steep cliffs. The north-most spur spawns a medial moraine indicating Devensian ice movement northwards along the east flank of Ben Hee, convergent with ice coming through the breach of Bealach nam Meirleach. Along the truncated east side, the cliffs are deeply fissured, with local toppling failures (Figure 2.51). Coire na Saidhe Duibhe has a remarkable midslope in-dipping open cleft approximately 250 m long and up to 10 m wide and 15 m deep. This heads a zone of fissuring and grabening with some toppling debris. The south flank of the bay is a 20 m crag heading an apparent rock slope failure cavity, with incipient fissuring. Above the great cleft, the slope is stable until a degraded short-travel debris-mass encroaches from a possible 40 m source scarp, which continues down the north-east ridge as a weak lineament defining a further large slope failure increment. This failure zone doubtless provided the large erratic on the spur above the medial moraine.

The breach north-west of Ben Hee (Bealach nam Meirleach) has a striking rock slope failure complex on its opposite flank. Torn vegetation and fresh debris indicate that creep and rockfall are still unusually active. Open fissures above the main rockslide suggest incipient encroachment into the residual summit plateau of Meall a' Chleirich (Figure 2.51).

Geologically, Ben Hee consists of Moine psammites with occasional pelitic schist bands (Johnstone and Mykura, 1989). The dip is rather regular, at 5°–15° to the east/ENE at An Gorm-choire, and a little steeper and to the south-east at the north end. Several prominent joint-sets are seen in the cliffs, and one inclined at about 25° may control the general slope to the ESE of the Ben Hee plateau.

The smoothness of the summit ridge, with shattered bedrock, 5–10 m deep, mantling most of it, suggests that it has not been vigorously glaciated (i.e. covered by active ice), and may even be a remnant of the pre-glacial land surface (cf. Hall, 1991). A periglacial trim-zone between 680 m and 750 m OD indicates the upper limit of the Late Devensian ice-sheet (Ballantyne *et al.*, 1998b). However, cold-based ice may have extended to higher levels, and also warm-based ice may have extended higher in earlier glaciations (Lukas, 2005).

Holmes (1984) records the main rock slope failure from aerial photographs as covering 0.36 km^2 (actually 0.40 km^2), the east cliff slope failure as 0.04 km^2, and the north-east corrie bay as having three slope failures totalling 0.14 km^2 (actually 0.09 km^2 for the upper and 0.17 km^2 for the lower failure). He gives the Meall a' Chleirich rock slope failure as 0.15 km^2 whereas it is part of a complex extending for over a kilometre and affecting approximately 0.5 km^2.

Interpretation

The main Ben Hee rock slope failure was first described by officers of the British Geological Survey (unpublished field slips, 1913–1926) as 'possibly not truly *in situ* but a whole crag slipped'. They noted 'scree has slipped away from the corrie edge in parallel ridges (leaving) a gully behind it'. Godard (1965) mis-interpreted these features as moraines, his Photo 17 showing the neat narrow ridge where the headscarp intersects the north corrie as a '*bourrelet morainique laissé par un petit glacier perché, tardiglaciaire*' (cf. Figure 2.55). Haynes (1977b) corrected this, and mapped the compound landslip with its two flank elements.

No geotechnical analysis has been made of Ben Hee. The simplest interpretation of its failure geometry invokes the joint-set dipping at 20°–25° south-east, which can be seen in the summit cliffs, and which forms the main head-scarp plane. A failure plane of approximately 18° is indicated by terrain reconstruction (Figure 2.56), and would just permit arrested translational sliding within the range of residual friction angles in the psammites (Watters, 1972). The tripartite scarp of the main failed mass (including the dry gully on its northern side) suggests an obtuse armchair-source configuration, with travel down the corrie axis plus

Figure 2.55 The source fracture for the main failed mass and north flank component, mis-interpreted by Godard (1965) as a glacial moraine. The slipped segment retains much of its pre-glacial character. (Photo: D. Jarman.)

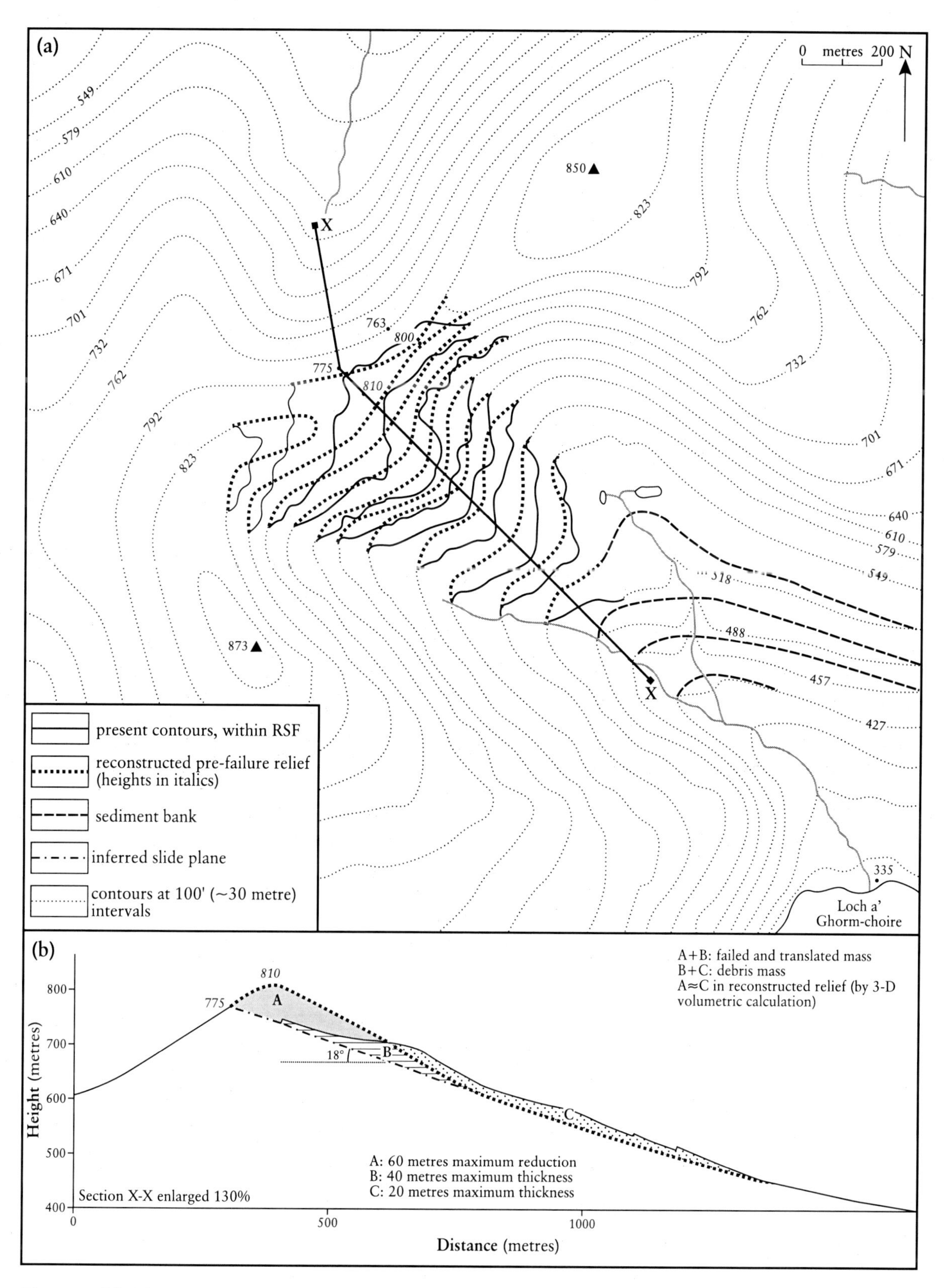

Figure 2.56 The reconstructed pre-failure topography of An Gorm-choire, with the long-section of the main rock slope failure showing reduction of the summit ridge by up to 35 m and the scale of mass displacement.

unconstrained spreading laterally out into the corrie head. This would account for the biaxial features in Tier 2, and the sideslip antiscarps on its steep south flank. A marked north–south joint-set seen in the summit cliffs may have encouraged disaggregation of the failed mass into several large slices, with Tier 1 remaining semi-intact above the traversing antiscarp. In Tier 2, these slices may be disguised by slumping of the deep weathering overlay into the tension hollows, hence the uncharacteristically random terrain. The small component to the north may have been an integral part of the main failure, comparable to Tier 1, from which Tier 2 has pulled away laterally, although whether at the same time or by later progression is unclear. The small component to the south is structurally separate and must have moved on a different failure plane, albeit sharing the same source fracture.

The extant rock slope failure complex clearly post-dates final deglaciation. Current research (Lukas, 2005; Lukas and Lukas, 2006) indicates that Gorm-choire was filled by a Loch Lomond Stadial glacier. However, the remarkably smooth debris-bank below the failure on the north side of the corrie was described in the unpublished British Geological Survey field slips as 'too regular to be a lateral moraine – ? a drumlin'. Lukas (2005) describes the feature as a stack of sub-glacial till sheets recording various ice-flow directions and possibly a succession of late Quaternary glaciations (Lukas, 2005). Haynes (1977b) noted that the quantity of drift is exceptional, and 'might well suggest that land-slipping has been progressing simultaneously with glaciation over a period, rather than merely just once'. This interpretation hints at the glacial–paraglacial rock slope failure cycling model now becoming recognized (Hall and Jarman, 2004; see **Trotternish Escarpment** GCR site report, Chapter 6). Similar exceptionally massive drift banks occur in corries on Maoile Lunndaidh in the Western Highlands (NH 140 450) and in Caenlochan (NO 170 765), with traces of rock slope failure in the headwalls. The officers of the British Geological Survey also noted the '3 or 4 ridges on the drift bank trailing like 'tail' from the 'crag' to the WNW', which Haynes (1977b) mapped as five ridges of 'streamlined drift' (Figure 2.54). These are probably sub-glacial flutes, as they are parallel to the palaeo-ice-flow direction indicated by the fabric analyses taken from the sub-glacial till accumulation they are placed on (Lukas, 2005). This might indicate that Gorm-choire had two ice sources, with the separating spur now destroyed by the rock slope failure that overruns them. Haynes (1977b) also observed the 'low hummocks' at the base of the slip. This anomalous lowest Tier 3 of the rock slope failure and the remarkable 'drift bank' merit closer investigation as to whether they stem from earlier episodes of rock slope failure, with differing degrees of glacial reworking.

Landscape Evolution

A reasonably constrained reconstruction of the pre-rock slope failure topography is possible (Figure 2.56). With contours extrapolated from both flanks, and with negative and positive pre- and post-rock slope failure relief in balance, it indicates a compound corrie with a large projecting nose on the headwall that has collapsed. The col between the summit and the north-east top has been lowered by up to 35 m as the failure-enlarged corrie has encroached by 80–120 m into the bowl of Garbh-choire Beag on the north. The failed mass may have reached 60 m thick above the slide plane at the former crest; it has spread out to depths reaching 20–40 m at an overall surface angle of only 18°, and has a total volume of 5–10 × 10^6 m^3. This is at least an order of magnitude larger than typical rock slope failures within corries estimated by Holmes (1984) and 2–3 times the size of the **Beinn Alligin** intra-corrie rock-avalanche.

The idea that rock slope failure may be a contributor to corrie initiation and development has been mooted since Clough (1897). However, only 18% of rock slope failures identified by Holmes (1984) are located within corries (broadly interpreted), most of them minor. It is difficult to point to any extant failure that is clearly seeding a new corrie, which is unsurprising given that the most promising localities have been selected over many glacial cycles. Ben Hee is therefore of exceptional interest in assessing the scale and *modus operandi* of this putative process. Here, it appears likely that the pre-glacial summit plateau of Ben Hee was originally continuous between the two present tops, and that the linking ridge has been lowered and displaced to the north-west by vigorous headward extension of An Gorm-choire, aided by repeated episodes of paraglacial rock slope failure on the favourable structural dip. Further episodes will lower the col to the point where breaching by transfluent ice may become possible, eliminating the corrie and converting Ben Hee into two separate mountains.

The north-east corrie bay offers an excellent comparator at an earlier stage of evolution, with the deeply fissured lower rock slope failure preparing the way for rapid headwall excavation, and the upper slope failure opening out a broader corrie bowl. The driver for this cluster of corrie-shaping rock slope failures appears to be basal over-steepening of the east flank of the Ben Hee massif by transfluent ice from the upper Loch Shin basin breaching north-east across the Reay Forest–Ben Klibreck divide (Sutherland, 1984). The pattern is repeated in the eastern corries of Ben Hope (NC 48 49), which contain Britain's northernmost montane rock slope failure. The relatively narrow breach of Bealach nam Meirleach on the north-west side of Ben Hee has evidently been cut or enlarged in the last main glaciation, to judge from the extensive destabilization and ensuing failure on Meall a' Chleirich.

Conclusions

The main Ben Hee rock slope failure is one of the largest and clearest examples to be found in a corrie. It is possible to reconstruct a pre-failure topography that confirms the scale on which such paraglacial mass movements can enlarge a corrie. It is then possible to project forwards a process whereby further rock slope failure increments will eliminate the corrie headwall and pave the way for dissection of the whole mountain block by glacial breaching. The adjacent slope failures on the north-east shoulder and on Meall a' Chleirich afford instructive comparators at different stages of corrie and breach initiation and evolution, including impressive evidence of incipient failure, which appears still to be actively propagating.

Ben Hee is a classic arrested translational slide. There is considerable scope here for detailed investigations into its geometry and mechanics, and into the sequence of rock slope failure activity probably over more than one glacial cycle. Finally, this isolated cluster of slope failures indicates that erosion both by corrie glaciers and by transfluent ice breaching has been sufficiently intense in Late Devensian times to destabilize mountain slopes even in the far north of Scotland.

CARN DUBH, BEN GULABIN, PERTHSHIRE (NO 113 721)

C.K. Ballantyne

Introduction

The landslide at Carn Dubh on Ben Gulabin is a translational failure in steeply dipping meta-sediments, a type of rockslide that is fairly widespread in the Scottish Highlands. The distinctiveness of this site arises from the unusual form of debris runout, which takes the form of two thick debris-tongues bounded by steep levées. There is no published account of this site; the observations below combine unpublished research by Cullum-Kenyon (1991) and the present author.

Description

Setting

Carn Dubh is the eastern spur of Ben Gulabin (806 m), 2 km north of Spittal of Glenshee in the south-east Grampian Highlands. The eastern slope of Carn Dubh descends towards the floor of the adjacent valley (Gleann Beag) at an average gradient of *c*. 33°. The slope is underlain by Dalradian schistose graphitic pelites and quartzites of Neoproterozoic age that dip moderately eastwards. At the last (Late Devensian) glacial maximum the area was completely over-ridden by SE-moving glacier ice that covered even the highest of the nearby summits (Glas Maol, 1068 m). During the final episode of glaciation, the Loch Lomond Stadial of *c*. 12.9–11.5 cal. ka BP, a small valley glacier advanced south down Gleann Beag, probably terminating near the valley mouth. The former extent of this glacier co-incides with a thick accumulation of till. On the west side of the valley the upper limit of till ascends northwards against the slopes of Ben Gulabin in the form of a discontinuous lateral moraine or drift limit that reaches an altitude of 490 m at the southern margin of the landslide, and is continued on the north side of the slide at an altitude of 520 m. Removal or burial of the intervening glacial drift confirms that the landslide occurred after glacier retreat. The failure scar occurs immediately upslope from the site of the former drift limit; and the debris tongues occur mainly downslope from this limit (Figure 2.57).

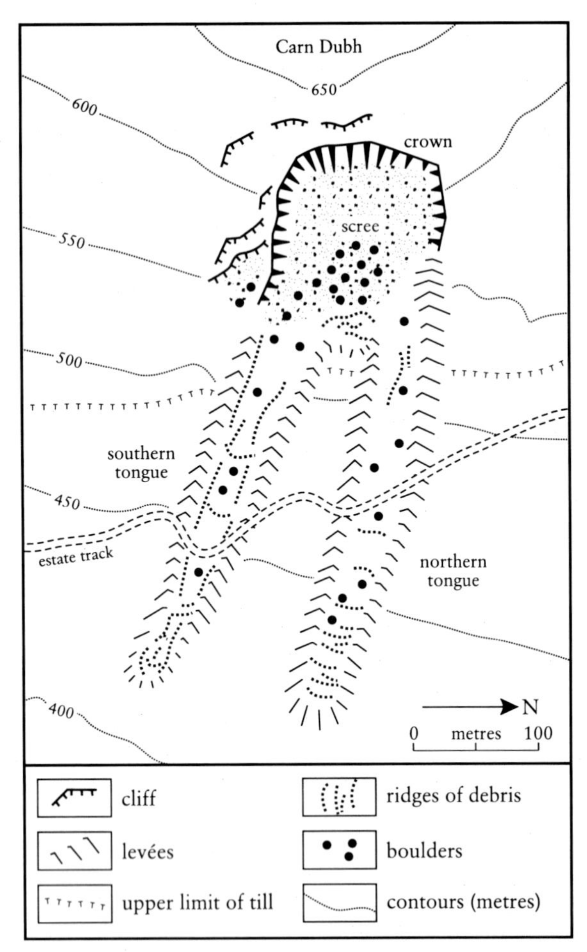

Figure 2.57 Map of the Carn Dubh rockslide scar and debris tongues, Ben Gulabin.

The failure scar

The failure scar has a roughly trapezoidal plan, but is extensively obscured by bouldery scree (Figure 2.57). Its southern margin is, however, defined by a partly vegetated cliff 10–15 m high, and a rockwall 20–30 m high crops out along the crown of the scar. The southern sidewall reveals a succession of well-jointed quartzites (Creag Leucach Quartzite Formation) in beds 1–4 m thick, interbedded with strongly foliated, weathered phyllitic semipelite and micaceous psammite beds generally less than 1 m thick. The bedding and foliation planes dip eastwards, sub-parallel with the slope at angles of 25° to 43°, forming part of a broad anticline. An exposure at the base of the southern sidewall suggests that rupture occurred (at least locally) along a 0.5 m-thick semipelite unit.

The debris lobes

Two thick tongues of debris extend downslope from the foot of the failure scar, and form the most conspicuous feature of the site (Figures 2.58 and 2.59). The two are separated just below the failure scar by a bulbous protrusion, probably underlain by bedrock, that split the mobile landslide debris into two separate flows. The top of the protrusion is covered by a wedge of angular bouldery landslide debris up to 7 m thick, formed into three transverse ridges.

The southern debris-tongue extends 320 m downslope from an altitude of 520 m to 415 m. It averages about 70 m in width and stands 3–10 m above the adjacent terrain. The gradient of the tongue declines downslope from 27° to 18°. The lobe margins are sharply defined by steep-sided boulder levées that decline in height downslope to terminate in overlapping ramp-like ridges 1–3 m high. Smaller longitudinal ridges occur inside the outer levées, suggesting that a pulse of flowing sediment occurred after the main body of the flow had stabilized.

The northern debris-tongue extends 380 m downslope from about 525 m to 420 m altitude, declining in gradient downslope from 31° to 15°. It averages about 80 m in width and 5 m in thickness, but is bounded by boulder levées up to 12 m high. Unlike its southern neighbour, it terminates abruptly downslope in a bold arcuate ridge that impounds a thick wedge or plug of debris, but lacks interior longitudinal ridges (Figure 2.58).

Between the outer levées, both tongues are crossed by arcuate transverse boulder ridges 1–3 m high. The central parts of the tongues are extensively vegetated, but numerous large angular boulders, mainly of quartzite, are scattered over the surface, and spreads of quartzite boulders mantle the higher parts of the levées. Sections in the levées occur where an estate track cuts across both debris-tongues (Figure 2.57). These reveal a predominantly clast-supported diamicton in which poorly sorted angular clasts, mainly of quartzite, are embedded in a compact, sandy matrix. Crude inverse grading is evident, with clasts increasing in both concentration and size towards the surface, forming a mantle of large (typically 0.3–2.0 m long) angular boulders. A small number of sub-angular facetted clasts, apparently derived from reworked till, are also exposed in the diamicton.

Figure 2.58 The Carn Dubh rockslide scar and debris tongues on Ben Gulabin. The southern debris tongue (left) peters out in low ridges and levées of vegetation-covered debris, whereas the northern tongue terminates in a bold bluff 5 m high. The inner levées of both lobes terminate upslope at the conspicuous bulbous protrusion that diverted flow of rockslide debris into two tongues. (Photo: C.K. Ballantyne.)

Figure 2.59 The Carn Dubh rockslide, Ben Gulabin, from above. (Photo: D. Jarman.)

Topographical survey of the site by Cullum-Kenyon (1991) indicates that the total volume of debris contained in the debris tongues and the debris on the intervening protrusion is *c.* $0.3 \times 10^6\,m^3$. Assuming an average rock density of $2550\,kg\,m^{-3}$ and allowing 20% for voids, this figure implies that the mass of failed rock was *c.* 600 000 tonnes. This figure is an upper estimate, as the debris lobes contain an unknown volume of reworked till.

Interpretation

Rock slope failure

Failure at this site took the form of a translational (planar) rockslide seated on quartzite strata that dip eastwards out of the slope at angles averaging around 34°. Tilt tests carried out by Cullum-Kenyon (1991) indicate an angle of plane sliding friction of 51° ± 4° for the quartzite and 36° ± 6° for the phyllite, confirming that rupture almost certainly occurred within the latter. The closeness of the friction angle of the phyllites to the dip of the bedding and foliation planes implies that the slope was in a state of conditional stability following deglaciation. The cause of failure is unknown. Progressive rock slope weakening is likely to have been caused by opening of stress-release joints following deglacial unloading and/or shearing of rock bridges and asperities as a result of rock-mass creep. Failure may ultimately have been triggered by build-up of joint-water pressures or possibly by a seismic event; the Glen Taitneach fault crosses Carn Dubh only 300 m from the crown of the failure scar, and may have been re-activated by differential glacio-isostatic uplift.

Debris flow

The two tongues have the morphological attributes of debris flow (sediment-gravity flow) deposits, namely elongate tongue-shaped plan-forms, steep bouldery lateral levées, near-surface inverse grading and longitudinal and transverse ridges indicative of flow surges (Van Steijn *et al.*, 1988; Coussot and Meunier, 1995; Corominas *et al.*, 1996). Morphologically, they resemble the hillslope debris-flow deposits that cover the lower slopes of many Scottish mountains (Ballantyne, 2002b,c), though the latter are produced by failure and flow within unconsolidated sediments and are generally at least an order of magnitude smaller than the debris tongues below Carn Dubh. Innes (1985), for example, found that 90% of hillslope flows in the Scottish Highlands have transported less than 60 m^3 of sediment, though large flows fed by gully systems may carry over 1000 m^3 of debris (Brazier and Ballantyne, 1989).

There are two competing models of flow movement. Some channelled flows move as Bingham flows, with a rigid plug of debris being transported by laminar shear of an underlying and surrounding mixture of sediment and water (Johnson and Rodine, 1984). Movement of most Scottish hillslope flows, however, appears to be dominated by cohesionless grainflow, in which momentum is maintained by inertial collisions, with boulders attaining partial buoyancy in a mobile mass of mud (Takahashi, 1981; Blikra and Nemec, 1998). The dispersive stresses implied by the latter mechanism account for movement of the coarsest debris to the top and sides of the flow, as observed in the sections cut through the levées of Carn Dubh debris-tongues.

Irrespective of the nature of movement, the flow of coarse debris below the Carn Dubh rock-slide implies a drastic reduction in viscosity. As the lower limit of the slide plane and the upper limit of the debris tongues co-incide approximately with the upper limit of thick till, it is tempting to relate the onset of flow to undrained loading of saturated till by cascading rock that raised porewater pressures in the till until the over-burden weight was transferred to the fluid, leading to liquifaction (Hutchinson and Bhandari, 1971; Bovis and Dagg, 1992). The alternative explanation appears to be that the high initial energy of the slide was sufficient to generate inertial grainflow or fragmental flow, partially buoyant in a mixture of expelled water, crushed phyllitic semipelite and possibly entrained till.

It is instructive to compare the characteristics of the Carn Dubh rockslide and debris-tongues with those of a rockslide in Gleann na Guiserein, Knoydart (NG 774 057; Bennett and Langridge, 1990), where planar sliding of psammitic meta-sediments over steeply dipping slabs resulted in the formation of a debris tongue very similar to those at Carn Dubh. The Guiserein tongue is approximately 440 m long, narrows downslope from 105 m to 40 m and is bounded by steep levées up to 10 m high. The adjacent slopes are underlain by bedrock with localized thin soil cover, implying that here the debris tongue developed without deformation of underlying sediment, and thus solely as a result of fragmentation and flow of rock debris. Ballantyne (1992) inferred that the Guiserein debris-tongue formed through inertial grainflow or fragmental flow, probably aided by reduction in effective normal stresses due to the presence of mud and water. Similarly, formation of the Carn Dubh debris-lobes may also have occurred independently of till cover downslope of the failure zone.

Flow of debris following rock slope failure is poorly documented in Scotland. An extreme example occurs at **Beinn Alligin**, where sliding of nearly 9×10^6 tonnes of rock along a steep (42°) failure plane resulted in movement of very coarse debris over a distance of 1.2 km along a corrie floor. Other sites occur along the **Trotternish Escarpment** on Skye, (Ballantyne, 1991a) and on the scarp face of the Lomond Hills in Fife (Ballantyne and Eckford, 1984) and below the basalt scarp of the Campsie Fells north of Glasgow (Evans and Hansom, 1998, 2003). None of these, however, have produced the elongate debris-tongues bounded by massive levées that characterize the Carn Dubh rockslide.

Conclusions

Sometime after the final deglaciation of Gleann Beag some 11 500–12 000 years ago, up to 600 000 tonnes of rock below Carn Dubh failed by sliding of interbedded quartzites and semipelites along bedding planes that dip towards the valley at angles of around 34°. Rupture occurred in the phyllites, which were probably weakened by post-glacial stress-release, though the failure trigger may have been high water-pressure or a seismic shock generated by reactivation of the nearby Gleann Taitneach fault.

The Carn Dubh landslide on Ben Gulabin represents an outstanding example of a rock slope failure where runout involved viscous flow, producing thick elongate debris-tongues that extended downslope from the foot of the failure scar. This phenomenon is very rare on the metamorphic rocks that underlie most of the Scottish Highlands; only two other examples are documented. Fragmentation and flow of mobile rockslide debris around a central protrusion resulted in the deposition of two thick debris-tongues, 320 m and 380 m long, flanked by massive bouldery levées up to 12 m high. These thick debris-tongues resemble those produced by hillslope debris-flows, but are an order of magnitude larger. Movement of the boulders in the debris tongues probably took the form of inertial grainflow sustained by the momentum of colliding boulders, with coarse debris partially buoyant in mobile mud. The degree to which movement was aided by loading and liquifaction of underlying glacial deposits is uncertain, though comparison with a similar site in Knoydart suggests that formation of massive debris-lobes such as those below the Carn Dubh rockslide was not dependent on deformation of underlying sediments. The internal composition of the massive flow-tongues suggests that flow generated by the momentum of the initial rockslide involved a mixture of expelled water, crushed phyllites, quartzite boulders and a subsidiary component of entrained till. Debris flow at this site was probably aided by focusing of runout debris around the central protrusion and runout on to initially steep gradients.

BEINN ALLIGIN, HIGHLAND (NG 867 603)

C.K. Ballantyne

Introduction

A corrie on the south-east side of Beinn Alligin (north-west Scotland) is the site of a major rock slope failure that probably represents the finest example of a rock avalanche in the British Isles. The term 'rock avalanche' is generally employed to describe the failure and rapid descent of large (> 500 000 m³) masses of rock from steep mountain walls. Rock avalanches occur when joints within a rockwall become progressively interconnected, reducing rock-mass strength until the rock fails under its own weight (Selby, 1993). They are particularly common in alpine environments where glacial erosion has steepened rockfaces, and where the stress field within the rock mass has altered in response to deglacial unloading. Although over 500 individual rock slope failures have been identified in the mainland Scottish Highlands (Ballantyne, 1986a), true rock avalanches are rare, probably because of the relatively modest relief. By far the most spectacular example is that at Beinn Alligin, which is particularly notable for the clarity of the failure scar and the exceptionally long runout of very coarse debris. The latter implies that the Beinn Alligin rock-avalanche may be classified as an excess-runout rock-avalanche or 'sturzstrom' (Ballantyne, 2003; Ballantyne and Stone, 2004).

Description

Beinn Alligin comprises two summits (Sgurr Mór, 985 m and Tom na Gruagaich, 922 m) joined by a narrow arête. On the south-east flank of the mountain, steep rockwalls of stepped Torridon sandstone strata rise 350–550 m above a deep corrie, Toll a'Mhadaidh Mór (Figure 2.60). The corrie floor is traversed by a tongue-shaped deposit of landslide runout debris composed of large Torridon sandstone boulders up to and occasionally exceeding 5 m in length, with no visible fine-grained interstitial sediment (Figure 2.61; see also fig. 6.12 in the *Quaternary of Scotland* GCR volume, Gordon and Sutherland, 1993). This deposit rises up to 15 m above the bedrock floor of the corrie, occupies an area of 0.38 km^2 and extends continuously downvalley for 1.25 km at an average gradient of 8°, from an altitude of 450 m at the base of the corrie headwall to 275 m. It tapers downvalley from a maximum width of 380 m to a width of 170 m at its terminus (Figure 2.60). The lateral margins of the runout deposit are sharply defined. Its surface relief consists of discontinuous and often poorly defined ridges and intervening depressions. The distal part of the debris tongue is dominated by arcuate-downvalley transverse ridges, but ridges in the upvalley part are transverse, sub-parallel and oblique to the downvalley trend of the deposit. The distal 360 m of the deposit is thinner than the remainder.

The source of this remarkable runout deposit is a deep failure scar on the northern wall of the

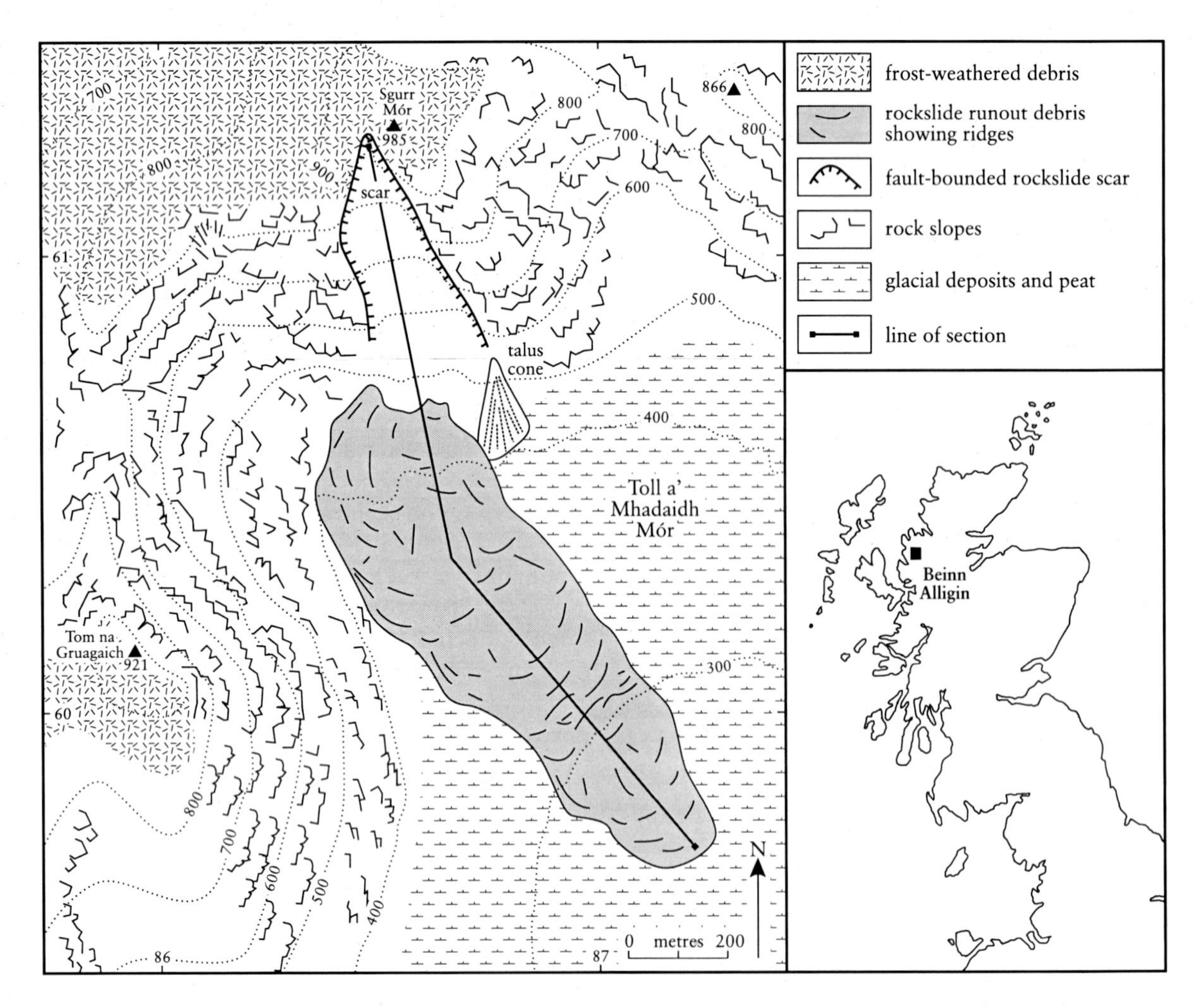

Figure 2.60 The Beinn Alligin rock-avalanche failure scar and runout deposit; map based on 1:25 000 aerial photographs and field mapping from adjacent summits.

Figure 2.61 The Beinn Alligin rock-avalanche failure scar and runout deposit photographed from the western end of the neighbouring mountain, Liathach. (Photo: C.K. Ballantyne.)

corrie. The scar is defined on both margins by near-vertical fault scarps up to 60 m high that converge near the summit of Sgurr Mór (Figures 2.60–2.62). The failure plane has an average gradient of 42°, and in comparison with the adjacent stepped rockwalls is relatively smooth, suggesting that failure was dominated by sliding. The planimetric area of the scar is *c*. 107 000 m^2, and its true area (taking gradient into account) is *c*. 144 000 m^2. Ballantyne and Stone (2004) estimated the volume of failed rock represented by the scar by interpolating the contours of the pre-failure rockface across the scar. Their calculations indicate that the failure involved 3.3–3.8 × 10^6 m^3 of rock, equivalent to a mass of 8.3–9.5 × 10^6 tonnes. At the east end of the foot of the scar, a small steep talus cone of very coarse rockslide debris abuts the main rockslide deposit on the corrie floor.

At the last (Late Devensian) glacial maximum, the site of the Beinn Alligin rock-avalanche was occupied by glacier ice to an altitude of *c*. 820 m, with the twin summits remaining above the ice as nunataks (Ballantyne *et al*., 1998a). The site was re-occupied by glacier ice during the Loch Lomond Stadial of *c*. 12.9–11.5 cal. ka BP, when a small corrie glacier, nourished in the corrie, fed a larger valley glacier to the south-east (Sissons, 1977).

Interpretation

The Beinn Alligin rock-avalanche has attracted considerable attention, particularly on account of the exceptionally long runout of debris along the corrie floor. This was initially explained in terms of downvalley transport of debris by remnant glacier ice at the end of the Loch Lomond Stadial, possibly in the form of a rock glacier (Sissons, 1975, 1976) or a supraglacial debris cover (Ballantyne, 1987a; Gordon, 1993). Whalley (1976), however, argued that the deposit could equally represent an excess-runout rock-avalanche, an interpretation also favoured by Fenton (1991). Ballantyne and Stone (2004) resolved the issue through cosmogenic radionuclide dating of the exposure age of the landslide debris. Three cosmogenic ^{10}Be ages obtained for the exposed upper surfaces of large boulders in the runout deposit yielded almost identical ages averaging 3950 ± 320 yr BP,

Figure 2.62 Rock-avalanche runout deposit with the failure scar in the background, showing the coarseness of the runout debris and the converging fault scarps that bound the scar. The vertical height of the fault scarp to the right of the failure scar increases upslope to nearly 60 m immediately below the summit of the mountain. (Photo: C.K. Ballantyne.)

implying that the rock avalanche occurred roughly 4000 years ago, and over 7000 years after the final disappearance of glacier ice. These findings demonstrate that the exceptional runout of the rock-avalanche debris cannot be attributed to transport by glacier ice, and must be related solely to landslide dynamics.

Causes of excess runout

Excess runout of rockslide debris has been defined as runout which exceeds that which might be expected from frictional sliding alone (Hsü, 1975). The expected travel distance of a rockslide can be estimated from the ratio H/L, where H and L are respectively the total vertical and horizontal distances between the top of the slide scar and the toe of the runout debris. For rockslides where runout distance is determined by frictional sliding, H/L typically has a value of about 0.6. The Beinn Alligin debris runout, however, yields an H/L ratio of 0.38, implying excess runout of *c*. 680 m (Ballantyne and Stone, 2004; Figure 2.63).

The phenomenon of excess runout appears to be related to the energy of the mobilized rock mass (Dade and Huppert, 1998; Kilburn and Sørensen, 1998), and the unusual long runout of the Beinn Alligin rock-avalanche in comparison with other rock slope failures in the Scottish Highlands probably reflects the exceptionally large mass and long vertical drop of failed rock at this site. Excess runout implies a reduction in the basal or internal friction of the mobile debris, and numerous theories have been proposed to account for this (Selby, 1993, pp. 316–19). At Beinn Alligin, the long-axis of the runout debris is oblique to that of the failure scar (Figure 2.63), implying that the mobile debris was re-oriented (by about 30°) during movement to follow the line of maximum slope along the corrie floor. When viewed from the mountain summit, it is clear that the debris surged a short distance up the slope opposite the failure scar then moved downslope along the corrie axis. Such re-orientation suggests that the debris moved as a grainflow or fragmental flow rather than a frictional slide. The abrupt margins of the deposit and the formation of arcuate transverse ridges are also consistent with this interpretation (Dawson *et al.*, 1986). Hsü (1975) argued that excess-runout landslides move as cohesionless grainflows driven by transfer of kinetic energy between colliding particles, and energy-balance calculations by Dade and Huppert (1998) are consistent with runout of densely concentrated debris in a state of granular flow. Alternatively, Kilburn and Sørensen (1998) have suggested that the upper parts of excess-runout landslides may be carried downslope as a result of fragmentation of clasts in a mobile basal boundary layer.

Cause of failure

Ballantyne (2003) and Ballantyne and Stone (2004) have suggested that the principal cause of the Beinn Alligin rock-avalanche was paraglacial (glacially conditioned) stress-release. Loading by glacier ice increases internal stresses

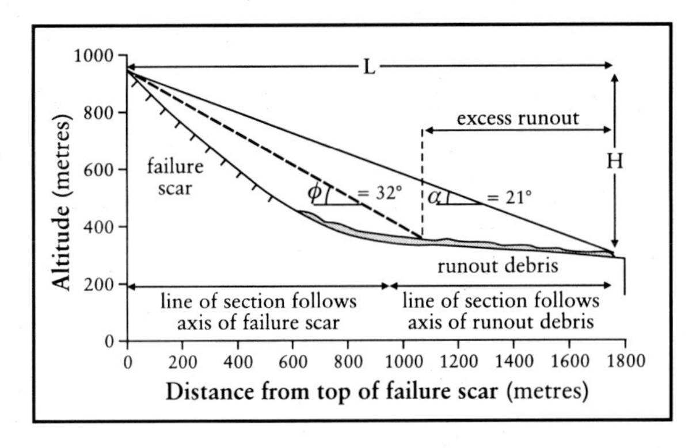

Figure 2.63 Long profile of the failure scar and runout tongue, projected along intersecting lines drawn down the axis of the scar and up the axis of the tongue (Figure 2.60). Excess runout is the difference between actual runout and the runout that is expected under frictional sliding alone. Under the latter condition H/L = 0.6 and $\tan^{-1}(H/L) = 32°$. At Beinn Alligin, H/L = 0.38 and $\tan^{-1}(H/L) = 21°$. Excess runout (Le) = (L – Htan 32°) = (1763 – 1080) m = 683 m, implying that the runout debris extended 683 m farther than would be expected under conditions of frictional sliding alone.

within underlying or adjacent rock masses, and part of the resulting ice-load deformation is stored as residual strain energy. During and after ice downwastage this strain energy is released, re-orienting the stress field within rock masses and resulting in the development of a region of tensile stress beneath rock slopes. Time-dependent relaxation of tensile stresses results in propagation of the internal joint network, ultimately reducing the strength of the rock mass to that of frictional contacts between joint-bound blocks. Depending on such factors as the steepness and height of the rockface and the orientation of joint-sets, this relaxation of internal stresses may lead to failure during or immediately after deglaciation, or, as in the case of the Beinn Alligin rock-avalanche, delayed failure conditioned by dissipation of residual stresses and consequent progressive joint propagation (Ballantyne, 2002a).

The actual trigger of failure, however, could have been an earthquake. Differential glacio-isostatic recovery during and after the downwastage and retreat of the last ice-sheet is believed to have re-activated ancient faults (Sissons and Cornish, 1982; Ringrose, 1989; Fenton, 1991). Although the diminishing rate of glacio-isostatic recovery following ice-sheet deglaciation implies a gradual decline in the magnitude of seismic activity, Davenport *et al.* (1989) estimated that the Western Highlands of Scotland may have experienced magnitude 5.0–6.0 events as late as *c.* 3.4 cal. ka BP. It is thus possible that even a fairly low-magnitude seismic event acting on a progressively weakening rock mass may have triggered failure on Beinn Alligin around 4000 years ago. The fact that the failure scar on Beinn Alligin is bounded by converging fault scarps (Figures 2.60 and 2.62) suggests that movement along one or both of these faults may have triggered the rock avalanche.

Wider significance

The Beinn Alligin rock-avalanche represents the largest documented rock slope failure on the Torridon sandstone terrain of north-west Scotland. Rock slope failures are rare in Torridonian rocks, despite the steepness of many corrie and valley rockwalls, suggesting that rock-mass strength in the gently dipping Torridonian arkosic sandstones is generally high. The closest analogue is a major rockslide at Creag an Fhithich, Baosbheinn (NN 856 676), 7 km north of the Beinn Alligin site. The Baosbheinn landslide, however, involved failure of an estimated 0.2×10^6 m^3 (0.5×10^6 tonnes) of rock (Ballantyne, 1986b), an order of magnitude less than that involved in the Beinn Alligin rock-avalanche. As a result, the mobilized rock mass at Baosbheinn had insufficient energy to generate excess runout, and was deposited as an arcuate boulder ridge at the foot of the failure scar. The exceptionally large scale of the Beinn Alligin failure reflects its unusual structural configuration, with two failure planes converging near the crest of a glacially steepened rock slope.

The Beinn Alligin landslide is also the largest known rock-avalanche in the Scottish Highlands and probably Great Britain. Other rock avalanches in the Highlands occur on a variety of lithologies, for example Tertiary basalts on Skye (see **Trotternish Escarpment** GCR site report, Chapter 6; Ballantyne 1991b), Devonian rhyolitic lavas and tuffs near Glencoe (see **Coire Gabhail** GCR site report, Chapter 4) and on Moine and Dalradian schistose rocks, for example at Carn Ghluasaid (NH 140 120) in Glen Cluanie, Beinn an Lochain (NN 217 083, Figure 2.11) in the south-west Grampians and Coire Ban (NN 618 447) in Glen Lyon. However, these rock avalanches are all roughly an order of magnitude smaller in terms of mass of failed rock, and runout distances are consequently much less (< 500 m), even where runout has been aided by moderate gradients. In a Scottish context, the phenomenon of pronounced 'excess runout' of debris is certainly best developed in the Beinn Alligin rock-avalanche.

The timing of the Beinn Alligin failure is also significant. Like The Storr landslide on Skye, which has been dated to *c*. 6.5 ± 0.5 cal. ka BP (Ballantyne *et al*., 1998b), the Beinn Alligin rock-avalanche occurred several millennia after deglaciation, demonstrating that major (paraglacial) rock slope failures were still occurring during Mid- and Late Holocene times in the Scottish Highlands and Hebrides. The long delay between deglaciation and failure at these sites suggests that the potential for major cataclasmic rock slope failures generated by deglacial unloading may not yet be exhausted in the Scottish Highlands.

Conclusions

Approximately 4000 years ago, roughly 9 million tonnes of rock became detached from the northern rockwall of a corrie (Toll a'Mhadaidh Mor) on the south-east side of Beinn Alligin and cascaded on to the corrie floor. The exceptionally large mass and height of fall of this landslide provided sufficient energy to cause boulder-sized debris to move as a grainflow or fragmental flow that surged up the opposite side of the corrie then moved downslope along the corrie floor. The debris came to rest as a tongue-shaped boulder deposit 1.25 km long, up to 380 m wide and up to 15 m thick. The site of rock failure is represented by a deep scar on the cliff-face, bounded on both sides by near-vertical fault scarps up to 60 m high that converge near the summit of the mountain.

The Beinn Alligin landslide is classified as an excess-runout rock-avalanche and is probably the largest and finest example of its type in Scotland. It is thought to have occurred due to 'rebound' or stress-release in the rock after it emerged from under the weight of the last ice-sheet. Stress-release resulted in the opening of joints (discontinuities) in the rock, so that the cliff became progressively weaker through time. The landslide may, however, have been triggered by movement along one or both of the faults that border the failure scar, causing the collapse of rock already weakened by stress-release. The Beinn Alligin rock-avalanche is the largest slope failure in Torridon sandstone bedrock. Its occurrence several millennia after final deglaciation suggests that the effects of unloading of rock from under the weight of the last ice-sheet may have continued to influence mountain-wall stability throughout most of the post-glacial period, and may persist to the present day.

Chapter 3

A mass-movement site in Silurian strata

R.G. Cooper

INTRODUCTION

Silurian strata are not well known for mass movements in Britain. Jones and Lee (1994) list only 322 sites, of which 86% fall into the 'unspecified' type category. Of those that have had their type identified, 30% (13) are compound, 28% (12) are translational, 14% (6) are single rotational and 12% (5) are topples. The site selected for the GCR is none of these. It was, rather, selected as an example of a site over which there is little agreement. Its location in an area of Silurian strata (Figure 3.1) rather than hard rock of some other geological system is, largely, incidental, as it is a product of the denudation system rather than a feature determined by age or rock-type.

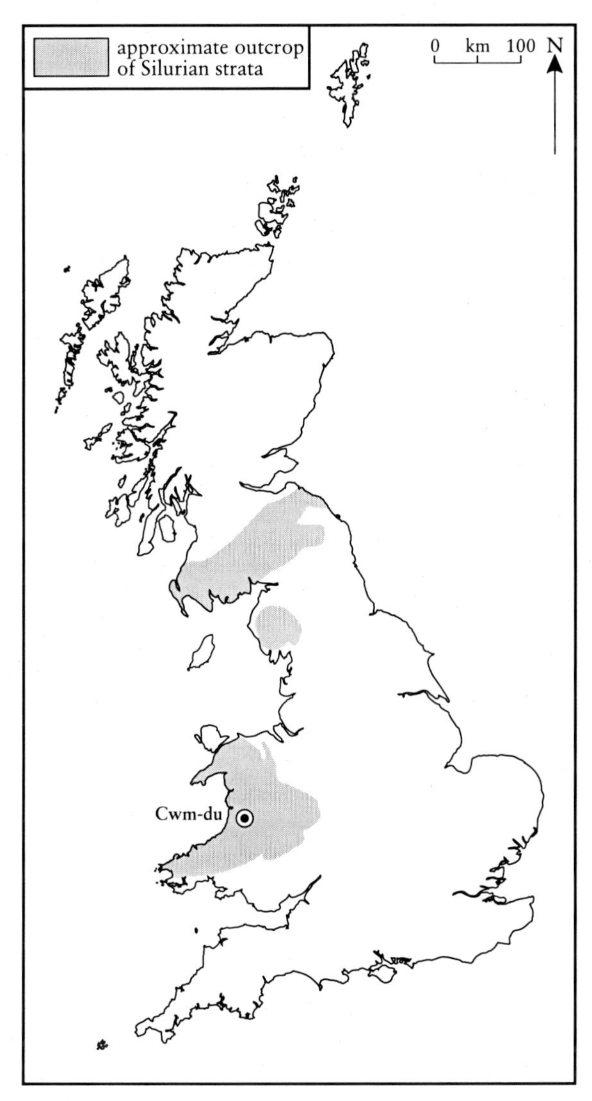

Figure 3.1 Areas of Silurian strata (shaded), and the location of the Cwm-du GCR site, described in the present chapter.

CWM-DU, CWMYSTWYTH, CEREDIGION (SN 813 740)

Introduction

Cwm-du is a large, north-facing hollow on the south side of the Afon (river) Ystwyth. It is fronted by a large fan of debris showing some stratification and a number of physiographical 'terraces'. Several types of mass movement have been involved in the production of these forms. Opinions differ as to their genesis, and concerning the relationship of the fan to the hollow which, in part, it occupies.

Some sites have been included in the GCR for mass movements because they have become classic sites for research and learning, and are, relatively speaking, well understood (e.g. **Folkestone Warren**, **Mam Tor**, **Black Ven**). Cwm-du is selected for a different reason: it is not well understood, and is the subject of continuing disagreement about the role of the mass-movement processes at work. Mass movements are certainly envisaged as having been instrumental in its development, but their precise nature, and their relationships to the surface forms present, remain unclear.

Description

General description

The large, north-facing hollow of Cwm-du, on the south side of the Afon Ystwyth (Figures 3.2 and 3.3) is 2 km east of Cwmystwyth. It is in the Aberystwyth Grits, a series of greywackes and mudstones of the Llandovery Series. The floor of Cwm-du is elongated like the floors of many glacially eroded basins (500 m × 300 m). The hollow has a spectacular backface (Graig Ddu) more than 50 m in height. It is 180 m lower in altitude than typical glacial hollows in the area and it shows none of the erosion features associated with them. It has a fan-shaped, terrace-like front without an enclosing moraine or rampart. It was first described by Keeping (1882) and has been studied in detail by Watson (1966, 1968, 1970, 1976) and Watson and Watson (1977).

Instead of a moraine, which would be expected if it were a glacial cwm, the basin of Cwm-du is fronted by a drift scarp 18 m high in the centre, which resembles a moraine when viewed from downslope but is in reality the front

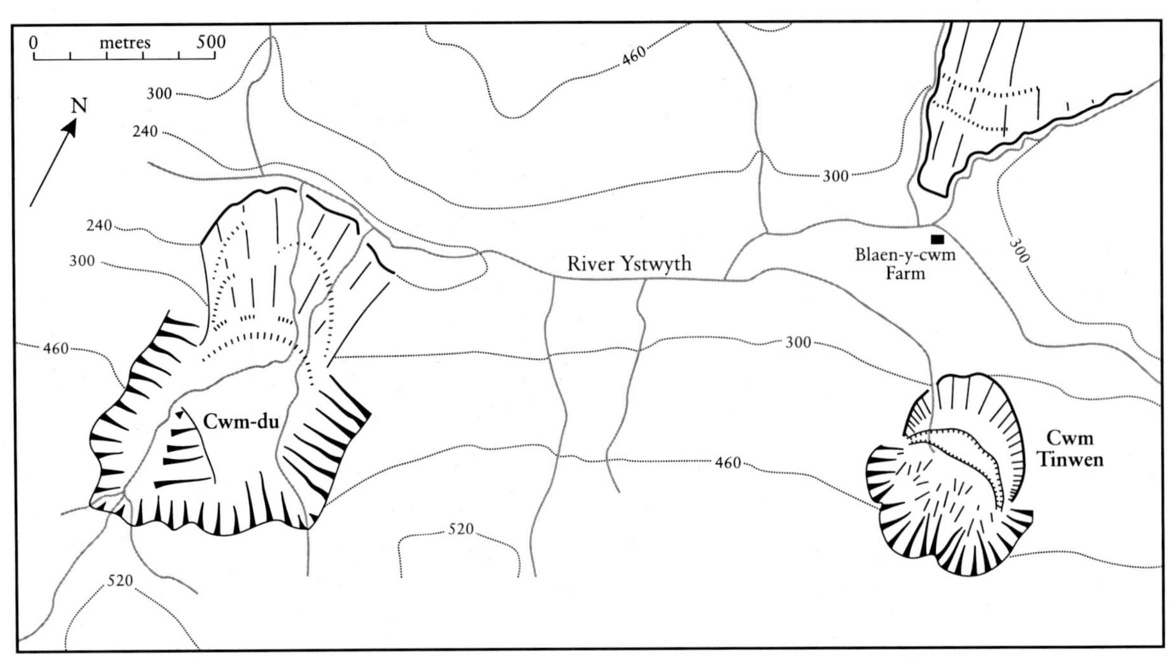

Figure 3.2 The location of Cwm-du in the Upper Ystwyth valley. Contours are in metres. After Watson (1966).

Figure 3.3 View across the Upper Ystwyth valley looking southwards into Cwm-du, up the extensive 'drift' or landslip debris incised by an 18 m-deep gully. (Photo: S. Campbell.)

of a terrace consisting of head deposits. The main stream in the hollow, Nant Cwm-du, has cut a deep gully parallel to this front but instrumental levelling in the area of the profile J–K (Figure 3.4a,b) shows that the top of this scarp is lower than the floor of the hollow (Watson, 1966). The gully reveals up to 18 m of head without the bedrock being exposed and there is a smooth slope rising up to the headwall of the hollow.

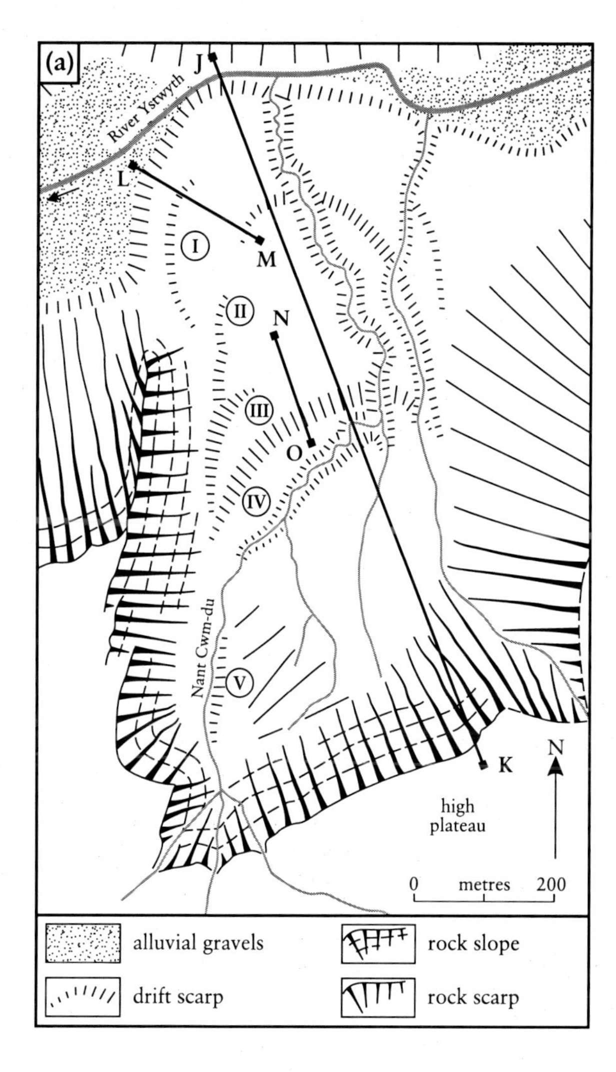

Figure 3.4 ◂(a) Field survey of Cwm-du and its associated debris fan. ▾(b) Surveyed sections, J–K, M–L, O–N. Scarps discussed in the text are numbered I–III. Scarp IV lies at the mount of the Cwm. Scarp V on the map is a protalus in the south-west corner. The dashed line is the profile of the main valley-side. After Watson (1966).

The long-axis of the hollow is south-west–north-east and its floor falls in the same direction. The angle of elevation from the front of the drift platform at the stream exit to the top of the backwall in its south-west corner is 13°. The smooth curve of the floor of the lower part of the hollow is replaced towards the head by a drift accumulation that fills the south-west corner of the hollow. The streams falling steeply into the hollow have filled in the area behind this accumulation at its southern end with bouldery alluvium, but downstream (at profile R–S, Figure 3.5) it is seen to be composed of head.

Surface topography

The south-west corner of the hollow contains an accumulation of drift, denoted 'Stage V' by Watson (1966) (see Figure 3.4). The drift encircling the basin of the hollow (IV, see Figure 3.4) appears to be the highest and most continuous member of a series that is developed across the

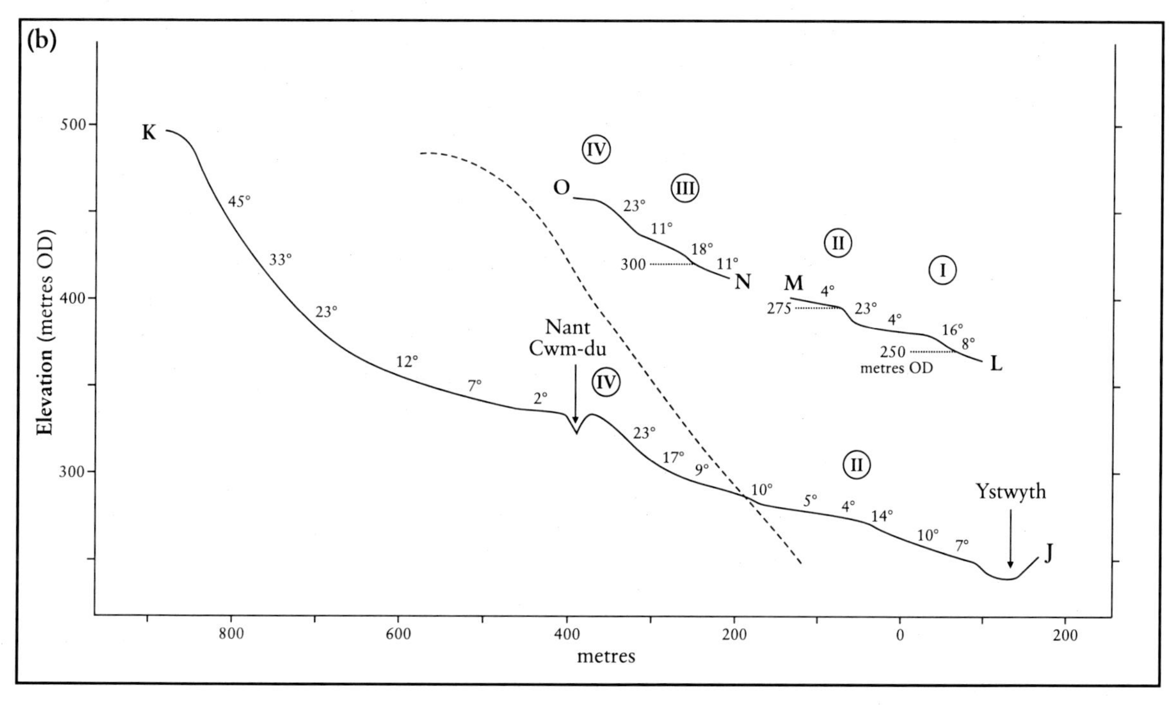

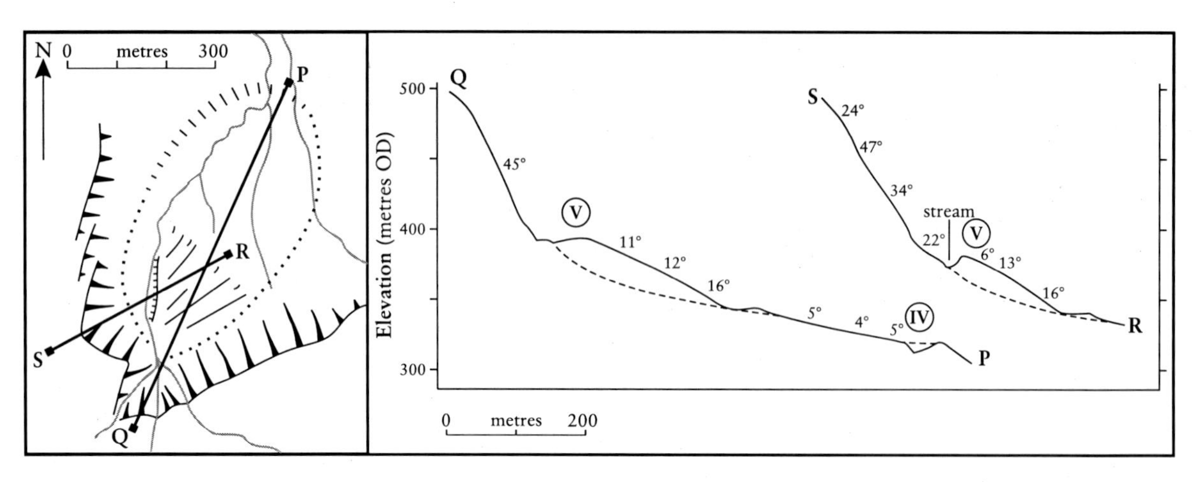

Figure 3.5 Profiles (P–Q, R–S) through Cwm-du. Roman numerals, IV, V, correspond to Figure 3.4. The dashed lines reconstruct the floor of the cwm. After Watson (1966).

surface of the drift fan below it. This series, marked 'I', 'II' and 'III' on Figure 3.4, gives the fan a stepped profile. These steps, or scarps, are not continuous across the fan so profile J–K does not give the complete series; profiles L–M and N–O restore the missing steps in the general picture. The upper scarp of the series, III, is parallel to the terrace front of the hollow, IV, and, like it, falls in elevation towards the east. Unlike IV, it appears to have been worn down at two points as if erosion had lowered the top of the scarp. Scarp II extends as a continuous scarp to the east of the Nant Cwm-du gully except where it is cut by the unnamed stream. West of this stream it appears to have been eroded as in the case of scarp III. Scarp I exists only on the flanks of the fan; in the central area the fan gives the appearance of having been built up so that there is a continuous slope from scarp II to the limiting bluffs of the fan, which are due to erosion by the Afon Ystwyth.

Deposits

The drifts in Cwm-du consist of two types of head (Watson, 1966). One is a tough bluish-grey silty deposit containing angular and sub-angular rock-fragments of all sizes from fine gravel to great boulders. The other is a yellowish-grey, loose deposit, containing similar debris but with a smaller proportion of finer material so that it may often be described as a muddy angular gravel. It often shows rusty mottling and, in the more open beds, manganese staining.

The exposures in the gully of Nant Cwm-du (15–18 m deep where it leaves the hollow) suggest that these beds form the whole of the deposits, occurring in distinct layers between 0.3 m and 1.0 m thick. The bluish-grey type is typical of the solifluction deposits on the greywackes and mudstones of the region, and the yellowish-grey type is probably basically the same except that it has suffered some degree of washing during deposition.

Stage V has many boulders scattered over its outer slope, and exposures show the two types of head; on the profile line R–S (Figure 3.5) 3.5 m of the loose yellowish-grey type overlies 0.6 m of the compact bluish-grey type, above 5.2 m of talus.

The main exposures of the deposits making up the Cwm-du fan are shown in Figure 3.6. The interbedding of tough bluish-grey head and loose yellowish-grey head seen in the floor of the hollow also occurs in the fan just behind scarp II at exposures 2 and 3. At other points, for example 11 m upstream of exposure 3, the west side of the gully shows only the blue-grey head and at exposure 5 only the blue-grey head is seen for 7 m above the stream.

In front of scarp II, small-calibre water-laid gravels (clast long-axes less than 5 cm) are interbedded with thin layers of grey silt at the top, from the lower 3 m of exposure 1. Again, in front of scarp III, exposure 4 shows blue-grey head overlain by 2 m of similar small gravels capped by 0.5 m of sand and silt, on top of which is blue-grey head. Exposure 5, in front of scarp IV, shows 0.7 m of water-laid sands, gravels and silts resting on the blue-grey head of scarp III. These are overlain by a stony, bouldery yellow-grey head (in places a muddy gravel)

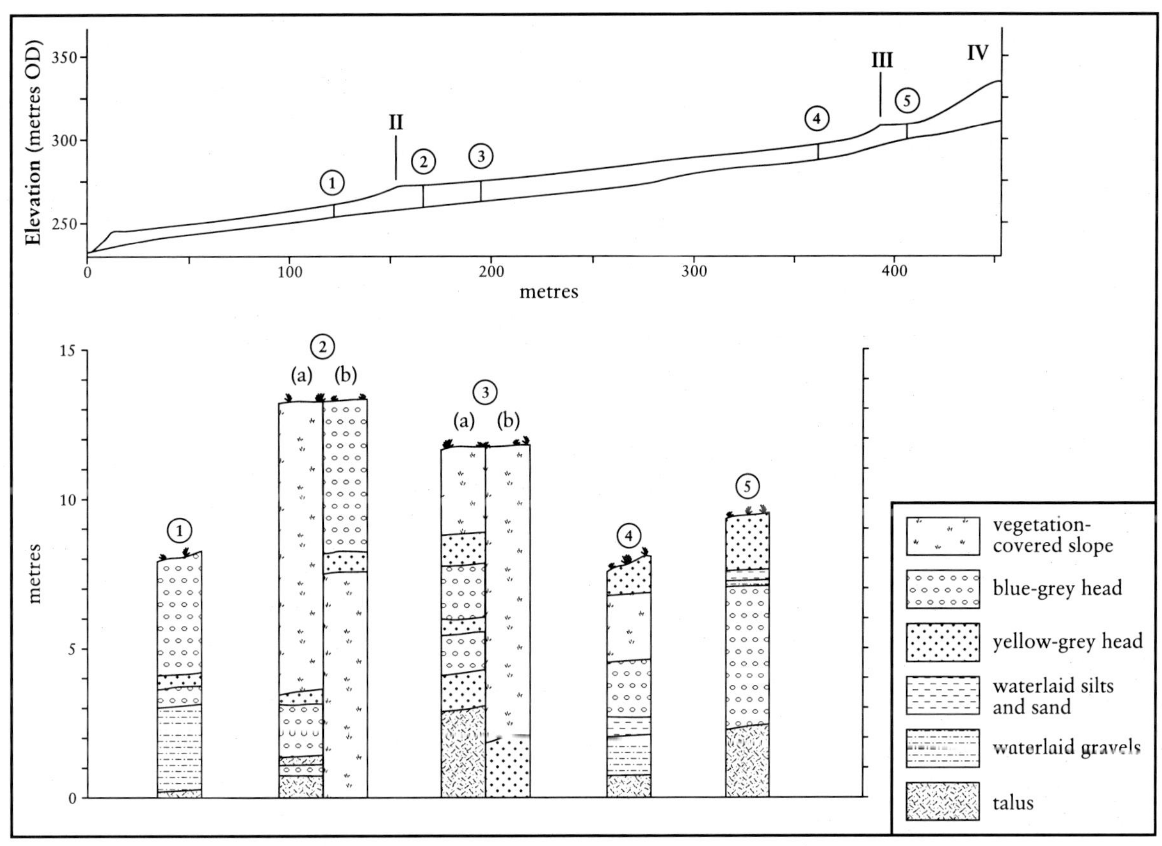

Figure 3.6 Exposures of the deposits in the Cwm-du fan. The profile is the west bank of the gully projected onto section J–K. After Watson (1966).

which thickens when followed upstream (Figure 3.7). Exposure 4 at the top of the sequence also shows a gravelly head, roughly stratified parallel to the present surface and showing similar conditions at stage II to those of stage IV.

An exposure new in 1974 (Watson and Watson, 1977), almost at the mouth of the gully, consists of grey clay and in much of it, especially near stream level, the maximum projection planes of the stones are tangential to the outer limit of the fan, indicating pressure along the fan axis. A bed of clay, sand and small gravel is torn out along a plane that curves upwards and outwards, suggesting thrusting.

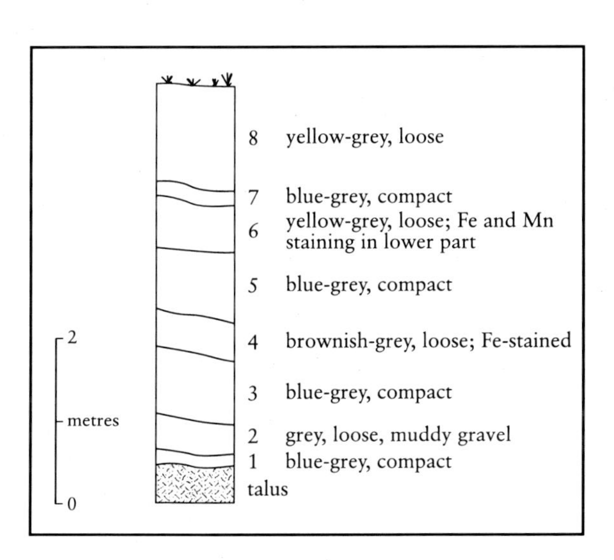

Interpretation

Watson (1966) considered the hollow to be due to nivation associated with inert masses of névé and ice that were not thick enough for plastic deformation and flow. This conclusion is supported by many of his detailed observations on the deposits present, since it provides interpretations of their genesis. For example, he attributes the yellowish tinge of the yellowish-grey deposits as being due to post-glacial weathering accompanying a more ready

◂**Figure 3.7** The sequence of head deposits in Nant Cwm-du gully, 120 m south-west of profile J–K. The scale begins at stream level. After Watson (1966).

percolation of water. He suggests that the water-laid sand and silt layers in the succession in front of scarp III might be older than the step behind, having been laid down during a recession phase when summer melting was more pronounced, and having then been overwhelmed by the solifluction deposits of the snowpatch of the succeeding cold phase. They might also represent meltwater deposits laid down while the step behind was being built up. The former explanation, he continues, supports the sequence at exposure 5, in front of scarp IV, where the bouldery head that tops the sequence would appear to be material that has been partly washed by snowmelt as it flowed down the face of the scarp.

The 1974 exposure caused him to revise his view. Interpreting it as a glacial deposit, he suggests that glacial ice played a part in the earlier stages of the development of the fan, incorporating this view by suggesting that an ice advance from the hollow was followed by a period of nivation that is responsible for the present surface form (Watson and Watson, 1977).

It is worth noting that he cites Flint (1957) to the effect that for slow plastic deformation 'the minimum thickness of ice and firn required is not known'. He envisaged a large snowpatch, covering the interior of the hollow from the backwall to the crest of scarp IV. This snowpatch would have sloped from the south-west corner towards the present stream exit. The upper limit of the snowpatch would be lower than the top of the backwall so that the surface of the snow would be less steep than the angle of 27° between the top of the drift accumulation which fills the south-west corner of the hollow, and the head of the backwall, measured along line R–S (Figure 3.5). The absence of a protalus suggests that no superficial debris reached the foot of the gently sloping snow surface in Cwm-du.

Botch (1946) produced a block-diagram (Figure 3.8) of the features associated with typical snowpatches in the Ural Mountains which, unusually, shows debris accumulating *under* the lower part of a snowpatch and moving on subaerially down the slope as a series of solifluction terraces. Watson (1966) suggests that the exposures of the floor of the hollow (stage IV) may have been built up in layers in the way shown in Botch's block-diagram. The exposures on the outer edge of the Cwm-du hollow may represent deposits laid down at the margin of the snowpatch, which might account for the increased washing of some of them, a further parallel with Botch's block-diagram.

The building up of such a platform of drift could be due to the fact that the ground below the snowpatch is affected by summer thaw only

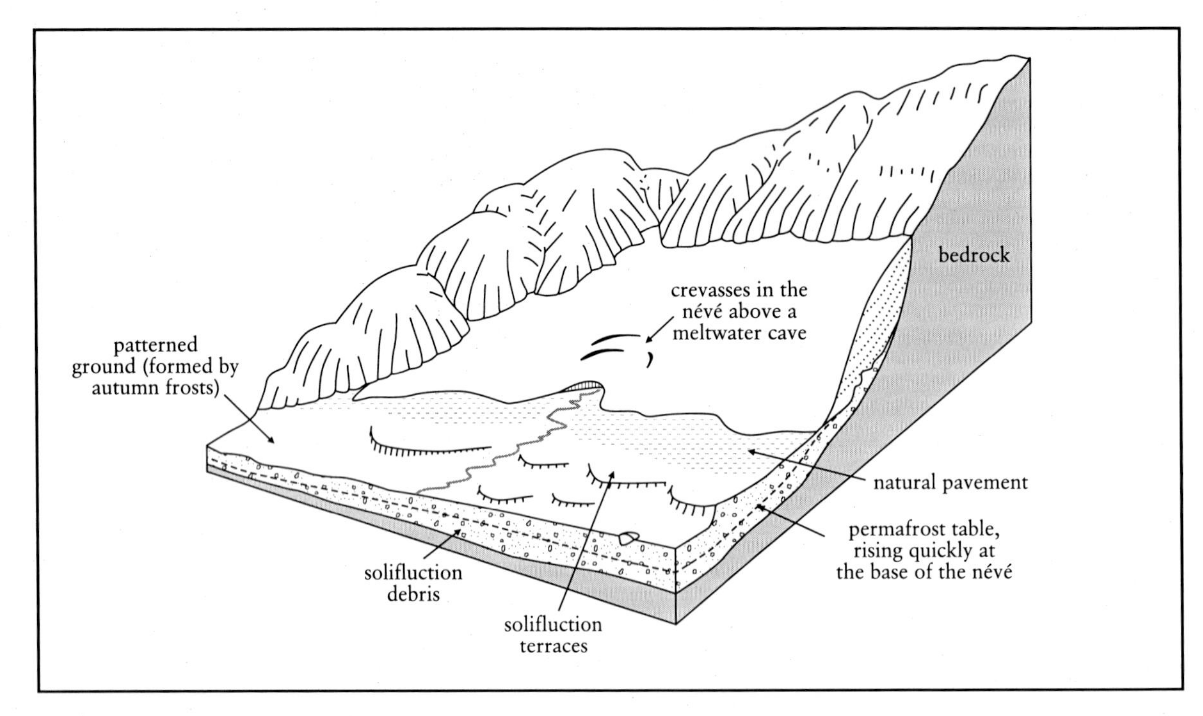

Figure 3.8 Block-diagram of a snowpatch during the melt season. Based on Botch (1946) and Watson (1966).

to a very shallow depth and that as the debris derived from the weathering of the backwall moves as a solifluction layer beneath the snow-patch it tends to thicken as the gradient lessens. In winter this muddy accretion is frozen to the permafrost below and never thaws out again fully so that summer after summer it is added to. The abrupt limit to the deposit, represented by the scarp 18 m high, seems to imply that the actual front of the snowpatch fluctuated relatively little, otherwise the thawing out of the uncovered ground must have been accompanied by soil flow.

The similarity between Botch's block-diagram and the situation at Cwm-du suggests that the steps of the fan may be solifluction terraces, but Watson (1966) came to believe that these scarps, I, II and III, are similar in origin to scarp IV at the mouth of the hollow and that they mark earlier snowpatch limits.

His argument against the view that the scarps are sub-snow solifluction terraces is their arrangement in plan. They do not form a series concentric with scarp IV but appear to enclose a series of lobes that have a progressively changing axis. The lobe enclosed by scarp II has an axis running west of north (about 350°); the axis of the higher lobes swings clockwise until in the hollow (the area enclosed by scarp IV) it runs NNE (about 025°). This would be consistent with an area of snow emerging from a gully which was being extended towards the south and west as freeze-thaw made its maximum attack on the backwall on these sides (Figure 3.9). However,

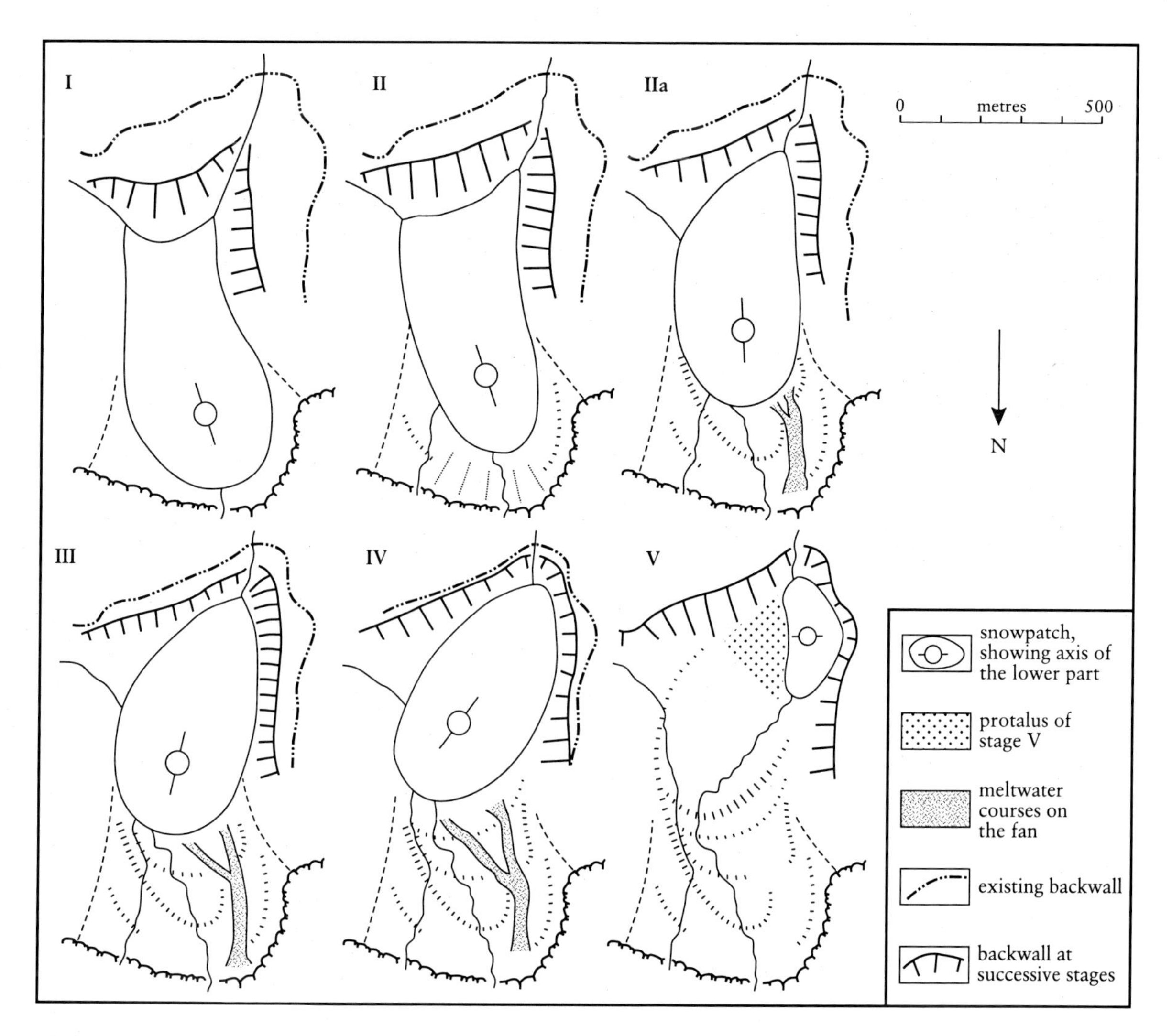

Figure 3.9 The evolution of Cwm-du. Step IIa is shown as marking a snow limit. If these steps represent climatic pauses, Watson (1966) suggested that research elsewhere in Wales may help to decide its status. After Watson (1966).

while he did not ascribe the surface form of the fan to solifluction, he believed that the 18 m minimum depth of head, and the smooth slope rising across the hollow from terrace IV to the headwall, point to the deposit as a whole being due to solifluction, accumulated beneath the putative snowpatch.

The position of the stream, Nant Cwm-du, on the centre line of a convex fan that is not of its own construction but predominantly of head, favours the view that the scarps of the fan mark snowpatch limits. Inside the hollow, Nant Cwm-du and the stream to the east occupy positions which would have been sub-marginal to the snowpatch of stage IV, and leave it at its lowest point, where one would expect meltwater to escape. The present stream appears to be a direct descendent of such a meltwater stream, having become entrenched in later times. Downstream of scarp IV the stream passes through the lowest point of each crescentic step so that the stream course cuts across the fan to a central position where it joins the Afon Ystwyth.

Watson (1966) made an attempt to reconstruct the evolution of Cwm-du and its fan by advancing the backwall of the hollow to compensate for the building of the fan, along the profile J–K of Figure 3.4. One of the difficulties he found was the fact that the hollow was not extending along the axis of the fan. Only a fraction of the material laid down at stage IV came from the backwall on the profile J–K. The bulk of it came from the south-west corner of the hollow. This is less true of the earlier stages, but the reconstruction in any case is only very approximate. The 'initial' profile shows a nick-point just above 275 m above OD. Though no rock is seen in the stream bed, the Nant Cwm-du gully is shallowest at this point, only 8 m deep compared to more than 16 m upstream and downstream of this, and the steepest stretch of the present stream profile in the fan occurs just below this.

From the elevation of the High Plateau here and the position of scarps I, II and III, it seems impossible that a snowpatch extending to them could fall into the steeply sloping class. On the reconstruction shown in Figure 3.10 the angles from the rim of the backwall for scarps I, II and III are similar to that for scarp IV, which is 18°. This is in harmony with the fact that each suggested snow limit is marked by a terrace front and not a protalus rampart.

Estimates of the thickness of snow at these stages, on the basis of Figure 3.9, show that the maximum thicknesses, if the hollow was filled to the top of the backwall, would be 82 m, 82 m and 72 m respectively, for stages I, II and III (75 m for stage IV). In this respect, the problem of stages I, II and III is the same as that of stage IV: the question of the thickness reached by snow and ice before it begins to behave as a hollow glacier.

With this escape of meltwater from the snow limit of stage IV, may be associated the fluting of the face of scarp IV suggesting wide shallow gutters leading down to scarp III. This drainage probably escaped by a shallow channel that breaches scarp II. It may be pointed out that this drainage and the deformation of the scarps is developed on the western, 'warmer' side of the fan.

The exposures in Nant Cwm-du gully show that the material in the fan has been laid down in several stages and that the scarps are not a series of terraces formed contemporaneously with the build-up of the platform of head at stage IV. The water-laid gravels and silts seen in three places, each a short distance from a scarp, suggest that the building of the steps may have followed on milder climatic interludes.

The drift accumulation, V, which fills the south-west corner of the hollow, has the stream passing behind it, between its rear and the backwall of the hollow, in a space which the stream has partly filled with bouldery alluvium. This suggests that after the snowpatch had disappeared from the hollow, it re-formed, filling only the south-west corner, and built up this drift as a protalus. The angle of elevation from the top of the protalus ridge to the rim of the backwall on profile R–S is 27°, indicating that this final snowpatch belonged to Botch's steeply sloping class.

The question arises as to whether the volume of the 'fan' is sufficient to account for the volume of material removed to create the hollow. Clearly, Watson believed this to be the case, so justifying his classification of Cwm-du as a 'nivation cirque' (i.e. a hollow formed by nivation). There is, however, a difference between a hollow formed by nivation, and a hollow (of glacial origin) in which nivation has taken place. There seems little reason to doubt Watson's argument that the fan at Cwm-du is a periglacial deposit, particularly in view of his many carefully reported observations at the site. His reconstruction of events may well be correct, but there is another possibility.

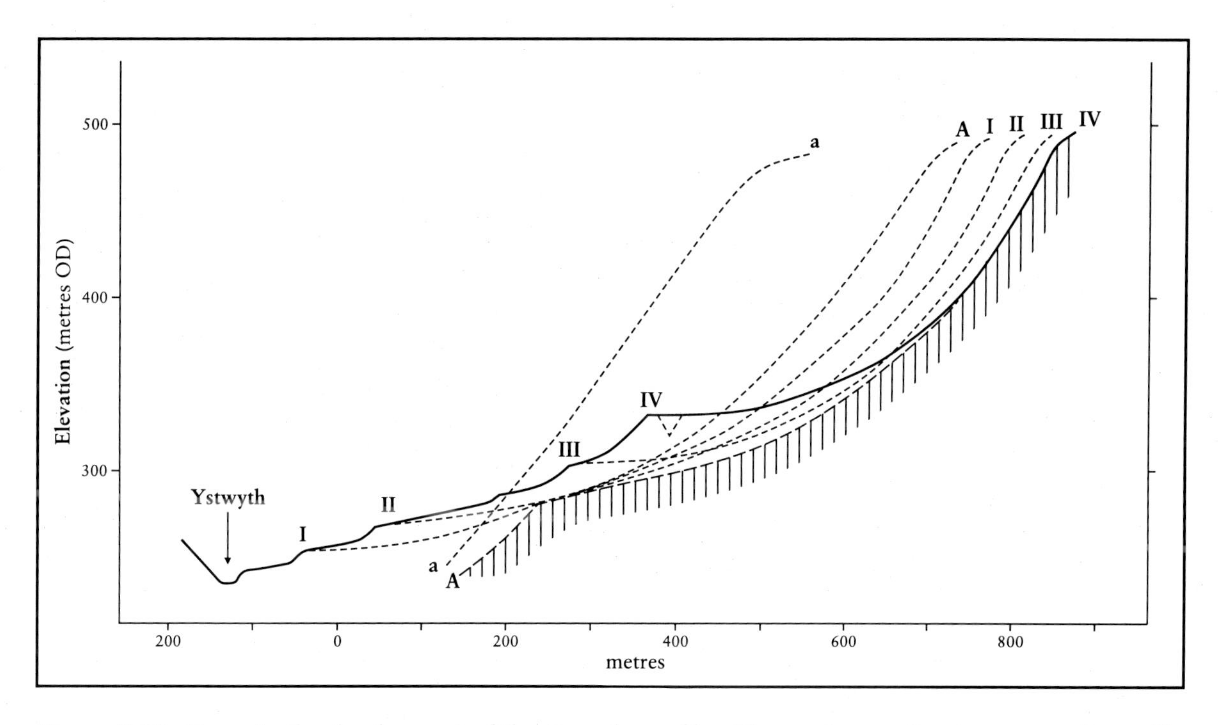

Figure 3.10 Stages in the development of the Cwm-du profile. The continuous line is a composite profile along section J–K. Bedrock is in part hypothetical. 'A' represents the initial profile; I–IV represent backwall positions corresponding to steps in the fan; 'a' is the profile of the main valley-side.

Thorn (1976, 1979, 1988) has questioned the efficacy of nivation alone to produce such large landforms. He has shown (Thorn, 1976; Thorn and Hall, 1980) that nivation processes operate at extremely slow rates, suggesting that large features may be in some sense 'inherited', and merely modified by subsequent nivation and frost-action processes. Thorn's work was concerned with large cryoplanation terraces, but similar considerations apply to the notion that nivation may form or develop large hollows ('nivation cirques').

Cwm-du seems a prime candidate for the situation envisaged in general terms by Thorn (1976): in this view, Cwm-du may be seen as an 'inherited' glacial hollow that has been modified by nivation. There is little evidence from which its original genesis may be deduced. It may be inferred that like some other hollows in the hills of mid-Wales, it originated as a glacial cirque. However, there is much evidence that hollows like this may be due to landslides.

In a detailed critique, Ballantyne and Harris (1994) point out that Watson's argument rests on the lack of evidence for glacial erosion in the hollows that he studied (Cwm-du and the nearby Cwm Tinwen). Instead, he invoked frost-sapping of headwalls, meltwater transport, sub-nival solifluction and the movement of debris across steep snowpatch surfaces to explain their development. Both hollows are fronted by thick drift accumulations up to 18 m deep. The drift is a tough diamicton with occasional looser layers deficient in fine-grained materials, and was interpreted by Watson as a solifluction deposit partly modified by meltwater eluviation. Watson's interpretation has attracted scepticism (see discussion in Watson, 1966). A particular difficulty (Ballantyne and Harris, 1994) is posed by the thickness of snow or névé required to produce a slope on the snow or névé steep enough to permit the sliding of debris over its surface. With gradients greater than 23° (Watson, 1966) the thickness would be such that unless densities markedly lower than that of ice are assumed for the accumulated névé, the basal shear stresses must have exceeded 1 bar, so that the 'snowpatch' would have moved as glacier ice. Although the snow surface gradients indicated by Watson are maximal, lower gradients would not have permitted movement of debris over the snow surface in the manner envisaged by Watson. Observations on active protalus ramparts (Ballantyne, 1987b) suggest that gradients of 23° and 25° are too low to permit this to occur (Ballantyne and Harris, 1994).

However, these strictures really apply only to that part of Watson's notion that the whole of Cwm-du owes its origin to nivation. The origin of the fan, the terraces upon its surface, and the distribution and properties of the head require explanation. At the very least, Watson's observations on the fan and its surface form and composition, pose a problem which is not solved by demonstrating that nivation cannot have been a major process in its genesis.

Conclusions

Two main means of mass movement have been invoked to explain, at Cwm-du, the transport downslope of material arriving as debris from rockfalls at the backwall of the hollow. The chief type of mass movement suggested by Watson (1966) is solifluction, unusually suggested to have taken place under a covering of snow. He also suggested sliding of debris over a snow surface; this has been proved not to be possible in this particular case, which leaves the apparent protalus in the south-west corner of Cwm-du unexplained. A further possibility is that the terraces could have reached their present positions as a result of landsliding, a process certainly capable of producing the features shown in Figure 3.9. This site's conservation value relies upon its protection so that it remains available for future scientific research.

Chapter 4

Mass-movement sites in Devonian strata

INTRODUCTION

R.G. Cooper

In Great Britain 1042 landslides are reported in Devonian strata by Jones and Lee (1994), of which almost three-quarters (75%) are of unspecified type. However, of those for which the type has been identified, the largest group is rockfalls, at 41%, followed by translational slides and debris flows at 31%. Two GCR sites have been selected in Devonian strata, **Coire Gabhail** in Scotland, and **Llyn-y-Fan Fâch** in Wales (see Figure 4.1).

Rockfalls are a commonly recognized form of mass movement, but there is usually little evidence of the processes that have brought them about. There are several possibilities, including:

(1) Frost-wedging in rock fissures: could cause a detached block to stand proud of a cliff-face to such an extent that its centre of gravity causes sliding under its own weight and eventual falling as the block becomes free of the restraint of the adjacent rock body. Dislodgement of one fragment could provide conditions for fragments above it, and possibly the whole cliff, to collapse.
(2) Weathering: could cause a decrease in cohesive strength so that a cliff-face could disintegrate into a large number of fragments.
(3) Undercutting of the cliff-face: could remove basal and lateral support.

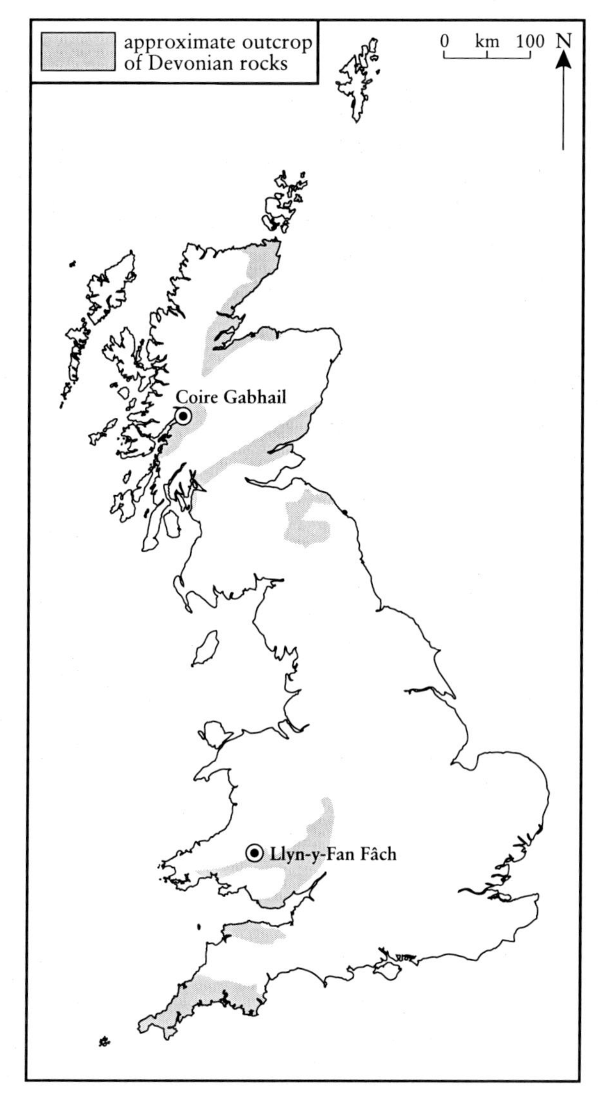

Figure 4.1 Areas of Devonian strata (shaded), and the locations of the GCR sites described in the present chapter.

Either way, the resulting form consists of a pile of fragments described as *talus*.

Where rockfall occurs on a rectilinear slope, a *talus sheet* is produced at its foot. However, where rockfall repeatedly takes place into a rock gully, the gully will constrain the path of the fallen debris, giving rise at the gully's mouth to a *talus cone*. Laterally adjacent cones may coalesce to form a sheet.

Ballantyne and Harris (1994) define a *talus slope* as a steep valley-side slope formed by the accumulation of debris at the foot of a rockwall. The term 'talus' is used to denote both the slope and its constituent material. *Scree* is taken to refer to any slope that is covered in coarse debris. Therefore 'scree' includes 'talus', which is more closely defined.

Rockfall talus characteristically may display fall-sorting, a process whereby larger fragments come to rest farther downslope than smaller fragments (Ballantyne and Harris, 1994). Grain size is commonly bi-modal, which can result in talus sliding, and sorting into garlands and stripes. This in turn imparts another characteristic to talus slopes: in profile, a talus slope lessens in steepness at its foot. This is because at the foot the talus slope consists of the largest fragments, which protrude most from whatever lies beneath the surface (usually talus from previous falls), although it can also be because the debris runs out and rests on the valley floor. This characteristic is absent from the famous talus of Wasdale, in the English Lake District, because the lower parts of these are underwater in the lake that occupies much of the glacially over-deepened valley.

Sliding may take place on a talus slope – dry and granular sliding by debris slide, or avalanche gullying. A widespread process by which debris is re-distributed on a talus slope is *debris flow*. The term is primarily used to refer to rapid downslope flow of poorly sorted debris mixed with water, but it is also used to refer to the suite of landforms produced by an individual flow. The overall effect of debris-flow activity on talus is that material is eroded from the upper part of the talus slope and deposited near or beyond the talus foot. In this way the overall gradient of the talus slope is reduced, and a long, sweeping basal concavity is produced. Debris flows disrupt the fall-sorting of undisturbed rockfall talus. Repeated flows on the same track produce a gully in the upper part of the talus slope, which is continued downslope by two parallel debris levées that mark the path of the flow and terminate downslope in one or more lobes of bouldery debris.

The GCR sites (Figure 4.1) chosen to represent some aspects of these features include the site of a very large post-glacial rockfall, **Coire Gabhail** in Glencoe, Scotland, for which the evidence is a talus of great coarseness, and **Llyn-y-Fan Fâch** in Carmarthenshire, Wales, where there are debris flows that have been investigated as a potential exemplar of an alternative explanation of the steepness of talus slopes (Statham, 1976).

COIRE GABHAIL, HIGHLAND (NN 166 557)

C.K. Ballantyne

Introduction

Of over 500 rock slope failures identified in the Scottish Highlands (Ballantyne, 1986a), relatively few involve total disintegration of the collapsed rock-mass. A double rock-avalanche site in lower Coire Gabhail, Glencoe (Figure 4.2), is of outstanding interest in several respects. First, the initial failed rock-mass completely disintegrated and accumulated as a massive talus cone resting on the valley floor. Second, a smaller talus cone partially overlaps the main cone, suggesting that a later rock-avalanche occurred after the initial failure. Third, the Coire Gabhail rock-avalanche site represents the largest failure on the Devonian volcanic rocks of the Western Highlands. Finally, the rock-avalanche debris has completely blocked the valley, forming a natural sediment trap and causing the accumulation of coarse alluvial deposits upstream (Werritty, 1997).

The larger Coire Gabhail rock-avalanche is also one of only a handful of Scottish rock slope failures to have been dated using cosmogenic isotope dating techniques.

Description

Setting

Coire Gabhail, popularly known as the 'Lost Valley', is a hanging valley on the south side of Glen Coe (Figure 4.2). The valley mouth is flanked by two truncated spurs, Geàrr Aonach (692 m) and **Beinn Fhada** (811 m). Gently dipping rhyolitic lavas of the Glencoe Volcanic Formation underlie the valley sides, forming steep stepped rockwalls, but the valley axis follows the line of a porphyritic dyke (Bailey and Maufe, 1960; Moore and Kokelaar, 1998). The area was completely over-ridden by westward-moving ice at the last (Late Devensian) glacial maximum (Thorp, 1987).

During the Loch Lomond Stade of *c.* 12.9–11.5 cal. ka BP, Coire Gabhail nourished a tributary valley glacier that fed the Glen Coe glacier; according to Thorp (1981), the surface of the Coire Gabhail glacier descended from 900 m at the head of the valley to about 560 m at the valley mouth. As the rock-avalanche runout debris at this site has not been modified by glacial erosion or transport, failure occurred after retreat of glacier ice at the end of the Loch Lomond Stade.

Failure scars

Two distinct failure scars are evident. Both are developed in flow-laminated rhyolites of the Upper Etive Rhyolite. The larger (south-west) scar comprises a steep 70 m-high backwall that rises above a bedrock ramp to the crest of the slope at 640–650 m (Figure 4.2). It has a maximum (across-slope) width of 150 m and is roughly trapezoidal in planform, with steep, lateral margins defined by near-vertical cliffs. At the foot of the failure scar is a large (*c.* 25 m-high) block of intact rock that has tilted outwards without toppling. A broad bedrock buttress separates the south-west failure site from the smaller failure scar to the north-east.

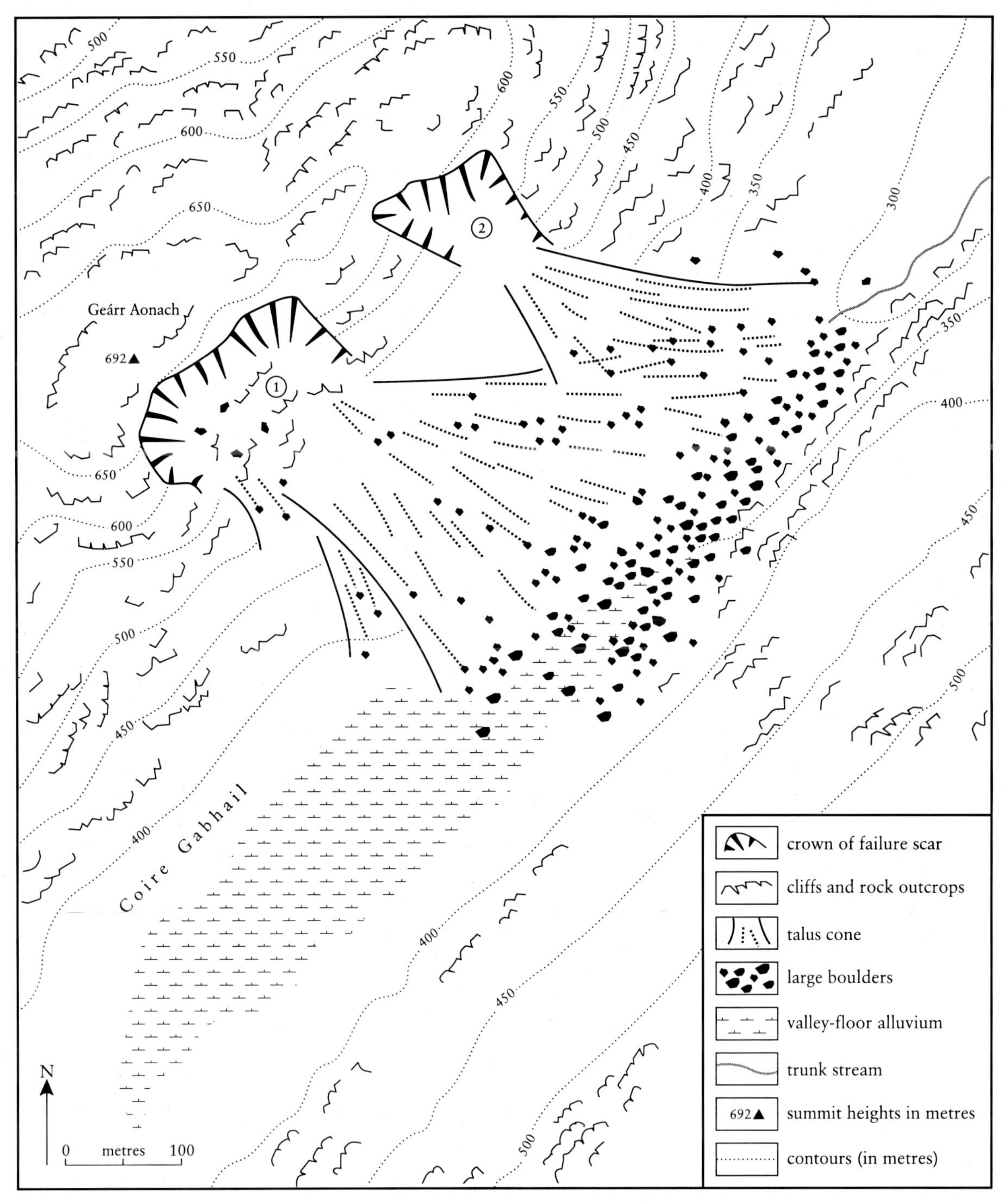

Figure 4.2 Geomorphological map of the Coire Gabhail rock-avalanches, showing the failure sites, the extent of the talus complex representing landslide runout and the area of alluvium deposited in the upper valley as a result of damming of the valley by runout debris. (1) site of initial rock-avalanche; (2) site of later rock-avalanche.

The latter takes the form of a broad funnel with a steep (50°–60°) basal failure plane, well-defined cliffed margins and a crown of pinnacled rock that extends to the crest of the ridge at 580–620 m. Under both failure sites and the adjacent rockwalls, stacked rhyolitic lava flows dip gently westwards, implying that failure was not seated on flow boundaries. The valley-side face is, however, seamed with vertical and near-vertical cooling and stress-release joints.

Debris accumulation

Most of the rock-avalanche debris has accumulated in two coalescing talus cones (Figure 4.2). The larger (south-west) cone extends from 470–490 m altitude downslope of the larger failure scar to the valley floor at 350–370 m and supports numerous large angular boulders exceeding 5 m in length. The southern part of the cone toe rests on the valley floor and is partly covered by alluvial gravels (Figures 4.2–4.4). Directly opposite this cone, numerous large (often > 5 m long) boulders have been thrown by impact on to a drift or bedrock bench on the far (south-east) side of the valley at an altitude of 385–395 m. These boulders extend up to about 30 m above the level of the adjacent valley floor and terminate at the foot of a steep, ice-moulded rockwall. Farther north-east, a jumble of huge angular boulders, many exceeding 1000 m^3 in size, completely fills the valley and abuts the opposite rockwall. Fractures in bedrock near the foot of the opposite rockwall suggest that some boulders impacted the cliff then rebounded. Some of the largest boulders have travelled over 250 m downvalley to an altitude of 300 m. Although most of the cone is mantled by coarse bouldery debris, shallow exposures suggest that fine-grained sediment occupies the interstices between clasts immediately beneath the surface boulder layer. The lower parts of the cone are extensively colonized by birch trees (Figure 4.3).

The smaller (north-west) cone consists of smaller boulders, is largely unvegetated, and presents a much fresher appearance. Unvegetated debris-flow tracks indicate recent reworking of debris. The south-west margin of this cone overlaps the larger cone, implying that this cone represents later (and possibly fairly recent) failure of the rockwall upslope.

The dimensions of the larger cone and associated runout debris were calculated from 1:5000 map data. The total planimetric area of debris cover is *c.* 23 000 m^2. Taking the average gradient of the cone surface (*c.* 33°) into account, this is equivalent to a real surface area of *c.* 27 400 m^2. The average depth of the

Figure 4.3 The complex talus accumulation formed by runout of the Coire Gabhail rock-avalanches, viewed from the south-west. The southern (left) edge of the larger talus cone is partly buried under alluvium. The conspicuous large boulder just beyond the toe of the cone rises about 10 m above the alluvium in which it is embedded. (Photo: C.K. Ballantyne.)

Figure 4.4 The south-west end of the talus complex, showing the fringe of large boulders. Alluvium deposited due to damming of the valley by the rock avalanche is visible in the foreground. (Photo: C.K. Ballantyne.)

runout accumulation, calculated by interpolating rockhead contours across the area occupied by runout debris and subtracting a grid of inferred rockhead altitudes from the debris surface altitude, is *c.* 11 m, with a maximum depth near the centre of debris accumulation of *c.* 30 m. Multiplying average depth by planimetric area yields a volume of approximately 300 000 m^3 of debris. Assuming that 20–30% of this volume represents voids, the implied volume of failed rock is 210 000–240 000 m^3, equivalent to a mass of $0.57–0.65 \times 10^6$ tonnes of rock for an assumed average density of 2.7 tonnes m^{-3}. The volume of the smaller cone could not be calculated, but appears to be about an order of magnitude smaller.

Interpretation

Mode of failure

The configuration of the failure scars suggests that at both failure sites thick slabs of rock failed along steep (50°–65°) basal shear planes bounded laterally by near-vertical rockwalls that represent the sites of stress-relief joints; open, near-vertical joints extend almost the full depth of the failure sites within adjacent intact rockwalls. Failure probably involved both sliding and toppling, leading to disintegration of the rock masses as they cascaded downslope. Travel of huge boulders across the valley floor and up to 250 m down the valley axis indicates that the earlier failure was characterized by extremely high energy; the large talus cone probably represents settlement of rock debris at the angle of residual shear in the final stages of movement. Only a very small component of the later failure reached the valley floor, however, with most debris accumulating in the smaller cone.

The immediate causes of the failures are unknown, but it is notable that the toes of both failure zones experienced debuttressing by retreat of glacial ice at the end of the Loch Lomond Stade, and the open near-vertical joints exposed in adjacent rockwalls imply rock-mass weakening as a result of joint extension due to deglacial stress-release. A zone of NE-trending vertical fractures followed by the Etive dyke swarm may have determined the loci of joint formation at the crown of both failures. The possible role of post-glacial seismic activity in triggering failure is difficult to assess. The larger failure scar lies roughly 1.4 km distant from two major NE-trending faults, the Ossian Fault to the north-west and Queen's Cairn Fault to the south-east, both of which originated during complex synvolcanic cauldron subsidence (Moore and Kokelaar, 1998). Neotectonic displacements due to differential glacio-isostatic adjustment along these or intervening fractures may have triggered failure, though there appears to be no evidence for post-volcanic dislocation.

Age of failure

The surface exposure age of a rock sample from the crest of a very large boulder deposited by the larger and earlier rock-avalanche has been subject to surface exposure dating by ^{36}Cl cosmogenic radionuclide assay. The provisional age obtained for this sample is 1.8 ± 0.33 ka BP, implying that failure occurred at least 9000 years after final deglaciation. The smaller and younger failure must have occurred after this date, and possibly within the last few centuries, consistent with its much fresher appearance. In common with landslide samples dated to *c.* 6.5 cal. ka BP for The Storr landslide (Ballantyne *et al.*, 1998b; see **Trotternish Escarpment** GCR site report, Chapter 6) and *c.* 4.0 cal. ka BP for the Beinn Alligin rock-avalanche (Ballantyne and Stone, 2004; see **Beinn Alligin** GCR site report, Chapter 2), this age determination implies that slope failure due to deglacial unloading and consequent stress-release has persisted well into Holocene times, and potentially may result in future failures from glacially steepened rock slopes in the Scottish Highlands.

Alluvial accumulation

According to Werritty (1997), the alluvial accumulation at Coire Gabhail is unique in Scotland, hence it is also selected as a GCR site for the Fluvial Geomorphology of Scotland GCR 'Block' (Werritty, 1997).

Following blockage of the lower valley by the earlier rock-avalanche, evacuation of coarse bedload sediment has been impeded, allowing progressive accumulation of alluvial gravels upstream. The alluvial basin is approximately 600 m long and 150 m wide, with a concave downvalley profile that is graded to the local base level created by sealing of the valley by landslide debris. The coarseness of surface gravels declines downvalley. The channel pattern is braided, but supports surface flow only following intense or long-duration rainstorms. Under normal flow conditions the river draining the valley (the Allt Coire Gabhail) sinks into the alluval deposits some distance upvalley from the rock-avalanche runout debris, and emerges near the north-east end of the boulder dam.

Conclusions

Roughly 1800 years ago, approximately 600 000 tonnes of rock failed near the mouth of Coire Gabhail, a steep-sided hanging valley cut in gently dipping rhyolites. The alignment of joints in the flank scarp of the failure zone suggests that failure was due to progressive joint extension and rock-mass weakening following debuttressing of the face as a result of glacier downwastage at the end of the Loch Lomond Stade, around 11 500 years ago. A smaller and apparently later failure occurred about 160 m north-east of the initial failure, depositing boulders as a talus cone on the flank of the debris deposited by the earlier event. This site is important for several reasons. It represents the finest example in Scotland of rock avalanches that have come to rest as massive talus cones, with debris resting at the angle of residual shear (*c.* 33°), though the high energy of the earlier failure drove boulders at least 30 m up the opposite slope and 250 m downvalley. It is also the largest rock slope failure on the Devonian volcanic rocks of the Western Highlands. This is one of the few ancient landslides in Scotland for which dating evidence is available, and the fact that failure occurred at least 9000 years after deglaciation implies that catastrophic failures due to paraglacial (glacially conditioned) stress-release persisted into late Holocene times. Finally, the site is unique in Scotland in that the rock-avalanche runout debris completely blocked the valley mouth, allowing sub-surface drainage through the runout zone but impounding coarse alluvial gravels. The alluvial floodplain that has developed upvalley as a result is without parallel in Scotland, with river runoff sinking into the alluvium near the valley head and emerging downvalley of the landslide runoff debris, except when exceptional flood events permit surface flow.

LLYN-Y-FAN FÂCH, CARMARTHENSHIRE (SN 801 215)

R.G. Cooper

Introduction

On the scarp face of the Black Mountain, facing northwards above Llyn-y-Fan Fâch, a number of deep gullies are cut through the vegetated scree surface (Figures 4.5 and 4.6). In some cases they reach the underlying bedrock for part of their length. Debris-flow activity is episodically taking place along the gullies' axes at the present time, transporting material derived from the gully sides by wash and other processes (Statham, 1976).

The Black Mountain is part of the north-facing Old Red Sandstone (Devonian) escarpment in south Wales. The scarp face above Llyn-y-Fan Fâch consists of a headwall of alternating hard sandstones and soft silty shales. The slope is about 60 m high, and stands at an overall angle of 45°–50°. Two prominent chutes/avalanche couloirs cut the upper slope of the headwall (Ellis-Gruffydd, 1972). Against the lower part of the headwall a scree-slope has accumulated, which is now well vegetated. The scree debris includes a significant proportion of coarse to fine sand-sized material, occupying the interstices between the boulders. Statham (1976) gives an average grading curve envelope for the scree material (Figure 4.7). Transport of debris down these gullies is resulting in the accumulation of debris-flow cones below the gully mouths. The cones are broadly concave

Figure 4.5 General view of the Llyn-y-Fan Fâch GCR site, showing the scarp face of the Black Mountain (Mynydd Du), and screes and gullies. (Photo: S. Campbell.)

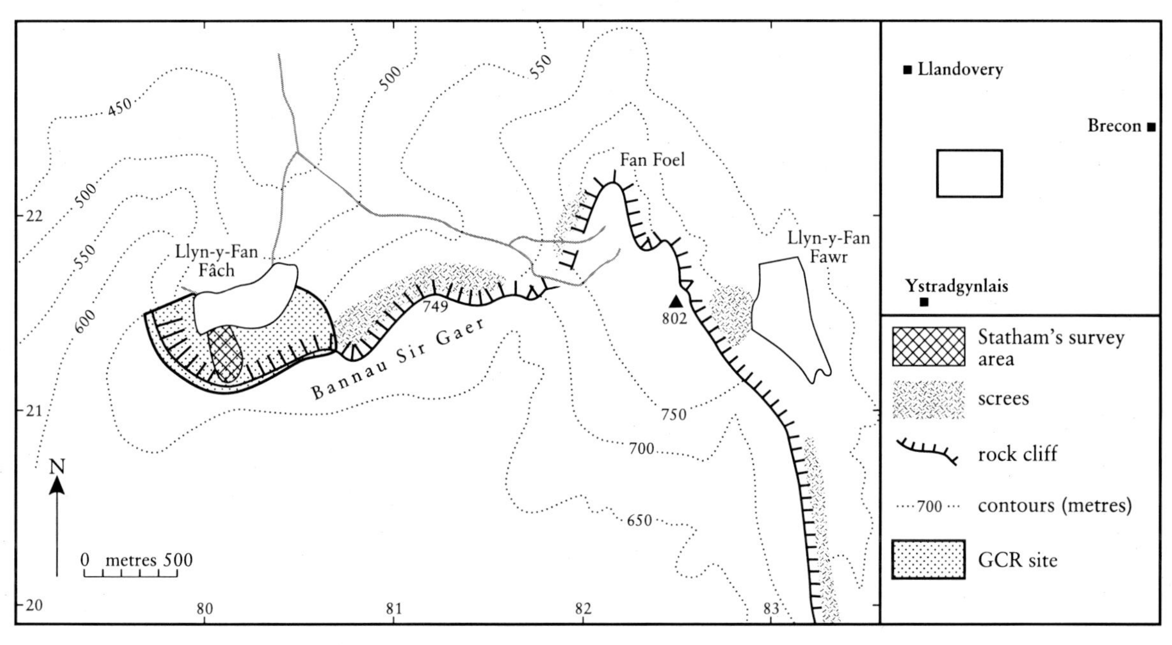

Figure 4.6 The location of the Llyn-y-Fan Fâch mass-movement site.

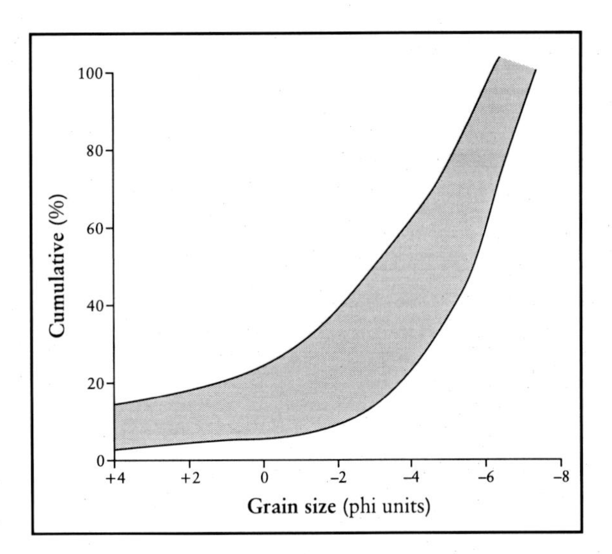

Figure 4.7 Range of sediment grain-size distribution of scree debris at Llyn-y-Fan Fâch. After Statham (1976).

in profile but irregular in detail as they are composed of a criss-cross accumulation of old debris-flow levées (Statham, 1976).

Debris flows are rapid mass movements of poorly sorted debris with a high water content. They are often associated with rockfalls from cliffs farther upslope, and tend to occur on talus that has accumulated from such rockfalls. Typically they follow a well-defined path as a fluid-like mass, leaving small linear ridges or levées on either side of their trail as they move. Sharp (1942) observed that quite rapid rates of movement (30–40 cm s^{-1}) are necessary for levées to be built. The levées have a high percentage of boulders, which are in a matrix of fine-grained sediments, and become concentrated at the head of the flow, partly blocking progress downslope until they are pushed aside by the advancing flow. The channel between the levées may or may not be eroded by passage of the flow, but if it is not it is usually free of any deposition.

Grass surfaces have been observed free of debris and entirely undamaged between levées (Rapp, 1960), indicating that basal shear, which may not always occur, is not great in such flows. However, many trails are not grassed over, because they erode their beds, and are, effectively, self-feeding. Trails tend to be a few centimetres to a few metres in width, with levées from about 1 cm to 1 m in height. Each flow generally transports anything from a few cm^3 to a few m^3 of material only. Many flows are single events, with no tendency for future flows to be concentrated along former lines, but more often flow activity is concentrated along a line to form a gully and a low-angled debris-flow cone of accumulation at its base (Johnson and Rahn, 1970). Debris flows almost always occur as a consequence of heavy rainfall or snowmelt. Mobilization may be due to the presence of a sub-surface, concentrated seepage line beneath the debris, which causes high porewater-pressure (Prior *et al.*, 1970). In the case of debris flows initiated in gullies, mobilization is a result of steady dilution of debris by water, rather than a steady increase in sediment content of a stream (Johnson and Rahn, 1970; see Iversen and Major, 1986; Addison, 1987). Movement probably begins as a slide but subsequent motion incorporates more water into the mass when it may behave as a fluid.

The Llyn-y-Fan Fâch GCR site was selected to represent well-documented debris-flows of a kind that is potentially atypical. Debris flows are a major and widespread type of mass movement in Great Britain, many occurring as relict periglacial forms (Ballantyne and Harris, 1994).

As well as being a mass-movement GCR site, the area was independently selected for the GCR for its Quaternary features of interest (Campbell and Bowen, 1989) and its fluvial geomorphology (Higgs, 1997).

Description

The Black Mountain debris-flows themselves are quite small, the tracks being from 1 m to 1.5 m in width with levées from 0.3 m to 0.4 m high (Figure 4.8). The uniformity of size, observable in recent flows and numerous old flows, is striking. On leaving the gully mouth, recent flows enter a short section undergoing erosion across the accumulation cone, but are then entirely depositional. Most have continued to move until nearly all of their load was deposited as levées, so that very little of the original mass can be found at the end of the flow track.

In profile, the steeper and straighter original scree has been replaced by a long, continuously concave profile along the gully axis. This concavity declines steadily in angle from about 40° at its top to about 8° at the base, and from the lowest bedrock exposure to the foot of the profile approximates well to a circular arc with radius of curvature about 310 m (Statham, 1976;

Figure 4.8 Detail of the gullying and mass-movement deposits at Llyn-y-Fan Fâch. (Photo: S. Campbell.)

Figure 4.9). There is no difference in curvature between the gully and the accumulation cone and so it seems reasonable to suggest that the entire form is continuous, controlled by the debris-flow process.

Debris flows are initiated in the gullies, at locations where the slope is between 27° and 37°. The gully sides attain a maximum stable angle at 43.5° when dry. Coarse debris with negligible clay- and silt-size percentages may

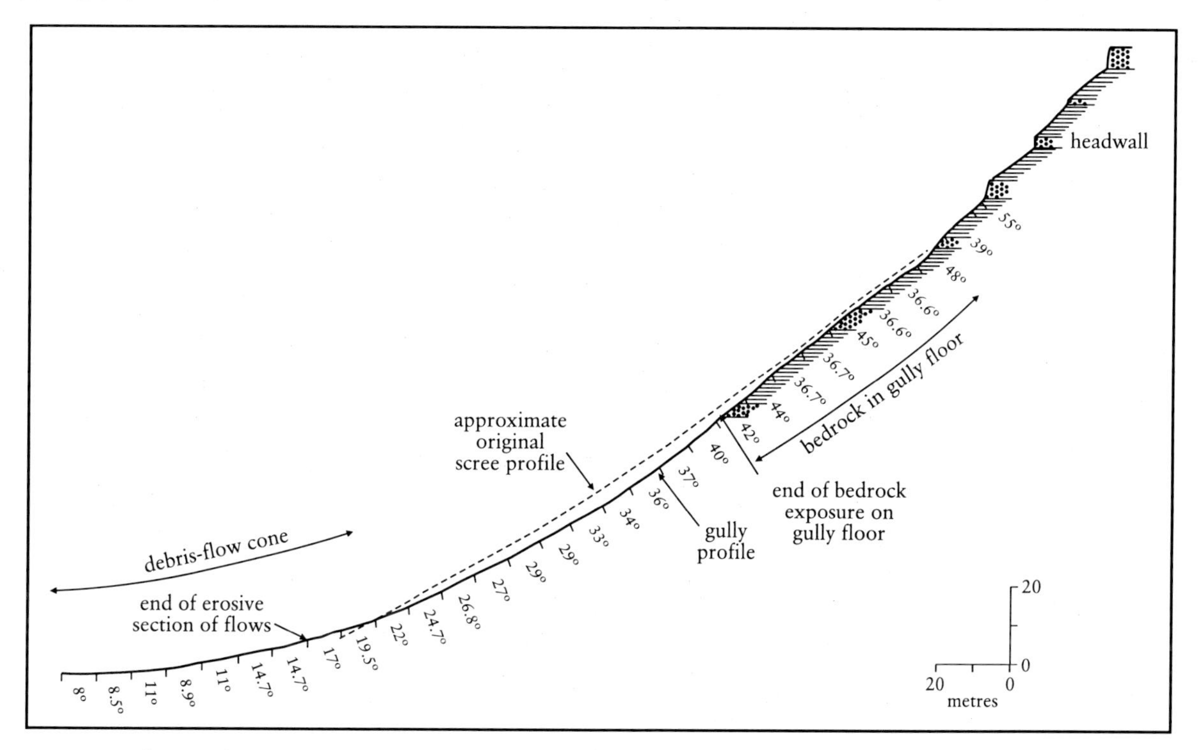

Figure 4.9 Profile of a typical gullied debris-flow cone system at Llyn-y-Fan Fâch. After Statham (1976).

be considered effectively cohesionless, and following this assumption Statham (1976) takes maximum gully-side angle as 'probably a reasonable estimate' of the lower repose angle, ϕ_r', for scree origins which involve a debris slide rather than gully flow and erosion.

Debris flows are easily formed by heavy rain and snowmelt yielding gully flow, which leads to erosion and gully-side collapse. Debris slides at the head increase the solids content, and when solids become 79% of the total by weight, the behaviour changes from mass transport to mass movement. The Statham (1976) model only considers the slide (the soil mechanics), not hydraulic or rheological models.

Assuming an infinite cohesionless slide analysis (after Skempton and DeLory, 1957) and taking ϕ_r' to be 43.5°, debris accumulating in the gully bottom due to erosion of the sides would be quite stable and would not begin to slide until the pore-pressure ratio r_u attained 0.15–0.4. For the debris to remain mobile after sliding r_u must increase steadily along the gully axis as slope angle declines. There are two mechanisms which might cause such an increase. Firstly, as suggested by Johnson and Rahn (1970), water may be added from rainfall and flow from the gully sides, which would lead to increasing water content and pore pressure as the debris moved down the gully. Secondly, accumulated material sliding from the upper part of the gully may over-ride already saturated debris in the lower part and cause undrained pore-pressures by rapid loading. Hutchinson and Bhandari (1971) have already suggested that undrained loading is important in the mobilization of coastal mudslides.

On surveying the cones illustrated in Figure 4.10, Statham (1976) observed that six recent debris-flow trails changed without exception from erosion to deposition when the slope fell below 16°. Contrary to the observations by Rapp (1960) already described, Statham (1976) found that, in the debris flows at Llyn-y-Fan Fâch, flows in the gully and steeper part of the cone increase in size by incorporation of debris at the base, whereas they decline by deposition of levées on the gentler section of the cone. Thus, movement over the lower part of the cone is not so much a reflection of very high pore-pressures but more of the inertia of the flow, as velocity declines on the lower angled section. The change from erosion to deposition implies that

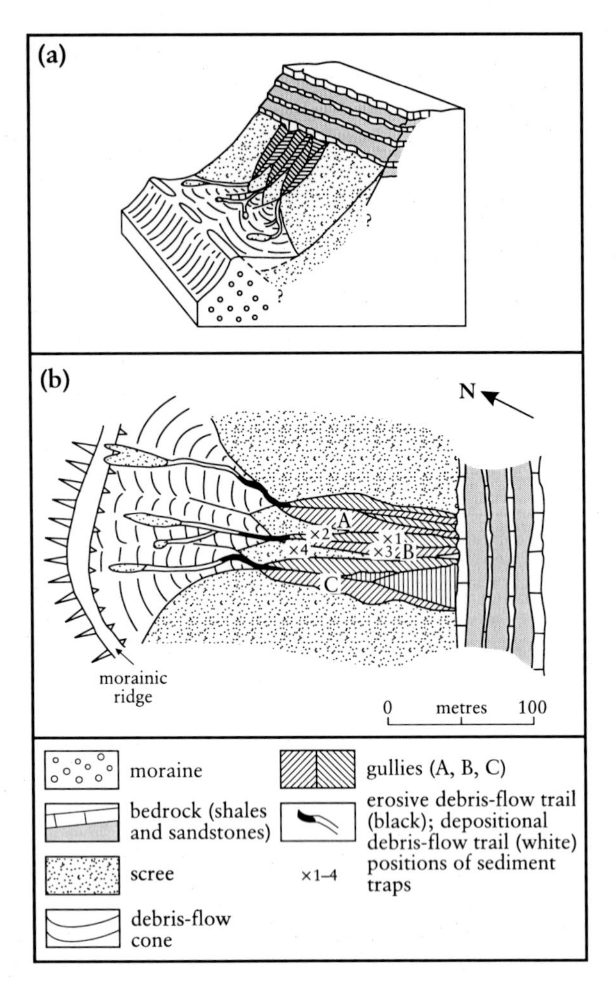

Figure 4.10 (a) Block-diagram of the Black Mountain scarp. (b) Plane table survey of three debris-flow gullies. After Statham (1976).

16° is the transit slope for these flows: the slope over which movement takes place without erosion or deposition. If this is the case it must also mean that flow velocity is such that drainage of porewater out of the debris is precluded and pore pressures are maintained. Assuming the flow behaves as a shallow infinite slide (which is open to question), a pore-pressure ratio of about 0.65 would be necessary to keep the slide going in this material. This is in excess of hydrostatic pore-pressure, which is about 0.5 for a material of unit weight about 2.0 tonnes m^{-3}.

Statham (1976) monitored the volume of sediment derived from the sides of one gully (gully B in Figure 4.10) and the volume transported to the accumulation cone of all three of the gullies in Figure 4.10. He found that the rate of lowering is much greater on the west-facing gully-sides than the east-facing sides. This has resulted in valley asymmetry in the gullies, the western slopes being both the steeper and the more unstable. Measuring the mean rate of surface lowering from the western- and eastern-facing gully-sides as 1.56 cm and 0.33 cm respectively, total yield of sediment to the gully bottom was calculated as 8.4 m^3 for the year. The volume shifted by one debris-flow from this gully was 9.8 m^3 and estimated volumes of another three flows from the three gullies (not necessarily in the observation year) were very similar, from 8.3 m^3 to 11.5 m^3. In consequence, the amount of sediment produced by surface lowering of the gully sides is roughly equal to that moved from the gully by debris-flow activity in one year. Naturally it is recognized that these quantities are only very approximate due to the methods used, and that the time-period of one year may not be sufficient to give a reliable overall picture. Nevertheless, it appears that input of sediment to the gully is balanced by output in debris flows, implying that sediment movement by other processes such as stream flow is negligible. This is supported by the fact that even in very heavy rain Statham (1976) did not observe any surface-water flow in the gullies except where bedrock was exposed in the base.

Although they were not observed, it seems likely that the flows took place on a day of heavy rain, since Prior *et al.* (1970) observed flows on days with 37 mm and 58 mm of rain in north Antrim. But although heavy rainfall was the most likely trigger mechanism, daily rainfalls of over 30 mm occurred on 16 occasions in the observation year, and on three occasions over 60 mm, with no associated debris-flow activity. Statham (1976) interpreted this as indicating that the debris-flow process is controlled by the rate of accumulation of loose sediment in the gully bottom and is not specifically a result of high-intensity rainfall. Thus, when there is sufficient material in the gully, a debris flow will occur. A storm is necessary to trigger the flow, but there is no shortage of storms of the necessary intensity and so the trigger is not an effective process control.

However, there is a substantial body of literature suggesting that availability of debris is the main control.

Interpretation

Given the occurrence of high-intensity rainfall, there seems to be very little climatic control on debris-flow activity (Statham, 1976), with a remarkable similarity of style and form of movement and of topographical situation in which flows are initiated. As the volume of the instrumented gully is 540 m^3, with removal of 8–10 m^3 per year, it cannot be more than 540–700 years old, assuming that the annual rate of removal has remained constant. Furthermore, there are no new gullies being initiated on the scarp and all the existing ones are in roughly the same state of development. It seems likely that some environmental change in the recent past was responsible for the initiation of debris-flow activity. Innes (1983) has noted similar initiation of debris flows in the recent past across the whole of upland Britain, and attributed this to environmental change, possibly deliberate fire-setting. Statham (1976) suggests that the causative environmental change at Llyn-y-Fan Fâch may have been the introduction of intensive sheep grazing in the area, resulting in damage to the vegetation surface and exposure of bare ground.

Therefore it seems that the progressive replacement of the straight scree-slopes of the Black Mountain scarp by a series of low-angled, concave debris-flow cones is a very recent change in process in the geomorphological timescale. Debris flows are initiated by heavy rainfall: this acts as a trigger, but does not control debris-flow activity. They probably require a minimum volume of material before mobilization can occur and are therefore controlled by the rate at which sediment is produced by gully-side lowering (Statham, 1976).

In a specific comparison with Statham's (1976) observations at Llyn-y-Fan Fâch, Ballantyne and Harris (1994) point out that Ballantyne (1981) observed at An Teallach in northern Scotland, that the transition at which deposition succeeds erosion occurs at 20°–28°, rather than the 16° observed by Statham. Similarly, while Statham observed the Black Mountain flows to come to rest on a slope of 8°, those at An Teallach stop on gradients of 11°–23°. They remark that these differences may reflect greater flow viscosity at An Teallach.

Conclusions

The Llyn-y-Fan Fâch GCR site is important in showing that the steeper and straighter scree-slope section of the Black Mountain scarp is being replaced by a series of low-angled, concave-upwards debris-flow cones. Rates of erosion in the debris-flow supply gullies suggest that this is a very recent change in process in the geomorphological timescale. Gully-side lowering produced about 8.4 m^3 of sediment from a monitored gully in one year and in the same year 9.8 m^3 of sediment was moved in a single debris-flow event. All of the sediment derived from the gullies is transported by debris flows, while other sediment transport processes, such as stream flow, are unimportant. Although debris flows are initiated by heavy rainfall, heavier storms occurr on other occasions, with no associated debris-flow activity. Since there is no shortage of large storms, they do not control debris-flow activity but merely act as a trigger. Debris flows probably require a minimum volume of material before mobilization can occur, and are therefore controlled by the availability of sediment, which in the Black Mountains is controlled in turn by the rate at which sediment is produced by gully-side lowering.

Chapter 5

Mass-movement sites in Carboniferous strata

For convenience, the GCR sites selected are discussed here in two sections, covering the Lower Carboniferous strata and the Upper Carboniferous strata respectively.

MASS-MOVEMENT SITES IN LOWER CARBONIFEROUS STRATA

EGLWYSEG SCARP (CREIGIAU EGLWYSEG), CLWYD (SJ 235 432–SJ 235 480)

R.G. Cooper

Introduction

Eglwyseg Scarp extends 8 km northwards from Castell Dinas Brân near Llangollen (see Figures 5.3 and 5.4). It faces westwards and is divided into a series of buttresses by deep gully dissection. The site has frequently been quoted as a type example for escarpments and screes, but it differs markedly in both form and history from many others in upland Britain.

Tinkler (1966) identified two distinct types of depositional slope below the free face. He used the term 'clitter' to describe a thin veneer of coarse rock fragments on a slope, the form of which is controlled by underlying structure, and reserved the term 'scree' for loose fragments in an accumulation of sufficient depth for the angle of repose to be determined by the physical characteristics of the fragments themselves, as distinct from its being determined by whatever lies beneath the accumulation of fragments.

The outcrop of the Lower Grey and Brown Limestone co-incides with that part of the escarpment below the most significant free face (Figure 5.5). It is characteristically composed of limestones 0.62–0.9 m thick, with intercalated shale beds (5–15 cm), and forms a stepped lower bedrock slope. The Middle White Limestone is lithologically distinct and forms the free faces. It is composed of three massive beds about 7.5 m, 13.5 m and 6 m thick respectively, separated by narrow shale beds that are locally absent. Lateral variations in thickness are considerable, with a general thinning towards the south. Above this are several low and degraded scarps in the Upper Grey Limestone and Sandy Limestone.

Description

The escarpment seems to have been initiated at a time when the River Eglwyseg joined the River Dee near Castell Dinas Brân, at about 300 m

Figure 5.3 View of Eglwyseg Scarp, surveyed by Tinkler (1966). (Photo: R.G. Cooper.)

Figure 5.4 Aerial phtograph of the scree-slopes at Eglwyseg Mountain, near Llangollen. (Photo: Cambridge University Collection of Air Photographs, Unit for Landscape Modelling.)

above OD. Local slopes indicate this drainage trend, and a terrace is preserved below the scarp at Craig Arthur. Some time in Early Pleistocene times the River Eglwyseg was diverted westwards, so that slopes south of the Dinbren Isaf col have since developed without a river to facilitate transportation or erosion (Tinkler, 1966). They appear to have declined, aided by the southward thinning of the Middle White Limestone. An extensive mantle of Devensian till covers the uplands and is also found in the deeper fissures on the scarps, on the inter-scarp ledges and on the main slopes below. During deglaciation, meltwater and periglacial activity re-deposited some of this as head, and upper deposits are common along the foot of the main escarpment (Figure 5.6).

Local slopes indicate the former drainage trend of the proto-Eglwyseg and a terrace is preserved below the scarp at Craig Arthur. Incision below this valley was considerable at the Dinas Brân and Dinbren Uchaf cols before diversion of the River Eglwyseg, an event still marked by an elbow bend in solid rock.

The slopes south of the Dinbren Isaf col below Creigiau Eglwyseg (Figure 5.6) are the closest to the line of the proto-Eglwyseg. To the north, where incision at the elbow of diversion is 90 m lower, the escarpment has retreated further, and the greater available height between the river and the slope crest (270 m compared with 150 m at Trefor Rocks) permits greater horizontal retreat of the upper cliff before complete decline. The latter stage is approached at Eglwyseg Mountain and Creigiau Eglwyseg. Pinfold Buttress, which shows least sign of decline, is opposite the elbow of diversion. At Trefor Rocks the slopes are in a degraded state, while Craig Arthur in the north has been protected in part from erosion by a terrace in front of it. This

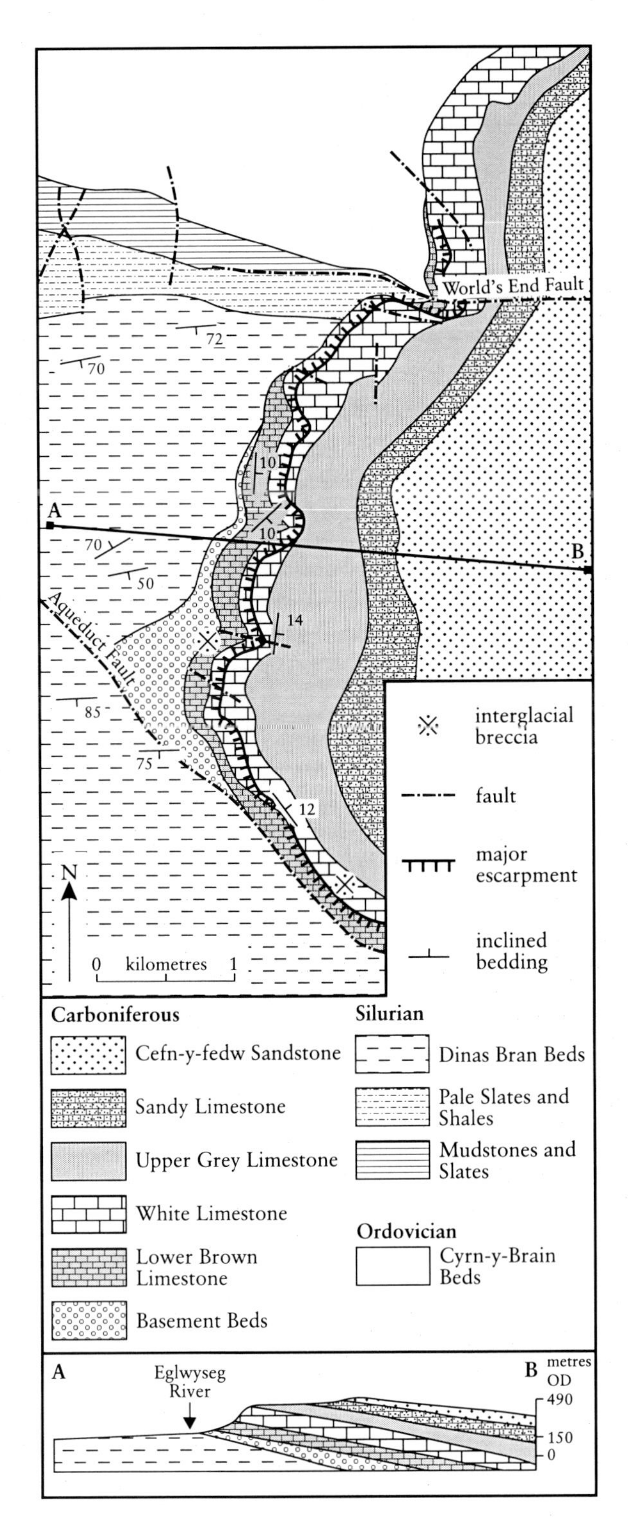

Figure 5.5 The geology of the Eglwyseg Valley, North Wales. After Tinkler (1966).

buttress lies to the front of the general line of the escarpment. The presence of till below scree and clitter on all parts of the escarpment indicates that the basic morphology dates at least to

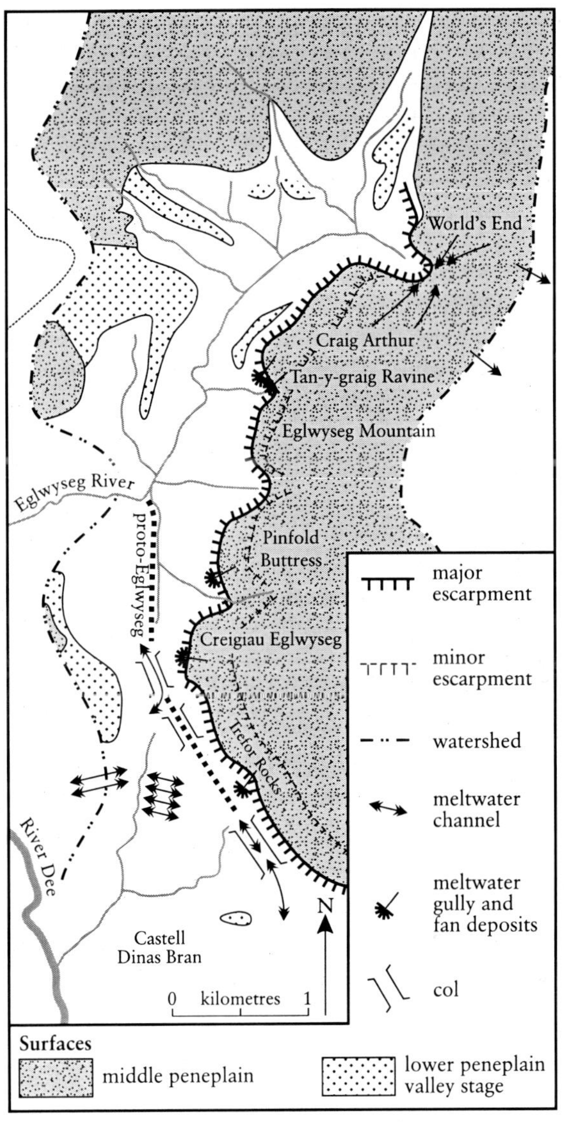

Figure 5.6 The geomorphology of the Eglwyseg Valley, North Wales. After Tinkler (1966).

the last (Ipswichian) interglacial, with only minor modification since then. The debris cover is therefore a shallow mantle on a fossil bedrock form partly buried by till. Its accumulation can have had little influence on the form of the bedrock slope below the free face or upon the free face itself (Tinkler, 1966). Jointing is variable, but deep cracks in the free face can produce huge tabular blocks that become embedded in the scree. Normally, fine surface cracks and joints in the free face have given rise to scree debris up to 30 cm maximum dimension.

The tallest free face is always in the White Limestone, and the scree, clitter and bedrock slopes below are developed in the Lower Grey and Brown Limestone, while the scree-slopes above are in the Upper Grey Limestone and the Sandy Limestone. Variations in lithology are minimal on different parts of the slope. Scree counts were made by Tinkler (1966) on the lines of profile at random intervals, and sizes refer to the maximum dimension of each of a sample of 100 pieces. The scree and slope type-data are restricted to the main slope of the scarp below the lowest free face, and the profiles are entirely limited to the limestone outcrop. The lower limit of profiles is that of loose debris, which is the upper limit of enclosure.

Tinkler (1966) surveyed 56 slope profiles at intervals along the length of the escarpment (Figures 5.7–5.9). Substantial scree-slopes are restricted to three localities: World's End, Craig Arthur and the south-west face of Pinfold Buttress. scree-slopes elsewhere are short and impersistent. In total the Eglwyseg scarp-face area is 28% scree, 11% bare bedrock, 11% grassed scree and bedrock, and 50% clitter (Tinkler, 1966). The term 'clitter' is used in vernacular English to describe either a slope composed entirely of rock clasts that litter the surface and have been derived from the runout of rockfall debris or a rock litter derived from rock weathering, the core stones being stripped of their matrix to leave the boulder field. Scree is present on the north side of the World's End valley, and the slope length increases 30 m to 75 m westwards along a baseline 129 m long. Nine profiles and 24 scree counts were made here. 54% were in the range 5–13 cm, 25% in the range 13–20 cm, and 3% were over 20 cm. 80% of the scree is between 5 cm and 20 cm in size. The percentage of the sample recorded at 5–13 cm decreases downslope, while the

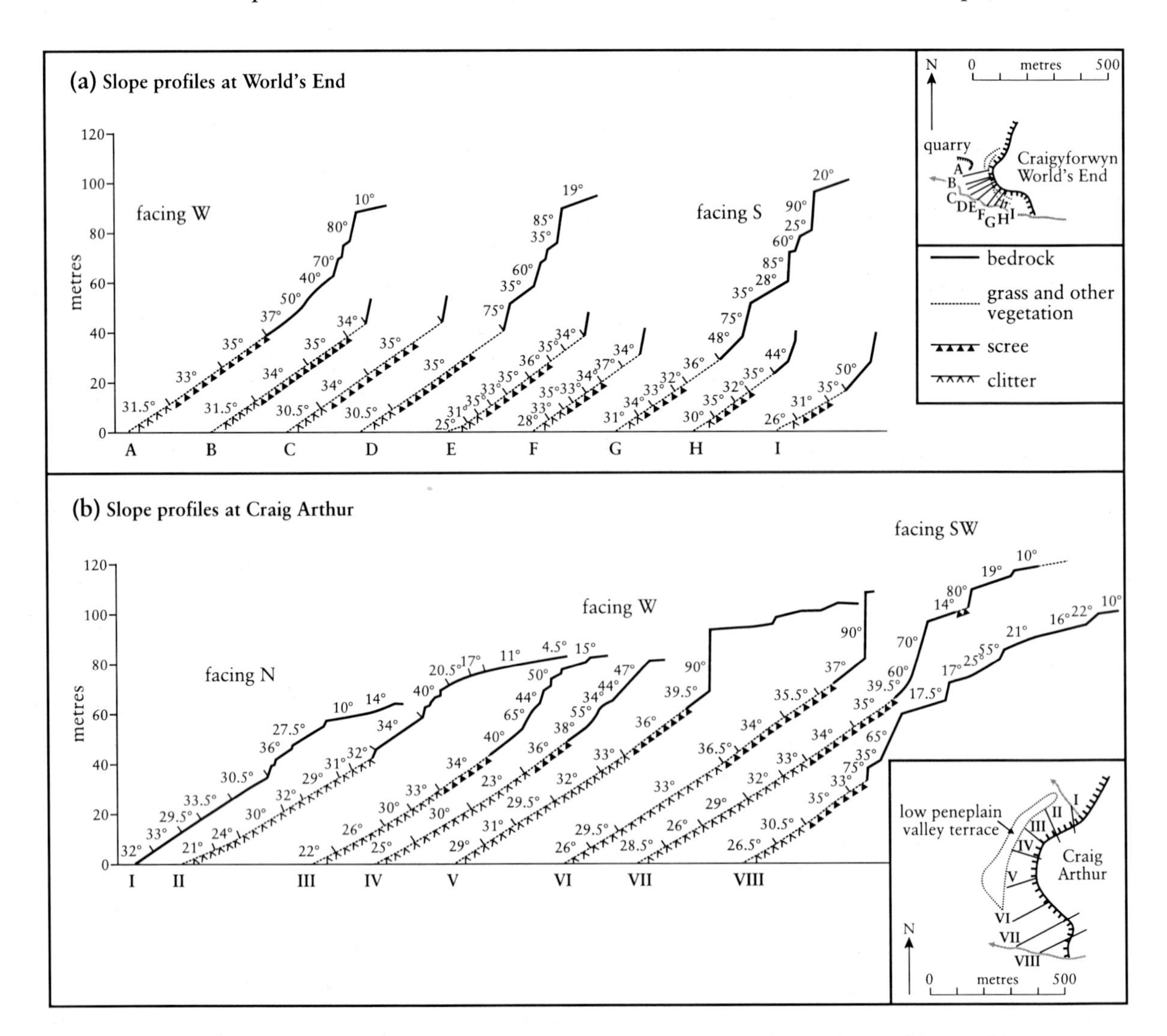

Figure 5.7 Slope profiles of (a) World's End, and (b) Craig Arthur. After Tinkler (1966).

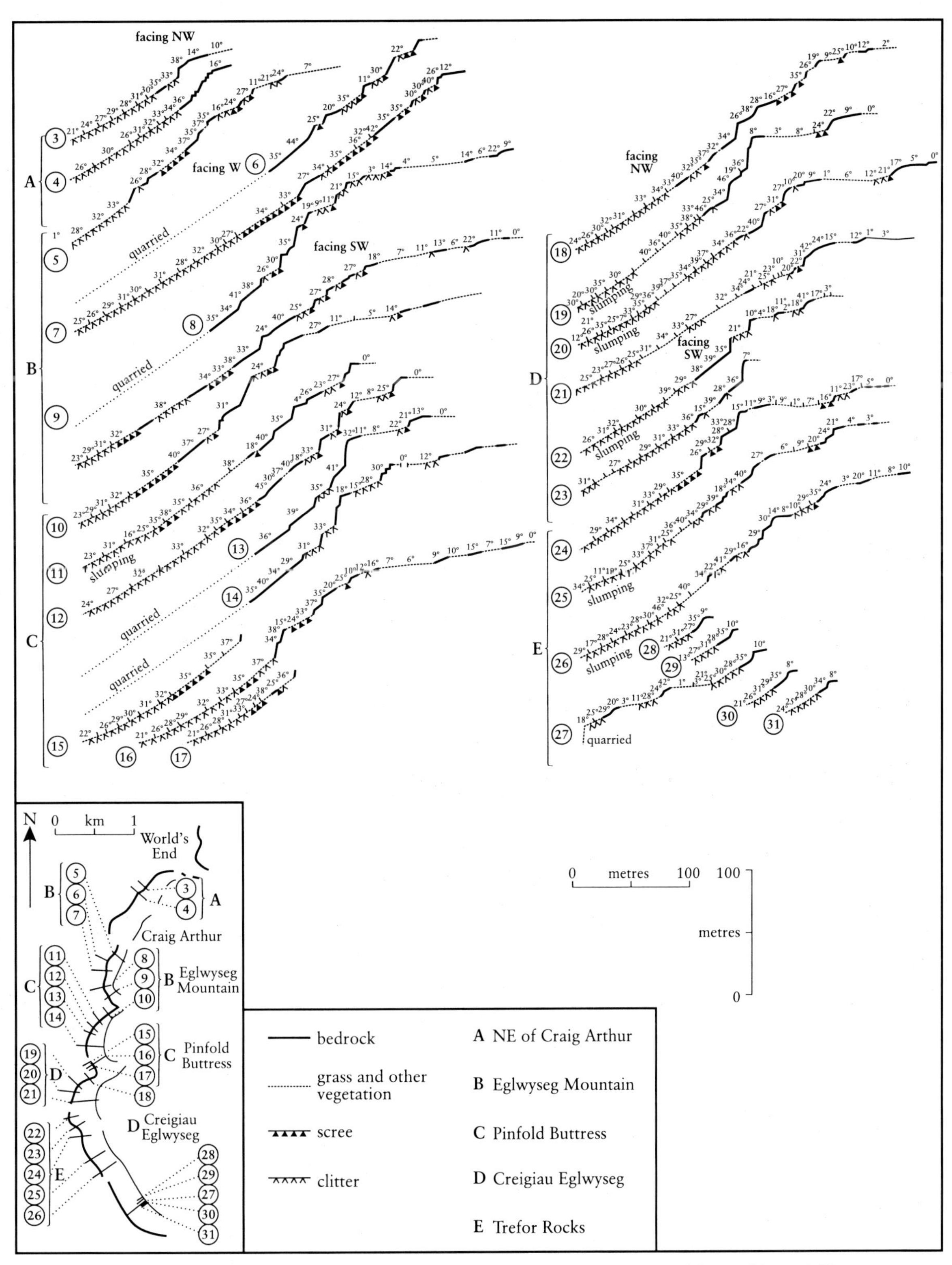

Figure 5.8 Slope profiles on Eglwyseg Scarp, North Wales, surveyed by Tinkler (1966).

percentage recorded at 13–20 cm increases downslope. Scree over 20 cm is limited to the bottom of the profile but is significantly related to slope position at the 1% level. Scree below 5 cm is barely represented in most of the samples. Coarser scree is present below the fine

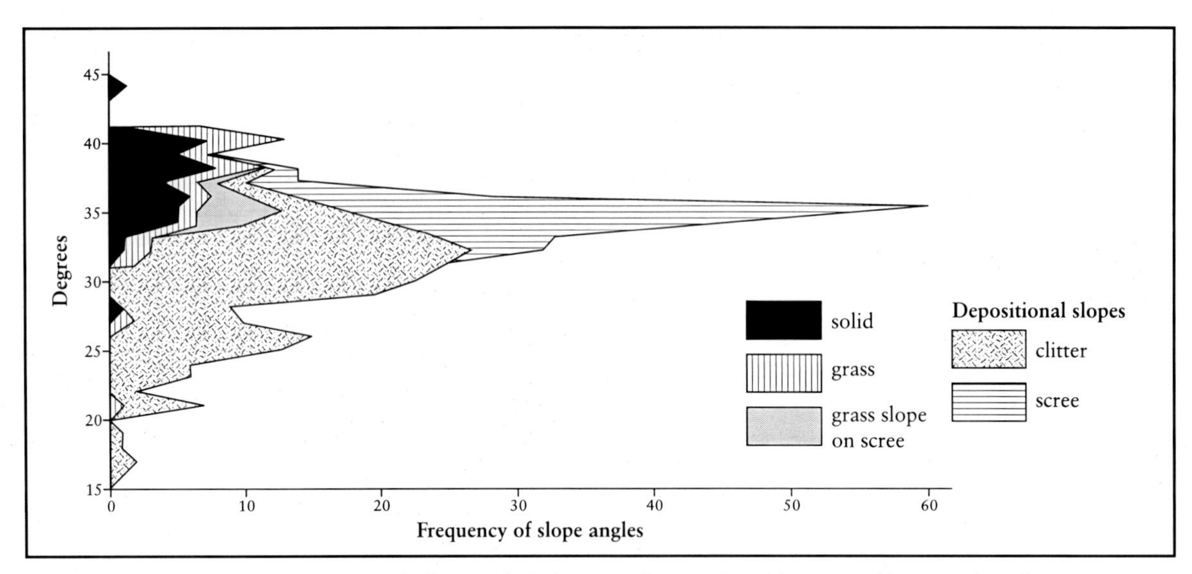

Figure 5.9 Histogram of all recorded slope angles on the Eglwyseg Valley, North Wales.

scree at a shallow depth, and occasionally overloading of fine scree at the top of the slope has caused it to spill downslope as a narrow trail, the lower end of which is built up at a slightly lower angle (32°–33°).

At Craig Arthur the slope cover of scree and clitter increases southwards, and on the most northerly facing slope there is no surface debris on a bedrock profile of 30°–33°. Elsewhere scree is present as a relatively narrow band above the lower clitter slopes. The increasing scree and clitter slope cover southwards, despite the almost constant height of the free face and increasing slope length, suggests differential weathering in the post-glacial period. Total scree percentages are similar to those at World's End: 18% with maximum diameter less than 5 cm, 66% between 5 cm and 13 cm, 13% between 13 cm and 20 cm, and 3% over 20 cm. However, at Craig Arthur much of the fine material low on the profiles is derived from the underlying till by surface washing. Clitter angles are all markedly lower than scree angles.

On the south-west side of Pinfold Buttress 11 scree counts were made, and similar proportions of scree sizes were recorded: 15% with maximum diameter less than 5 cm, 63% between 5 cm and 13 cm, 19% between 13 cm and 20 cm, and 3% over 20 cm. The fairly constant proportions at three different sites seem to highlight the constant lithology of the free face.

The distribution of bedrock angles is significantly higher than the distributions of scree or clitter, and this partly depends on the masking effect of the scree and clitter on the lower-angled bedrock slopes. However, where exposed on the lower slopes, bedrock slope is nevertheless steep (over 35°) (see Figure 5.9).

Interpretation

In general the pattern is for a clitter slope to occur below a free face of bare bedrock. Parts of the clitter slope may be grassed over. At World's End, Craig Arthur and Pinfold Buttress, a scree-slope is interposed between the free face and the clitter slope. The order 'free face–scree–clitter' applies to many of the smaller free faces above and set back from the main free face. The upper levels of the scree-slopes may be grassed.

Particle-size counts (Tinkler, 1966) of scree from the three main sites indicate that about 80% of scree particles have sizes (presumably *b*-axis) between 5 cm and 20 cm, and that particle-size proportions are fairly constant between sites. At all three sites, fall-sorting is evident, with the smallest particles most frequent at the top of each scree run, and the largest at its foot. Talus creep and surface wash also affect the distribution, particularly where long clitter slopes are present and the till is near the surface, as at Craig Arthur and Pinfold Buttress. scree-slopes have only developed where there is a substantial free face. They are currently active, and only stabilize where a thin soil covers the uppermost part, as at World's End.

The range of slope angles recorded on the screes is only 6°–7°, clustering around a modal value of 35° (Figure 5.9). In contrast, the modal angle on bedrock (excluding the free face, which generally stands at more than 50°) is 38°, with a very definite upper limit of 40° (upper semi-quartile). This limit is taken as an indication (Tinkler, 1966) that the bedrock slopes may have developed as a Richter slope in relation to a debris cover which no longer exists. They clearly represent the 'buried face' of Wood (1942), but the morphology is not always clear: the form is essentially exhumed, with only very minor convexity. Till is always found at shallow depth beneath the clitter on the clitter slopes, and the angles on it reflect this: the modal angle is about 32°. This suggests that the clitter may be a residual deposit resulting from the washing out of till. Clitter can grade upslope into scree but the junction is usually sharp.

Conclusions

As noted by Tinkler (1966), post-glacial erosion and deposition at Eglwyseg Scarp has been a mere etching upon a morphological framework inherited from late Tertiary and Pleistocene times. For this reason, expressed mainly through the prevalence of clitter slopes, the depositional slopes at Eglwyseg cannot be regarded as true scree-slopes like those of Snowdonia, the English Lake District or the Cuillins. It is this unusual aspect of their nature that makes them particularly appropriate for conservation.

HOB'S HOUSE, MONSAL DALE, DERBYSHIRE (SK 173 710)

Introduction

R.G. Cooper

Hob's House (Figure 5.10) is a rare example of a large-scale rotational slip in the Dinantian limestones of the southern Pennines. The sliding has taken place over a weathered horizon of lava.

Hob's House consists of a group of about seven large blocks of Carboniferous limestone standing on a low-angled shelf halfway down the otherwise steep northern slope of Fin Cop, at Monsal Dale, in the valley of the River Wye, Derbyshire (Figures 5.10 and 5.11).

Figure 5.10 The backscar and transported blocks of the Hob's House landslide. (Photo: S. Graham, English Nature/Natural England.)

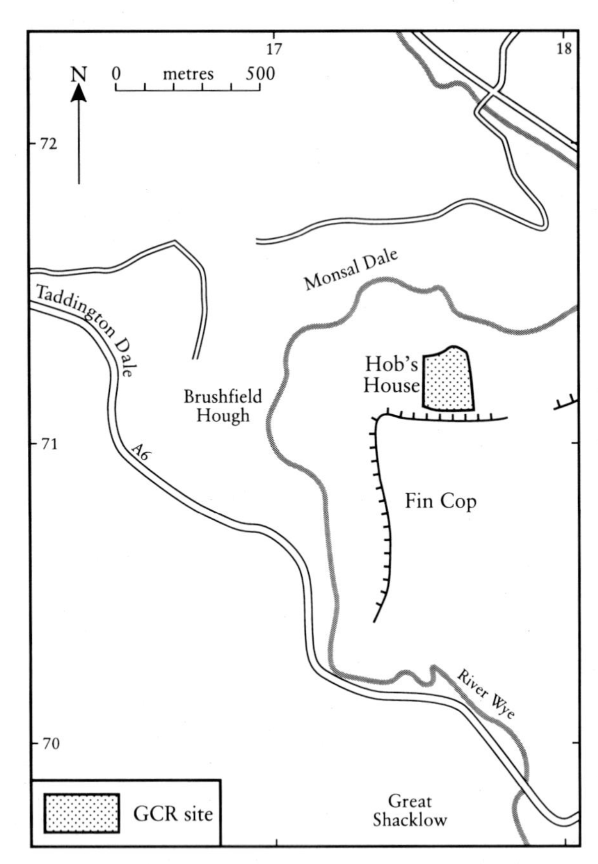

Figure 5.11 Location of Hob's House GCR site.

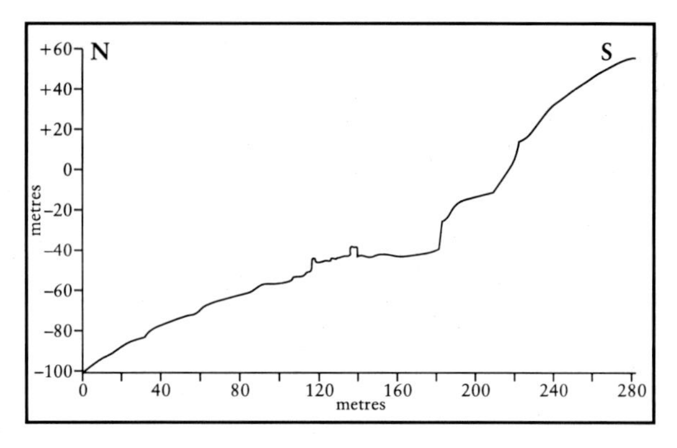

Figure 5.12 Slope profile at Hob's House.

The Hob's House mass-movement site at Monsal Dale should not be confused with Hob Hurst's House, a Bronze Age round barrow on Baslow Moor 10 km to the east, which was excavated by Thomas Bateman in 1853 (Bateman, 1861, pp. 87–88).

Description

The blocks at Hob's House are in the dark lithofacies of the Monsal Dale Limestones (Dinantian, Lower Carboniferous). The slope is about 330 m long and stands at an overall angle of about 35° (Figure 5.12). The vertical-sided blocks, standing on an approximately horizontal boulder-strewn shelf 65 m wide, are up to 7 m high and 20 m long and broad. They are backed by a 12 m-high vertical cliff-face in the limestone bedrock, above which the slope continues 150 m to the crest of the hill, at about 40°. Below the shelf the slope runs down 130 m to the river, at an angle of about 20°. The cliff contains an enterable fissure which has been penetrated for 20 m, indicating that it is in a shattered condition, and has itself been engaged in some mass movement (Figure 5.13). Coarse limestone scree mantles the shelf between the blocks and the cliff, and around and between the blocks. On either side, and uphill and downhill from these features, the slope is grassed over.

Some indication of the degree of displacement can be gained from the stratigraphical levels. Aitkenhead *et al.* (1985) point out that

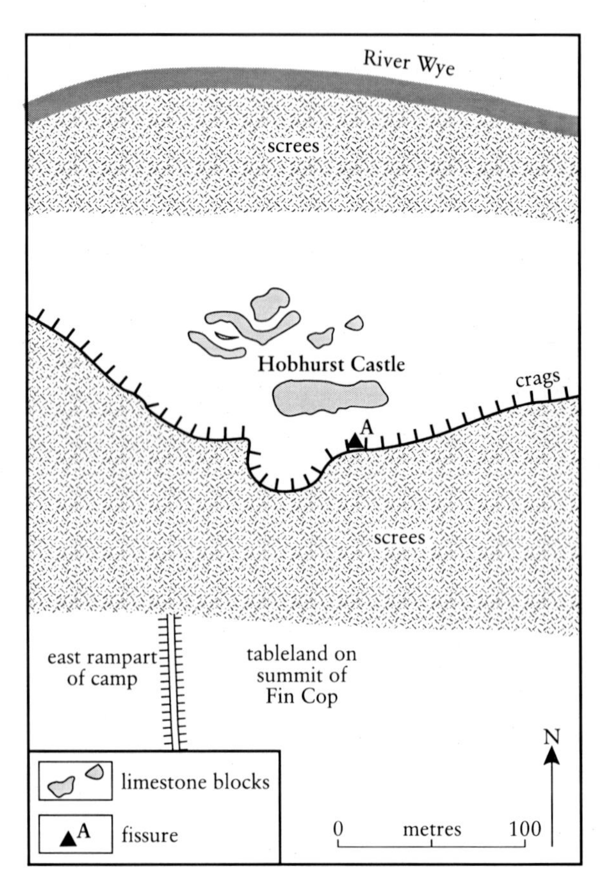

Figure 5.13 Sketch plan of Hob's House (Hobhurst Castle). 'A' is the mouth of the fissure in the cliff-top, probably a camber structure or landslide 'labyrinth'.

the 51.7 m section at Hob's House contains the Hob's House Coral Band, which is 0.4 m thick. Brown (1973) identified the cliff, and the 'towers' as containing his 'Hob's House Coral', and that this coral has dropped 15 m in the towers compared with the cliff. Furthermore, the coral is at different heights in the different towers.

Interpretation

The drop in altitude suggests that the blocks lie within the upper zone of a large-scale rotational slip. Aitkenhead *et al.* (1985) remark that landslips are not widespread on the Dinantian limestone outcrop in Derbyshire. They cite Hob's House as a rare example, where rotational slipping has occurred due to movement on softer-weathered igneous rocks, in particular the Shacklow Wood Lava. Such lavas are discussed by Ford (1977): characteristically, in the Derbyshire outcrop, their original minerals (from basalt, tuffs and dolerites) have broken down under chemical attack, usually resulting in clay minerals. The process forms a soft, green clay at the top of the chemically altered lavas, which are known locally as 'toadstones'.

Conclusions

At Hob's House an unusual set of circumstances has led to a landslip which has resulted in several block-shaped limestone 'towers' of large size having become separated from the cliff behind them. Their situation, halfway down the slope, is also unusual.

MASS-MOVEMENT SITES IN UPPER CARBONIFEROUS STRATA

ALPORT CASTLES, DERBYSHIRE (SK 142 914)

R.G. Cooper and D. Jarman

Introduction

The Alport Castles are a massive landslide complex, the most prominent elements of an extensive landslip complex affecting at least 0.85 km^2 along the eastern (west-facing) side of Alport Dale near Ladybower in the Peak District (Figure 5.14). Alport Dale is one of several valleys incised in the sandstone plateau of the Dark Peak, here by up to 200 m. Landslipping is very extensive in this part of the Pennines (cf. Johnson, 1965; Stevenson and Gaunt, 1971; Johnson and Walthall, 1979), and the geological controls on its incidence are particularly evident here. This site is notable for its array of distinct slip-masses and slumps in varying stages of intactness, attitude and distance travelled, extending in places to the River Alport.

Description

Landslipping on the steep valley-sides in the 'Millstone Grit' areas of the Pennines commonly occurs where competent sandstones overlie less competent shales. Here, the River Alport rises on the Bleaklow plateau of Shale Grit (Kinderscoutian R1 stage of the Namurian, Upper Carboniferous). About 1.5 km above this site, it begins to cut down through the Mam Tor sandstones (Kinderscoutian). It eventually reaches the underlying Edale Shales (Alportian (H2) stage of the Namurian Period). These are predominantly mudstones, though sandstones occur, and include exceptionally weak pyritic shales studied at **Mam Tor** (Vear and Curtis, 1981). Immediately the river enters them, the valley-sides 'become covered with huge landslips formed of masses of the Shale Grit which have slid down from the hilltops above' (Green *et al.*, 1887; Figure 5.14). The weakness of these Edale

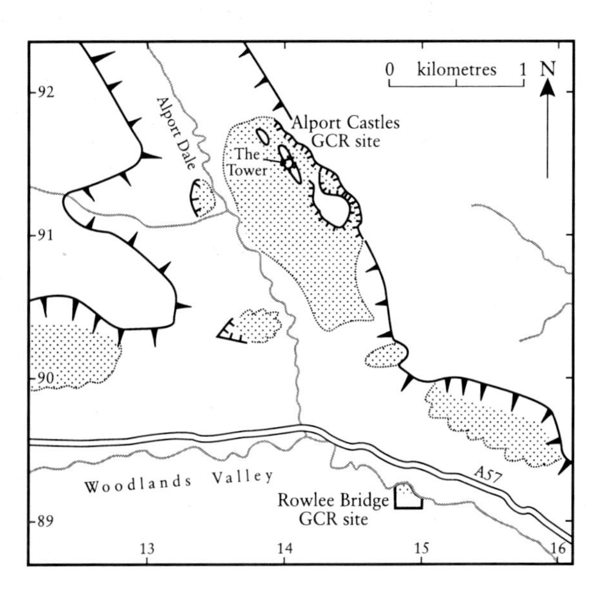

Figure 5.14 Location of the Alport Castles and **Rowlee Bridge** GCR sites, showing other landslips (stippled) and scars ('spiked' lines) in the vicinity.

Shales is evidenced by deep crumpling revealed in dam trenches in the adjacent valleys (Thompson, 1949) and at **Rowlee Bridge** nearby. The strata are here nearly level, with a slight eastward tilt.

The Alport Castles site is one of the largest landslip complexes in the district (Figure 5.15), and has been studied by Johnson and Vaughan (1983). It divides into two sectors (Figure 5.16): a main (northern) sector, where landslipping encroaches into the plateau rim along a bold craggy scar, and where the prominent 'Castles' are located; and a secondary (southern) sector, where the source scar is a less significant feature running across the upper valley-side.

Northern sector

Three main rock-masses have broken away from the rim of Alport Moor (Units A–C), leaving a scar which above Unit B reaches 68 m to the narrow boulder-filled trench floor. The scar here comprises a 30 m vertical sandstone crag above a talus slope and the rift-trench, which is of unknown original depth (Figure 5.17). These units are substantially intact, and prominent

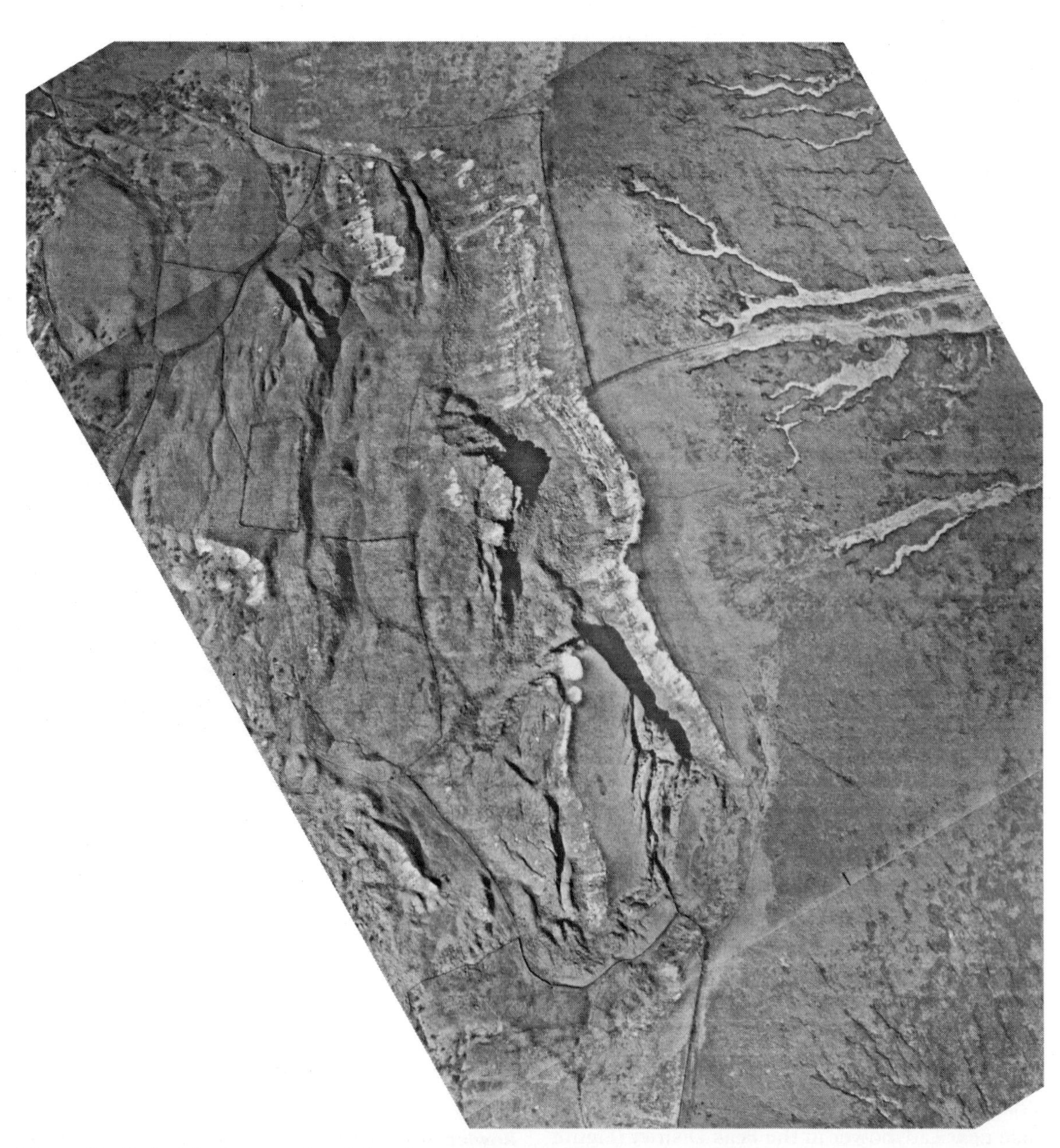

Figure 5.15 Aerial photographs of the Alport Castles landslip complex. (Photos: © Crown Copyright/MOD. Reproduced with the permission of the Controller of Her Majesty's Stationary Office.)

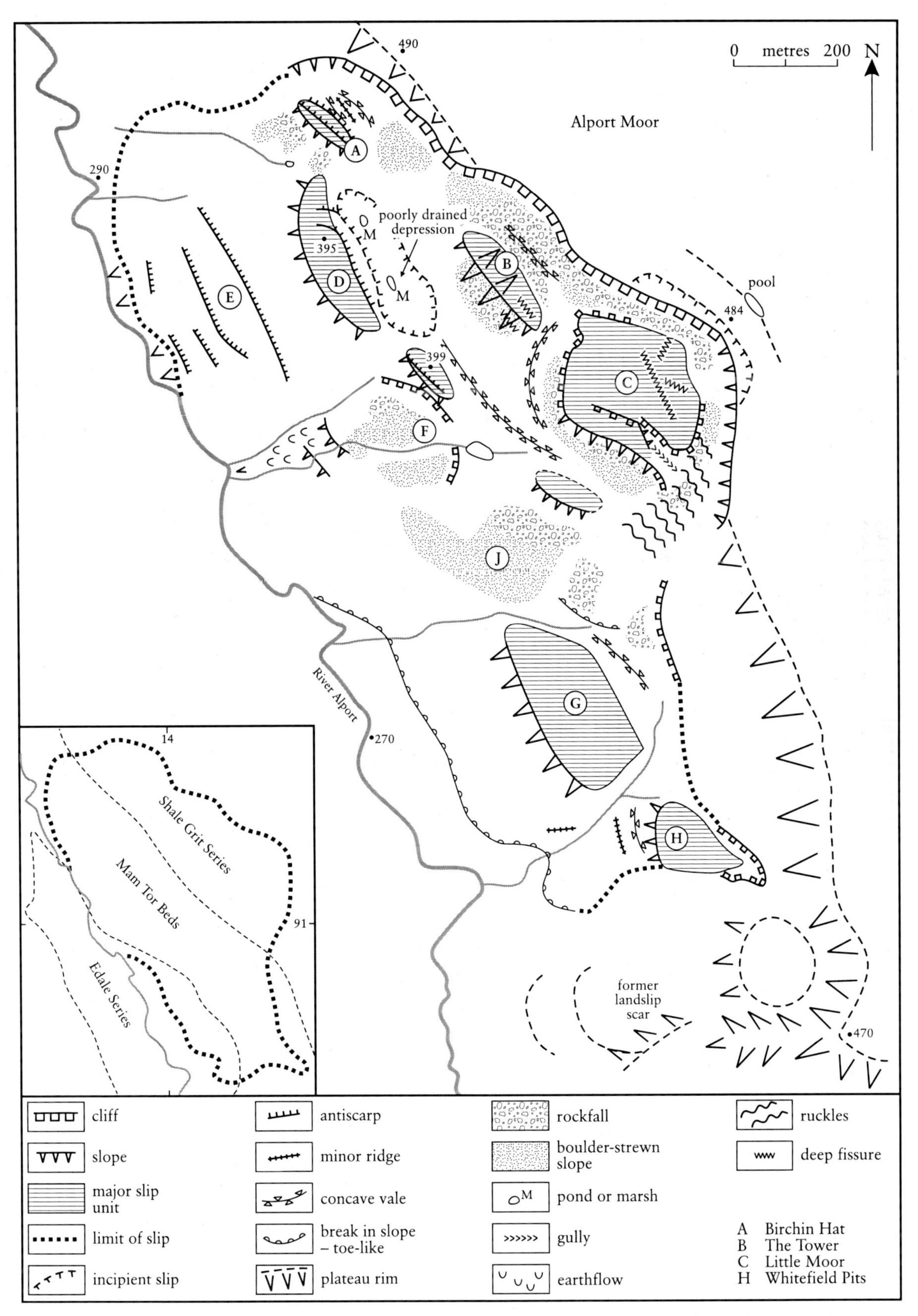

Figure 5.16 Morphological map of the Alport Castles landslip complex, identifying the main slip units described in the text, and indicative geology. The source scar transgresses the original valley rim above Units B and C, but daylights below it elsewhere. After Johnson and Vaughan (1983).

Figure 5.17 The main source scar at its greatest above The Tower (Unit B). The sandstone cliff has a degraded upper slope to the plateau rim and a long talus slope with abundant coarse debris in the trench. The rounded and split slip of Birchin Hat (Unit A) beyond contrasts with the ruggedness of The Tower (Unit B). Far below Unit A, the lower parts of the failure bulge into the broad trough of Alport Dale, where it widens out from a narrow, V-shaped valley. (Photo: R.G. Cooper.)

enough to be named on the 1:25 000 map as Birchin Hat, The Tower, and Little Moor. The extent of downslope movement and tilting varies considerably (see Figure 5.18). Unit A displays 100 m travel and forward rotation by 7°; Unit B displays 130 m travel and considerable backward rotation to leave a sharp crest; while Unit C has only descended 5 m, with a slight valley-ward tilt.

Little Moor (Unit C) is much the largest, about 300 m long (valley-parallel), leaving a deep wedge-shaped bay encroaching into the plateau by some 200 m from the inferred original valley rim. It has moved out by about 70 m, creating a trench 18 m deep which contains several back-tilted slices of rock. Some of these appear to be small rotational slips off the scar, with others produced by backward rotation from the rear of the slipped mass. Open tension fissures up to 8 m deep have begun to dissect the mass, which on its southern side has disintegrated into low chaotic ruckles below a degraded grassy return scarp. A ridge halfway down the steep outer slope appears to be formed by forward rotation, and with similarly tilted blocks on the outer face of Unit B suggests tensional stress on the plateau rim either before or during mass movement. The Tower (Unit B) is almost as long at 200 m, but has left a much shallower encroachment into the plateau. Its pinnacle is 30 m high, the crest reducing southwards to 10 m high as it splits into triple fins with 3 m deep trenches. Birchin Hat (Unit A) is only 100 m long, with a more rounded character; it has split to give 3–5 m anti-scarps (Figure 5.17). The source scar is here well below the valley rim, so that there is a steep grassy slope above the crag.

The rim is essentially intact except above Little Moor, where a 1.5 m furrow runs up to 15 m behind the plateau rim, a proto-slip extending 16 m into the plateau has dropped down by 0.5 m, and a lineament with linear pool lies 45 m in. These indications of incipient or latent failure extend over a length of 300 m.

Below the three 'castellations' a broad swale crosses most of the northern sector. Now partly infilled, it is partly drained by small streams, but below The Tower there is a large depression with ponds. This depression is impounded by another slipped mass, Unit D, which is about 300 m long. Its summit is of the Shale Grit, which constitutes the rim, and it presents an

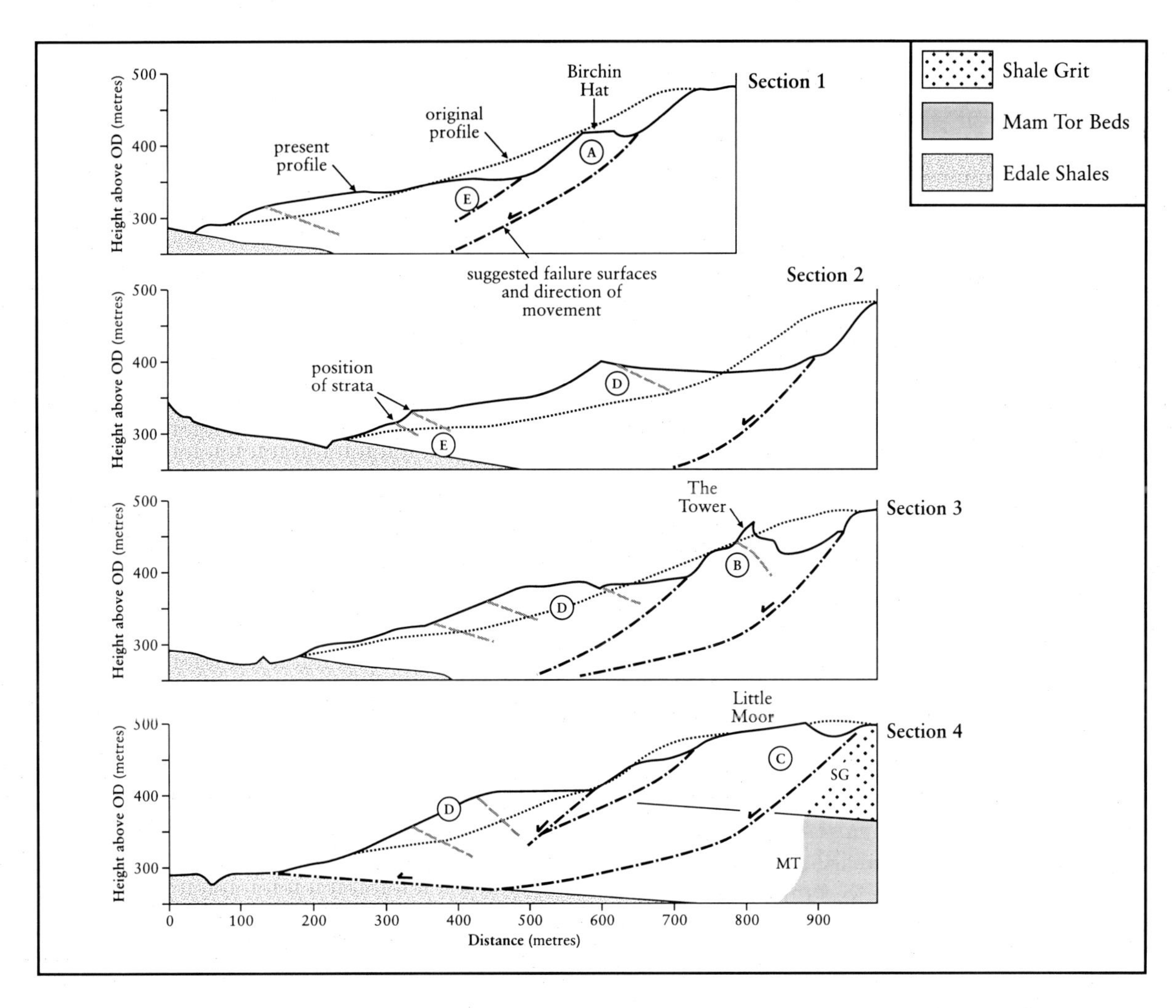

Figure 5.18 Profiles showing the varying arrangements of slump units at Alport Castles (letters refer to slump units in Figure 5.16). Note that the depth and nature of a failure surface or zone are not surmised. After Johnson and Vaughan (1983).

uphill edge 15 m high to the swale. There are tension hollows across its top, and multiple sub-metric antiscarplets down its dipslope.

The lower slopes (Area E) comprise a semi-coherent slip mass which presents an uphill edge 5 m high to a shallow transverse depression. Its surface is generally smooth, with several 1 m antiscarps. It bulges to narrow the valley appreciably, but although the slide toe steepens into a 15–25 m bluff above the river, which meanders in a flood plain upstream of this site, the river does not appear to have been significantly dammed or displaced. Conversely, fluvial erosion of the toe has caused minor undercutting without removing much of the material or re-activating sliding. Mam Tor strata are exposed on the surface and in the river cuts, and while individual elements appear intact, their varying inclinations show them to be much deformed by slide movements.

On the south flank of this slip complex (Units D and E), there is a distinctly different zone (F) headed by a small secondary scar. This is the product of more recent slumping, where saturation by impeded drainage has led to disintegration and in one part of the lower slope an 'earthflow'. The scar reveals Mam Tor beds which dip steeply backwards, suggesting that prior failure of the midslope had predisposed it to more complete collapse. The slump reaches the river, where erosion has produced an 8 m cliff in stiff structureless debris containing < 300 mm sandstone clasts, which contrast with the very weak thin flaggy mudstones revealed in the 'slipped bedrock' cuts, immediately upstream.

Southern sector

The wedge scar above Little Moor (Unit C) angles down the valley slope and then wanders across it discontinuously and at much reduced height. The main slipped mass (Unit G) is 300 m long at the scar foot, from which it is separated by a 60 m-wide swale. Streams rising in this depression descend either side of the slip mass, which broadens to over 400 m at its steep outer edge, from where minor earthflows or mud-slides have descended. The surface of Unit G is undulating, with several ridges and vales parallel to the contours. A further failure increment, Unit H, has descended a short distance from the acute wedge scar of Whitefield Pits (similar to that on the opposite side of Alport Dale – Figure 5.14). This forms the southern limit of the extant landslip complex, but immediately to its south the plateau rim is indented by a bowl which sharpens the angle of Alport Dale and Woodlands Valley, and appears to be an older landslip cavity with irregular terrain below.

Between areas F and G there is an amorphous zone of disturbed ground with blocky debris strews (area J), not recognized by Johnson and Vaughan (1983) as a distinct slip mass, but which appears to have descended from the rhomboidal space defined by the outer edge of Little Moor and the degraded angle of the source scar. The lower slopes beneath Units G–J are largely concealed by forestry, but their hummocky character suggests considerable disintegration and mobilization, followed by substantial consolidation. Only below area J does the slippage reach (and possibly deflect) the river. At Unit H slide debris is seen to over-run periglacial 'head' deposits.

Interpretation

The extensive landslipping here clearly reflects a combination of deep valley incision and geology, with weak strata exposed in the valley floor at the base of the failed slope. The preservation of very large masses of former plateau in varying degrees of intactness reflects the competence of the upper sandstone strata and the gradual nature of their translation. Little Moor (Unit C) is one of the largest known individual slipped masses, with an area of 0.06 km^2 that despite deep fissuring is substantially coherent. Its rhomboidal shape probably relates to a source cavity controlled by near-vertical joint orientations diagonal to the trend of the escarpment. Those units that have travelled further tend to be more disintegrated, but all bear signs of splitting along quite closely spaced NW–SE-trending joints.

Sequence of failure

The sequence in which the various units failed and separated from each other is unknown, and several scenarios and permutations can credibly be proposed. Johnson and Vaughan (1983) suggest that the failure began with displacement by rock-mass creep in the lower slope, with landslipping developing retrogressively toward the plateau edge as successive displacements took place. Similarly, landslipping would have extended laterally as failure in one part of the hillside removed support from the adjacent slopes. The varying locations of the source scars at different elevations up the valley side, on the rim, and encroaching into the plateau, lend weight to this incremental view, as do the distinctly separate Units A and H on either flank. However, the process may not simply be one of upward and lateral propagation. For example, the apparent 'fit' of Unit D between Units A (Birchin Hat) and B (The Tower), and the continuity of the scar above and between them, suggests that this unit is a much longer travelled slip mass. Unit D may thus have released at the same time as A and B and merely travelled further, with Unit E below being a subsequent subdivision. Alternatively, the swale across the northern sector, which presents a fairly continuous uphill scarp, could indicate an initial midslope rupture embracing Units D and F, which then provoked upward propagation. This interpretation could indeed be extended to embrace the entire suite of mid-slope units from D to G, without implying that all commenced moving in unison. However, the greater degradation of the most enigmatic area J might suggest that failure originated here, with the outer face of Little Moor either being intact plateau rim at that time, or part of a whole central sector (Units C/F/J) which failed at depth *en masse*, with the lower parts becoming more disaggregated, breaking away, and slipping and slumping to the slope foot.

Depth and mode of failure

The depth to which failure extends is equally unclear. The size of the coherent masses, and the heights of the scar and the trench walls, suggest depths certainly reaching 30–50 m (allowing for trench infilling) and possibly 60–80 m, a scale comparable with large slope deformations in the Highlands (see Chapter 2). However, the lateral margins are low, although Johnson and Vaughan (1983) suggest this is because of outward, as well as downward, spreading of the lower parts. Neither the position nor the nature of the basal failure surface can be readily determined, without geotechnical investigation. It seems unlikely that concave sliding surfaces could readily develop in the sandstones which comprise most of the landslip complex, and even less than a through-going planar surface could shear cleanly across the grain of joints and bedding. Although concave rotational failure is more feasible in the weak shales, these only crop out at the very foot of the slope and dip gently into it; while they may have helped mobilize the lower slopes, they seem unlikely to have influenced the higher parts of the complex some 150–200 m above (cf. **Mam Tor**, where they extend more than halfway up the slope). If mass movement is predominantly within the Mam Tor sandstones and the base of the Shale Grits (Figure 5.16), a zone of crush and deformation stepping down through the strata might be envisaged rather than a simple shear surface; this can more readily develop where weaker and stronger strata are intercalated. Above this zone of weakness, tension stresses would develop until rock masses gradually parted from the plateau along sub-vertical joints and slipped away. This process would account for the remarkable intactness of such large translated masses, and for the highly variable degree of both backward and forward rotation. Indeed, Johnson and Vaughan (1983) single out the gentle dip of the strata into the hillside, which normally predisposes against failure, as the main reason for the scale of the movement units.

They also divide the landslip complex into zones of depletion and accumulation, following Varnes (1978), whereby the latter zone stands proud at a higher level than the original (pre-failure) ground surface (Figure 5.18). This can arise either by debris over-running the intact lower slope, or by the landslip mass bulging out under compression. The latter must apply here, if failure has propagated from the slope foot, leaving no original ground surface in place; the antiscarped character of Area E attests to such compression. Johnson and Vaughan (1983) place the transition along the 'swale', such that Unit D lies within the accumulation zone. Indeed, this is clearly seen on the north flank, where the source scarp turns downhill beside Birchin Hat and neatly transmutes into a flank rampart near the forest edge (cf. **Benvane** GCR site report, Chapter 2).

Groundwater and failure morphology

Johnson and Vaughan (1983) recognize a strong morphological contrast between the upper and lower slopes, but suggest that this is a geological difference between massive sandstones above, giving rise to angular masses with castellated crags and scarps, and mudrock below, with smooth rounded ridges and wide troughs up to 100 m in amplitude. However, if most of the slip complex is in Mam Tor sandstones and Shale Grits, other factors must be found. The emergence of numerous streams from springs and seeps along the midslope (Figure 5.16) indicates that the lower valley-side is not free-draining despite rock-mass failure extending to the slope foot. The failed masses in the midslope area would become saturated, and thus liable both to superficial slumping and flowing, and to more pervasive degradation (even so, they have barely reached the slope foot, and have not gained sufficient momentum to become a landslide dam). By contrast, the upper units are dry today, and their arrested descent may indicate rapid dewatering at the time of failure. The band of incipient failure along the rim indicates where upward propagation had initiated vertical fracturing, with some slight settlement but with insufficient lubrication for movement.

Age of failure

It is reasonable to infer a Holocene age for most if not all of this complex. The relative freshness of much of the upper morphology implies lack of periglacial attrition, although the top 5–8 m of the scarp above The Tower is a battered grass slope in thinner or deep-weathered strata (an unusual hazard requiring fencing). The over-riding of periglacial head by Unit H has been noted, and pollen from a small peat lens in the slide toe suggests that the flows are not more than 8300 years old (Johnson and Vaughan, 1983). However, this need not preclude a history of landslipping here and in the vicinity earlier in the Quaternary. Alport Dale has a fluvially incised character in its upper reaches on the Bleaklow moors, but widens and straightens at the slide locus; the extent to which erosion by local glacier ice has played any part in slope destabilization merits further exploration in the Pennines.

Conclusions

Alport Castles is one of the largest landslip complexes on the sedimentary lithologies of inland Britain. It is particularly remarkable for the size and relative intactness of its individual movement units, some of which are striking and well-known landscape features. It clearly displays geological controls on both its location and its topography. The depth to which failure extends, the nature of the translation surface or zone, and the sequence of evolution are largely unknown, and Alport Castles presents excellent opportunities for further research of wider relevance in the Pennines. The model of upward propagation, after rupturing in weak strata exposed by valley incision, has been applied here and may account for the freshness of the uppermost units and the boldness of the main scar, which attains an exceptional 60 m plus in height. The scale of encroachment into the plateau by up to 200 m at Little Moor, with signs of further incipient extension, exemplifies the contribution of bedrock mass movement to valley widening, with local rates of scarp retreat vastly in excess of those yielded by all other slope processes. This is far from being an isolated case (cf. **Beinn Fhada**, Chapter 2; **Trotternish Escarpment**, Chapter 6), and represents an extensive suite of such slope failures in the vicinity and in similar lithological contexts across the Dark Peak and farther north in the Pennines.

CANYARDS HILLS, SHEFFIELD (SK 250 948)

R.G. Cooper

Introduction

The Canyards Hills GCR site is an area of irregularly ridged ground downslope from a 10 m-high vertical scar, in the Ewden Valley, south of Broomhead Reservoir, near Bradfield in South Yorkshire. The ridges are the complex physiographical expression of a large landslip, 1 km long from west to east, extending downslope from the scar for at least 0.4 km (Figure 5.19).

Canyards Hills is in the upper part of the Millstone Grit succession (Namurian, Upper Carboniferous), but higher in the succession than at **Mam Tor** or **Alport Castles**. The site is formed in Beacon Hill Flags and the Huddersfield White Rock, with a thick series of shales in between (Elliott, 1979). South-east of Wigtwizzle (Wightwizzle) the Huddersfield White Rock forms a gently sloping plateau with a steep northern scarp face. The north-easterly dip is causing the rock to slip over the underlying shales, and great masses of slipped material cover the slopes below (Figure 5.20). The western part of the plateau has been reduced by this process of denudation to a tongue of high ground only 180 m wide (Bromehead *et al.*, 1933). The Huddersfield White Rock exposed in the scar consists of massive well-bedded and open-jointed sandstones less than 30 m thick, dipping 6° north-eastwards, i.e. towards the river. Overall, the slope is concave in profile. The shale outcrop occupies the longest downslope segment of the slope in the western part of the area, and the profile is most concave there also. The landslide area also occupies the longest segment of the slope in the west, but does not reach the river until the centre and east (Bass, 1954).

Hunter (1869, writing in 1819) described the landslide as 'The Canyers, a range of conical hills

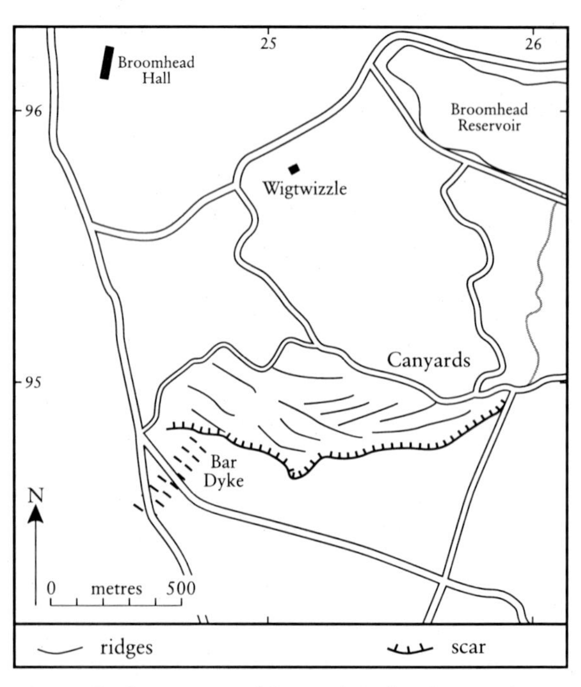

Figure 5.19 Location of Canyards Hills, showing linear features below the rockface and the upper slope.

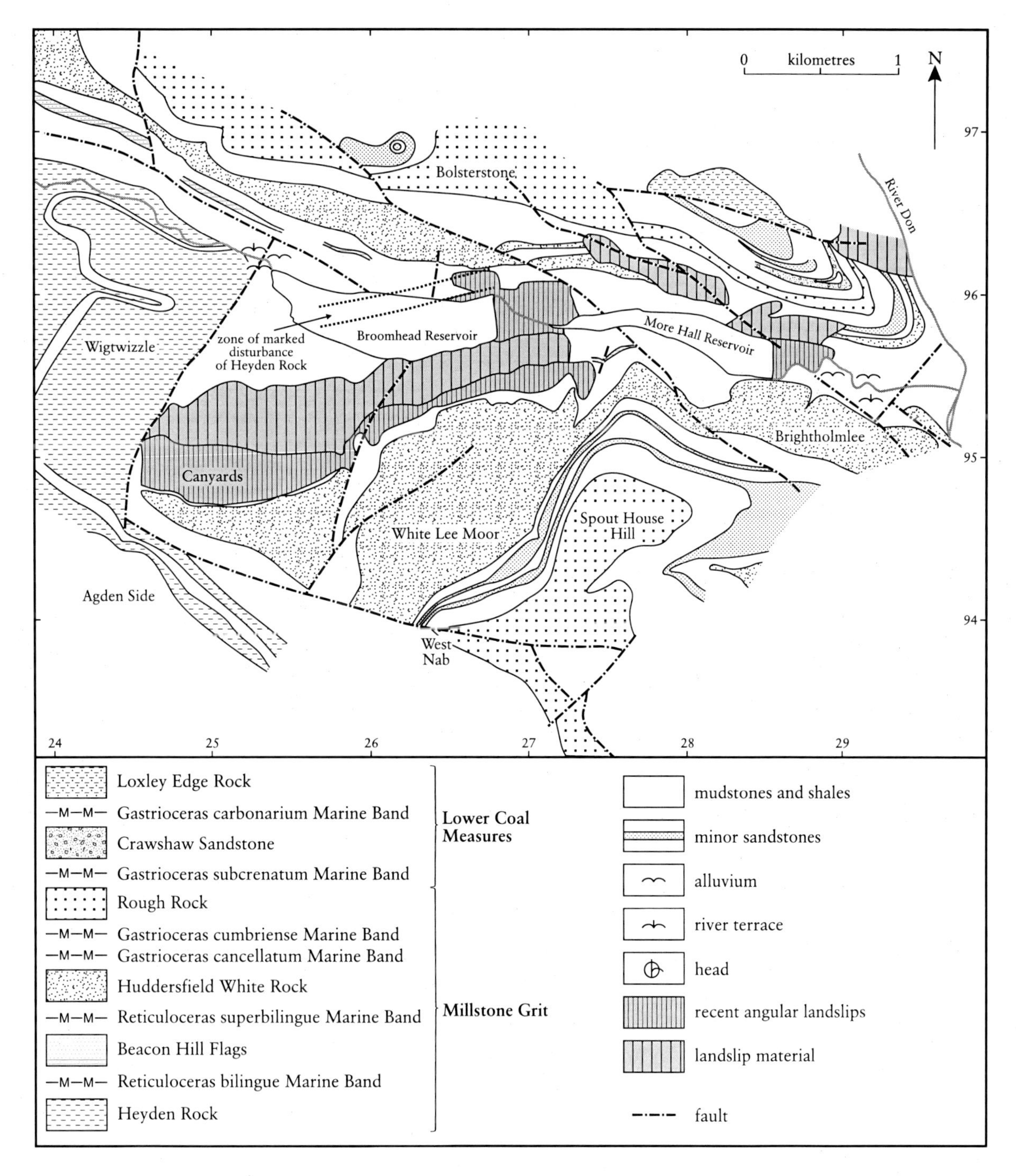

Figure 5.20 Geological map of the setting of the Canyard Hills landslide complex south of the Broomhead Reservoir.

stretching about a mile', while Hepworth (1954) remarks, 'Canyard Hills were formerly called 'Kenhere' or Kenyer Hills'. These variations of name (now fixed by the Ordnance Survey as 'Canyards') probably result from the remote location, but the landslip certainly does not consist of 'conical' hills.

Description

The site is chiefly remarkable for the very large number of irregular ridges running along the slope approximately parallel to the cliff-face (Figure 5.21). They enclose numerous poorly drained and often marshy elongate troughs.

Figure 5.21 General view of the Canyards Hills landslide complex from the south-west, showing the elongate ridges and troughs of the upper slope and the lateral extension of the slope. (Photo: R.G. Cooper.)

Transects downslope may cross as many as eight ridge-and-trough pairs, with amplitudes of 4–5 m, but generally decreasing in amplitude with distance downslope (Figure 5.22). In the west the area has separate ridge-like masses with breadths of up to 90 m, rising in some cases to over 15 m. They are rotational and appear to be aligned along curves which become shallower closer to the scar. They are covered with grass and bracken, and have steep sides. A few in the west are still partly attached to the scar. To the east of Canyards Brook the width of the area

Figure 5.22 Canyards Hills landslide complex viewed from ENE – the uppermost failure blocks and arcuate scar. (Photo: R.G. Cooper.)

decreases and the landslides have more bench-like features. A few are ridge-shaped, but most have wide and steep downslope sides and little or no backslope (Bass, 1954).

Interpretation

The features are thought to result from major break-up of the sliding mass during a single movement, as there is little to suggest that a succession of upper slides has taken place causing gradual cliff recession. The ridged physiography is probably due to break-up along the lines of fairly closely spaced pre-existing joints in the Huddersfield White Rock. This physiography is similar to that developed in shales, Shale Grit, siltstones and sandy shales of the Namurian at Bretton Clough, 18 km to the south (Boggett, 1989). Wood (1949) records that construction of the Broomhead Reservoir at the foot of the slope was begun in 1913 but not brought into service until 1936, in part because remedial works were necessary to stabilize 'hillside ground movements'. Therefore it would seem that all or part of the Canyards slope movements have been subject to artificial stabilization.

Conclusions

Canyards Hills is probably the largest site, and has the most pronounced examples in England and Wales, of closely spaced hillslope ridges with intervening troughs, which are both sub-parallel and sub-regular in form. As such they represent an unusual form of lateral extension failure caused by retrogressive unloading along a weak Namurian shale layer. The associated ridge-trough form is due to the coherence of the Huddersfield White Rock and the joint spacing.

LUD'S CHURCH, NORTH STAFFORDSHIRE (SJ 987 656)

R.G. Cooper

Introduction

Lud's Church is a vertical fissure in Roaches Grit (Namurian R_2b; C.M. Jones, 1980). It lies on a north-facing slope in Back Forest, overlooking the River Dane, in the Staffordshire moorlands 10 km to the north of Leek (Figure 5.23).

Description

The Lud's Church fissure is remarkable for its size: it measures about 165 m from end to end, and including all its side passages its length totals 220 m. For much of its length it is 4–5 m wide, and up to 18 m deep (Figures 5.23 and 5.24).

Associated with the fissure are hillside trenches and their associated intervening ridges ('ridge-and-trough' features) and a curious tor known as 'Castle Cliff Rocks'. This is sited 70 m from the fissure, and at a similar position about halfway down the slope. It rises 4 m above the surrounding soil surface. Its location on the slope is unusual in the Pennines (Palmer and Radley, 1961) and probably relates to its position on the surface of a slipped mass. There is a short steeper section upslope from Lud's Church, which could be interpreted as the degraded upper part of a landslip scar, but there is no trace of a toe farther downslope. These factors suggest that the slip which opened the Lud's Church fissure may be of some antiquity, perhaps immediately post-glacial. The toe may co-incide with the river bed or bank, and material pushed forward may have been washed away by the stream, or may have diverted the stream northwards into its present northward-arcing course (Figure 5.23).

Interpretation

Lud's Church was described by Hull and Green (1866), who noted that 'it gives the idea that the front of the hill has parted bodily from the main mass, and slipped a little forward, leaving this fissure along the line of fracture', (if this is correct the fissure is a tension crack or 'gull' marking the backscar of a landslide). More recently Millward and Robinson (1975) ascribed its origins to post-glacial incision of the River Dane. Lud's Church appears to have been formed as a result of the detachment of a large sliding mass as it began to move valley-ward over a possibly irregularly shaped slip-plane. The possible backface scar on the slope profile suggests that the Lud's Church fissure may be *within* the slipped mass, both of its walls having moved with the main slipped mass, followed or accompanied by a more surficial movement (at least 18 m thick) as the fissure opened. Cooper *et al.* (1977) noted the presence within it of fissures that are roofed-over by fallen boulders, forming covered tunnels up to 12 m deep.

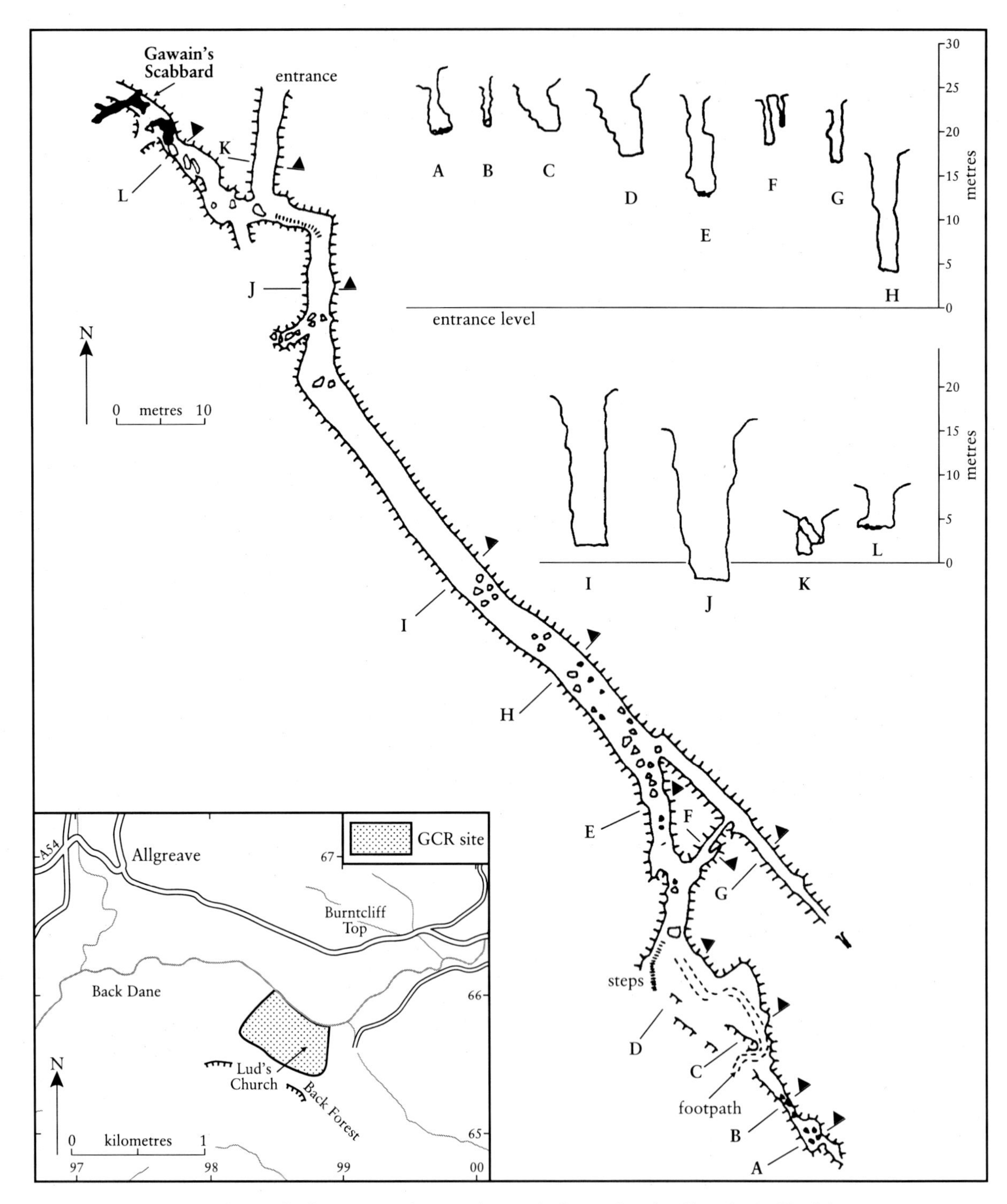

Figure 5.23 The location and general morphology of Lud's Church, Staffordshire.

Aitkenhead *et al.* (1985) describe it as 'a spectacular example in sandstone, of bedding-plane slip, which is common in major sand-stone units and in mudstone-with-sandstone sequences.' However, there is no feature lower down the same slope corresponding to the lower end of the mass which has slipped on the bedding plane. While Aitkenhead *et al.* (1985) seem to describe an essentially translational movement, it is uncertain whether the main movement, which opened the Lud's Church fissure, was translational, or whether material within a rotationally upper mass involving the whole of the slope from crest to stream, underwent a small

Figure 5.24 View along the 'labyrinth' of the Lud's Church fissure. (Photo: R.G. Cooper.)

translational movement equal to the width of Lud's Church.

Elliott's (1977) meticulous identification of Lud's Church with the 'Green Chapel' of the medieval alliterative poem *Sir Gawain and the Green Knight* (Tolkein and Gordon, 1967; Stone, 1974) may provide some indication concerning its age. The unknown *Gawain*-poet wrote in a north Midlands dialect which has been identified as late 14th century. It may be concluded, tentatively, that the Lud's Church fissure was both open, and wide enough to allow axe-swinging men to fight in, more than six hundred years ago.

Conclusions

Lud's Church is of educational importance because of the unique opportunity it provides to walk on a reasonably easy footpath through the interior of a large-scale 'detaching' landslide, and examine it from within. It is by far the largest such fissure within a landslipped mass in Great Britain, and may be the best example of a 'rock labyrinth' or 'lattice' structure formed by unloading.

MAM TOR, DERBYSHIRE (SK 130 836)

R.G. Cooper and D. Jarman

Introduction

Mam Tor is the locus for one of the most conspicuous and active landslides in the inland sedimentary strata of Britain. It has long been well known as the 'shivering mountain'. The head of the Hope Valley in the Peak District (Figure 5.25) rises steeply to the little plateau top of Mam Tor (517 m), the side of which has sheared away leaving a bold rock scar 80 m high above talus and the top of the slip mass (Figure 5.26). Failure probably occurred as a single event in about 3600 BP, an unusually recent date for inland areas. It is inferred that the initial slide came to rest at a marginally unstable angle, rendering it liable to re-activation after periods of heavy rain. It is believed to have extended downslope by over 500 m at very slow and gradually reducing rates of intermittent creep, but whereas almost all similar failures in the Pennines are now inactive, Mam Tor is exceptional in being predicted to continue moving for a long time to come. This reflects the unusually weak pyritic shales that both underlie the slide and comprise a proportion of the debris.

It is unusual for an inland slide in Britain to interfere with important communication routes. Here, the original direct road between Manchester and Sheffield (A625) became so affected by repeated slippage (Figure 5.27) that it was closed in 1979, necessitating unprecedentedly lengthy detours for trunk traffic. This has prompted a series of detailed geotechnical studies including Lounsbury (1962), Brown (1966, 1977), Stevenson and Gaunt (1971), Lant (1973), Lupini (1980), Al-Dabbagh (1985), and notably Skempton *et al.* (1989), making Mam Tor probably the best understood mass movement of natural origin in inland Britain. It provides an invaluable point of reference for interpreting similar and contrasting sites. It also has the benefit of a guide for visitors interested in its geology (Cripps and Hird, 1992), while Derbyshire County Council has placed an informative board at the foot of the closed road.

Although the Mam Tor slide is of only medium extent, affecting 0.35 km² (Figures 5.28 and 5.29), it is of wider significance as part of a cluster of four major extant landslips which collectively

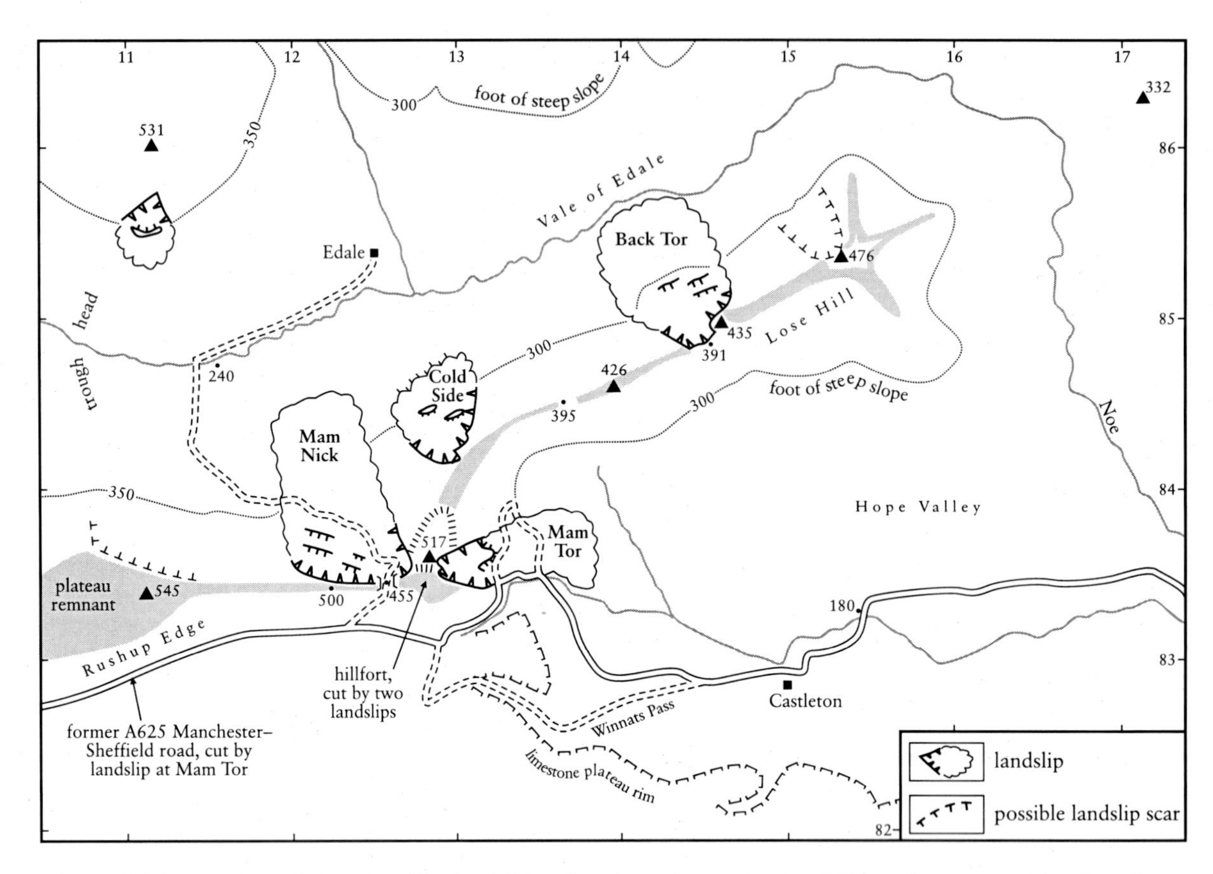

Figure 5.25 Location of the Mam Tor landslide, showing other major landslides also encroaching into Mam Tor hillfort and Rushup–Lose Hill ridge, and the former trunk road severed by it.

Figure 5.26 The Mam Tor landslide scar from the top of the upper slump sector. (Photo: M. Murphy, English Nature/Natural England.)

Figure 5.27 Oblique aerial photograph of the Mam Tor area, showing the old road which was closed in 1979 owing to repeated slippage. The scar of the landslip is clearly visible, as well as a slumped mass. (Photo: © National Trust/High Peak.)

have encroached into the Rushup Edge–Mam Tor–Lose Hill ridge (Figure 5.25). The summit of Mam Tor is ringed by the earthworks of a hill-fort, except for two sections where landslip scars cut into the hill from opposite sides, adding an archaeological dimension to site interpretation.

Description

As with many landslip locations in the 'Millstone Grit' areas of the Pennines, the slopes of Mam Tor are predisposed to mass movement by virtue of successions of weak strata underlying more competent rocks. Here, the top 100 m are of Mam Tor Beds in which micaceous sandstones alternate with siltstone and shale (Spears and Amin, 1981; Figure 5.30). Beneath them, the Edale Shales are hard mudstones with occasional bands of siltstone and ironstone. Pyrite occurs at several horizons, generally as scattered crystalline aggregations. Below some metres of weathered material, the mudstones are weakened by fissuring to about 10 m depth, probably resulting from stress-release after valley erosion, accentuated by Pleistocene permafrost action (Skempton *et al.*, 1989; cf. **Rowlee Bridge** GCR site report, this chapter). The strata dip NNE at about 8°.

The scar itself stands at an average angle of 45° (range 40°–51°) with a crest at 510 m above OD, and a free-face height of 80 m, entirely in Mam Tor Beds (Figure 5.29). It is an asymmetrical rectilinear wedge, but with little evidence of strong joint control. Talus extends about 30 m down from the scar foot at an average slope of 23° (range 18°–28°). The partially evacuated cavity has minor rockfalls and slumps within it. The debris mass has extended to 1000 m long at an average gradient of 12°; it is gently convex and attains 450 m wide, although the scar itself is only about 200 m wide.

Skempton *et al.* (1989) divide the landslide debris into three sectors (Figure 5.28).

(A) Upper 'slump' sector

The upper slump sector is that part of the initial slide mass that has travelled a relatively short distance, and remains largely where it first came to rest before subsequent extension of its toe. It extends from about 370 m above OD down to

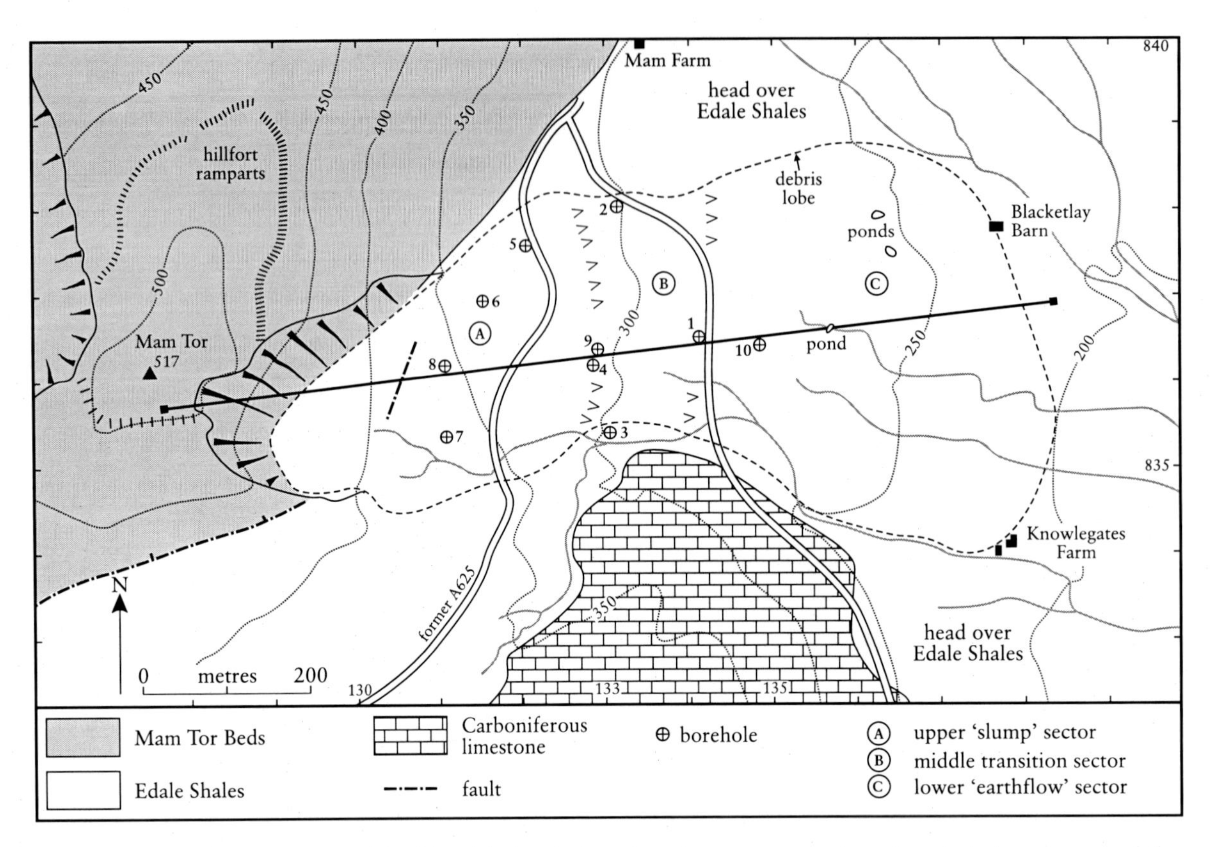

Figure 5.28 Schematic plan of the Mam Tor landslide, showing sectors, geology, borehole locations, and the former trunk road. The line running almost west–east is the line-of-section shown in Figure 5.29. After Skempton *et al.* (1989).

about 310 m above OD. Geological markers indicate that the slump mass has moved about 160 m (Figure 5.31), although the actual distance from the centre of the cavity to the lower contour is over 400 m (Figure 5.29). The slump has a very irregular surface, initially rather flat-topped then steepening where traversed by the upper leg of the former road. Borehole 8

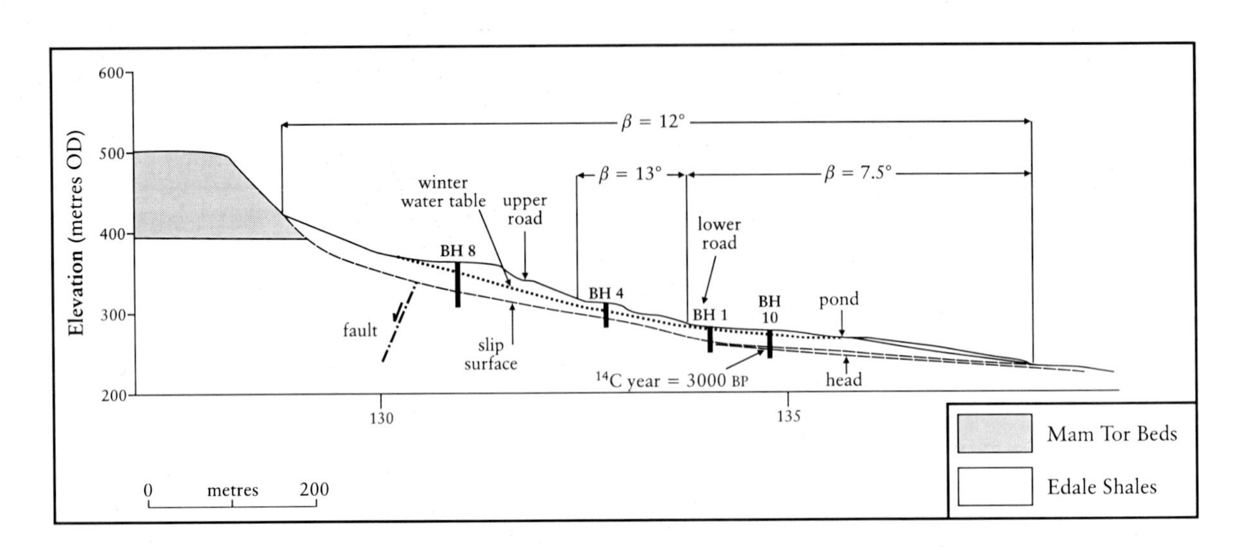

Figure 5.29 Longitudinal section and location of boreholes through the Mam Tor landslide. After Skempton *et al.* (1989). The section line is shown in Figure 5.28.

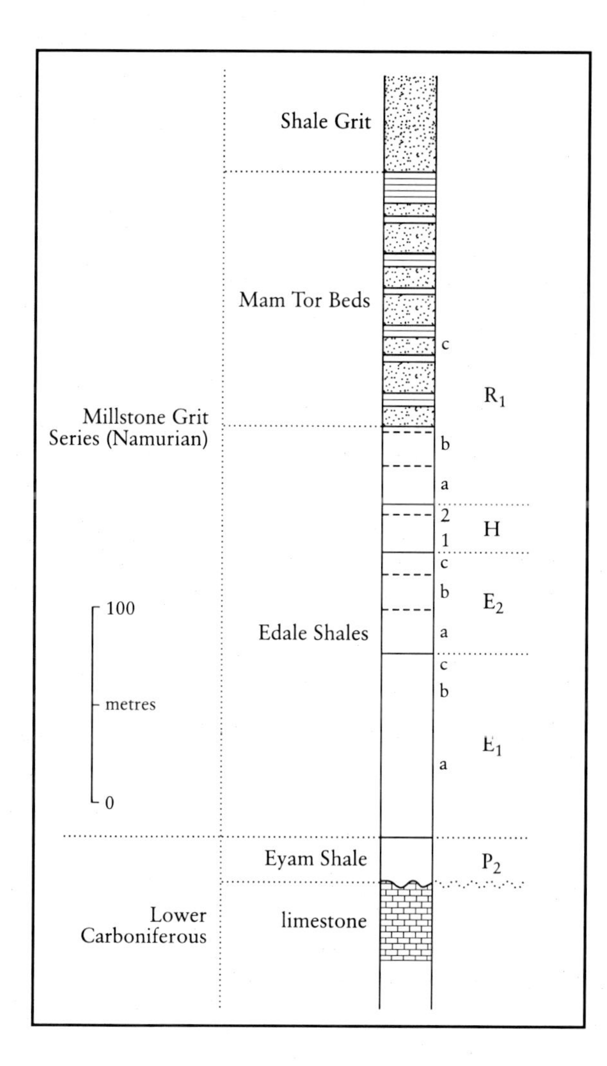

◂**Figure 5.30** Stratigraphical section at Mam Tor. After Skempton *et al.* (1989).

(Figures 5.28 and 5.29) proved a slide thickness of 32 m here, with two polished and striated slip-surfaces, 10 cm apart, near the base of 2.6 m of brecciated clay. This shear zone cuts through unweathered Edale mudstone, of which at least 8 m has been removed (in addition to weathered material and superficial deposits). Landslide debris includes pieces of mudstone from the upper Edale Shales (Zone R_{1a}) in the base, and 10 m above there are blocks of micaceous sandstone from the Mam Tor Beds, back-tilted at 40°–45°. Though broken and distorted, and missing a portion of Zone R_{1b}, the rocks are recognizably in their correct sequence, and show a displacement along the slip-surface here of about 160 m. Similar features were found in several other boreholes (Skempton *et al.*, 1989).

(B) Middle transition sector

The middle transition sector extends for about 150 m down to the lower leg of the road at about 280 m above OD. The terrain is less irregular but still quite steep, ranging between 9°–16°, and with transverse compression ridges. The debris is generally less than 20 m thick

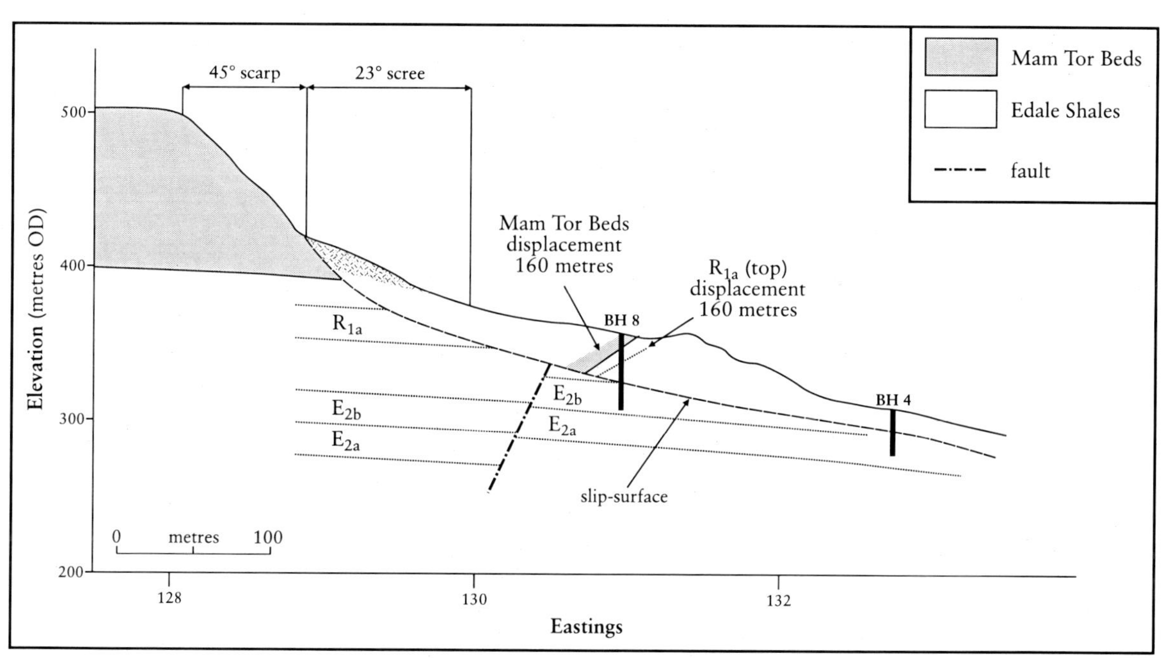

Figure 5.31 Geological and morphological details of the scar and the upper slump sector of the Mam Tor landslide, showing the locations of boreholes 4 and 8 (BH 4 and BH 8). After Skempton *et al.* (1989).

(Figure 5.32), and becomes very thin at the road. Borehole 4 proved a slide thickness of 11.8 m to a 1.8 m shear zone of brecciated clay with a thin layer of intensely sheared clay at its base on or just below the top of the weathered mudstone, which was disrupted for a further 2.2 m. This indicates that by this point no intact bedrock was incorporated in the slide.

(C) Lower 'earthflow' sector

The lower earthflow sector extends for about 420 m, down to a present lowest point of 220 m above OD. The slope varies between 6°–9°, and the deposit is generally less than 10 m thick. This part of the landslide has moved in almost pure translation by sliding on or just below the original ground surface, which is here little disturbed. The flow stands above the adjacent ground with clearly defined flanks and toe. The surface is hummocky, with transverse ridges in the upper parts, and contrasts strongly with the smooth fields alongside. The slide tongue has spread laterally, widening to 450 m compared with 250 m in the upper sector. Ponds and marshes indicate high winter groundwater levels, while there are perennial surface streams.

Near the head of this 'earthflow' Borehole 10 proved a slide thickness of 19.3 m, with a slip-surface 10 cm above the base of 1.2 m of brecciated clay. Immediately beneath this shear zone, 15 cm of peaty material contained small pieces of wood and an alder (*Alnus glutinosa*) root, above 5 cm of structureless grey clay. Together these form a fossil topsoil buried beneath the landslide.

Records of landsliding

In the historical period, records of awareness of Mam Tor's crumbling character date back to Michael Drayton's *Poly-Olbion* (1622). Charles Cotton's *The Wonders of the Peake* (1685) described seven wonders, one of which was Mam Tor:

> 'To the South-East is a great Precipice,
> Not of firm Rock...
> But a shaly Earth, that from the Crown
> With a continual motion mouldring down
> Spawns a less Hill, of loose mould below...'

The Wonders of the Peake was influential because all later visitors felt obliged to see the seven wonders described by Cotton. The most graphic description is probably that given by Celia Fiennes in the late 18th century (Fiennes, 1947):

> 'The fifth Wonder is Mamtour which is a high hill ... next Castleton...on that side its all broken that it looks just in resemblance as a great Hay-Ricke thats cut down one halfe, on one side that describes it most naturall, this is all sand, and on that broken side the sand keeps trickling down allwayes ... [it is] ... very dangerous to ascend and none does attempt it, the sand being loose slips the foote back againe.'

However, these accounts appear to refer to the scar itself being active, and a rare and conspicuous exposure of friable, layer-cake bedrock, rather than movement of the debris lobe.

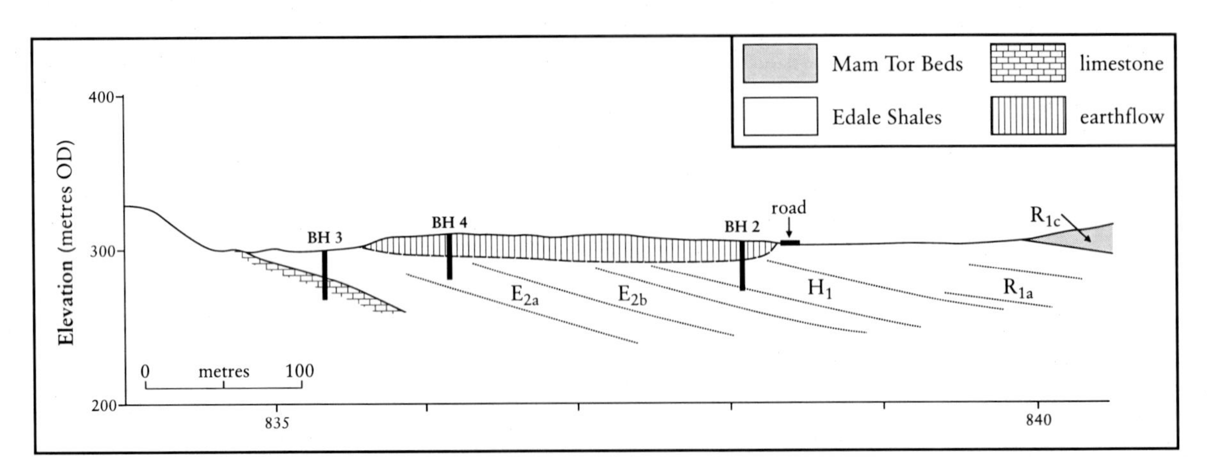

Figure 5.32 Transverse section of the middle transition sector of the Mam Tor landslide (Ordnance Survey, gridline 133). The shaded area is the landslide; borehole (BH) positions are marked. After Skempton *et al.* (1989).

The coach road was constructed in 1810, its hairpin actually taking advantage of the landslip to gain height and surmount the 100 m headwall of the Hope trough, which is otherwise only negotiable by the narrow fluvial channel of Winnats Pass (Figure 5.25). Evidently at that time, the terrain was not seen as hazardous.

Since 1907, notes of maintenance to the road after slips have been kept. This need not imply some general re-activation after long quiescence – this date co-incides with the advent of motor traffic, heavier vehicle loadings, and an expectation of a smooth, bound and better-drained surface. Until then, it is probable that cracks and minor slips would have been infilled and regraded as with any other road after a winter's attrition. Table 5.1 (Skempton *et al.*, 1989) shows a summary of such information since 1915, when local rainfall records became available. Slips of varying magnitude have been noted on 16 occasions during 66 years, on average at four-year intervals. Movements usually arise from re-activation of the transitional and lower sectors of the landslide, revealed by tension cracks in or above the upper leg of the road, accompanied by subsidence and outward displacements. In some cases there is upheaval on the edge of the slide.

In winter 1965–66 almost the entire landslide lobe re-activated. On 10 December 1965 cracks appeared following 120 mm of rain in six days. Abrupt movements were noticed on 18, 23 and 29 December, in each case within a day of further rainy spells. By mid-January, when movement had practically ceased, the total displacement in the upper sector amounted to 0.7 m, and shear displacements of about 0.4 m were observed where the road crosses the flanks of the slide (Brown, 1966). Most of this movement would have taken place during the last 20 days of December at an average rate of around 30 mm per day. The upper road subsided by as much as 1.5 m in places and a local 'confined' slip developed over a short width below the road. Activity renewed in February, mainly in response to 100 mm of rain in 10 days towards the end of the month. The rate at that period was about 15 mm per day and diminished almost to zero by mid-March (Figure 5.33).

Movements at the lower road are less than at the upper road, an observation consistent with the existence of compression ridges; indeed, it remains open for access to Mam Farm. Forward movements of the toe are therefore smaller than those at the upper road, though on a long timescale the difference cannot be great, and clear proof of advance at the toe in recent times is provided by slide debris encroaching on Blacketlay barn (see Figure 5.28).

Table 5.1 Records of movement and rainfall at Mam Tor, 1915–1977. After Skempton *et al.* (1989).

Slip	Date	Movements	Monthly rainfall (mm)
1	Jan 1915	Crack 30 m long	200
2	Dec 1918	slip, 0.3 m subsidence	240
	Jan 1919	movements continue	140
3	Dec 1919	steady movement	280
	Jan 1920	movements continue	200
4	Dec 1929	serious slip	300
	Jan 1930	movements continue	180
5	Jan 1931	slip, 60 m crack	210
	Feb 1931	movements continue	190
6	Feb 1937	considerable subsidence	220
7	Jan 1939	100 m crack, 0.25 m subsidence	210
8	Oct 1942	30 m crack, 0.1 m subsidence	160
9	Feb 1946	extensive slip	240
10	Nov 1946	new movements	230
11	Feb 1948	subsidence on 200 m length (preceded by 280 mm rain in Jan)	100
12	Dec 1949	slip (no details)	230
13	Jan 1952	large slip (preceded by 400 mm rain in November and December)	150
14	Dec 1965	serious slip, 0.7 m displacement	320
15	Feb 1966	renewed movement, 0.3 m displacement (preceded by 385 mm rain in December and January)	190
16	Feb 1977	large slip; 0.4 m subsidence (average)	230

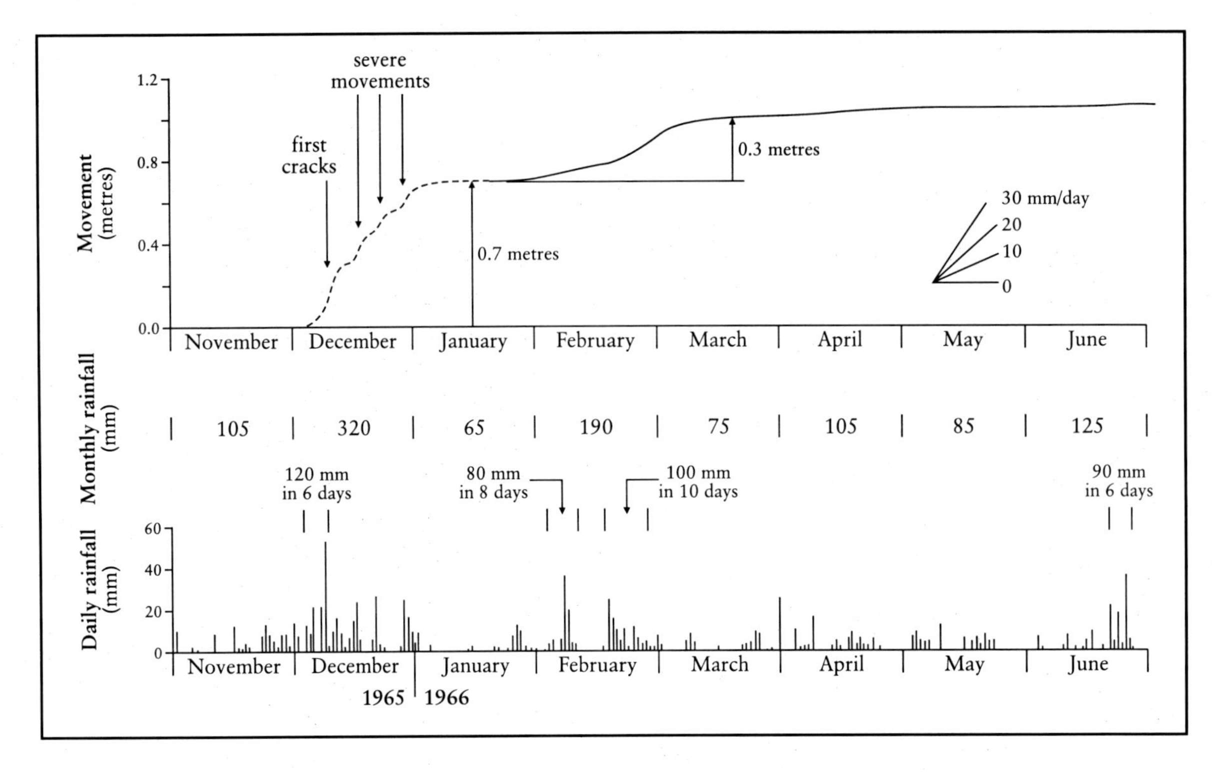

Figure 5.33 The relationship between rainfall and landslide movements, winter 1956–1966. Recorded by Brown (1966), updated by Skempton *et al.* (1989).

From observations in 1918, 1939, 1965, 1966 and 1977 the average displacement of the upper road when a re-activation slip occurs is about 0.3 m, allowing for the fact that not all slips involve the full width. With a return period of four years this is equivalent to 7.5 m per century, and because the toe of the landslip will have moved by a rather smaller amount the present rate of advance is likely to be about 7 m per century.

Rainfall and groundwater

As is commonly reported in Britain and elsewhere, landslide mobilization can relate closely to groundwater availability, both seasonally and in relation to peak rainfall events. The unusually detailed records for Mam Tor exemplify this relationship. Remobilization events are most frequent in December–February, while rainfall is heaviest from October to February (Figure 5.34). Records from local rainfall stations were analysed by Skempton *et al.* (1989) in terms of the return periods of rainfall amounts over 3-day, 6-day, 10-day, and 1-month periods. Comparing these with the situation at Mam Tor in December 1965 shows that instability was almost certain to occur at this time, and in the case of the February 1966 situation (Figure 5.33) when the 10-day rainfall was around 100 mm, the analysis showed that for every ten such events about five may be expected to result in a slip. When no slip occurs, this is likely to be because winter groundwater has been lower than average.

Observations from piezometers installed at or near the shear zone for short periods in 1977 and 1978 give an indication of the seasonal response of the landslide's groundwater levels to rainfall (Figures 5.35 and 5.36). In winter months, when the soil moisture deficit is effectively zero, the greater part of the rainfall not lost as runoff penetrates to augment groundwater. Under such conditions the 'storm response' in groundwater level is more or less directly proportional to the rainfall. While the ratio may vary locally with permeability, slope angle, depth to water table and intensity of rainfall, in uniform strata the response, seasonal as well as short term, is practically the same at different depths. Thus for winter rainstorms capable of causing substantial movements, the corresponding transient groundwater rise in the lower sector is about 0.5 m. In the upper sector,

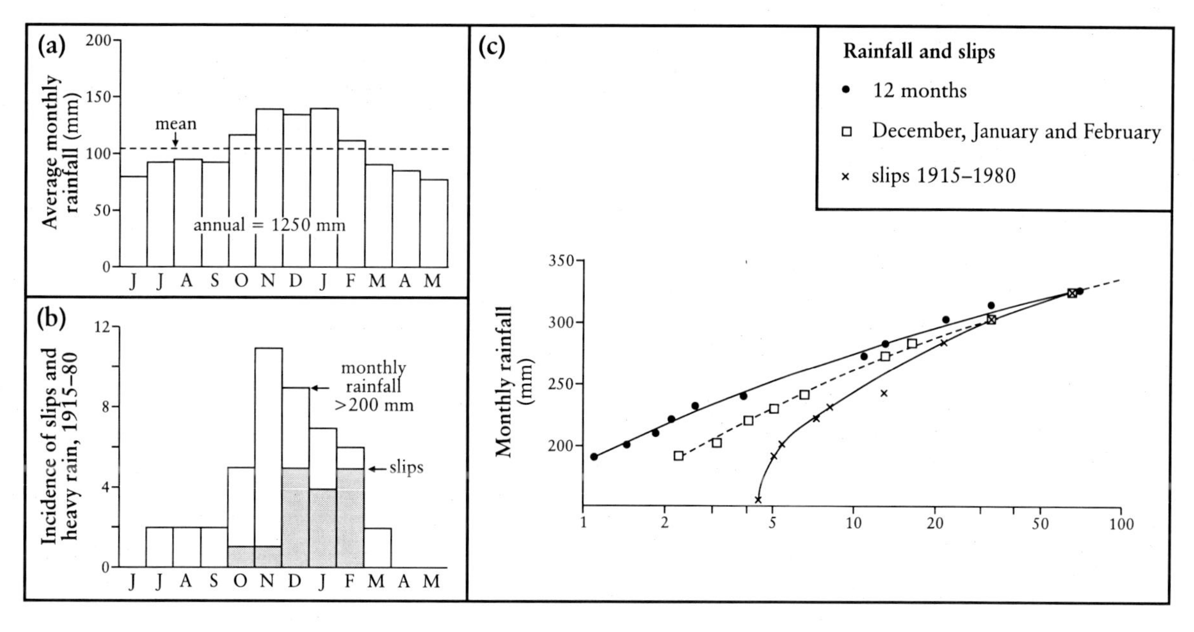

Figure 5.34 The relationship between average monthly rainfall and the incidence of heavy rain (1915–1980) and the return periods of monthly rainfall on an annual and winter basis. After Skempton *et al.* (1989).

permeability is higher, where broken sandstone exists in the debris, and winter groundwater is at a greater depth; the average slope is steeper, but runoff from the scarp face will contribute a throughflow component. Therefore storm response in this part of the landslide is not very different from that in the lower sector, and in any case it is unlikely, even as an upper limit, to exceed the seasonal response of 0.7 m measured in a borehole in the upper sector.

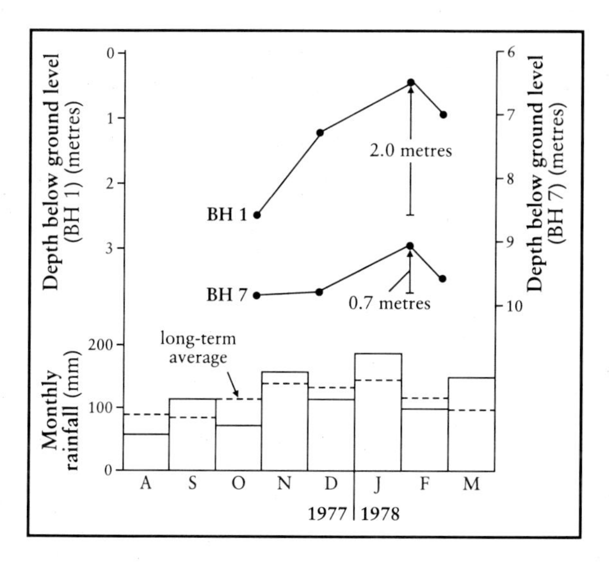

Figure 5.35 Piezometric levels recorded at Mam Tor during 1977 and 1978. After Skempton *et al.* (1989).

The chemistry of the groundwater at Mam Tor has been studied, with results of great significance. This is probably the classic discovery of the importance of the role of pyrite weathering reactions in lowering the residual shear-strength of the mobile horizon of a large landslide. Oxidation of pyrite (FeS_2) within the mudstones forms sulphuric acid, which liberates iron, calcium and other elements into solution. Chemical analyses have shown this process to be operating in the landslide (Vear and Curtis, 1981) because water issuing from springs and seeping off the lower sector in winter is acidic with an ion concentration much in excess of that in runoff from adjacent slopes. Steward and Cripps (1983) have shown that the residual shear-strength of the pyritic Edale Shale near Castleton is sensitive to modification of both mineralogical and porewater composition. At Mam Tor, weathering solutions penetrate deeply into the landslide and may attack fresh shale below the slip-surface, thus reducing its strength. Over a period of years, this would then reduce the factor of safety to a value close to unity so that other destabilizing effects could initiate a major failure event. Of course, the effect of this leaching on the strength of the landslide materials is likely to be small in the short term; strength and other geotechnical properties determined by Skempton *et al.* (1989) are those resulting from at least 3000

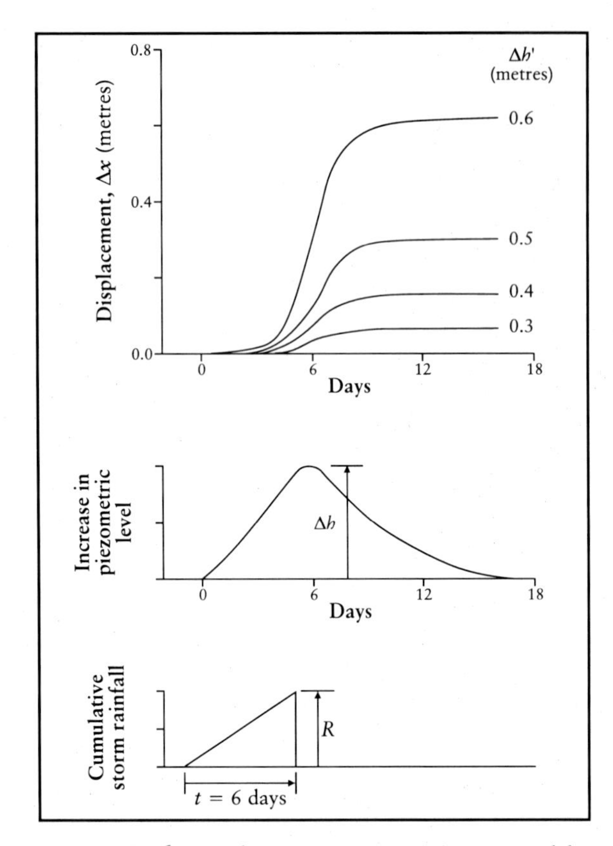

Figure 5.36 Displacements at Mam Tor caused by winter rainstorms. Note: movement begins at a piezometric level that is lower than the level at which movement ceases. This may imply that there has been a strength gain that may be of chemical origin in which movement releases cations. After Skempton *et al.* (1989).

years of debris-lobe activity. The importance of geochemical research here is in demonstrating that residual shear-strength is dynamic over the long term, and that there will be small seasonal variations in strength.

Geotechnical analyses

Skempton *et al.* (1989) carried out detailed geotechnical analyses to establish the index properties of the slide debris, its residual strength and stability, and the mechanics of storm-response movements, leading to a comprehensive back-analysis of the whole sequence of past and contemporary movement. Note that what follows is a much-simplified account, extracting key findings relevant to the Mam Tor failure; reference should be made to the original study and standard engineering geology texts.

The slide debris can generally be classified as clays of medium plasticity, although sand particle content and sandstone block incorporation can considerably affect its properties. The average water content is about 21% of the dry mass. It has a porosity of 36%.

The peak friction angle in intact rock, and its reduction after failure to a residual friction angle, is essential to understanding landslide activation. Here, the differences between the overlying Mam Tor Beds and the underlying Edale Shales are demonstrated to be very appreciable:

Mam Tor Beds	Peak strength	37	Residual strength	30
Edale Shales	Peak strength	30	Residual strength	14

The post-failure drop in strength therefore amounts to about 30% in sandstone and 60% in mudstone. At Mam Tor, where slip-surface testing was not feasible, estimates were obtained from the index properties. The value of 14°– 15° deduced for the shear zone compares well with test results of slip-surfaces developed in compacted mudstone at other sites.

Stability analyses demonstrate that in high winter groundwater conditions, as studied in February 1978, the landslide is close to limiting equilibrium, with the factor of safety at 1.0, because it can be substantially re-activated by a transient rise in water level of 0.5 m. Moreover, because large displacements have occurred in the past, the strength along a slip-surface must be at the residual.

Skempton *et al.* (1989) considered four cases involving re-activation of different parts of the slide mass (Figure 5.37):

Case 1: Re-activation of the whole transitional and lower sectors of the landslide, below tension cracks at the upper road (between points j–e–g), and sliding on a slip-surface passing through slide debris and along the basal shear zone. The best result was obtained with residual friction angles (RFA) of 18° (in slide debris) and 14° (shear zone) when the calculated value of factor of safety (*F*) is 1.02.

Case 2: As a variant of case 1 the slip-surface is assumed to thrust upward through the slide debris at f–k. This simulates a 'confined' slip

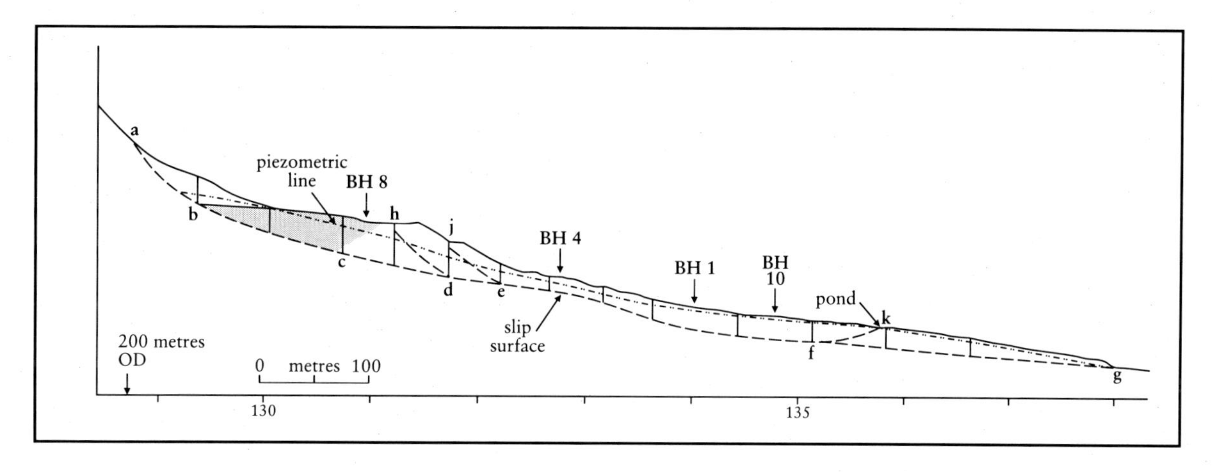

Figure 5.37 The stability analysis used by Skempton *et al.* (1989) at Mam Tor. Case 1: j–e–g; Case 2: j–e–f–k; Case 3: h–d; Case 4: b–c. The uppermost part of the failed mass is shaded.

not extending to the toe. With the same RFAs, $F = 1.00$, confirming that these two modes of failure have very similar probabilities.

Case 3: In the slip of February 1977 cracks additional to those at the upper road were seen above it, as at point h. On the slip-surface h–d a higher RFA value may be taken because the slide debris here probably includes a considerable proportion of sandstone, as observed in nearby Borehole 6. A reasonable assumption is that about one-third of the debris is sandstone, with RFA of 30°, and the rest is clayey debris with RFA of 18°. The average residual angle on h–d is then 22°, and, still using an RFA of 14° in the basal shear zone, $F = 0.99$. Had it been assumed, as perhaps an upper limit, that half of the debris consisted of sandstone, the average RFA value on h–d would be 24° and the corresponding factor of safety would increase by 1.7% to $F = 1.02$ (i.e. the higher the sandstone content, the less prone to re-activation).

Case 4: Exceptionally heavy rain in December 1965 led to re-activation of practically the entire landslide up to and including the talus. The talus is granular material with little if any clay fraction, in which the RFA can be taken as 30°. The uppermost part of the failed mass (toned in Figure 5.37) will contain a rather high proportion of sand, derived from the overlying predominantly sandstone debris in this upper part of the upper sector. The RFA in sandy clays is very dependent on relatively small changes in plasticity index and clay fraction. The procedure in this case was therefore to determine, by back-analysis, an RFA value in the shear zone b–c. This gave a factor of safety not less than 1.0, with an RFA of 14° in the slip below point c. The result is an RFA value lying between 23°, which gives $F = 1.0$ exactly, and 24° which gives $F = 1.02$. An RFA of 24° corresponds to a clay fraction of approximately 20%. This is an acceptable result, as compared to 35% in the shear zone further down the landslide where the debris consists chiefly or entirely of degraded mudstone and clay matrix.

These four cases demonstrate that no difficulty exists in showing that the landslide as a whole, and various parts of it, are delicately balanced in a state close to limiting equilibrium with groundwater level at about the normal winter maximum.

Skempton *et al.* (1989) conclude their stability analyses by examining the effects of variations in the parameters and assumptions used, including lateral confining pressures and their interaction with internal forces. A change of RFA in the basal shear zone of 1° leads to an increase or decrease in F by 5–7%. Changes of this magnitude they describe as not admissible, indicating that 14° is the correct value for the RFA within narrow limits. Finally, a change in groundwater level of 0.5 m results in a change in factor of safety of 2–3%. Although this only affects the RFA in the basal shear zone by ± 0.4°, such a rise in groundwater level would reduce the factor of safety by 3% which is sufficient to cause substantial re-activation of a landslide previously existing in a state of limiting equilibrium.

Storm-response movements are controlled by an apparent paradox. A rise in water table will cause an increase in pore pressure in the shear zone, and therefore a decrease in shear strength. Consequently the factor of safety (F) falls below 1.0 and movement takes place. However, in clays of medium to high plasticity the shear strength increases with rate of displacement and with changes in the soil chemistry; movement is therefore restricted to a finite amount. The rates of movement involved are sufficiently low (about 100 mm per day) for inertial forces to be negligible. Thus although a rise in water table causes an initial drop in factor of safety, the consequent remobilization leads to an increase in strength and cessation of movement. This is why displacements are chiefly concentrated within a few days of peak rainfall events, and why they become self-limiting regardless of continued high water-table conditions (in other words, movement does not continue indefinitely, or accelerate into a mudflow).

The total advance of the slide mass at Mam Tor by these episodic storm-response movements is currently less than 10 m in a century. This has produced a very small change in overall geometry of the slide mass, and therefore a correspondingly small change in static factor of safety under normal winter groundwater conditions. Consequently the process can be repeated many times without any considerable change in parameters. Nevertheless, on a long timescale the cumulative effect must be to bring the slide mass into a more stable configuration: F becomes marginally greater and a larger water-table rise is required to produce a given displacement. The return period for re-activation is therefore longer and the movement per century is smaller. Eventually, a state will be reached in which F is sufficiently high for the landslide to remain stable under the heaviest winter rainstorms, and this may be defined as the condition of permanent equilibrium, under present climatic and geomorphological conditions.

Interpretation

Although the nature and behaviour of the Mam Tor slide can be described in unusual detail, there still remain matters of interpretation.

Age of the landslide

Several lines of dating evidence point to a relatively young (mid-Holocene) age for the initial event. The fossil topsoil beneath the lower slide sector yields an age in pollen zone VIIb, which agrees with a radiocarbon date of 3900 BP for the fine-grained fraction. However, the *Alnus* root within it dated to 3000 +/–150 radiocarbon years BP, which is in agreement with other wood fragments sieved out of the buried topsoil. From correlations between radiocarbon and tree-ring dating (Pearson and Stuiver, 1986) the absolute age of the *Alnus* root is about 3200 ± 200 calendar years BP.

Skempton *et al.* (1989) estimated the date of inception of the landslide by reasoning as follows (Figure 5.38):

If at a time T in the past the toe of the lower sector was at a distance X from its present position, any curve relating X and T would have to satisfy these conditions:

(a) At the toe, $X = 0$, $T = 0$ and dX/dT is the present rate of movement (7 m per century).
(b) At the borehole containing the dated Alder root, $X = 320$ m and $T = 3200$ years.

By extrapolation to point B, where $X = 440$ m, the time T_1 to the initiation of lower sector movement can be found. For the time-displacement curve shown in Figure 5.38, $T_1 = 3600$ years, and dX/dT at that time is 0.5 m per annum. The initial slip, by comparison, would have been a sudden event. Therefore T_1 is the estimated date of origin of the Mam Tor landslide. This is about 3600 ± 400 calendar years BP, which agrees with the upper limit of the radiocarbon dates described above.

However the radiocarbon dates relate to material beneath the lower slide toe, and Figure 5.38 relates to re-activation of an initial slide which may previously have come to rest. Climatic changes to wetter conditions after 4000 BP could have contributed to re-activation, as could consolidation of the debris to elevating water tables. The possibility of an earlier initial event cannot readily be ruled out.

The archaeological evidence is also equivocal. The rampart is generally agreed to date from the Iron Age (flourished *c.* 2500 years BP), although remains of Bronze Age (*c.* 4000 years BP)

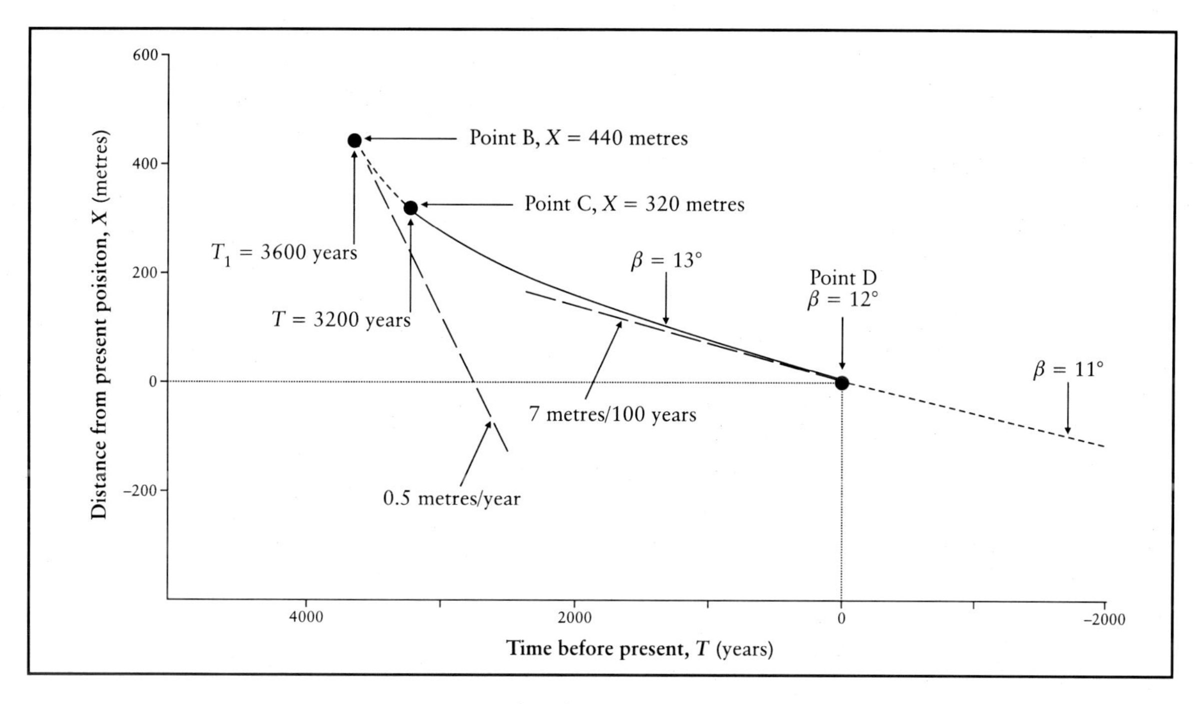

Figure 5.38 Method of determining age of Mam Tor landslide by projecting back from current configuration and rate of movement (points B–D as Figure 5.39). $T = 0$ corresponds to 1950 AD After Skempton *et al.* (1989).

dwellings have been found within it. Both the Mam Tor landslide and the Mam Nick landslide on the opposite side interrupt the rampart (Figure 5.25). Archaeological opinion is that the rampart was originally continuous, although both scars are so steep as to make construction of a rampart superfluous. Although the rampart appears to have been breached by a subsequent landslide, this could simply be the product of the later attrition which has created the large talus bank. The Mam Nick scar is however grassy, and a sample at the toe gives an age of 5900 radiocarbon years BP, suggesting that its flank scarp might have been incorporated in the defences.

Failure character and geometry

Mam Tor is described by Skempton *et al.* (1989) as a 'massive example of a slump-earthflow'. It is certainly a 'composite landslip' (WP/WLI, 1993). Following Hutchinson (1988; see Chapter 1), it can be classified as H4 (Landslides breaking down into mudslides or flows at the toe) and D3 (Compound failures – markedly non-circular, with listric or bi-planar slip-surfaces).

Although borehole evidence indicates relatively limited initial displacement, this may relate to particular back-tilted masses originating near the base of the scar cavity. It is possible that material released near the rim travelled further, over-riding the basal material to form the forward part of the initial toe. Further investigation of the rapidity of the initial movement, its degree of disintegration, and its trajectory is merited; Mam Tor has some of the topographical characteristics of **Beinn Alligin** (see GCR site report, Chapter 2), also in near-horizontal sandstones, if in miniature. The role of faulting in facilitating failure here might also be examined, along with proximity to the formerly mined zone of mineralization in the adjacent limestone.

The actual failure surface has been described as a 'concave-upwards curved slip-surface' evidenced by the back-tilted strata (Skempton *et al.*, 1989). Figure 5.31 clarifies that the upper sector is a listric failure, i.e. a spoon-shaped surface that is here essentially bi-planar (scar plane and basal plane) linked by a curve. In detail, this curve may take place within a shear zone rather than as a discrete smooth concavity, given that in this sector it is in unweathered bedrock. From borehole evidence a marked convexo-concave failure surface step-down in the transition zone to the lower sector is interpolated. This could imply that the weight of the initial failure mass surcharged the weak weathered shales and triggered a secondary failure with its own listric profile. The transition between the upper and the lower listric surfaces is poorly understood, but may incorporate some bedrock at the head of the lower sector. The existence of this transition, at the points selected for construction of the hairpin road, may account for the tensional dislocations that ultimately closed it.

Stages in Development

Skempton *et al.* (1989) interpret the probable evolution of the landslide in four stages as inferred in Figure 5.39a–d:

(a) The initial event was a single large slip, rather than several relatively small slips, because:

1. About 520 m from the present toe, in the transitional zone, the slip-surface is at a very shallow depth below the original ground level. Almost all of the slide debris east of this point, including the lower sector, must therefore have derived from material to the west.
2. For the same reason the initial slip or slips must have been to the west of this point, i.e. in the upper sector.
3. The volume of the initial slip or slips must equal that of the slide debris, after allowing for expansion due to softening and degradation. The bulking factor is about 20% in mudstone, 40% in clay matrix and (say) 10% in sandstone.
4. The original profile cannot have been much steeper than the steepest slopes currently existing adjacent to the landslide.
5. The basal slip-surface must pass through the points where it was observed in boreholes, and through the foot of the scarp.
6. Given the contrasting peak and residual strengths for the stronger overlying and weaker underlying strata (described above), the slip mass will undergo large and rapid displacements before reaching a position of (temporary) equilibrium.

(b) Trial-and-error solutions lead to a spreading displacement of the initial Mam Tor slip represented by point B. The slide debris is taken as having a volume 15% larger than that of the initial slip, to allow for bulking without a substantial increase in water content.

(c) As a result of degradation and softening, secondary slips occur in the lower part of the mass of the upper sector, leading to the development of a lower sector. Comparative studies (see below) indicate that the rate of advance of the lower sector would initially have been far greater, by roughly one order of magnitude, than the average figure in recent times. The advancing lower sector reached point C,

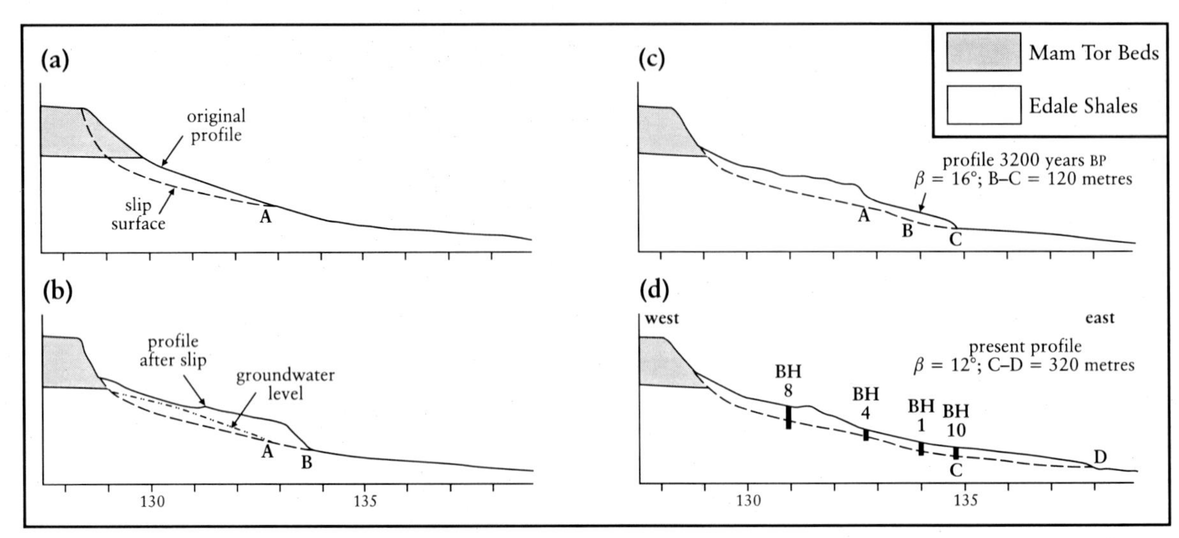

Figure 5.39 Inferred evolution of the Mam Tor landslide (ß = slope angle at toe): (a) initial failed mass; (b) arrested slide immediately after initial landslide event; (c) early stage in progressive advance of toe; (d) present profile. After Skempton *et al.* (1989).

320 m back from the present position of the toe, about 3200 years ago, as demonstrated by the age of the *Alnus* root in the buried fossil soil. At that time the debris slope angle would have been about 16°. Meanwhile, the upper sector itself had been moving, partly as a result of additional weight imposed by talus eroded off the scarp face, and in response to rainfall, but partly also as a result of material lost from its front edge, due to the retrogressing secondary slips removing material at a faster rate than could be supplied by forward movement of the upper sector.

(d) These processes are still continuing today. Their resultant effect is to bring the landslide into a more stable configuration, and the rate of movement per century will therefore be decreasing. At 7 m per century the present rate of advance is substantially less than the average of 10 m per century for the past three millenia. However it is clear that a state of permanent equilibrium has not yet been reached, and the present shape of the landslide Figure 5.39d, with a debris slope angle of 12°, is simply the latest stage in a development that will continue for a very long time.

Comparisons

The southern Pennines contain many other large landslides in Namurian strata. These have been described by Johnson and co-workers (Johnson 1965, Franks and Johnson 1964, Johnson and Walthall 1979, Tallis and Johnson 1980; cf. the **Alport Castles** and **Canyard Hills** GCR sites). Pollen analysis and radiocarbon dating show that some of these are older than Mam Tor, and, unlike Mam Tor, apparently stable (as at Alport Castles). Skempton *et al.* (1989) made map measurements of the of the debris slope angle at four other landslips to compare with Mam Tor:

Coombes Tor	9.5°	Rough Rock cap over shales
Millstone Rocks	12.5°	Kinderscout grit over Grindslow Shale
Didsburk Intake	11°	Kinderscout grit over Grindslow Shale
Mam Nick	10°	Mam Tor Beds over Edale Shales
Mam Tor	12°	Mam Tor Beds over Edale Shales

Collating these measurements with the datings, they showed that large landslides in Namurian mudstones remain unstable if the slope exceeds about 10–11°, and that a period of the order of 8000 years is required for such landslides to attain a state of permanent equilibrium. Thus the by-road which traverses the Mam Nick landslip (Figure 5.25) does not appear to have suffered any serious disruption.

Landslipping and the shaping of the Mam Tor ridge

This conspicuous landslide is only one of a cluster which significantly shapes the ridge on the south side of Edale (Figure 5.25). Most landslips in the Peak and Pennines are on plateau rims, and their main geomorphological contribution to landscape evolution is simply one of valley widening (as is well seen at **Alport Castles**). Here, the south wall of Edale commences as a plateau, but from the summit of Rushup Edge eastwards for 4 km to its terminus at Lose Hill it has a well-defined crest. Only the summit of Mam Tor itself broadens out, as a residual of the former plateau ridge.

The Mam Tor slide is the only one on the southern aspect of this ridge. On the north side, three major extant slips occur, although the terrain suggests that earlier events have embayed the ridge, the failed material having been evacuated by subsequent valley glaciers:

(a) Mam Nick (0.60 km^2): this landslide has twice the extent of the Mam Tor event, and narrows the crest of Rushup Edge to a half-arête for 500 m (Figure 5.40). The source configuration is an obtuse wedge. In its south-east corner, it breaks through the ridge to create the 'Nick' followed by the minor road over to Edale. Here it has lowered the ridge by about 40 m, and truncates the west flank of Mam Tor, including its hillfort rampart (Figure 5.41). The main headscarp is of steep grass approximately 25 m high, with a clutch of short-travel sharp-crested slip masses having bold antiscarps 3–5 m high. Beneath these, the apron of the main failed mass is crossed by the road, below which an amorphous 'earthflow' extends at a lesser gradient for 500 m down to an 8–15 m-high toe bank above Greenhill Farm (Figure 5.40). This two-tier configuration is very similar to Mam Tor.

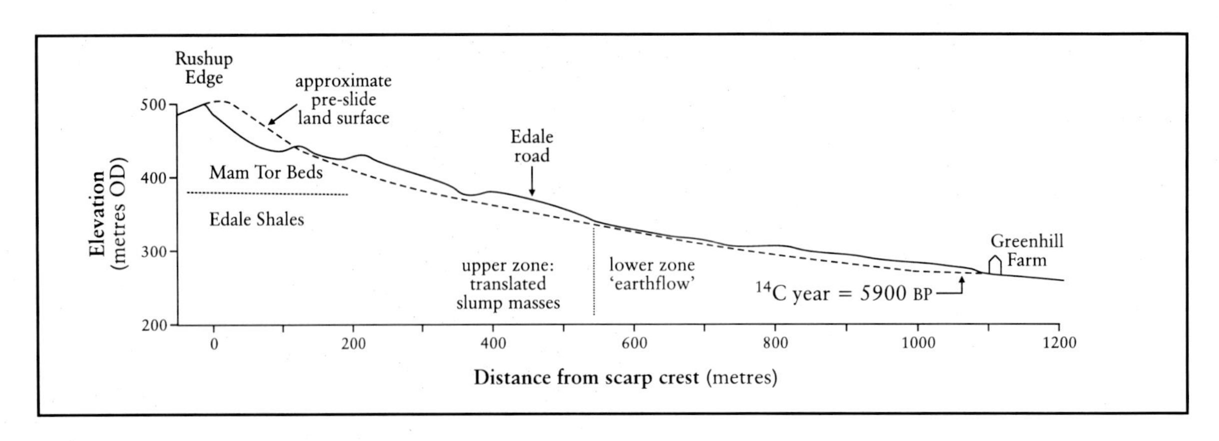

Figure 5.40 Landslide profiles at Mam Nick, where ß = 10°. After Skempton *et al.* (1989).

Figure 5.41 Mam Nick landslide source cavity and upper slump zone, from Rushup Edge looking east. Note the far flank scar interrupting Mam Tor hillfort rampart, and the headscarp narrowing the former plateau ridge to a crest. Edale road passes through Mam Nick and descends across the slump, which has grassy slip-masses presenting uphill-facing scarps. (Photo: D. Jarman.)

(b) Cold Side (0.25 km^2): the grassy source scars are exposed just below the crest north-east of Mam Tor (cf. **Alport Castles**). They are up to 32 m in height, in a double-wedge obtuse splay. The main slip mass has a striking antiscarp 8 m high impounding a pond, while the toe is a steep rampart 10–15 m high.

(c) Back Tor (0.50 km^2): this dramatic landslide bites right through the ridge west of Lose Hill for 200 m, lowering it by up to approximately 50 m (Figure 5.42). The main failed mass has slumped almost to the floor of Edale, possibly deflecting the river slightly. A more recent increment on the east side yields impenetrable antiscarped terrain colonized by Backtor Wood, below a 60 m sandstone crag comparable in scale with that on Mam Tor.

Figure 5.42 Back Tor landside from the east. The 60 m main crag is a source scar comparable to Mam Tor. The intervening ridge has been lowered by some 50 m by virtue of the slide surface here exposed behind the crest. (Photo: D. Jarman.)

Whereas these three slips are metastable, and the Mam Tor slide is evolving towards that condition, future re-activation in response to fluvial or glacial valley incision will tend to see these and other slope failures coalesce, first eliminating Mam Tor as a separate hill, and then reducing the whole ridge to a rump.

Conclusions

The landslide at Mam Tor is not unusually large or articulated, by comparison with (for example) the **Alport Castles** GCR site. In many ways, it represents the typical Pennine slump-flow where competent rocks overlie weaker strata that are less permeable and prone to deformation. However, Mam Tor is important for the high degree of knowledge of its recent evolution, including quantitative assessment of the relationships between movements, ground-water levels, storm response, and properties of the clay/mudstone (notably its geochemistry) of which the slide is largely composed. This is because it was crossed by a trunk road that eventually had to be closed because of continuing and irremediable slippage – the only such case affecting a major transport artery in inland Britain.

Mam Tor is unusual in still being an active landslide, with spasmodic advances associated with peak rainfall and raised water-table conditions. The current recession rate is 7 m per century, which, while gradually diminishing, has no foreseeable end-date. The nexus of factors that sustain instability are finely balanced, and this site is of great comparative importance. Mam Tor is also a conspicuous and readily accessible site, with a bold scar, and the 'shivering mountain' has long been known as a 'Wonder of the Peak'. It is of considerable interest to archaeologists, since this and the Mam Nick slide interrupt an Iron Age hillfort rampart. It is also important in studies of land-scape evolution, since this and several even larger sites have made substantial inroads into the ridge separating Edale from the Hope Valley. It is thought to be a relatively young feature, with an inferred mid-Holocene date of initial failure at around 3600 BP, although an older date remains possible.

ROWLEE BRIDGE, ASHOP VALLEY, DERBYSHIRE (SK 150 894)

R.G. Cooper

Introduction

On the right (southern) bank of the River Ashop, 130 m upstream from Rowlee Bridge (Figures 5.13 and 5.43) and 2 km before the east-flowing river enters the Ladybower Reservoir in Derbyshire, there is an exposure, 3 m high, of the highest Edale Shales (Alportian (H_2)) with the lowest Mam Tor Beds (Kinderscoutian (R_1)) of Namurian, (Upper Carboniferous) age.

Description

The exposure shows sharp, symmetrical, straight-limbed folds. Similar but less well-exposed structures exist in the Edale Shales and Mam Tor Beds in Edale, the Derwent valley and its tributary valleys including Abbey Brook and Ouzelden Clough, and in the Hope valley. Trenches excavated for the foundations of the Howden and Derwent dams exposed a single large fold in each case, decreasing in magnitude with depth (Lapworth, 1911; Sandeman, 1918; Fearnsides *et al.*, 1932) (Figure 5.44). Photographs of the trenches excavated during Sandeman's (pre-1910) construction of these dams and exhibited by Thompson in 1949 show that at the Howden Dam this fold is a 'crumple' in the form of a 'double V' fault in the strata. This extends to at least 15 m below the ground surface, but reduces with depth. At the Derwent Dam there is a similar crumple, but the crumpling is less complex: it takes the form of a simple 'V'; and at the bottom of the trench (about 15 m deep) the movement had largely died out (Thompson, 1949).

A more detailed investigation was undertaken on construction of the Ladybower Dam in the 1930s and 1940s, with a large number of vertical borings across the valley along the line of the dam (Hill, 1949). These revealed the presence of an enormous crumple extending down from the valley bottom for at least 58 m vertically (Figure 5.44). Thompson (1949) commented that the crumples occur throughout the Derwent and Ashop valleys.

Figure 5.43 Valley-bulge structures exposed in the Rowlee Bridge section, in the Ashop Valley, Derbyshire. The sharp, symmetrical folds are one of the most remarkable examples of compressional folding ever recorded. The folds are due to the extrusion of clays and ductile flow in bedded strata, often called valley-bulging. (Photo: R.G. Cooper.)

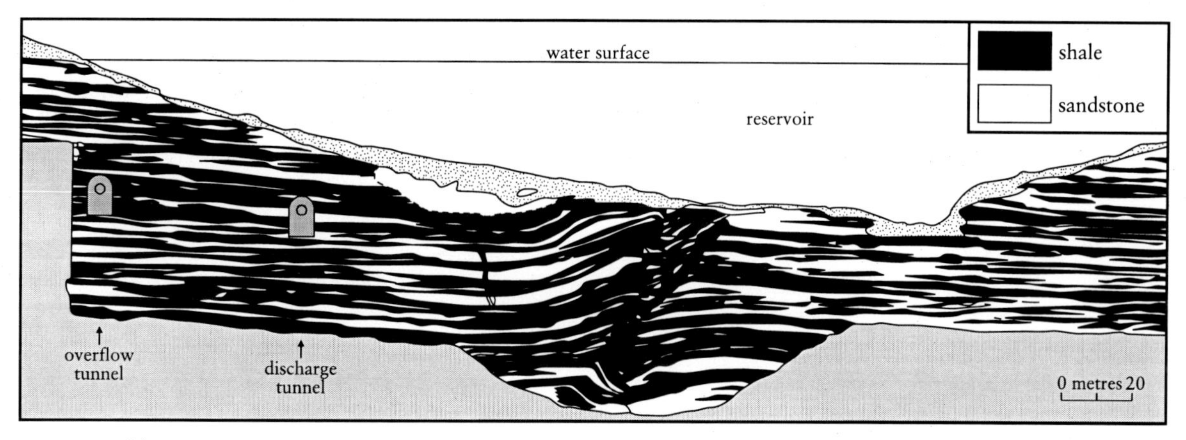

Figure 5.44 Section through valley-floor crumple (according to Thompson (1949)), at Ladybower Reservoir. After Hill (1949).

Interpretation

It may be inferred, therefore, that a similar crumple lies beneath the Ashop valley at Rowlee Bridge, in which case the folding observed at the site represents accommodation of the surface strata to a larger event that took place at greater depth.

Thompson's (1949) explanation for the crumples was that since the strata in the hillsides and at the bottom of the valleys consists of a succession of beds of shale and sandstone, the weight of the hillsides had compressed the shale and had caused it to flow towards the valleys, where the stress had been relieved by crumpling. In the case of the Ladybower Dam site the process had gone a little further, and the fold had become an overthrust fault. The reduction of the structure with depth distinguishes it from an ordinary geological fault. Stevenson and Gaunt (1971) came to broadly similar conclusions.

These structures have many similarities to those described in Northamptonshire by Hollingworth *et al.* (1944) and by Hollingworth and Taylor (1951); they are believed to have resulted from the pressure exerted on predominantly argillaceous strata by superimposed beds inducing lateral movements or 'squeezing out' of the argillaceous strata into the adjacent load-free valley areas, producing compression folds and thrusts (Stevenson and Gaunt, 1971). Movements caused in this way would be possible wherever valley-deepening exposed the top of any substantial thickness of shale, but would be greatly facilitated by thawing of ground-ice following a glacial or periglacial phase.

Conclusions

The site is at some risk from normal fluvial erosion processes. The River Ashop runs at the foot of the exposure itself, and clearly it is eroding the exposure. This would not matter but for the position of the exposure on a short promontory 3 m high, standing in the floodplain of the river. The promontory is a salient from what appears to be a river terrace, but it is only 5 m across, and when the river has eroded 5 m into it, the promontory and the site will have ceased to exist. The River Ashop does not, however, experience continuous natural flow levels at this point; about 200 m upvalley a variable proportion of its flow is from time-to-time taken into a concrete channel and thence by tunnel to the Rivelin valley as part of the water supply for Sheffield (Wood, 1949). This may be responsible for the site's survival thus far.

In the absence of a clear exposure associating valley-bulging with cambers and gulls in the Northampton ironstone field (the good ones have all been backfilled), the Rowlee Bridge exposure provides the clearest example of valley-bulge structures currently accessible in Great Britain.

Chapter 6

Mass-movement sites in Jurassic strata

INTRODUCTION

R.G. Cooper

The Jurassic strata in Britain (Figure 6.1) are susceptible to large-scale mass movements. Of the ten sites chosen for evidence of deep-seated slipping in Lower Jurassic strata, four have a substantial over-burden of non-Jurassic rocks that are also involved in the slides: in the Axmouth to Lyme Regis Undercliffs Nature Reserve (**Axmouth–Lyme Regis** GCR site) these are Cretaceous sediments, whereas at mass-movement sites on the **Trotternish Escarpment** on the Isle of Skye Paleocene lavas form a dramatic caprock. All of these sites are examples of the classic cases of failure where thick permeable strata overlay relatively impermeable argillaceous strata. The exception is at **Hallaig** on the Isle of Raasay, where the landslip is entirely in Jurassic strata, but at this site other factors may have had a greater influence than stratigraphy or lithology.

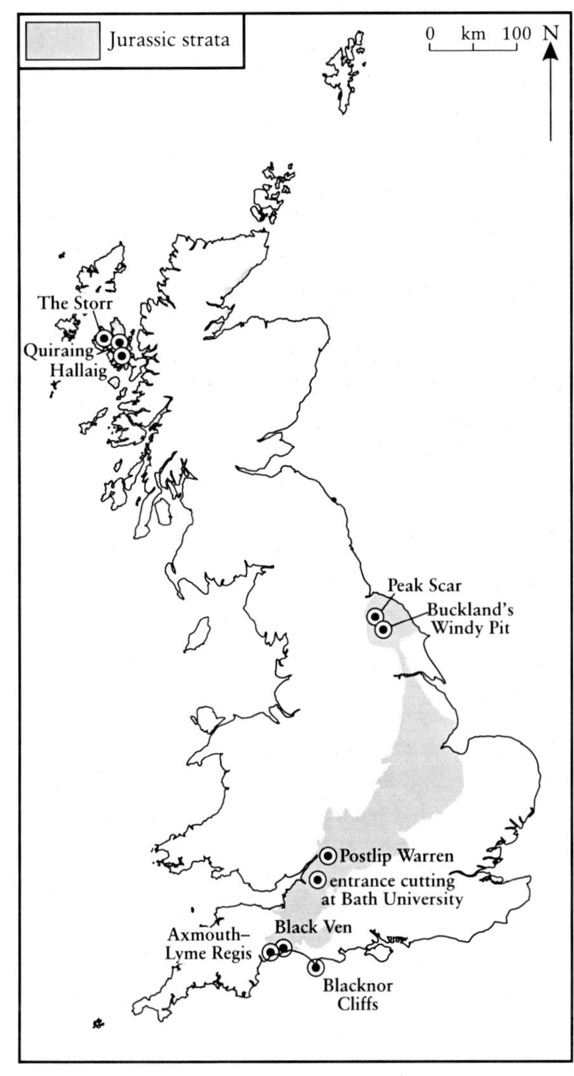

Figure 6.1 Areas of Jurassic strata (shaded) and the locations of the GCR sites described in the present chapter. The Storr and Quiraing lie within the larger **Trotternish Escarpment** GCR site.

The Lower Jurassic scarps through central England have been subject to cambering, valley-bulging and an attendant suite of features. These were first recognized in the Jurassic strata of the Northamptonshire Ironstone Field (Lapworth, 1911; Hollingworth *et al.*, 1944), but no Northamptonshire sites now provide good exposures of these features. Dip-and-fault structures can be seen in Jurassic strata at the **Entrance Cutting at Bath University** GCR site, and ridge-and-trough features can be seen in Jurassic strata at **Postlip Warren**, near Cheltenham. The site chosen to illustrate valley-bulge structures (**Rowlee Bridge**) has been described in Chapter 5, as it is in Carboniferous strata.

Illustrating the effects of the Lower Jurassic consisting largely of highly mobile clays, **Black Ven** is a landslide that is in the process of being actively degraded by mudslides, but episodically rejuvenated by other processes. It is formed in the Lower Jurassic argillaceous beds overlain by the Cretaceous Upper Greensand.

Large-scale superficial structures are also found in Upper Jurassic strata. Three sites, **Buckland's Windypit** and **Peak Scar** in North Yorkshire and **Blacknor Cliffs** on the Isle of Portland, give a broader impression of the types of mass movement in Jurassic strata. At Blacknor Cliffs clay extrusion and deep-seated settlement followed by rotation is typical. At Peak Scar the main mechanism is toppling.

The survey of landsliding (Jones and Lee, 1994) records 2236 examples in Jurassic strata, of which 55% are of unspecified type. The most common types are successive rotational slips (21% of those where the type is specified) and cambered/foundered strata (20%). The portmanteau classification type 'complex' accounts for 17% and single rotational slips for 13%. At the opposite extreme, 0.5% are topples, and no records of sags were obtained from Jurassic strata. The latter point is surprising: the Jurassic rocks in and around Dundry Hill, south of Bristol, are recorded by the British Geological Survey as 'foundered strata' because of the complexity of their superficial structures; this term suggests that the superficial structures are likely to include sags.

Thirty percent of the slides in Jurassic strata are recorded as having taken place in the Upper Lias, with the Inferior Oolite at 18%, the Middle Lias at 15% and the Lower Lias at 14%. Of the Upper Lias slides, 48% are of unspecified type, but of those where the type is specified, two types predominate: successive rotational slips (28%) and cambered slopes (24%) (Jones and Lee, 1994).

For convenience, the GCR sites selected are discussed here in two sections, covering the Lower Jurassic strata and the Upper Jurassic strata respectively.

MASS-MOVEMENT SITES IN LOWER JURASSIC STRATA

POSTLIP WARREN, GLOUCESTERSHIRE (SO 997 265)

R.G. Cooper

Introduction

Postlip Warren is an area at the top of the Cotswold cuesta, on Cleeve Hill, north-east of Cheltenham. Cleeve Hill is the highest part (300 m above OD) of the dissected Jurassic limestone escarpment, in a region of intra-Jurassic subsidence remarkable for its considerable thickness of Inferior Oolite (*c.* 107 m). The Inferior Oolite is composed mainly of limestones, with occasional sandy beds, for example the Harford Sands. The hill is deeply dissected by dendritic dry-valley systems that feed the dip-slope River Coln and the scarp-face river, the Isbourne. There are also two groups of anomalous troughs, one group occurring along the main scarp face above Prestbury and Southam, and another group truncating spurs near Postlip Warren (Figure 6.2).

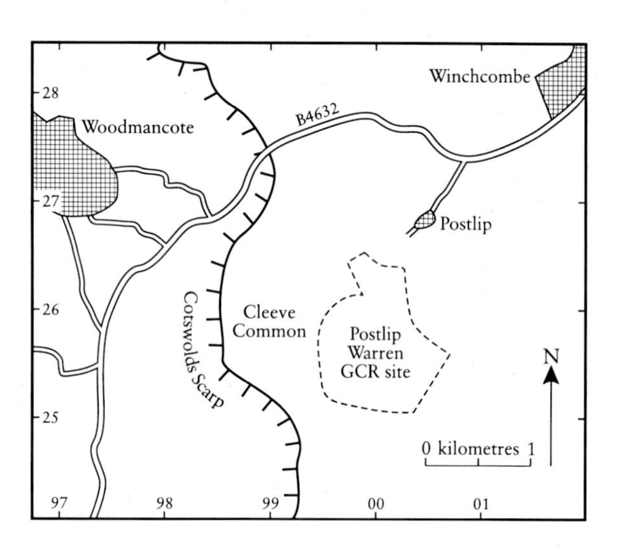

Figure 6.2 Location of the Postlip Warren GCR site.

Description

The troughs are dry and grass-covered, occurring at or near the crest of the escarpment, at heights exceeding 240 m above OD (Figure 6.3). The Postlip Warren group consists of three features truncating the spur between the Postlip

Figure 6.3 View across Cleeve Common showing the deep dissection of the escarpment and the setting of Postlip Warren. (Photo: Gloucestershire Geology Trust.)

and Corndean valleys. They are approximately parallel, and 12–15 m deep. They are aligned perpendicular to the major valleys. They have convex longitudinal profiles, with irregularities, rather than the normal concave longitudinal profile of most stream valleys. In places the maximum angle of the longitudinal profile approaches 10°. The bottoms of the troughs are broad and flat (Figures 6.4 and 6.5), and typically they have a width of 50–70 m from one break of slope to the other. From crest to crest, the widths are of the order of 150 m. They are characterized by asymmetry, with a tendency for the slopes on the plateau side to have a maximum steepness of 18°–21°, and those on the embayment side to stand at 10°–14.5° (Goudie and Hart, 1976). A further group of small depressions runs parallel to the Postlip Warren troughs. Most of them have a dominantly linear form, but they are in essence closed depressions. They contain a relatively deep fill of dark-brown clayey material, attain depths of 3–5 m and tend to follow the contours.

One of the main troughs ('Trough 3' of Goudie and Hart, 1976) contains more than 4.9 m of dark-brown clayey fill with oolitic fragments. The content of coarse oolitic material increases with depth.

Interpretation

The troughs are a distinctive type of landform because:

- they possess a constant asymmetry;
- they possess irregular or closed longitudinal profiles;
- they contain, in at least some cases, a deep, non-alluvial fill;
- they run parallel to the main relief trends;
- they truncate major drainage lines; and
- in some cases they rise where there is little or no catchment area.

Goudie and Hart (1976) argue that they are neither solutional nor glacial features. They point out that the deep fill and the closed nature of some of the features is consistent with a solutional origin, but that a solutional origin does not fully explain either the asymmetry of the cross-profiles of the troughs or the way in which the troughs run parallel to the main relief trends.

Another possible hypothesis for the origin of the troughs is that they are some type of glacial form. The up-and-down longitudinal profiles could have been formed by sub-glacial streams

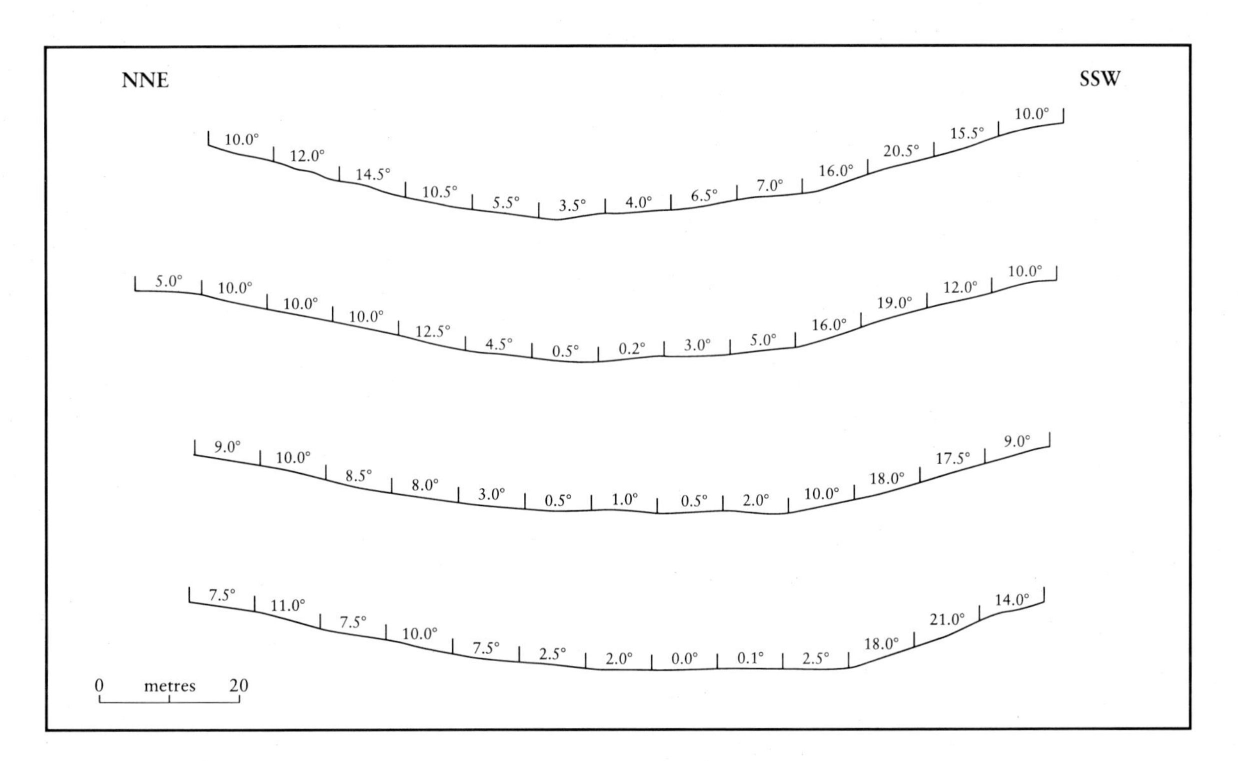

Figure 6.4 Representative slope profiles at Postlip Warren to show distribution of maximum angles. After Goudie and Hart (1976).

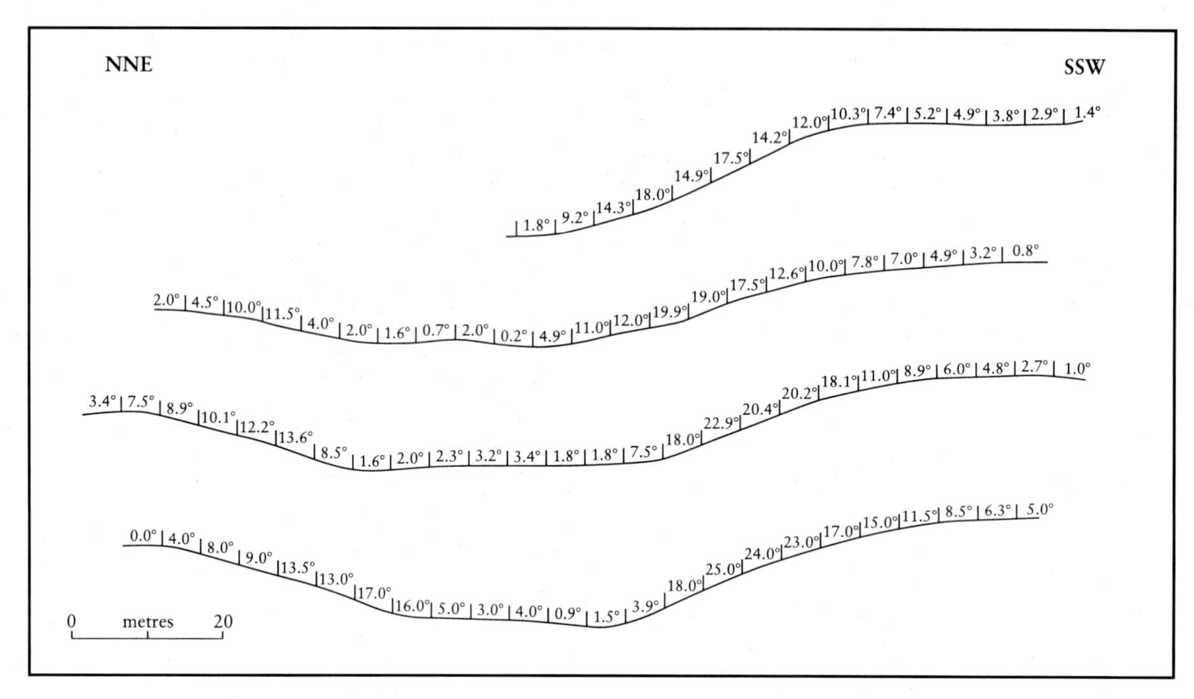

Figure 6.5 Representative profile at Postlip Warren to show broad and flat valley floors.

under hydrostatic pressure. The absence of a normal catchment for the troughs lends support to this idea. Many sub-glacial channels also have flat bottoms. However, there is little evidence for glaciation in this part of the Cotswolds. Also, the sub-glacial meltwater hypothesis fails to account for the orientation of the troughs along the scarp face, rather than down it, and the striking asymmetry of their cross-profiles.

Goudie and Hart (1976) conclude that the most satisfactory hypothesis explains the troughs as large-scale gravitational slip features produced by the foundering of large masses of oolitic limestone over less-competent Liassic clays and marlstones. It may be expected that the face along which slip took place would be relatively steep and that the opposite slope would be relatively gentle. Likewise it may be expected that the troughs would develop parallel to the edges of either the escarpment or embayments within it. Kellaway (1972) has suggested that over-steepening of the escarpment by ice coming down the Severn Vale is a possible contributing factor, though accelerated spring sapping or periglacial cambering could have similar effects. Small-scale superficial cambering is evident in many quarry sections on Cleeve Hill (notably at SO 987 272). Once the depressions were formed, water may have flowed along them under nival conditions (Beckinsale, 1970) and solutional activity may have accentuated initial irregularities.

The deep fill of the troughs is explained as soil that was washed down into the troughs from the slopes on the embayment side (Goudie and Hart, 1976). Before movement took place, the original land surface would have been approximately flat. The asymmetry may be explained by the development of shallow rotational movements of underlying blocks of thick Inferior Oolite as they foundered or cambered into the Liassic clays. The flat surface became inclined towards the developing trough, and with this change in slope angle the soil cover became unstable and sludged or washed down. On this hypothesis there need be no catchment area and the depressions might be expected to truncate the main dendritic dry-valley systems of the area.

The troughs bear comparison with the 'ridge-and-trough' features at Lower Slaughter in the north Cotswolds and are similar to features described generally in the literature as gulls, vents, grabens, dip-and-fault structures, rock labyrinths or camber crests (see Brunsden, 1996b).

Conclusions

Postlip Warren provides very clear physiographical evidence of large-scale gravitational slip processes. It exhibits the longest, deepest, and most-pronounced gravitational troughs in Great Britain, clearly displaying asymmetry and deep fill.

Although very little is known about this site, the dates or the mechanism involved – there is no sub-surface evidence – there is the opportunity to study the subject of material spreading at this site. There is both academic and economic importance in the subject because the forms imply that the whole hill may be in a 'residual strength' condition and might easily be re-activated by inappropriate civil engineering activity.

ENTRANCE CUTTING AT BATH UNIVERSITY, AVON (ST 767 645)

R.G. Cooper

Introduction

Bath University of Technology is sited on Bathampton Hill to the east of Bath at 204 m OD. It has two main roadway entrances, one 400 m down the hill on North Road, and the other on the top of the hill where North Road meets The Avenue. The first of these runs through a cutting in limestones of the Great Oolite Series (Lower Jurassic) (Figure 6.6). In the cutting, the rockface on the western and north-western side of the road shows very clear examples of dip-and-fault structure (Figures 6.7 and 6.8).

The City of Bath is situated in one of the most intense zones of landslipping in Great Britain. This is largely due to the presence of Lower Jurassic limestones overlying incompetent Lias Clays, in an area where the River Avon and its tributaries are deeply incised in steep-sided valleys (Kellaway and Taylor, 1968; Chandler *et al.*, 1976). The largest-scale movements took place during Pleistocene times. The Great Oolite is cambered on the western, northern and eastern slopes with downslope dips of up to 37° (Hawkins and Kellaway, 1971). Gulls that result from this cambering have been described by Hawkins (1977).

Description

Until 1967 the only outcrop at the site was an old quarry. The roadway to the university was then cut through the Great Oolite, revealing that the strata are in a disturbed state (Hawkins, 1977). The basal 2 m seen in the northern part of the quarry section are generally medium to thickly bedded oobiosparites (the terminology used to describe limestones in this section is

Figure 6.6 Location of the Entrance Cutting at Bath University GCR site. (Photo: English Nature/Natural England.)

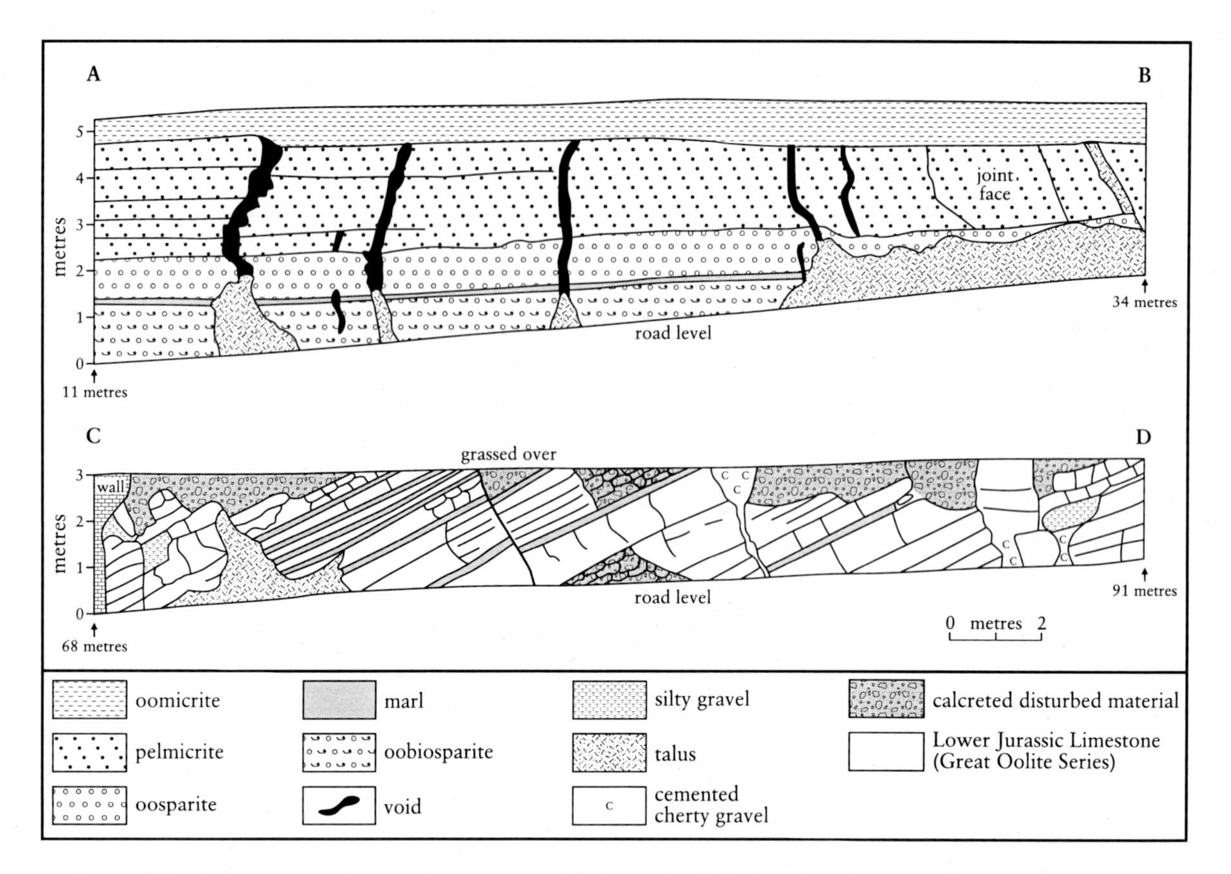

Figure 6.7 Geological sections at the Entrance Cutting at Bath University GCR site. After Hawkins (1977).

taken from the classificatory/descriptive system of Folk (1959)). Above this is another massive bed 1.9 m thick, which, especially near the road cutting, has weathered to show thickly bedded strata of oobiosparite grading upwards into oosparite. A poorly seen 0.1–0.2 m-thick band of marl in the northern cambered and collapsed section of the quarry is more clearly seen on the left of the cut entrance. In the old quarry section, this marl has frequently penetrated up into the disturbed, thinly bedded 1 m-thick overlying pelmicrite bed. Above this is a 0.4 m-thick bed of oomicrite which has a thin band of pelmicrite, and contains rounded gravel-sized clasts of micrite. In the quarry section, this intensively bored, more resistant band acts as a type of roof bed: gulls in the lower strata, often infilled with calcreted limestone fragments, frequently do not penetrate upwards through it. It is overlain by 1.25 m of cross-bedded biosparrudites, with the foresets dipping westwards. These cross-bedded rocks are overlain by a 0.35 m bed of biomicrudite, which, in the left-hand side of the cutting, is capped by 0.7 m of oosparite (Hawkins, 1977).

The structural disturbance of the beds, as revealed in the road cutting, is of considerable interest. It has not yet been possible to explain satisfactorily all of the structures, or to determine accurately the geological succession further than that given in the paragraph above. The gulls, up to 0.4 m wide on the left-hand side of the cutting 15 m from the entrance, are good examples of the way that tension applied to massive beds causes a complete fracture to open; yet in the 2 m thinly bedded upper horizon, bed-by-bed slip means that the tensional strain is taken up by many small movements and no fracture penetrates through to the surface (cf. Hawkins and Privett, 1979). The fact that these cavities exist, yet are not visible at the surface, is obviously of importance in the construction of buildings on the plateau surface (Hawkins, 1977).

Entering the cutting, one of the first things noticed is the almost horizontal bedding to the left, yet the beds on the right have a northerly dip of 26°. This face has not been sampled in detail (Hawkins, 1977). It exposes an oosparite bed, over 1 m thick with a highly bored and

Figure 6.8 The section at the Entrance Cutting at Bath University GCR site showing the dip-and-fault structures. (Photo: R. Wright, English Nature/Natural England.)

oyster-covered surface. This distinct bed, when followed laterally is displaced in several places at the position of old gulls, now largely infilled with travertine. Ascending the incline, the sub-horizontal beds on the left-hand side suddenly begin to dip northwards, and by the bridge they have a dip of 30°. Between the sub-horizontal and inclined beds is a 0.3–0.5 m zone of disturbance, possibly representing a fault or gull breccia. Just east of the bridge, dip-and-fault structures can be seen displacing by 0.82 m and 1.7 m respectively, the 0.43 m bed of bio-sparrudite with irregular borings overlain by oosparite with fine, generally vertical, boring.

Limestones of the Great Oolite Series are seen in the old quarry at the lower end of the cutting. In the cutting section deposits of chert and flint gravel with loamy clay fill solution cavities in the limestone. Locally, one band of limestone has been dissolved out and the space infilled with gravel overlain by bedded loam. Both the lime-stone and the cave filling had subsequently been strongly tilted and faulted as a result of cambering and the formation of dip-and-fault structures. It therefore follows that at the time when caves were formed and the overlying drift deposits were washed into the horizontal solution cavities, the camber slope on the west side of Bathampton Down did not exist. At that time, even after the deposition of the re-sorted high-level drift deposits, this section of the Great Oolite Limestone of Bathampton Down was virtually undisturbed (Hawkins and Kellaway, 1971).

During the construction of reservoirs on the top of Bathampton Hill in 1955, some large gulls were seen, in which masses of stiff plastic clay with pebbles of patinated flint, chert and lime-stone, and roughly bedded loamy gravel, were enclosed in limestone fissures. In 1969 a second set of reservoirs was built on the flat plateau top. These showed large caverns, pipes and swallow holes, some having been developed by the differential solution of individual limestone bands. In these excavations the Great Oolite limestones were not cambered, the bedding of the rocks being horizontal (Hawkins and Kellaway, 1971).

Interpretation

It is clear that the gull-bounded blocks of Great Oolite have settled on the underlying Lias Clay in such a way that their downslope dips are greater than would have been expected as a result of differential lowering by the cambering process. Each individual block has rotated in the downslope direction, as well as being lowered.

Each may also have settled into the underlying clay by different amounts.

Where, as here, the downslope dip on cambered blocks exceeds the dip of the camber itself, irrespective of slope angle, the structure is termed 'dip-and-fault' (Hollingworth *et al.*, 1944). Contacts between adjacent blocks, displayed particularly well in the section at Bath, appear as normal faults heading upslope, with small downthrow on the upslope side.

Conclusions

Dip-and-fault structures can be seen at many sites in Great Britain. They are, however, displayed with particular clarity at the Entrance Cutting at Bath University GCR site.

TROTTERNISH ESCARPMENT, ISLE OF SKYE, HIGHLAND (NG 450 717–NG 481 494)

C.K. Ballantyne

Introduction

The Trotternish peninsula is the northernmost part of the Isle of Skye, the largest island in the Inner Hebrides. The peninsula is dominated by the Trotternish escarpment, a bold cliff of Paleocene lavas up to 200 m high that extends almost continuously for 23 km along almost the full length of the peninsula. Foundering of the lava scarp over the underlying Jurassic sedimentary rocks has produced the largest continuous area of landslide terrain in the British Isles, covering nearly 40 km^2 (Ordnance Survey, 1964), and two of the most spectacular post-glacial landslides in Scotland, at The Storr and Quiraing. The Trotternish landslides were first described in detail by Godard (1965), who identified an inner zone of post-glacial failures and an outer zone of subdued, hummocky slipped rock that he inferred to have been over-ridden by glacier ice. Anderson and Dunham (1966) mapped the extent of landslide terrain, all of which they regarded as post-dating the last ice-sheet, and proposed that failure had been dominated by successive deep rotational slides seated on dolerite sills. Ballantyne (1990, 1991a,b) re-evaluated the age of the Trotternish landslides in the light of a revised glacial history of the escarpment, and described the characteristics of individual landslide blocks and other landforms produced by rock slope failure along the escarpment. The age and cause of post-glacial failure at The Storr has been investigated by Ballantyne *et al.* (1998b) using cosmogenic radionuclide dating of landslide pinnacles, and the history and processes of rockfall accumulation along part of the escarpment have been reconstructed by Hinchliffe *et al.* (1998) and Hinchliffe and Ballantyne (1999) on the basis of talus stratigraphy. These investigations have thrown light on the post-glacial evolution of the Trotternish escarpment, but the deep structure and mechanics of slope failure remain more speculative.

Description

Geology and topography

During the Palaeogene Period (65–23 Ma), the opening of the North Atlantic Ocean was accompanied by prodigious volcanic activity along the western seaboard of the Scottish Highlands. This volcanicity focused between 61 Ma and 56 Ma at the end of the Paleocene Epoch. Volcanoes on Skye, Mull and elsewhere discharged large quantities of fluid, primarily basaltic lavas that buried the underlying rocks and accumulated progressively to form extensive, near-horizontal lava plateaux (Bell and Williamson, 2002; Emeleus and Bell, 2005). In the area of Trotternish, a thick succession of lava flows buried a great thickness of relatively weak sedimentary rocks (mudstones, sandstones and limestones) of Jurassic age (Bell and Harris, 1986; Hallam, 1991; Hudson and Trewin, 2002). Paleocene volcanicity was also accompanied by the intrusion of dolerite sills deep within the Jurassic strata. Jurassic sedimentary rocks and dolerite sills now underlie the low ground east of the Trotternish Escarpment, whereas west of the escarpment crest these rocks are buried under *c.* 300 m of basalts that dip gently westwards (Figure 6.9).

Anderson and Dunham (1966) have suggested that the original scarp face in Trotternish formed as a result of tilting and faulting in Neogene times. Such scarp development probably exposed the full thickness of lavas and the

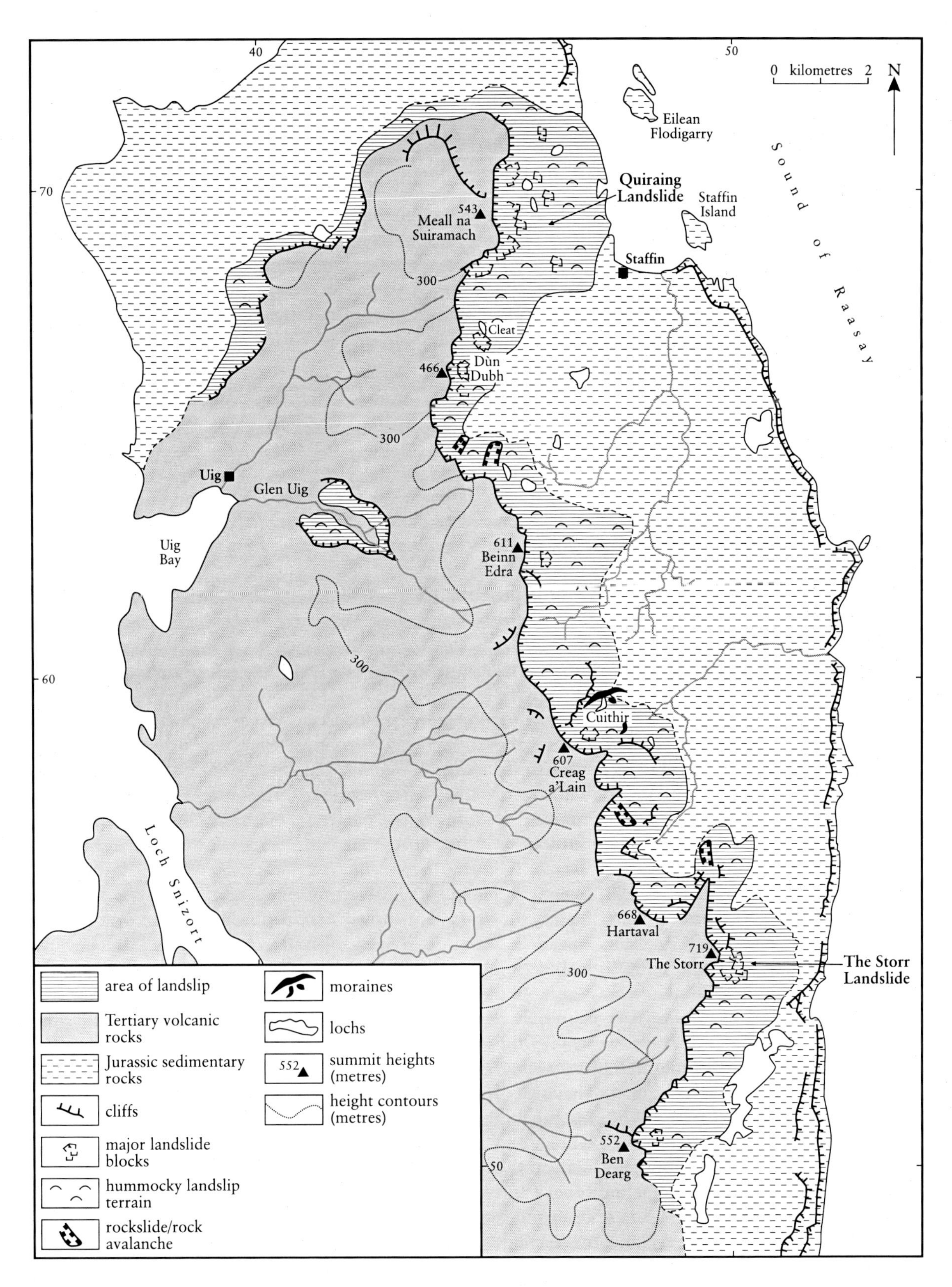

Figure 6.9 Geological and geomorphological map of the Trotternish Escarpment GCR site. The extent of landslide terrain is based partly on British Geological Survey mapping. Modified after Ordnance Survey (1964).

uppermost underlying sedimentary rocks, and initiated a long-term process of scarp retreat through failure of the latter under the weight of the former. The present escarpment exhibits a scalloped planform attributed by Anderson and Dunham to intersection of arcuate failure planes. Scarp retreat has cut backward into a gentle dipslope of broad valleys and rounded spurs, so that the crest of the present scarp consists of alternating cols and summits, the latter including eight peaks over 500 m and culminating in The Storr (719 m).

Geomorphological mapping indicates that the last (Late Devensian) ice-sheet moved northwards across the Trotternish peninsula. Ballantyne (1990) identified a periglacial trimline that descends northwards from 580–610 m to 440–470 m along the escarpment, and proposed that the higher peaks remained above the ice as nunataks, a proposition supported by X-ray diffraction analysis of clay minerals and cosmogenic nuclide dating of bedrock surfaces above and below the trimline; cosmogenic isotope dating also indicates emergence of the escarpment from under the last ice-sheet at *c.* 17.5 cal. ka BP (Ballantyne, 1994; Stone *et al.*, 1998). As a result of renewed cooling during the Loch Lomond Stade of *c.* 12.9–11.5 cal. ka BP, two small corrie glaciers developed east of the scarp, at Coire Cuithir (NG 470 592) and Coire Scamadal (NG 498 552), but the remainder of the escarpment appears to have been ice-free and exposed to severe periglacial conditions at this time (Ballantyne, 1990).

Figure 6.10 Detached lava blocks at Dùn Dubh (NG 441 666). Since deglaciation, these displaced blocks have foundered and moved laterally away from the scarp face, but without the back-tilting of lava flows characteristic of rotational sliding. (Photo: C.K. Ballantyne.)

The Trotternish landslides

Slipped rock-masses extend continuously along the foot of the escarpment over a horizontal distance of 23 km, and for a further 9 km below the basalt cliff that extends south-west from the northern end of the escarpment. A further small area of slipped rock fringes the basalt cliffs of Glen Uig on the west side of the peninsula (Figure 6.9). Two distinct zones of landsliding can be identified: an inner zone, adjacent to the scarp face, of bold angular detached blocks and pinnacles (Figures 6.10 and 6.11) and an outer zone of subdued, rounded landslip terrain, extensively covered by peat (Ordnance Survey, 1964). Anderson and Dunham (1966) suggested that erosion by the last ice-sheet removed all evidence of previous landslides, implying that all of the present area of slipped rock represents failure following ice-sheet retreat. Godard (1965), however, proposed that the outer zone represents slipped rock-masses that were subsequently over-ridden by glacier ice. Various lines of evidence favour Godard's interpretation, particularly northwards-oriented ice moulding of landslide blocks, the presence of till deposits in hollows, and over-riding of landslide blocks by lateral moraines deposited by the two corrie glaciers that developed east of the escarpment during the Loch Lomond Stade (Ballantyne, 1990). Conversely, angular blocks and delicate pinnacles of the inner zone show no evidence of glacial modification and appear to represent failure and sliding of rock masses after ice-sheet retreat.

Several different landslide forms occur within the inner zone adjacent to the basalt scarp. Adjacent to the escarpment crest are a number of incipient failures, where basalt blocks have

Figure 6.11 Pinnancles of shattered basalt at The Storr landslide. The highest pinnacle is the Old Man of Storr. Note the eastwards (forwards) tilt of the slipped mass, away from the escarpment. (Photo: C.K. Ballantyne.)

become detached from the crest along master joints aligned northwards, but have experienced little or no displacement. Resting against or detached a short distance from the scarp are foundered blocks of intact rock, for example at Baca Ruadh (NG 478 576), Coire Cuithir (NG 470 586) and Dùn Dubh (NG 441 666). Farther out from the crest are isolated tabular blocks such as Cleat (NG 447 669) and shattered rock pinnacles such as the Old Man of Storr (Figure 6.11) and the Quiraing needle. Along many stretches of the escarpment, however, evidence for major post-glacial rockslides is absent, and the scarp face is fringed with relict talus slopes, now extensively gullied and eroded (Hinchliffe, 1998, 1999; Hinchliffe *et al.*, 1998).

The Storr and Quiraing

The most impressive areas of landsliding occur at The Storr (NG 485 540; see also Emeleus and Gyopari, 1992) and Quiraing (NG 455 692). The entire south-east face of The Storr has collapsed to form a great hollow, Coire Faoin (NG 497 537), that is bounded to the south-west and north-west by sheer lava cliffs 200 m high. The undercliff zone of the landslide is a labyrinth of lava-capped blocks, narrow defiles and pinnacles of shattered rock, of which the 49 m-high Old Man of Storr is the largest (Figure 6.11). Below the eastern threshold of Coire Faoin, however, the landslipped blocks are subdued and rounded by glacial erosion, indicating that post-glacial landsliding was confined to an area of about 0.25 km^2 above the 350 m contour. Older, glacially modified slipped rock, however, extends down to 200 m, up to 1.5 km from the present cliff-face.

The Quiraing landslide at the northern end of the escarpment is one of the largest landslides in Great Britain. It occupies an area of *c.* 8.5 km^2 and extends 2.2 km eastwards from the scarp crest to the coastline. Like The Storr landslide, it consists of an inner (post-glacial) zone of tabular and toppled landslip blocks, pinnacles and deep clefts, and an outer zone of more subdued landslide terrain representing remnants of ancient landslides that occurred before the last and possibly earlier ice-sheets crossed the area. In the central part of the slide area, landslip blocks up to 70 m high have dammed a chain of small lakes, of which Loch Fada (NG 458 698) is the longest, with a length of over 300 m.

Interpretation

Structure and mode of landsliding

Interpretation of the deep structure of the Trotternish landslides has focused on The Storr and Quiraing slides. Anderson and Dunham (1966) estimated that prior to the failure of slipped rock at The Storr, the scarp crest lay about 600 m east of its present position. They observed repetitive outcrop of steeply dipping Jurassic sediments, palagonite tuffs and lavas in the lower parts of the slide mass, and inferred that the landslide represents successive rotational failures of a thickness of up to 300 m of sedimentary rocks and tuff under a similar thickness of basalt (Figure 6.12). They concluded that the most recent (i.e. proximal) failures were seated on an upper sill, the Creag Langall Sill, but that earler failures took place over a thicker lower sill, the Armishader Sill (Figure 6.12). The outer part of the Quiraing slide also appears to exhibit cyclic outcrop of Jurassic sediments, tuffs and lavas. Here Anderson and Dunham (1966) inferred that a thickness of *c.* 200 m of sedimentary rocks and tuff had failed under the weight of *c.* 300 m of lavas, again in the form of successive deep rotational slides, here seated on a transgressive, westward-dipping dolerite sill that crops out in nearshore islands.

Although the cyclic repetition of Jurassic sedimentary rocks, palagonite tuffs and lavas identified by Anderson and Dunham (1966) appears consistent with their model of rotational sliding and consequent back-tilting of slipped blocks (Figure 6.12), other evidence suggests that it represents over-simplification of the nature of rock displacement. It requires rotation of slipped masses against the regional westwards dip of the strata, until the outmost slipped masses are resting at improbably low angles, and infilling of the huge gaps between slipped blocks with 'rock debris' of unspecified origin. Moreover, some landslide blocks, for example at The Storr, are tilted forward (eastwards) and not backward as the rotational model suggests (Figure 6.11). The model, moreover, does not account for the detachment of large intact landslide blocks without back-tilting, notably Cleat (NG 447 669), Leac nan Fionn (NG 453 704) and Dùn Dubh (NG 459 687). These suggest a different interpretation, involving planar sliding or gliding of thick lava blocks over the weaker sedimentary rocks, probably the Upper Jurassic Staffin Shales, which in Trotternish achieve a thickness of over 117 m (Hallam, 1991). Lateral displacement and subsidence of thick lava blocks would have been accompanied by pronounced deformation of underlying sedimentary strata under the weight of the lavas; this, rather than block rotation, may account for the outcrops of steeply dipping sedimentary rocks and tuffs between the outermost lava blocks. By contrast,

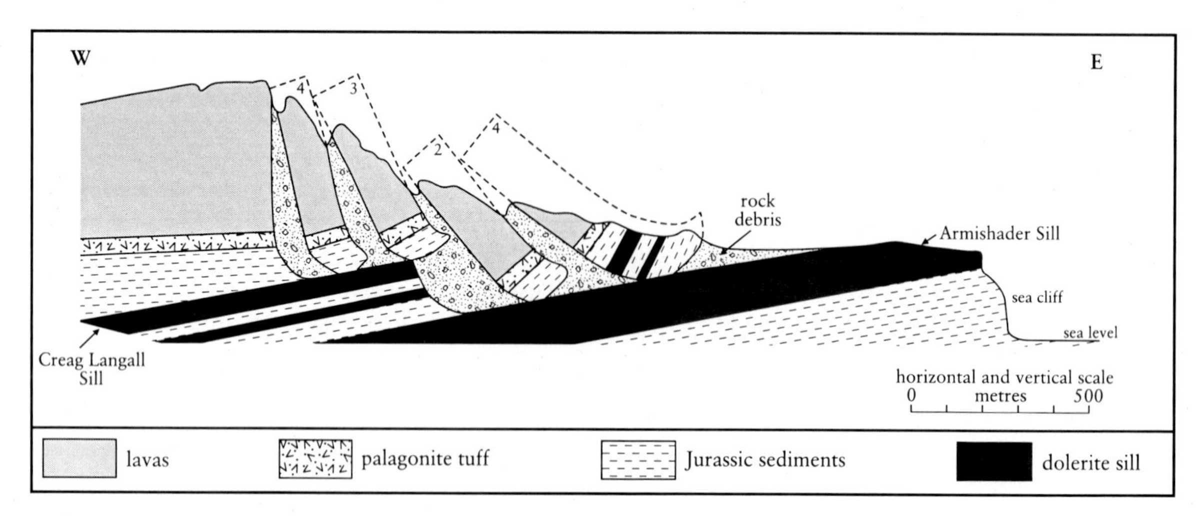

Figure 6.12 A previous interpretation of The Storr landslide, redrawn from Anderson and Dunham (1966; The geology of northern Skye, Memoir of the Geological Survey of Great Britain, p. 191, fig. 23), by permission of the Director, British Geological Survey. Their model depicts slide evolution as a sequence of successive deep rotational slides in the sedimentary rocks and palagonite tuffs underlying the basalt scarp, and involves back-titling of lava blocks. Compare with Figures 6.10 and 6.11.

backward rotation is evident in a similar context nearby at **Hallaig** (Isle of Raasay – see GCR site report, this chapter) where active landslipping is entirely in Jurassic sedimentary rocks (Russell, 1985).

The present position and altitude of such intact blocks implies displacement and subsidence without fragmentation of the lava caprock, possibly indicating gradual movement analogous to that proposed for cambering of limestone caprocks over Jurassic mudstones in England (see **Postlip Warren** GCR site report, this chapter). Like these, block movement may relate to reduction of shear strength in underlying argillaceous strata during thaw of ice-rich permafrost (Parks, 1991; Hutchinson, 1991; Ballantyne and Harris, 1994). In contrast, the shattered lava pinnacles of the post-glacial slides at The Storr and Quiraing imply catastrophic failure after deglaciation. It thus appears likely that no single mode of failure and block movement accounts for all attributes of the slipped rock-masses.

Not all rock slope failures along the Trotternish Escarpment have involved block displacement. Ballantyne (1990) mapped four smaller post-deglaciation failures, probably rock topples or translational slides, that have produced runout of coarse debris below the scarp face. South of Baca Ruadh (at NG 476 568), a tongue of vegetated debris terminates downslope in a *c*. 100 m-wide zone of bouldery mounds and hummocks, and a similar but smaller deposit occurs at NG 441 651. At NG 449 646, a jumble of limestone boulders covers an area of nearly 0.2 km^2 and records collapse of the cliff upslope. At Carn Liath (NG 464 593) a tongue of rock debris 260 m wide and 500 m long descends from 400 m to 260 m. The visible debris consists of large angular boulders, many exceeding 2 m in length. The lower part of the debris tongue extends for 250 m over a slope of only 8°–9°, and appears to be about 10 m thick. The debris tongue reflects collapse of the lava scarp upslope, probably as a small rock-avalanche, and its extended runout over gentle gradients suggests flow (debris flow or grainflow) of the landslide debris (Ballantyne, 1991b).

Timing and activity

The lateral extent of landslipped rock along the Trotternish Escarpment implies a long history of slope instability and failure, with evidence of post-deglaciation sliding confined to a few sites adjacent to the present scarp. Although the age of failures that pre-date the advance of the last ice-sheet over the area is unknown, the evidence for multiple landslide events suggests a probable link with glaciation. Successive Pleistocene ice advances northwards along the escarpment are likely not only to have steepened the scarp face and removed earlier landslide debris from the footslope, but may also have induced scarp failure due to paraglacial stress-release (Ballantyne, 2002a). It is possible that deglaciation and consequent debuttressing of the scarp face created conditions favourable for failure on several different occasions throughout the Pleistocene Epoch. Hence the broad zone of ice-moulded landslide terrain along the length of the escarpment represents a sequence of landslides that occurred in response to several episodes of deglaciation, much as the post-glacial slides at The Storr and Quiraing represent response to Late Devensian deglaciation. The long-term evolution of the escarpment may thus be envisaged in terms of alternating glacial erosional and interglacial (paraglacial) landsliding episodes.

Godard (1965) proposed that most post-glacial (inner zone) failures from the Trotternish Escarpment occurred soon after deglaciation, suggesting that they represent collapse of glacially steepened cliffs following the withdrawal of a supporting buttress of glacier ice. Cosmogenic ^{36}Cl radionuclide dating of the exposure age of two separate basalt pinnacles at The Storr landslide, however, yielded virtually identical ages averaging 6.5 ± 0.5 cal. ka BP (Ballantyne *et al*., 1998b), consistent with radiocarbon-dated evidence for exposure of the present cliff at this time (Ballantyne, 1998). These dates indicate that the landslide occurred 7–10 ka after ice-sheet deglaciation, and at least 4 ka after the end of the Loch Lomond Stade, the final period of permafrost conditions in Scotland. Ballantyne *et al*. (1998b) inferred that progressive rock-mass weakening due to stress-release had been critical in conditioning failure, but noted that a seismic trigger could not be discounted. They also observed the significance of the age of failure in terms of indicating persistence of major landslide events well into the Holocene Epoch, an observation subsequently re-inforced by the even more recent date (*c*. 4 cal. ka BP) obtained for the **Beinn**

Alligin rock-avalanche, 30 km east of The Storr (Ballantyne and Stone, 2004). The rock avalanche at Carn Liath (NG 464 693) on the Trotternish escarpment appears to have occurred even more recently, as limited lichen cover on the source rockwall suggests failure within the past few centuries (Ballantyne, 1991b).

Although the slipped rock-masses at Trotternish appear to be stable at present, Anderson and Dunham (1966) suggested that where the toe of the Quiraing slide is being eroded by the sea there is 'continuous though not extensive movement' and that the main road near Flodigarry (NG 465 716) is 'frequently dislocated'. However, deformed sediments at the landside toe are overlain by undisturbed raised beach deposits and cut into by Holocene raised shorelines, suggesting that recent movement has been slight (Ballantyne, 1991a).

Talus accumulations

Since deglaciation, only limited areas of the Trotternish Escarpment have been affected by major rock slope failures. Much of the remainder, however, has experienced post-glacial cliff recession due to intermittent rockfall, with concomitant accumulation of talus at the scarp foot. Such talus accumulations are now essentially relict (Figure 6.13). They support an almost continuous vegetation cover and are deeply dissected by gullies, many of which exhibit evidence of recent reworking of talus sediments by debris flows. The morphology, sedimentology and stratigraphy of a section of talus near the southern end of the escarpment (NG 492 533; Figure 6.13) have been investigated by Hinchliffe (1998, 1999), Hinchliffe *et al.* (1998) and Hinchliffe and Ballantyne (1999). Calculations based on the volume of talus accumulations at the foot of the basalt cliff indicate that 4.3–7.8 m of rockwall recession has occurred since deglaciation at *c.* 17.5 cal. ka BP. The composition of the talus sediments indicates that approximately 70% of overall rockwall retreat has been due to rockfall, and that the remainder reflects granular weathering of the cliff-face. Radiocarbon-dated soil horizons within the talus imply that rockfall inputs have been very limited since the end of the Loch

Figure 6.13 Relict talus accumulations at the southern end of the Trotternish Escarpment. The talus slopes are now vegetated and deeply dissected by active gullies. The main period of talus accumulation occurred prior to 11.5 cal. ka BP. (Photo: C.K. Ballantyne.)

Lomond Stade at *c.* 11.5 cal. ka BP, and suggest that about 80% of total rockwall retreat occurred between 17.5 cal. ka BP and 11.5 cal. ka BP, during which period rockwall retreat rate averaged about 0.75 m ka^{-1}. Hinchliffe and Ballantyne (1999) attributed this high rate of late-glacial cliff recession to stress-release following ice-sheet deglaciation, and/or frost wedging under severe periglacial conditions. The stratigraphy of the upper parts of the talus accumulations reveals stacked debris-flow horizons intercalated with occasional slopewash horizons and buried organic soils, implying that rockfall debris has been extensively reworked by intermittent debris-flows throughout much of Holocene time.

Wider significance

The assemblage of landslide features represented along the Trotternish Escarpment is characteristic of those of basaltic successions not only along the western seaboard of Scotland, but also of areas of similar geological configuration in the wider North Atlantic Igneous Superprovince (Emeleus and Bell, 2005) and elsewhere (Evans, 1984). In Scotland, less extensive but equally spectacular areas of slipped basalts overlying Mesozoic sedimentary sequences occur at several locations in the Inner Hebrides, such as Score Horan (NG 285 594) and Ben Tianavaig (NG 517 410) on Skye, and at Gribun on Mull (Bailey and Anderson, 1925; Godard, 1965; Anderson and Dunham, 1966; Richards, 1971; Ballantyne, 1986a, 1991b). In all these cases the long-term history of scarp retreat probably reflects alternating periods of glacial steepening and loading, and intervening episodes of paraglacial stress-release, rock-mass weakening, rockfall and slope failure. Many of these occur at coastal locations, raising the possibility that coastal erosion may have triggered and accelerated post-glacial failure. Some exhibit signs of recent movement (Anderson and Dunham, 1966). These landslides constitute a family of structurally and lithologically conditioned slope failures quite distinct from those found on the Precambrian or Palaeozoic rocks of the Scottish mainland (cf. 'Introduction', Chapter 2). As at Trotternish, however, the deep structure of these landslides remains uncertain.

Although much of the literature on the Trotternish Escarpment focuses on post-glacial rock slope failures and slope modification, the outer zone of older, ice-moulded landslide terrain not only provides evidence for previous episodes of slope failure during past interglacials, but also forms the most extensive area of glaciated landslide topography in Great Britain. Most pre-glacial landslides in Scotland can be identified only as modified failure scars, as the failed rock-masses and runout debris have been removed by glacier ice (Clough, 1897; Ballantyne, 2002a). Only on Trotternish have extensive areas of slipped blocks survived the passage of one or more ice-sheets, leaving a landscape of ice-moulded lava blocks and intervening lochans or peat-filled hollows. Such terrain is particularly well exemplified by the outer (eastern) parts of the Quiraing landslide.

Conclusions

The Trotternish peninsula in northern Skye includes probably the largest continuous area of landslide terrain in Great Britain (totalling nearly 40 km^2), together with the largest individual rockslide (the Quiraing slide, 8.5 km^2) and some of the finest examples of landslides associated with Paleocene volcanic rocks. These distinctions reflect the geological configuration of Trotternish, where a succession of basalt lava flows roughly 300 m thick overlies a similar thickness of Jurassic mudstones, sandstones and limestones into which have been intruded thick dolerite sills. Westward tilting and faulting of this sequence, probably in Neogene time, created a steep escarpment of lavas overlying relatively weak sedimentary rocks. Throughout the Quaternary Period, this scarp face has retreated westwards through a combination of alternating episodes of glacial erosion and interglacial landsliding and rockfall. Trotternish provides probably the clearest example in Britain of this glacial–paraglacial cycling process.

The landslide terrain east of the Trotternish Escarpment can be subdivided into two zones: an outer zone of subdued, ice-moulded landslide blocks and an inner zone of fresh tabular landslide blocks, shattered basalt pinnacles and talus accumulations. The outer zone represents landslide terrain that was over-ridden and modified by the last (and possibly earlier) ice-sheets, and represents the most extensive area of glaciated landslip terrain in Great Britain. The inner zone represents the products of scarp failure, landsliding and rockfall since ice-sheet deglaciation, which probably occurred about 17 000 years ago. Scarp failure was probably caused by steepening of the basalt cliff by glacial erosion and stress-release within the rock mass caused by unloading as the last ice-sheet down-wasted. Stress-release results in slow opening of joints within the rock, resulting in progressive loss of strength. Although this effect often causes slope failure soon after deglaciation, failure may be delayed for millennia as the joint network propagates. The major landslide at The Storr occurred between 7000 and 6000 years ago, some 10 000 years after deglaciation.

Various modes of slope failure and landsliding are evident along the escarpment, but the structure of the Trotternish landslides remains speculative. Anderson and Dunham (1966) proposed that scarp retreat has been dominated by deep rotational slides in the sedimentary rocks underlying the lavas, but the configuration of some intact landslide blocks suggests planar sliding or gliding of the lava caprocks over a layer of deforming shale, possibly when the strength of the latter was reduced by thaw of ice-rich permafrost. A small number of shallow rockslides or topples have also occurred since deglaciation, and extensive areas of talus have accumulated as a result of rockfall from lava cliffs. In the southern part of the escarpment, rockfall and weathering have resulted in 4.3–7.8 m of cliff retreat since deglaciation, most of which occurred in the interval between ice-sheet deglaciation and the end of severe periglacial conditions around 11 500 years ago. The resulting talus slopes were subsequently modified by intermittent debris-flows, and are now essentially relict and extensively eroded.

Much greater amounts of post-deglaciation cliff retreat have occurred at major landslide sites, notably at The Storr and Quiraing. Both of these famous landslides consist of a chaotic inner zone of tilted lava blocks, deep defiles and shattered lava pinnacles, and an outer zone of ancient landslide blocks modified by the passage of the last and probably earlier ice-sheets. The Storr landslide extends 1.5 km from scarp crest to toe, and Quiraing landslide reaches the sea 2.2 km from the escarpment crest. The toe of the latter is reported to have experienced recent movement, possibly due to coastal erosion.

HALLAIG, ISLE OF RAASAY, HIGHLAND (NG 588 387)

R.G. Cooper

Introduction

The Isle of Raasay lies between the Isle of Skye and the Applecross peninsula of the Scottish mainland. On its eastern side, at Hallaig, is a large landslip (Figure 6.14), which has moved in recent times (a slip was recorded in 1934). It is

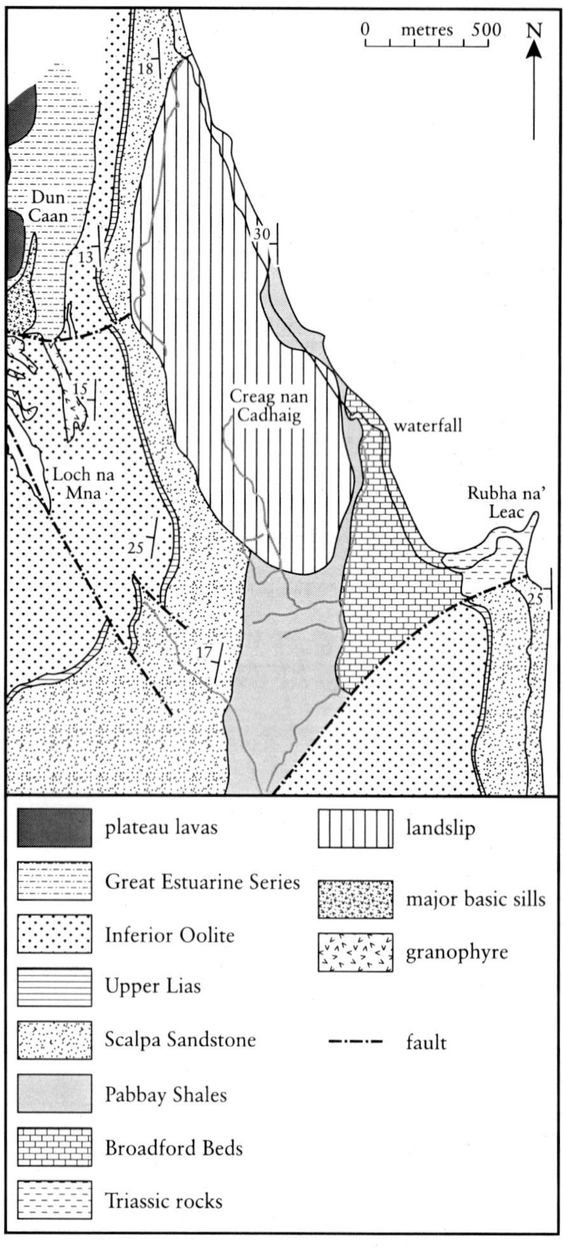

Figure 6.14 The geological setting of the Hallaig landslide on the coast of the Isle of Raasay.

about 1.8 km long from north to south, and extends from the Cadha Carnach cliff on the east side of Dun Caan (NG 583 384) for 800 m to the coast. The landslip has been mapped in detail by Russell (1985, see Figure 6.15), and much of his work is utilized in this account.

Description

The Hallaig landslip lies on the eastern side of Raasay, immediately south-east of Dun Caan (NG 579 395), adjacent to the Inner Sound. The landslip forms a large crescent-shaped topographical feature extending from 290 m above OD to sea level. The main backslope to the slip is itself 150 m high in places (e.g. Cadha Carnach) and a major bench at *c.* 150 m above OD represents the top of the slipped mass. Loch a' Chada-charnaich infills a hollow on this bench. The landslipped material forms a steep slope from *c.* 150 m above OD to sea level,

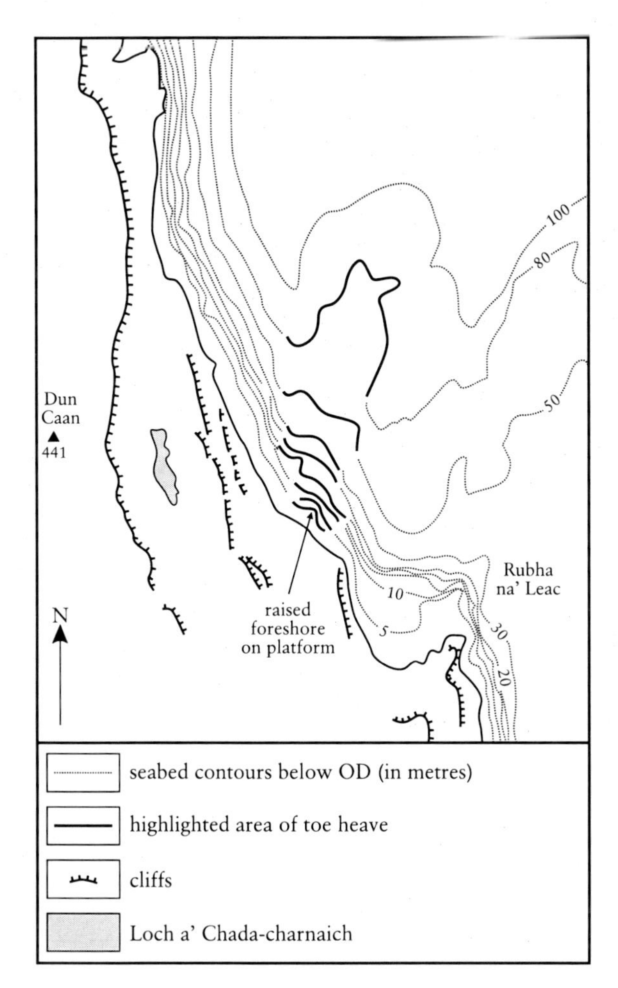

Figure 6.15 Slipped masses of the Hallaig landslips and the offshore features. After Russell (1985).

constituting a very large mass of failed material. The crags of Creag nan Cadhaig are prominent on this lower slope. They lie immediately to the north of the deserted village of Hallaig and are formed of Jurassic limestone, containing fissures (probably widened joints) of unknown depth, which are partly open to the surface. The fissures are up to 4 m wide at the surface, and appear to have opened in response to internal deformation of the slipped rock mass.

The most significant aspect of Russell's (1985) mapping of the landslide was the recognition, in the field, of the stratigraphical sequence on the backface of the landslip, and its repetition in the slipped mass. The critical sequence is:

Bearreraig Sandstone Formation
Raasay Ironstone Formation
Scalpa Sandstone Formation

In the field he mapped this sequence both along the backface and again across the landslip. Allied with measurements of dip, this provided data for five cross-sections (Figure 6.16). In the backface, he found the in-situ bedding dipping between 12° and 15° west, while in the slipped mass the bedding strikes north–south with dips between 44° and 58° west.

In the vicinity of the sea cliffs he found that the Pabbay Shale, which underlies the Scalpa Sandstone, has exactly the changes in dip shown in Figure 6.16. One small exposure of Pabbay Shale exhibits sheared surfaces and slickensided surfaces, the latter cutting one another. In the same area the underlying Broadford Beds have the normal tectonic dip, and hence were not involved in the slip at this location. Russell (1985) pointed out that near the foreshore the Pabbay Shale is raised, indicating an area of toe heave. The Admiralty Chart of the area (Chart No. 2480) shows tongue-like sea-floor features suggesting that this area of toe heave continues offshore.

Russell (1985) observed that open joints are a distinctive feature of the Raasay landslides, and accordingly made a study of them. In the area of the landslide itself, they are to be found only in the Bearreraig Sandstone Formation (here, a series of coarse-grained calcareous and non-calcareous sandstones). The joints generally have an east–west trend, with a few in the south of the landslide block trending north–south.

To the west of the southern part of the landslip block, widened joints in the Scalpa Sandstone have grassy bottoms and show no

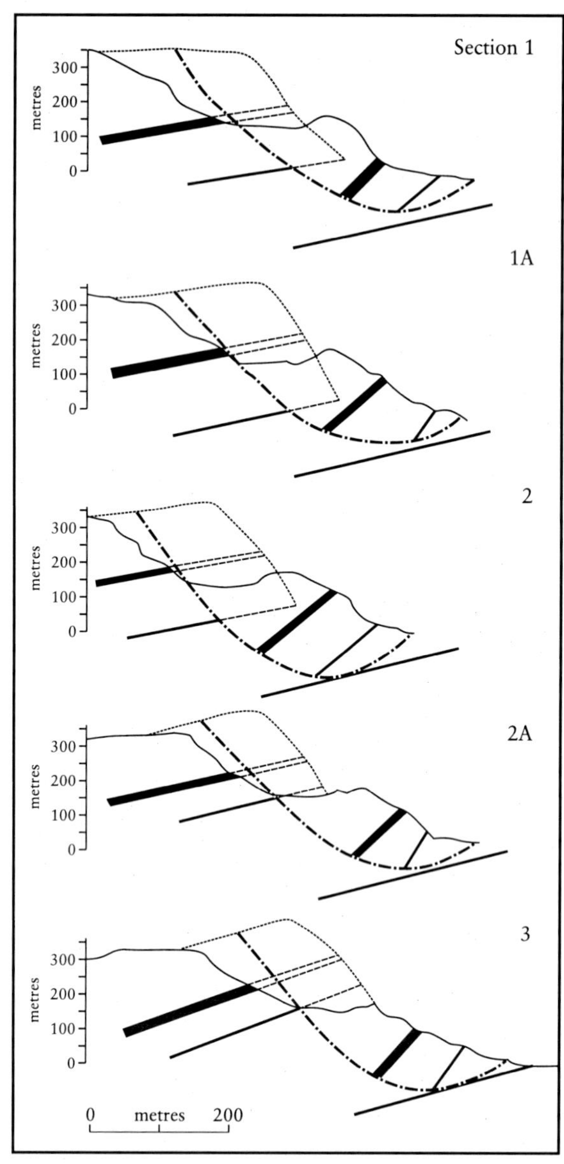

Figure 6.16 Cross-sections of the Hallaig landslip. After Russell (1985).

displacement across bedding units. On average they are 6.65 m wide, which Russell (1985) attributes in large part to subaerial erosion of their walls. On average they are 1.5 times as deep as they are wide. Their average weighted trend is 343°, approximately parallel to the slope trend. Many of the widened joints in the Bearreraig Sandstone Formation in the same area are narrower and deeper than those in the Scalpa Sandstone, being on average 5.4 times as deep as they are wide. Their average weighted trend is 349°, again running approximately parallel to the slope. Those close to the crest of the landslide's backface show some downward vertical displacement of their eastern walls (Russell, 1985, p. 59).

Interpretation

Despite the similarity of its east-facing position, the landslip at Hallaig on the Isle of Raasay differs greatly in its geological setting from the **Trotternish Escarpment** landslips on the Isle of Skye. Anderson and Dunham (1966) suggest that within the 145 m pile of sedimentary rocks exposed in the backface of the landslip, the first movement was that of the Portree Shale and Raasay Ironstone formations strata overlying the competent Scalpa Sandstone. This was followed by slipping of the Pabbay Shale with the Scalpa Sandstone acting as over-burden, and then by failure of the Great Estuarine Series and overlying Paleocene lavas.

Such a sequence of events may be necessary to explain the development of the entire slope eastwards from the summit of Dun Caan, which is the highest point of the island, but there is little evidence of it on the ground. Russell (1985) found no slipped lavas. He considers the large landslip block at Hallaig to have moved by rotational slip with the failure surface located in the Pabbay Shale. Its limit is controlled by the stronger Broadford Beds. He measured the strength of four samples of Pabbay Shale using a portable shear box, and carried out a stability analysis to assess the validity of the circular slip model depicted in the cross-sections in Figure 6.16. The average peak shear-strength obtained was c = 590 $\mathrm{kNm^{-3}}$, ø = 35°. Taking the slope height (H) to be 300–350 m, and the rock density (γ) as 21.6 $\mathrm{kNm^{-3}}$, he used Hoek and Bray's (1981) circular failure charts and their ratio $c'/(\gamma H \tan ø)$ to find the factor of safety at various degrees of slope steepness (Table 6.1). The use of 21.6 $\mathrm{kNm^{-3}}$ for the value of rock density is precisely the value of 137 $\mathrm{lb/ft^3}$ used by Hoek and Bray (1981, p. 223) in their practical example of the method. It equates to a density of 2200 $\mathrm{kNm^{-3}}$, which converts to 2.2 $\mathrm{g\,cm^{-3}}$. According to Farmer (1968, p. 15), this is a value typical of sandstones, mudstones and limestones; it is therefore a valid approximation to use. Russell (1985) concluded that failure could occur at an angle of 64° with a slope height of 300 m, and at an angle of 58° with a slope height of 350 m. This is in good agreement with slope gradient reconstructions shown on the cross-sections (Figure 6.16).

Table 6.1 Factors of Safety (*F*) for the Hallaig landslip. After Russell (1985).

Slope height (m)	*Slope angle (degrees)*	*F*
300	80	0.75
300	70	0.90
300	60	1.06
350	80	0.69
350	70	0.84
350	60	0.97
350	50	1.12

The Hallaig landslip also differs in its geological setting from the **Trotternish Escarpment** landslips in that features in the immediate vicinity suggest that recent tectonic movements have taken place on the east side of Raasay. Such movements are likely to have had some effect on the landslip. The most apparent sign of such movement is a regular ridge of sandstone blocks running along the north-western side of Beinn na' Leac, a flat-topped hill of Bearreraig Sandstone (Aalenian and Bajocian, Mid-Jurassic, age) which rises to 319 m above OD, immediately south of the landslip. First described by Lee in 1920, the ridge is up to 4 m high along most of its length (Figure 6.17), and appears to mark the position of the fault which throws down this Jurassic outlier of Bearreraig Sandstone to the south-east. The fault is arcuate, and where it crosses the coast at the small bay south of Rudha na' Leac, the Jurassic succession from the Scalpa Sandstone (Upper Pliensbachian) to the Bearreraig Sandstone to the south is juxtaposed against the Broadford Beds (Sinemurian and Hettangian) to the north; the throw of the fault is of the order of 300 m (Morton, 1969). Lee (1920) comments that the materials forming the ridge must be derived from the crags immediately upslope, and interprets the feature as scree accumulations now separated from the slope by movement on the fault. Since the ridge is too fragile to have survived glaciation, this movement must have been post-glacial, and the ridge is so fresh in appearance that the movement appears to have occurred recently. It may be a late expression of isostatic re-adjustment to the removal of the Devensian glaciers. Further evidence of recent movement is provided by features of Beinn na' Leac itself, and by reports quoted by Anderson and Dunham (1966). The east side of Beinn na' Leac is made up of

Figure 6.17 The Hallaig landslide – Beinn na' Leac ridge. (Photo: R.G. Cooper.)

landslips, to the extent that they greatly increase the apparent thickness of the Scalpa Sandstone that crops out there (Lee, 1920). Many vertical fissures of great depth occur on the surface of Beinn na' Leac. The latter are wide enough to be entered, and have been partly explored by cavers.

Evidence of recent movement, as adduced by Anderson and Dunham (1966), includes the local newspapers for 7 August 1934, which reported that twice within a period of six weeks a volcanic eruption had taken place in Raasay. Rising steam and smoke, showers of stones and a loud rumbling noise were reported by local inhabitants. They also reported that Professor A.D. Peacock, who visited Raasay shortly after the most recently reported occurrence, attributed these phenomena to stones falling down one of the very extensive fissures backing the latest slipped mass. Anderson and Dunham (1966) comment that it is 'more probable that such a spectacular disturbance was due to movement on a larger scale, i.e. to renewed slipping of the unstable mass'. It can be added that renewed movement on the fault could also give rise to such phenomena, and indeed to renewed slipping of the unstable mass.

It may be worthy of note that Musson *et al.* (1984) have traced evidence of an earthquake that took place on the nearby mainland on 16th August 1934, nine days after the newspaper reports noted by Anderson and Dunham (1966). They placed its epicentre in Strathconon Forest. In the light of additional evidence, Musson (1989) placed the epicentre farther west, in the Torridon area, although the position is still poorly determined. The earthquake was felt over a wide area, and Neilson and Burton (1985) list it as one of the larger British earthquakes of the 20th century, with an instrumental magnitude of 3.8 ML, which Musson (1989) regards as a rather small estimate considering the large area over which it was felt. Torridon village, assumed by Musson (1989) to be close to the epicentre, is 35 km from the Hallaig landslip on the Isle of Raasay. Anderson and Dunham (1966) note that in the mid-1950s many new fissures could be seen, as could evidence of 'considerable and recent movement', and Russell (1985) provides photographs of manifestly recent shallow slides (his plates 8 and 9). In this connection it is relevant to record that Ballantyne (1997) in a review of rock slope failures in the Scottish Highlands noted that Holmes (1984) found that most translational rockslides had taken place over failure planes inclined below the residual angle of friction, the lower threshold angle for rock masses to slide under their own weight. After considering and eliminating glacial over-steepening, progressive failure, and high cleft-water pressures as causes of rock slope failures under these conditions, Ballantyne (1997) concluded that seismic activity was a likely triggering factor.

One further circumstance may be significant in relation to the Hallaig landslip. This is the remarkable depth of the sea-channels which flank Raasay. In particular, the Inner Sound, which lies between Raasay and the Applecross peninsula of mainland Scotland, has a channel which includes the deepest submarine hollows in the British sector of the continental shelf (Whittow, 1977). These extend down to more than 300 m below sea-level. The submarine slope is steep, with an average angle of about 19°, although the eastern side of the trench is seldom steeper than about 8° (Robinson, 1949; confirmed by the bathymetric survey of Chesher *et al.*, 1983). Thus the steeper side of the trench lies adjacent to the east coast of Raasay, and the Hallaig landslip. The asymmetry of the trench led Sissons (1967) to write: 'a fault control of this trench seems likely and it may well be that the fault has caused these relatively soft rocks [i.e. the Jurassic sedimentary rocks] to form the floor of part of the Inner Sound and so aid its excavation [to these great depths] by glacier ice', even though they have been strengthened by the injection of Paleocene igneous rocks, e.g. dolerite sills and dykes. Whittow (1977, p. 276) put forward the opinion that the presence of major faults on Raasay suggests that the entire chain of islands (Scalpay, Raasay, Rona) may be fault controlled: 'Their entire eastern shores may have been carved from faultline scarps, and the neighbouring ocean deeps excavated along the relatively soft rocks of the down-faulted [Jurassic] sedimentary basins'.

Conclusions

The Hallaig landslip on Raasay is a very large crescentic failure of Jurassic sedimentary rocks that shows evidence of a classic circular failure. Unlike the **Trotternish Escarpment** landslips of northern Skye, Paleocene lavas were not of importance here, at least in the first two of the landsliding stages postulated. The landslip is lent particular interest by the manifestly unusual events (for Great Britain) apparently taking place at Beinn na' Leac, immediately to the south. Here, inferred fault movement, detachment of scree, and widening of joints all seem to point to some kind of post-glacial flexure and faulting of the sandstones which make up Beinn na' Leac itself. These tectonic events may be related to the great depth of the Inner Sound, enhanced by glacial excavation. These factors and the various landslips of the area seem intimately related, in ways still to be evaluated in detail. However, the Hallaig landslip is the only British mass-movement site with good evidence for neotectonic activity.

Shocks accompanying submarine slumping on the 19° underwater slope might have been responsible for the phenomena observed in Raasay in 1934, and could also have acted as an episodic trigger of continued movements of the Hallaig landslip. Both subaerial movement of the landslip, and submarine slumping, could be triggered by fault movement, and it is known that the general area was seismically active in 1934.

AXMOUTH–LYME REGIS, DEVON–DORSET (SY 257 897–SY 333 915)

R.G. Cooper

Introduction

The Axmouth–Lyme Regis stretch of the south coast (Figure 6.18) comprises one of the best known areas of landslipping in Great Britain: it includes the site of arguably the first large-scale landslide ever to have been the subject of detailed scientific description by geologists, and it was the mass-movement site most widely suggested for inclusion in the Geological Conservation Review (Cooper, 1982). It is a National Nature Reserve (declared in 1955), selected primarily for its geological interest, especially its landslides. There has been considerable debate about the mechanisms responsible for the landslides and the develop-

Figure 6.18 The Axmouth to Lyme Regis Undercliffs region. This photograph shows the famous Bindon Landslide that took place on Christmas Eve 1839. It is probable that this is the first landslide to be fully described in a scientific memoir. (Photo: courtesy of http://www.ukaerialphotography.co.uk.)

ment of the complex of landforms found there at the present day.

The site is about 9.5 km in length from west to east, and generally extends inland from the high-water mark for about 500 m. The strata involved consist of a series of easterly dipping early Mesozoic argillites and 'limestones' (Keuper and Rhaetic (Upper Triassic) and Lower Lias (Lower Jurassic)), successively exposed by faulting. These are overlain unconformably by arenaceous and calcareous sediments (Gault and Upper Greensand (Lower Cretaceous) and Chalk (Upper Cretaceous)). The plane of unconformity dips just east of south at about 5°. The present in-situ sea cliffs on the coastal boundary of the area are formed of a more limited range of strata: Keuper Marls, the White Lias division of the Rhaetic, and the Blue Lias division of the Lower Lias (Figure 6.19; Pitts, 1979).

Description

(a) General

There is a continuous series of landslips along the coast from Axmouth to Lyme Regis, named successively Haven Cliff, Culverhole Cliffs, Bindon Cliffs, Dowlands, Rousdon Cliff, Charton Bay, Whitlands, Pinhay Bay and Ware Cliffs (Figure 6.20). All have histories of large-scale landslipping throughout post-glacial times, and have displayed similar features (Pitts, 1982, 1983a). Although the major component in most or many of the slips has probably been rotational, detailed examination by Pitts (1979) has shown that a wide variety of mass-movement types are present. These include:

1. Rockfalls caused by undermining of relatively competent rocks by erosion of relatively incompetent horizons, typified by falls of Blue Lias calcarenites in the sea cliffs at Pinhay Bay.
2. Rockfalls and clayfalls caused by frost, water or desiccation in multi-jointed or fissured materials. The scale of the falls varies with the frequency of discontinuities, between the relatively widely spaced major fractures of the Chalk and the indurated Upper Greensand facies in the cliffs in the back of the undercliff, to the closely fissured Keuper Marls of Haven Cliff and Culverhole Cliffs.
3. Gully enlargement associated with cliff-top seepage points, as in the Keuper Marls of Culverhole Cliffs.
4. Forward toppling of columns of rock bounded by approximately vertical, continuous fractures, on the seaward edge of Goat Island.
5. Successive rotational slips, as in the Chalk–Upper Greensand–Lower Lias succession of Pinhay Bay, or the Chalk–Upper Greensand–Lias–Rhaetic succession of Charton Bay. These are all renewed movements of the original slipped masses.
6. Retrogressive slips, as in the Chalk–Upper Greensand succession in The Chasm at Bindon Cliffs.
7. Non-circular to translational slides, as at Bindon Cliffs, leaving a relatively undisturbed slipped mass. These represent re-activation of slipped masses.
8. Debris slides with weathering or depositional discontinuities and a mainly

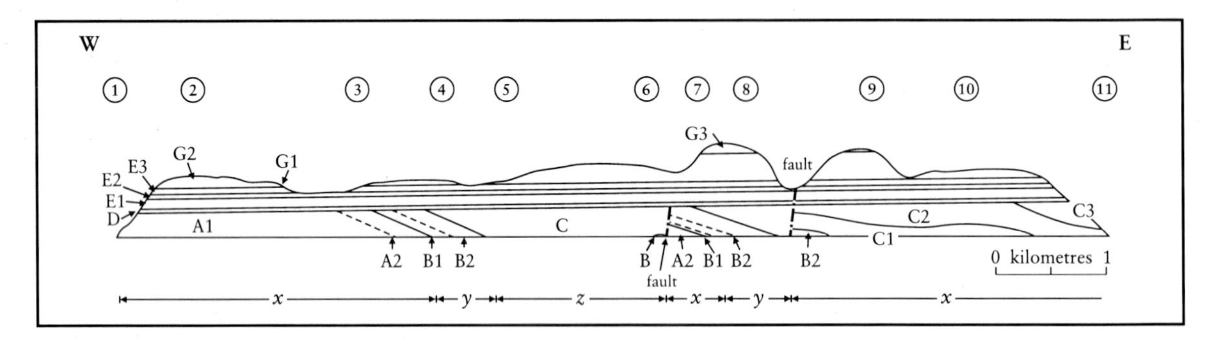

Figure 6.19 Schematic geological section of the coast between Axmouth and Lyme Regis. (1) River Axe; (2) Haven Cliff; (3) Culverhole Cliffs; (4) Bindon Cliffs; (5) Dowlands; (6) Rousdon Cliff; (7) Charton Bay; (8) Humble Point; (9) Pinhay Bay; (10) Ware Cliffs; (11) Lyme Regis. Geological succession: G3 – Upper Chalk; G2 – Middle Chalk; G1 – Cenomanian limestone; E3 – Phosphatic Upper Greensand; E2 – Cherty Upper Greensand; E1– Foxmould; D – Gault; C2 – Shales-with-Beef; C1 – Blue Lias; B2 – Lilstock Formation; B1 – Westbury Formation; A2 – Blue Anchor Formation; A1 – Red and Variegated Marls of the Mercia Mudstones Group. After Pitts (1979).

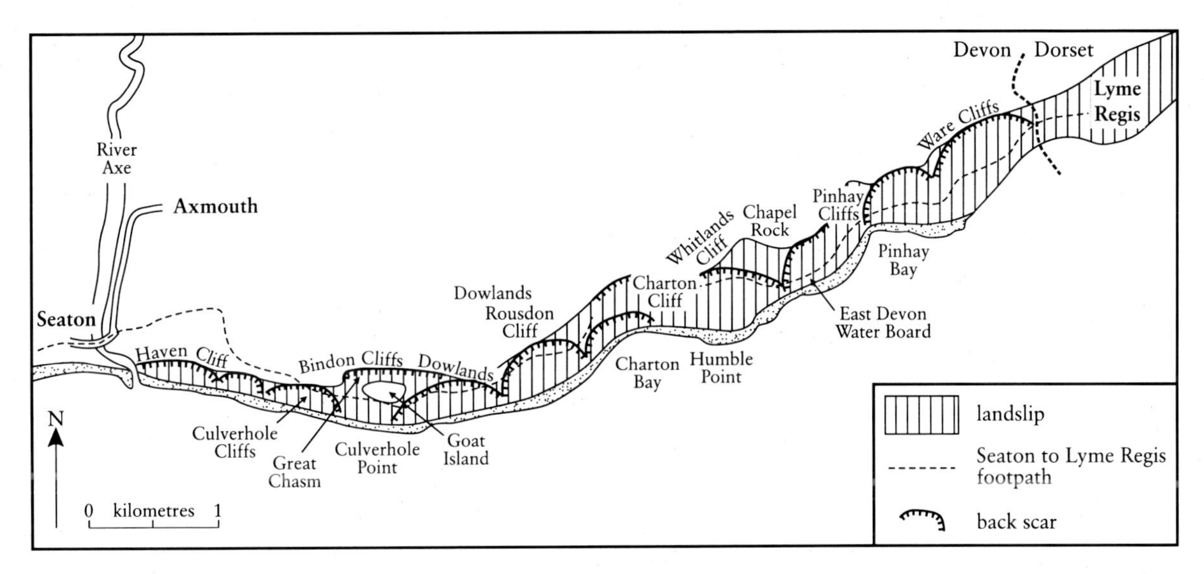

Figure 6.20 The landslides of the Axmouth to Lyme Regis Undercliffs National Nature Reserve. After Pitts and Brunsden (1987).

disturbed slipped mass, as at the toe failures of Haven Cliff, and the scree-slope failures in front of slipped Chalk and Upper Greensand blocks at Charton Bay.

9. Mudslides: these are mainly toe features, being the terminal stage of successive rotational slipping where adequate comminution of debris, clay bedrock and seepage tend to occur together, as at the cliff-top at Pinhay Bay, and at beach level at Dowlands.
10. Radial heave and slow, creep deformation at the toes of mudslides at both Dowlands and Pinhay; and small cliff-foot flows from saturated talus beneath seepage points, as at Haven Cliff and Ware Cliffs.
11. Liquefaction: structural collapse of Foxmould and subsidence of overlying strata, as at depth beneath Goat Island in the Bindon slip.
12. Sand-runs: collapse of dried-out non-cohesive arenaceous deposits, especially Foxmould, as at Charton Bay.

At Charton Bay the landform is unusual: the undercliff is two-tiered, i.e. there have been two separate landslips, one above the other, within the 127 m-high cliffs (Pitts, 1986). The upper undercliff is probably of great antiquity, while the lower undercliff formed as recently as 1969.

Two large-scale slips are very well-documented; the slip at Bindon in 1839, and the slip at Whitlands in 1840.

(b) The 1839 slip at Bindon

The Bindon area has a long and very complex history of landslipping, which has been pieced together in great detail by Pitts (1982, 1983a), using documentary sources. Probably the best-documented event, because it has been the most spectacular to take place in modern times, was the landslip at Dowlands and Bindon in 1839 (Pitts, 1974, 1982). Numerous eyewitness accounts of this slip have been documented (e.g. Roberts, 1840), and many illustrations are still extant (Figures 6.21–6.24).

The slip is particularly remarkable for its cross-sectional shape. The new main cliff-face at the rear (landward) side was then up to 64 m high, and has in front of it a large depression ('The Chasm') into which about 8 ha of land had subsided. The length of The Chasm is about 800 m, while its breadth increases from 60 m in the east to 120 m in the west. The amount of foundered material, of which some is back-tilted, was estimated at 4.2×10^6 m^3, weighing nearly 8.1×10^9 kg. Most of this is broken into a jumble of small rifted masses and pinnacles of rock, but in places blocks of about one hectare remain intact though tilted. Beyond The Chasm a counterscarp of Chalk borders an isolated upstanding area of 6 ha, which soon became known as 'Goat Island'; it had moved seawards and subsided to some extent, but corn- and turnip-fields and hedges survived (Hutchinson, 1840) (Figure 6.25).

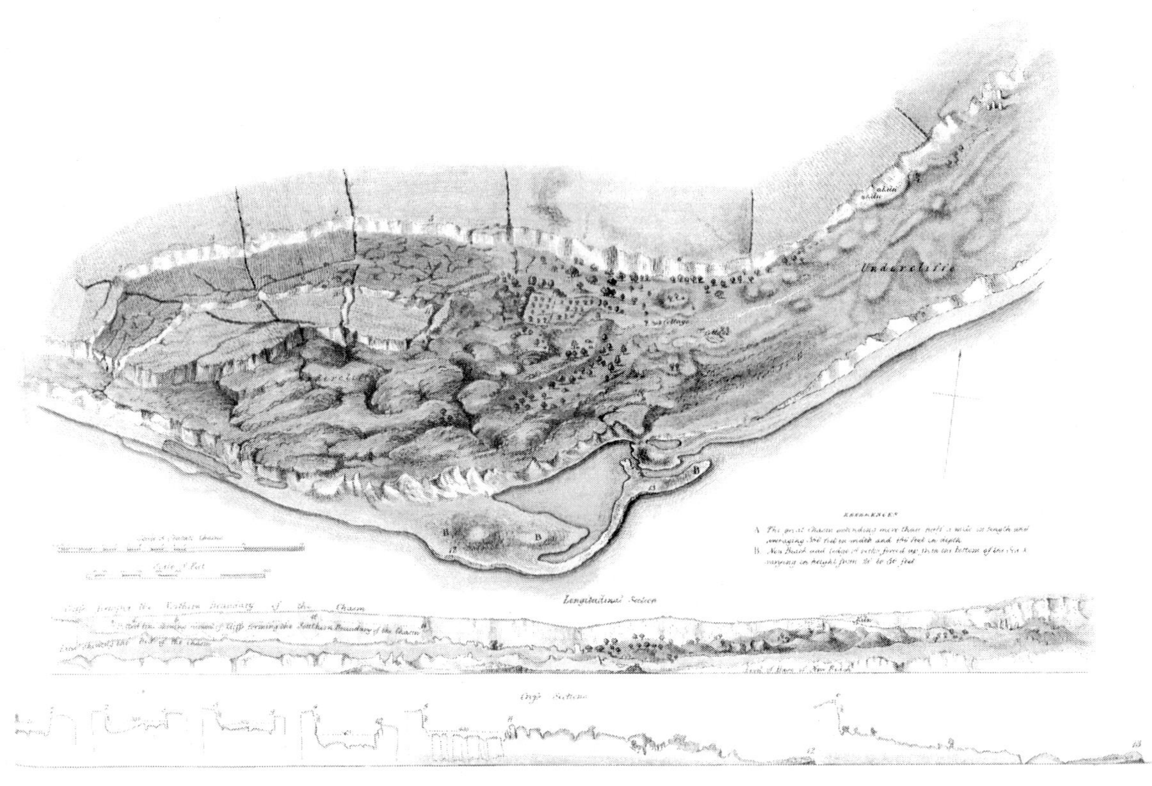

Figure 6.21 Ground plan and section of the Bindon Landslip (1839). From Conybeare *et al.* (1840), reproduced with permission of Lyme Regis Museum.

Figure 6.22 A view of 'The Chasm' looking west. From Conybeare *et al.* (1840), reproduced with permission of Lyme Regis Museum.

Figure 6.23 A view of 'The Chasm' looking west. From Roberts (1840), reproduced with permission of Lyme Regis Museum.

Figure 6.24 The reef and lagoon at Culverhole Point looking east. An engraving on stone by G. Hawkins Jr, reproduced with permission of Lyme Regis Museum.

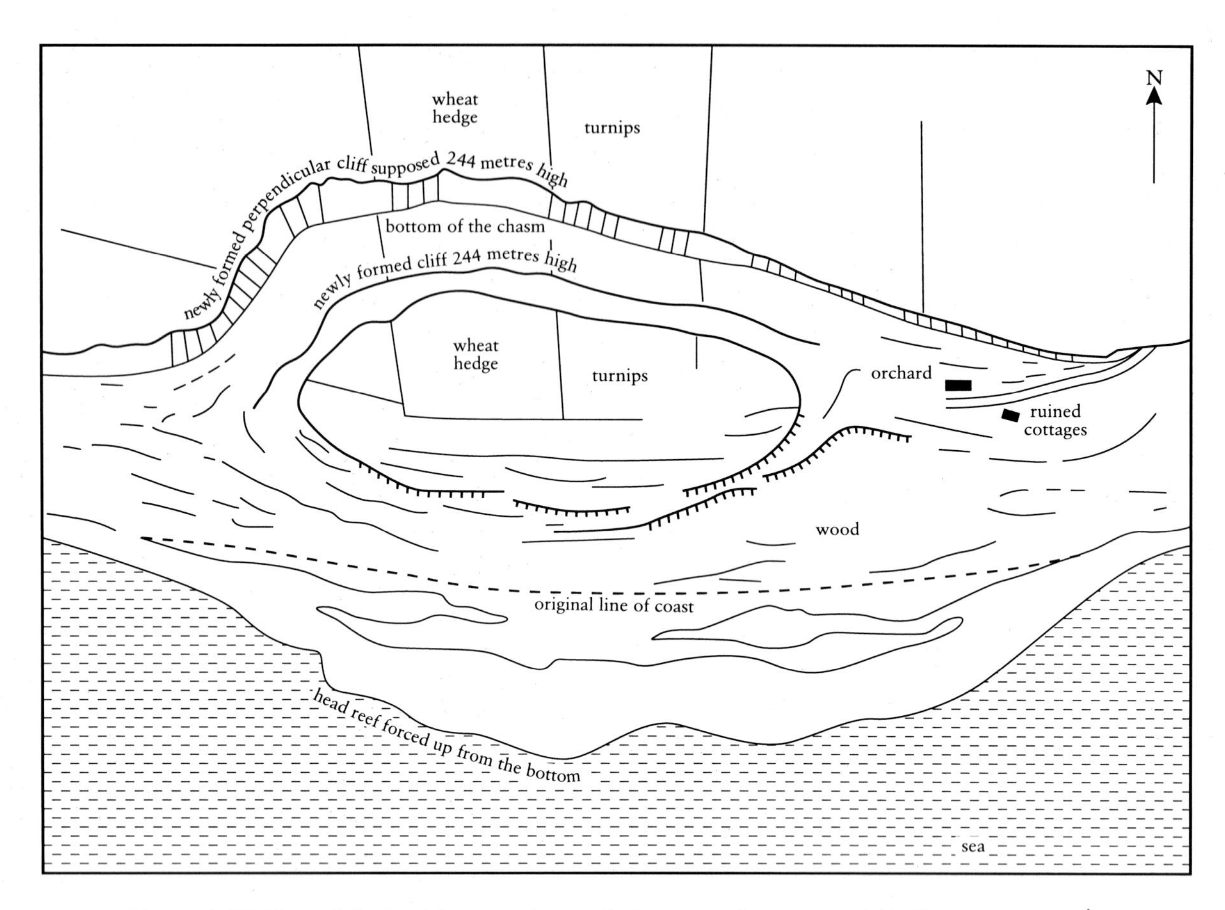

Figure 6.25 Plan of the landslip near Axmouth, Devon. After Anon (1840), from Pitts (1974).

The sea cliffs of the displaced Chalk and Upper Greensand, which had previously stood 15 m to 30 m in height, were now broken and lowered, and thrust 15 m toward the sea. A ridge of the sea shore was pushed up in front of the slip, forming a reef of Upper Greensand. This stretched laterally for nearly 1.2 km, with its outer edge 90 m to 150 m seaward of the previous high-water mark. The beds were much broken, and now dipped inland at angles varying from 30° to 45°, while the surface, which previously had been at least 3 m underwater at low tide, was now raised in places to 12 m above high-water level. The middle of the reef was joined to the mainland by shingle, but one arm extended freely at the western end, and the other to the east enclosed a lagoon which formed a natural harbour. This reef persisted for several years.

Pitts and Brunsden (1987) give a geomorphological map and a geological section of the Bindon slip (Figures 6.26 and 6.27). They made an examination of the groundwater conditions at Bindon, but unfortunately the quality and quantity of groundwater data are very poor for all parts of the site, including Bindon. Only one well exists for which records are available and that is at Dowlands Farm about 1.0 km to the east and 0.5 km inland from the cliff-top. The water level in the well varies very little. Records cited by Roberts (1840) show that the latter half of 1839 was particularly wet, contributing notably to the Bindon slip: at least 50% more rainfall than average was experienced during that period (Arber, 1939).

Pitts and Brunsden (1987) carried out laboratory investigations as follows: samples of the black shales of the Westbury Formation were obtained for determination of residual shear-strength parameters. The samples were obtained from the slipped block at Culverhole Point but were too weathered and disturbed for peak strength determinations to be reliably undertaken.

Residual shear-strength parameters were determined using a 100 mm square shear box. Reconstituted samples of the Westbury Formation shales were formed by consolidation under high loads. A plane was then cut in the

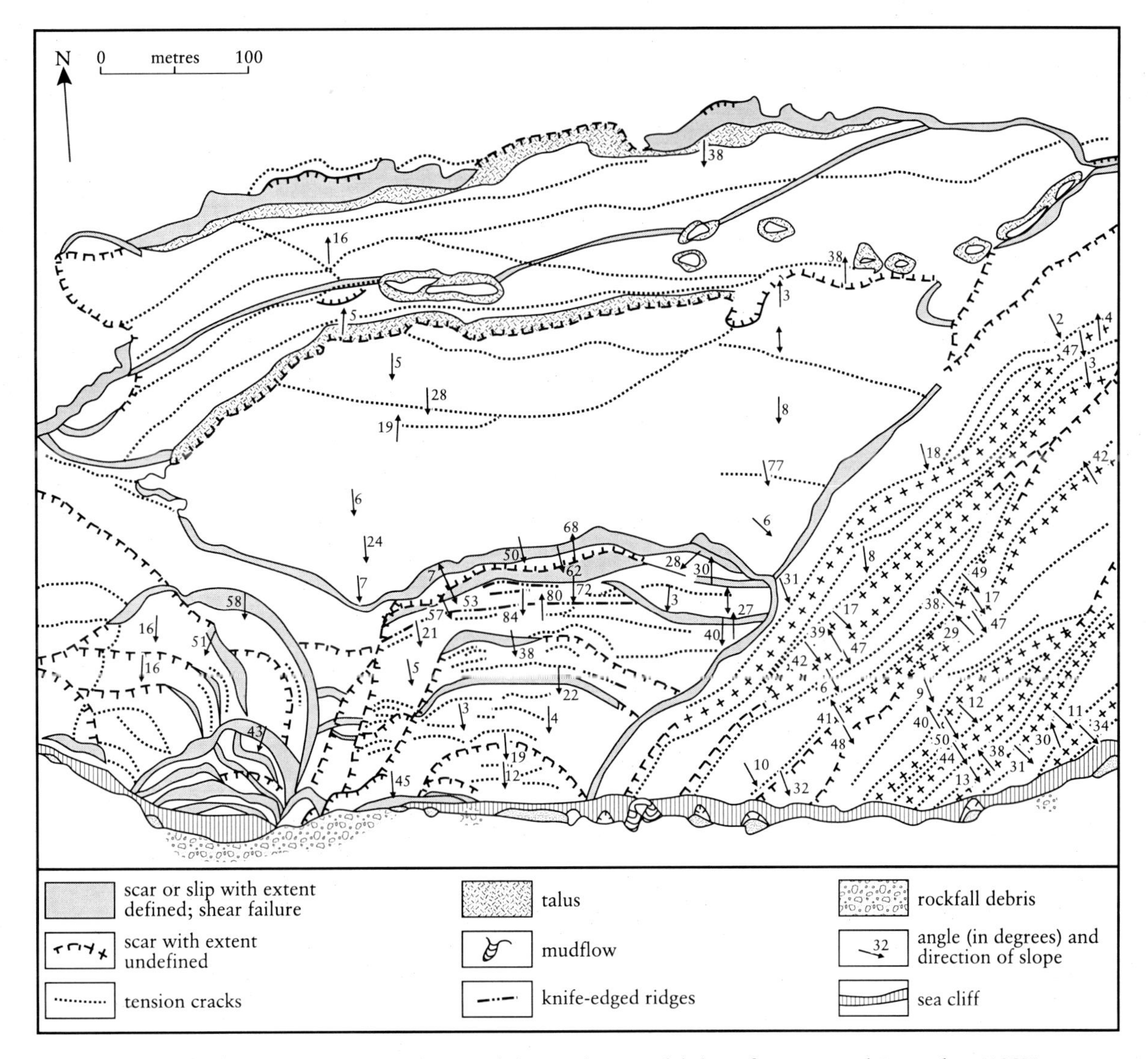

Figure 6.26 Geomorphological map of the Bindon Landslide. After Pitts and Brunsden (1987).

sample and the faces of the cut plane smoothed and polished on a glass plate. The sample was then re-consolidated and sheared several times until a fairly consistent value was obtained. The final pass was made at a low rate of shear. Values of normal stress were increased and the shearing process was repeated. The results obtained for the effective residual cohesion were $c_r = 4$ kN m^{-2}, and for the effective residual angle of shearing resistance, $\phi = 4.5°$. There must be some doubts as to the validity of such an unusually low value for the angle of shearing resistance, and the rate of shear may have been too great to generate truly drained conditions. Nevertheless, similarly low values were obtained for the Westbury Formation in Somerset by Hawkins and Privett (1985).

Pitts and Brunsden (1987) carried out stability analyses for the 1839 slide at Bindon. The analysis was carried out in terms of effective stresses using the method of Janbu (1973), and Hoek and Bray (1981) for a block slide on a clay layer. Values of pore pressures on the slip-surface were based on water levels recorded in the well at Dowlands Farm. Using considerations detailed by Pitts and Brunsden (1987), this enabled an average value of unit weight, of 20 kN m^{-3}, to be used in analysing first-time slides. In analysis of slips incorporating previously slipped material, a 10% reduction in density was assumed.

The slope profile before the 1839 slip was reconstructed using data from contemporary sources, and changes in slope facet positions

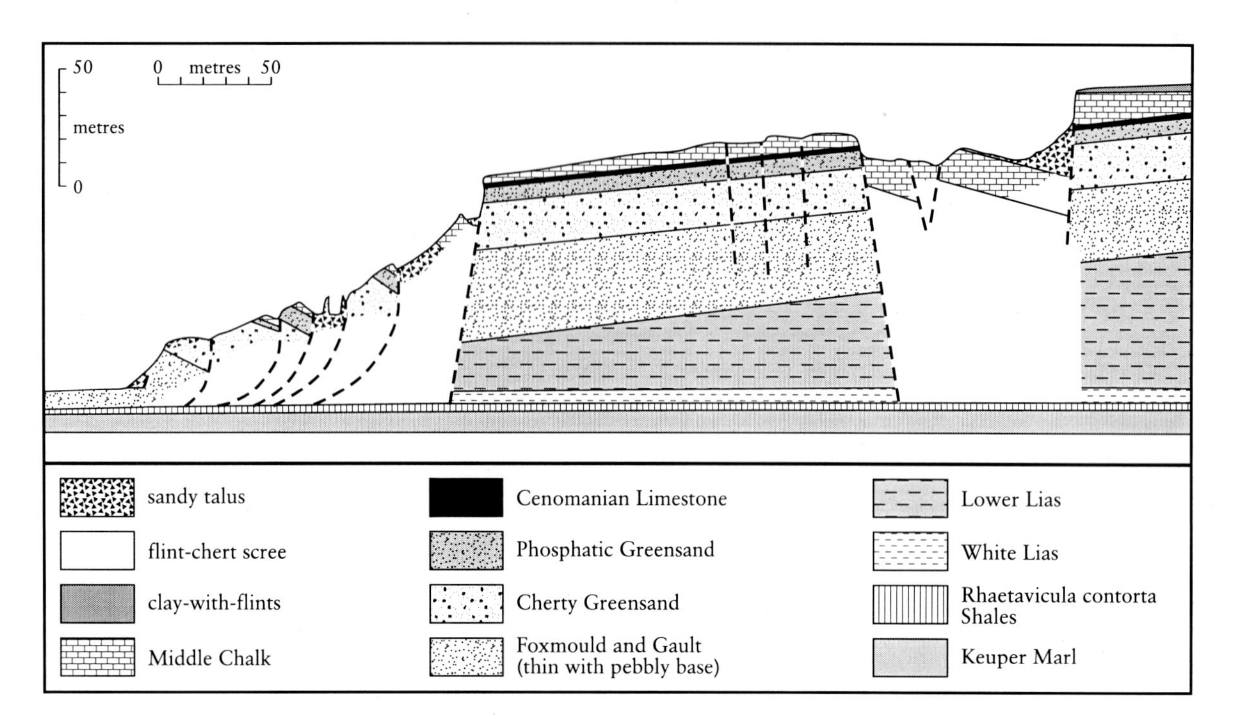

Figure 6.27 Geological section of the Bindon Landslide. The Rhaetavicula contorta Shales are the Westbury Formation and the Keuper Marl is the Mercia Mudstones Group in the modern terminology of Warrington (1980). After Pitts and Brunsden (1987). There are no accurate sub-surface data for this slide.

based on rates of erosion determined from the recession of high-water marks between 1888 (the date of the first 1:2500 plan) and 1972 (the date of the latest aerial photographs of the area) (Figure 6.28). It was assumed that this part of the Axmouth–Lyme Regis Undercliffs area was previously occupied by a small coastal slip extending offshore. The landward extent became the seaward edge of Goat Island in its pre-slipped position. The shear surface was taken to be within the Westbury Formation, and was assumed to develop on bedding partings, a situation in which cohesion values would be very low.

In order to obtain more realistic values of the shear strength of the Westbury Formation, a back-analysis was performed on the simple planar slide which took place at Charton Bay in June 1969 (Pitts, 1986), for which conditions are quite well established. The original profile of this slide was reconstructed from aerial photographs, providing a peak value for the angle of shearing resistance of the bedding of $\phi = 13°$. This value was then used in the analysis for Bindon where peak strengths were required, the cohesion being considered to be zero.

The use of zero cohesive strength throughout the analysis of the failure of Goat Island reflects the probability of sliding along a discontinuity, of progressive strength loss within the Westbury Formation shales by shear creep during the pre-failure period, perhaps during progressive erosion of the toe, and the use of drained strength parameters.

The stability of the slope in front of Goat Island was then investigated using a present-day profile, and a single unit weight was used to represent the slipped material, in conjunction with the residual strength of the Westbury Formation shales along the shear surface. Finally, analyses were undertaken to investigate the state of the stability of the current slope. This considered the three main components of the modern slope: The Chasm, Goat Island, and the slipped mass in front of Goat Island.

A slab slide was assumed despite the existence of some inconsistently back-tilted blocks in the floor of The Chasm. A perfectly planar shear-surface at the base of a deep tension crack sited at the rear of The Chasm was assumed. The results of the back-analysis for the failure of Goat Island and The Chasm produced a value for ϕ' of 20.8° mobilized at failure, a much higher value than that obtained at Charton Bay for the same material. This disparity may relate to the proposed trigger of the failure.

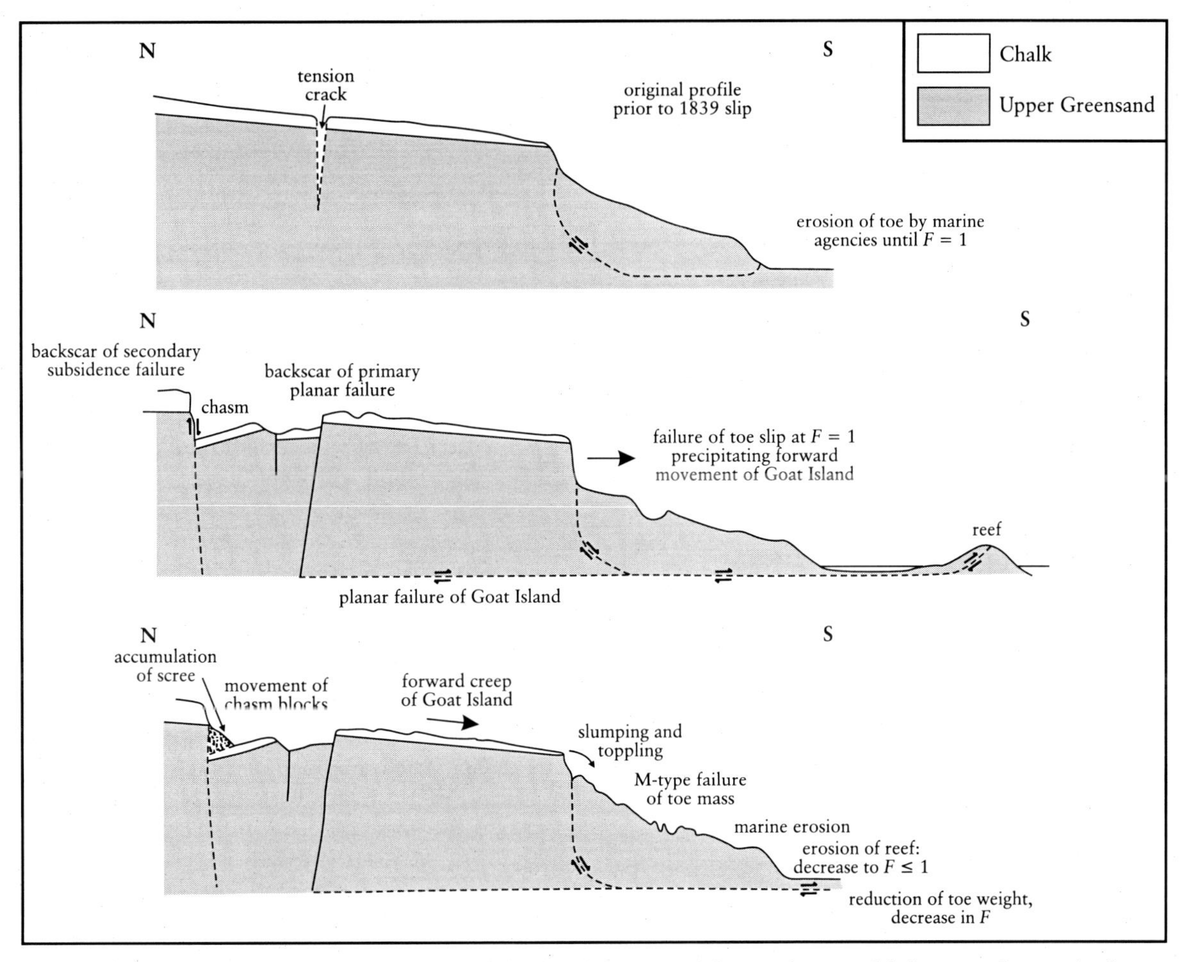

Figure 6.28 Diagrammatic reconstruction of the development of the Bindon Landslide. '*F*' refers to the factor of safety against landsliding. An M-type failure is a multi-rotational slide from the classification of Skempton and Hutchinson (1969). After Pitts and Brunsden (1987).

Several simple stability analyses were carried out to investigate further the order of failure at Bindon. This particularly concerned the hypothesis that the trigger of the main failure of the Goat Island mass was a non-circular rotational failure in front (seaward) of that mass. An analysis of the forces acting on Goat Island when it was unsupported seawards, using the shear-strength parameters from the back-analysis of the Charton Bay slip and the water levels in Dowlands Farm well, in each case produced factors of safety (*F*) of less than 1.0.

An attempt was therefore made to analyse the contribution of the seaward support to the stability of Goat Island. No direct method of analysis seemed to exist that dealt with this contingency. A method was adopted which had been outlined by Hoek and Bray (1981) as a part of a stability analysis procedure for rock masses subject to toppling failure. The formula presented by Hoek and Bray (1981) for calculating the propensity of any of the blocks to slide rather than topple is;

$$P_{n-1} = P_n[W_n(\tan\phi \cos\alpha \sin\alpha)]/(1 - \tan^2\phi)$$

where ϕ is the angle of shearing resistance and the various forces acting on the block are as shown in Figure 6.29. For the situation prior to the 1839 failure, a factor of safety of 1.15 was obtained for Goat Island.

It is difficult to be sure at what stage of failure the toe block was required to be in order to produce a factor of safety (*F*) of 1.0, that is, the factor of safety at which sliding just begins to occur, for Goat Island. The indication is, however, that the failure of the toe mass would have been almost completed before the slip of Goat Island occurred, a factor of safety of 0.99 being obtained for the pre-failure geometry.

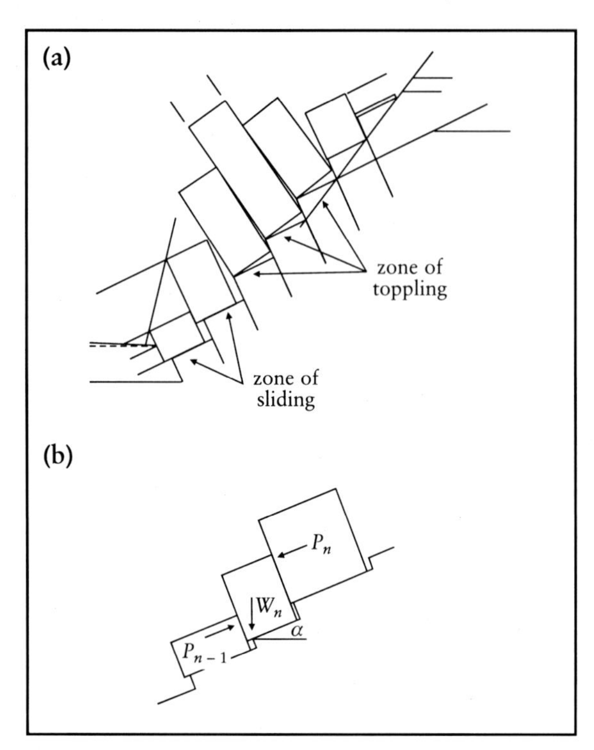

Figure 6.29 (a) Model for limiting equilibrium analysis of toppling failure on a stepped base. (b) Forces acting on a toe block liable to failure by basal shearing. After Hoek and Bray (1981).

Unfortunately no calculation was attempted to assess the effect of the subsiding masses in The Chasm on the stability of Goat Island. Too little is known about the precise course of events to quantify their effects adequately.

An analysis was carried out on the slope geometry to determine the gain in the factor of safety resulting from the lengthening of the profile and the formation of the offshore reef. Although residual strengths operated throughout the length of the failure surface, the factor of safety (F) increased to 1.49.

The slope geometry in front of Goat Island as it exists today has suffered erosion of the reef, about 140 years of marine erosion, and crown loading from degradation of the seaward-facing slope of Goat Island. The factor of safety is very low, around 1.0, and the slope is in a quasi-stable state, if the assumptions made, particularly about shear strengths and groundwater, are realistic.

Finally an analysis was undertaken of the whole slope using the present-day geometry. A value of $F = 1.23$ was obtained, compared to that of $F = 1.49$ for the immediately post-failure situation, a reduction of 17.5% in about 140 years. In view of the apparently less-stable situation in front of Goat Island, some decrease in support by a failure of the toe mass may dramatically decrease the stability of Goat Island in a way similar to 1839, except that now, lower strengths obtain below Goat Island.

(c) The 1840 slip at Whitlands

Buckland (1840) and Conybeare *et al.* (1940) also provide accounts of possibly the largest slip between Axmouth and Lyme Regis: the Whitlands slip which took place on 3rd February 1840. Pitts (1982) has assembled evidence of movements at this site over a long period, and remarks that it has probably the longest period of landslipping between Axmouth and Lyme Regis. He suggests that the extensive slips of 1689, described simply as 'West of Lyme' (Roberts, 1840) may have been at Whitlands. Wanklyn (1927) attributes a description of the effects of this slip to William Pitt the Younger. Records cited by Roberts (1840) show the period 1764–1765 to have been particularly wet. Buckland (1840) makes it clear that the high cliff was not affected, while the undercliff, a mass of Chalk and Greensand 'which had descended in former ages, began gradually to sink downwards'. The scar was over 18 m high and over 400 m long (Conybeare *et al.*, 1840). The slipped material was split into a series of 'irregular ridges and furrows' (Buckland, 1840). Houses on the slipped mass had their floors squeezed upward and their walls were tilted. A nearby garden was converted into a pond of water.

Two reefs close to the shore were seen to 'rise slowly and simultaneously with the slow descent of the subsiding portion of the adjacent undercliff' (Buckland, 1840). This extended about 0.8 km and 30 m seaward of and parallel to the old sea cliff. Buckland (1840) also suggested that 'the bottom of the sea, for a great distance from the present shore, is composed of large fragmentary masses of subsided Chalk and cherty sandstone brought thither by the destructive action of the sea and of land springs in former ages upon ancient undercliffs'. Pitts (1982) points out that this is supported by Conybeare *et al.*'s description of the seaward of the two reefs which was capped by 'a stratum of chalk capped with angular flint gravel, exactly as would be found on the summit of the chalk downs above the undercliff' (Conybeare *et al.*, 1840).

Before the main landslip on 3rd February, the undercliff at Whitlands was 'broken up into great cracks and fissures' (Roberts, 1840) at Christmas 1839. After the 'continued rains of January 1840' (Roberts, 1840) the main slip occurred on 3rd February.

The most recent major landslip at Whitlands occurred in 1981, starting on 28th February and continuing until about 7th March. A mass of black plastic Lias clay (Macfadyen, 1970), was squeezed up under the beach for a distance of about 450 m. The cobbles disappeared from the foreshore, which was raised by 3–4.5 m. The larger boulders formed a soft cliff varying between 1.8 m and 4.5 m in height on the foreshore (Wallace in Pitts, 1982). In the area of Whitland Cliff Pools, 46–229 m inland from the shore, the cliff was found to have subsided by 6 m (Macfadyen, 1970) and the pools to have lost their water. During the succeeding months, the reef was eroded, but the foreshore boulders were left unstable and in disarray for a long period afterwards (Wallace in Pitts, 1982).

Pitts (1983b) has used Ordnance Survey maps and aerial photographs to examine recent landsliding along the Axmouth–Lyme Regis coastline. He was able to show that, in general, present movements follow long-established patterns. At Haven Cliffs, it is clear that there have been major phases of block disruption, a process whereby large blocks are successively broken up into smaller blocks as material moves downslope, as described by Brunsden and Jones (1976). Colluvium has accumulated at the foot of the sea cliff. In the eastern part of Haven Cliffs, this has reduced movement almost to a standstill: scars are degrading and becoming vegetated.

At Culverhole Cliffs, activity has increased since 1905. There are many fresh scars and increased activity around the backscar.

At Bindon, rapid erosion of the reef and the extended toe of 1839 have resulted in steady forward creep of slope elements in front of Goat Island, at an average of 0.3 m a^{-1}, as these elements have been loaded by the degradation of the seaward face of Goat Island by slumping and toppling. As a result, the high-water mark has migrated seawards. Goat Island itself has steadily moved downslope, but the spatial pattern of movement is not uniform: there is a relative lack of movement in the eastern part. This may be due to stabilizing effects of slope movements in the adjacent Dowlands Cliff which took place subsequent to 1839 (Pitts, 1983b).

At Dowlands Cliff there had been relatively little change over the period of analysis. The main activities were rockfalls and rockslides at the rear. The toe block of Chalk is undergoing parallel retreat at 0.1–0.25 m a^{-1}. The main elements of the landslide have descended at 0.075–0.1 m a^{-1}.

At Rousdon the west part of the toe of the landslide is occupied by a mudslide. The toe of the mudslide seems fairly stable in position, so production of mudslide material by surging, and removal of material by the sea, may be presumed to be roughly in balance. At the toe, blocks have been moving downslope at 0.14 m a^{-1}.

At Charton Bay the lower part of the undercliff slipped in June 1969. This undermined the slip in the higher undercliff. Marine erosion of the toe, however, is inhibited by the high cobble beach, and by a large accumulation of colluvium against the sea cliff. Pitts (1986) reported a stability analysis using the method of Janbu (1973): shear-strength parameters were determined for the Shales-with-Beef at Charton Bay. Using drained reversed shear-box tests, values of residual cohesion, $c_r' = 2$ kN m^{-3} and angle of internal friction, $\phi_r' = 14.5°$ were obtained; the stability analysis gave a factor of safety (*F*) of 1.06.

At Humble Point the slope failed in March 1961. There was an apparent advance of the shoreline between 1888 and 1904, which may be an effect of block disruption in the undercliffs, unrecorded elsewhere (Pitts, 1983b). Pitts (1983b) remarks that this slope seems to be in a pattern of evolution towards large-scale slipping. He notes that at Pinhay Bay there was increased mudslide activity in the 1970s, and recession at the toe resulted in 'a substantial re-failure of the face of one of the main blocks within the undercliff. The highly comminuted and weakened nature of much of the slipped material makes its incorporation into the toe mudslide relatively rapid where major seepage from the face of the slope becomes an influence. Major block disruption events at increasingly high upslope positions appear inevitable. The broad, wet, low-angled toe area is being fed by debris which is causing rapid undermining of the upslope blocks. At the same time the crest of the sea cliff continues to recede' (Pitts, 1983b).

Further light is shed on this by Grainger *et al.* (1985; Grainger and Kalaugher, 1995), who report shallow landslide activity that poses a long-term threat to the Hart's Tongue Spring, a large spring of clean water, which issued from the base of a large slipped block of Chalk and Chert Beds, at about 30 m OD and 160 m from the beach, between Whitlands and Pinhay. This source was tapped in 1935 and remains the sole source of water supply to Lyme Regis. Their investigations (survey pegs, temporary bench marks at the pumping station and on large concrete blocks on the beach, continuous-flight auger holes, geomorphological mapping, electrical resistivity and seismic refraction techniques) enabled them to draw a cross-section and reconstruction (Figure 6.30), actual surges of movement being clearly related to rain-

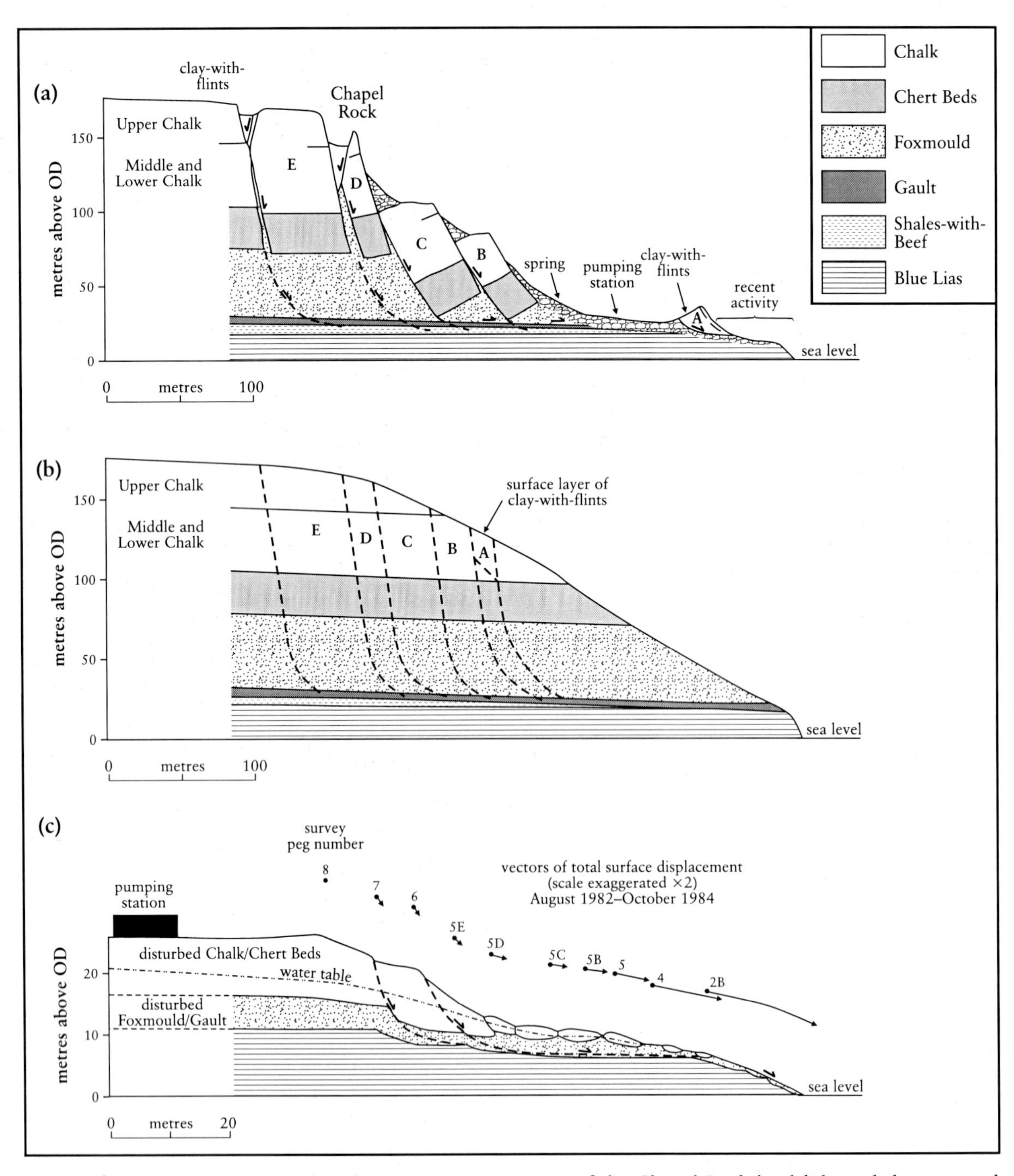

Figure 6.30 Cross-sections and evolutionary reconstruction of the Chapel Rock landslide and the surveyed movements at the undercliff water pumping-station. Note the loss of the Foxmould by flow or extrusion. After Grainger *et al.* (1985).

fall events (Figure 6.31). They were able to conclude that the cliff above the source is currently in a stable condition, but the zone below the pumping station is unstable and unlikely to achieve stability. In this zone retrogressive development of backscarps in the degraded Chalk blocks continues, and at its present rate will reach the pumping station in a few tens of years.

At Ware Cliffs, major block disruption in the lower parts of the slope has been very active since the 1950s. These events have extended the zone of major activity as far back as the public footpath that runs through the Nature Reserve. There is a noticeable boundary between the basically dry and wet parts of the Ware Cliffs slope at a position just downslope of the footpath, from which large amounts of seepage are discharged. Extensive seepage near the toe of the slope and from points in the lower parts of the undercliff are also making the lower slopes unstable. Much material has slipped over the edge of the sea cliff, the effects of which have

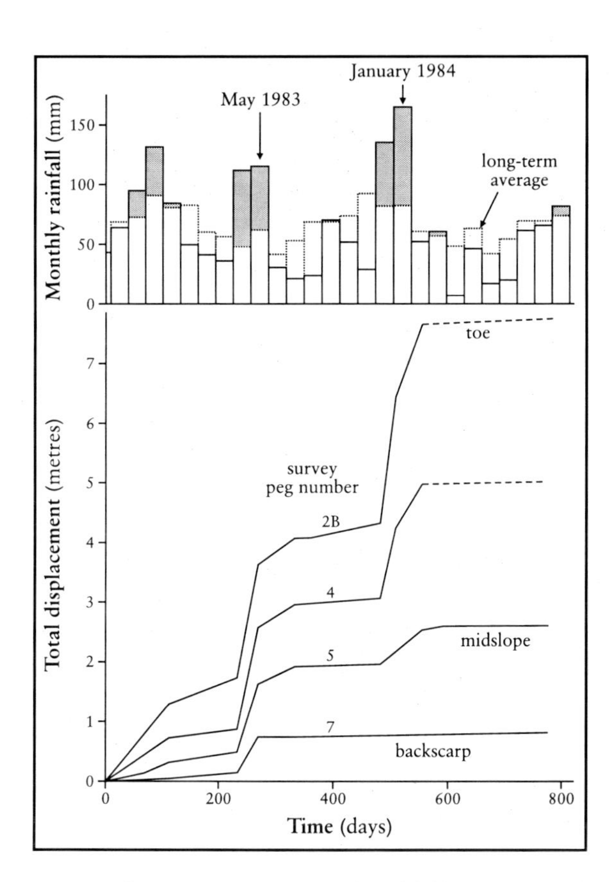

Figure 6.31 Comparison of rainfall and ground movement of the lower slopes of the landslide between Humble Point and Pinhay. After Grainger *et al.* (1985).

been transferred upslope and which have resulted in re-routings of the public footpath to positions progressively farther upslope. This pattern seems likely to continue, resulting in the progressive undermining of the major blocks of Chalk and Greensand in the upper half of the undercliff.

Interpretation

Conybeare (1840; and in Conybeare *et al.*, 1840) and Buckland (1840) suggested that the relatively undisturbed state of Goat Island indicated translational seaward sliding over some saturated horizon. They identified the Foxmould (glauconitic sands, lower division of the Upper Greensand) as the most likely candidate, and postulated that it had been reduced to the condition of a quicksand and 'washed out' by heavy rains, causing undermining of the superincumbent strata, and seaward slipping over the underlying argillaceous Lias and Gault. This is probably the first description of liquefaction and metastable sands in the scientific literature. Arber (1940) suggested that the slip is associated with the Cretaceous overstep, and supported the notion of 'washing away' of the Foxmould leading to undermining and translational sliding. A major (but erroneous) challenge to this view came from Ward (1945) who stated categorically that the main movement was rotational and similar to that at **Folkestone Warren**. This was questioned by Arber (1962), in view of the relatively undisturbed nature of Goat Island, with its (if anything) slightly seaward tilt. The strongest expression of this view is a diagram by Macfadyen (1970; Figure 6.32). However, a close examination of a series of ridges running parallel to the edge of Goat Island on its seaward side unfortunately led her to a later acceptance of Ward's view (Arber, 1971). Arber has since pointed out (1973) that the back-tilting of blocks in The Chasm, and the pushing up of Lower Greensand strata in the reef, support this view, and has suggested that Goat Island may not have moved at all. However, morphological mapping by Pitts (1979) suggests that some foundering and seaward slipping have in fact taken place. It is now widely recognized that the mechanism of 'graben' or 'chasm' formation requires forward movement of the foundered blocks and is a diagnostic feature of non-circular failure. The Macfadyen (1970) model should be discarded.

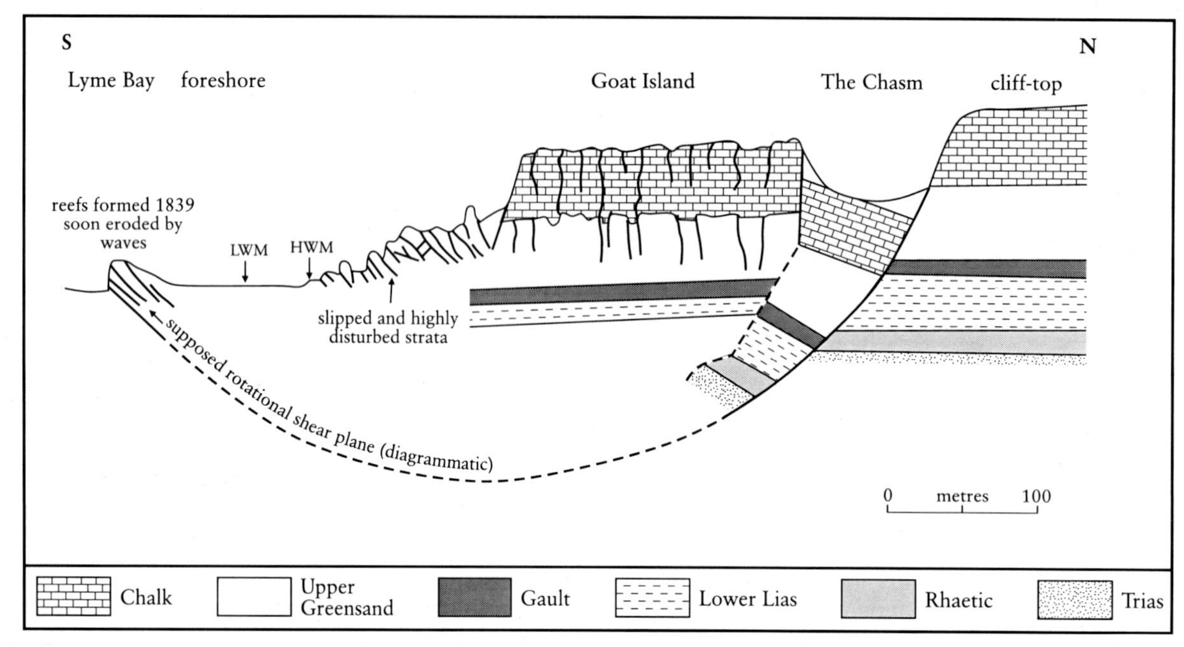

Figure 6.32 Hypothetical section through the Bindon Landslide showing a rotational failure mechanism (after Macfadyen, 1970, from Pitts, 1974). The model is not substantiated by sub-surface information. Note that the model does not explain the toe slips, nor why the strata in Goat Island remain horizontal when subject to rotational movement. The graben is diagnostic of a non-circular failure on the bedding.

Pitts (1983b) concludes from his study of recent movements that present-day developments depend primarily on conditions at the toe. The removal of the toe of a landslide, landslide debris, slipped blocks, or even offshore reefs subsequent to a slope failure or the gradual recession of sea cliffs upon which landslides have developed, result in critical fore-shortening of the profiles and a reduction in the factor of safety. Progressive downslope movement of the large, more-competent slipped blocks behind then takes place more easily. Processes at the rear, notably rockfalls from the backing cliffs (rear scarp), and the general accumulation of scree at their foot seem to result in crown loading but this appears to have only a marginal influence on the destabilization of the remainder of the slope. As Pitts (1983b) remarks, this is borne out by the rarity of whole-slope failures which are not demonstrably preceded by a series of increasingly large failures developing rearwards from the toe. He concludes by observing that over the time period which map, aerial photograph and field observation cover, there seems little reason to believe that instability following the currently established patterns will not continue.

With respect to the 1839 Bindon landslide, Pitts and Brunsden (1987) propose that the mechanisms were as follows:

1. An initial non-circular deep-seated failure of the pre-existing undercliff, which reduced the toe support of the block behind.
2. Resultant planar slippage on the underlying Westbury Formation, of the mass known as Goat Island.
3. Subsidence of pinnacle-like masses of Middle Chalk and indurated Upper Greensand rocks into The Chasm from both the rear of Goat Island and the rear cliff.

Additional points which may have a bearing on the interpretations are:

1. Geophysical investigations in Lyme Bay (Darton *et al.*, 1981) have revealed many 'axes' trending roughly parallel to the coast; these were interpreted as being of structural origin. However, as the main structural axes in the pre-Cretaceous strata in the area trend approximately normal to the coast, Pitts (1983b) suggests that the 'axes' are the eroded remains of multiple rotational slips, repeatedly planed off by a transgressing sea in immediately post-glacial times.
2. Alternatively (Kellaway *et al.*, 1975) the landslips may have resulted from glacial oversteepening and dissection by meltwaters, but evidence for glaciation is inconclusive.

Conclusions

The Axmouth–Lyme Regis site is worthy of GCR status for several reasons, and has additional historical importance. Owing to the extraordinary co-incidence of two eminent geologists (Conybeare and Buckland) being in the vicinity at the time of the major 1839 landslip, the site was meticulously examined and recorded at the time, the first occasion that this had happened with any landslip, worldwide. The results of the slip were scenically so spectacular that many pictures of it were drawn and painted and accounts of the events collected. Conybeare's explanation represents an important landmark in the development of attempts to explain and understand mass movements. Secondly, the existence of Goat Island beyond The Chasm at Bindon has led to an instructive controversy over the nature of the 1839 slip. Thirdly, the entire stretch of cliffs (the Ware Cliffs slides are currently extending eastwards into the gardens of Lyme Regis), shows an astonishing and quite exceptional variety and richness of mass-movement types, as documented above. As noted by Arber (1973), this is chiefly in terms of large-scale, deep-seated slumps, in which many large slices of rock have maintained something of their original shape, and form discontinuous ridges paralleling the inland cliff-line that forms the backface of the slips. It should be noted that at the present day very few of the characteristics and features described 150 years ago can be observed with ease at the site, as the vegetation cover is very thick and in many places virtually or actually impenetrable.

BLACK VEN, DORSET (SY 347 927–SY 363 931)

R.G. Cooper

Introduction

The Jurassic outcrop in Great Britain terminates on the south coast in Dorset, where some of the finest landslides in Britain are located. At many locations along the Dorset coast, a permeable caprock overlies an impermeable clay. The structural relationships of the Jurassic strata are such that a variety of different subjacent beds form caprock-and-aquiclude pairs, and within this range of landslide generators, each has different properties and styles of failure (Brunsden, 1996b).

Among these, Black Ven (Figures 6.33 and 6.34), located between Charmouth and Lyme Regis, is of particular interest because of its active mudslides. Indeed, it is the most active and complex landslide site in the British Isles. It comprises rotational slides, topples, rockfalls and slumps in Upper Greensand, above mud-slides, mudflows and sandflows, which feed down to the beach across Liassic materials. It has a long history (Lang, 1928; Arber, 1941,

Figure 6.33 The Black Ven landslide. (Photo: R. Edmonds, Dorset County Council.)

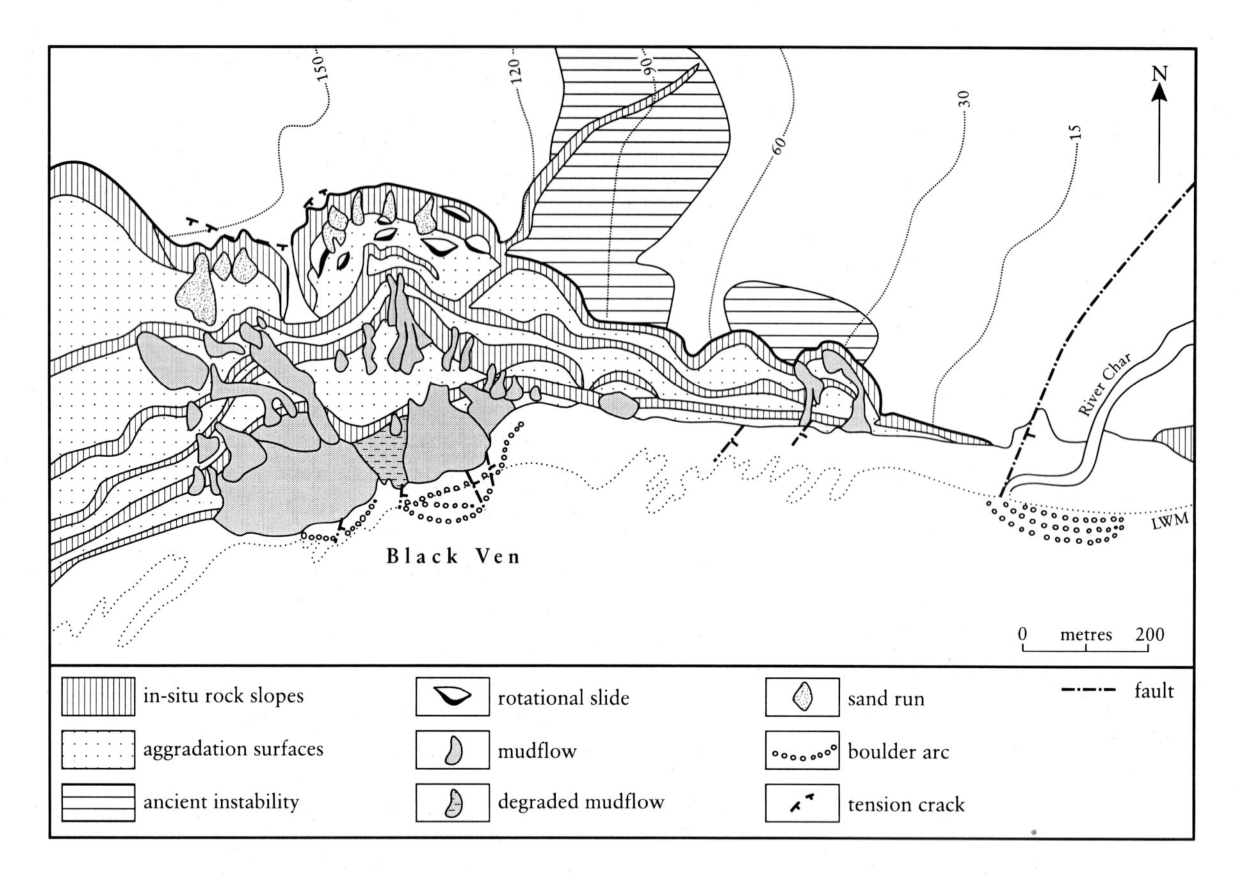

Figure 6.34 The mass-movement complex at Black Ven as it appeared in 1974. After Conway (1974).

1973; Wilson *et al.*, 1958; Brunsden, 1969, 1984; Brunsden and Goudie, 1981; Chandler and Cooper, 1988, 1989; Chandler and Brunsden, 1995; Brunsden and Chandler, 1996), but its present character was established by major movements that took place in 1956–1957, 1958 (Conway, 1974) and more recently by renewed activity in the 1980s and 1990s (Figure 6.35). The cliffs reach 150 m above OD. Cliff retreat in the order of 5 $m\,a^{-1}$ to 30 $m\,a^{-1}$ is typical during periods of activity. Between major events, erosion at the toe is around 15 $m\,a^{-1}$ to 40 $m\,a^{-1}$. During periods when detached material at the head of the slope is highly saturated, debris tumbles off the edge of an upper bench and drops 20 m onto the middle of three terraces. The upper segment of the Black Ven slope is therefore loaded at the head of each bench. The change begins at the rear scar, where the cycle of primary instability is generated. The original road along the coast was destroyed by landslips in the 18th century (Koh, 1990). A cart track running parallel to the road 100 m farther inland disappeared in 1965 and a section of the Heritage Coast Path collapsed in 1985; it was renewed, but lost again in 1994.

By the time the moving material has reached the middle terrace, mudslides and mudflows have developed. The debris is funnelled into large mudslide tracks, which pour across the cliff separating the middle and lower terraces. The process repeats as the material moves across the top of the Blue Lias towards the beach where the mudslides merge to form large composite fans and toe lobes.

Description

The Black Ven cliff is composed of the Blue Lias, Shales-with-Beef, Black Ven Marls and Belemnite Marl divisions of the Lower Lias (Lower Jurassic), overlain unconformably by Gault Clay and then by the Foxmould and Chert Beds divisions of the Upper Greensand (Lower Cretaceous) (Figure 6.36).

Dips in the Jurassic beds are about 2°–3° south-east or ESE, and the plane of unconformity at the base of the Cretaceous strata dips 1°–2° south or SSW. The cliff profile (Figure 6.37) shows well-developed terraces at the levels of the base of the Black Ven Marls (Conway,

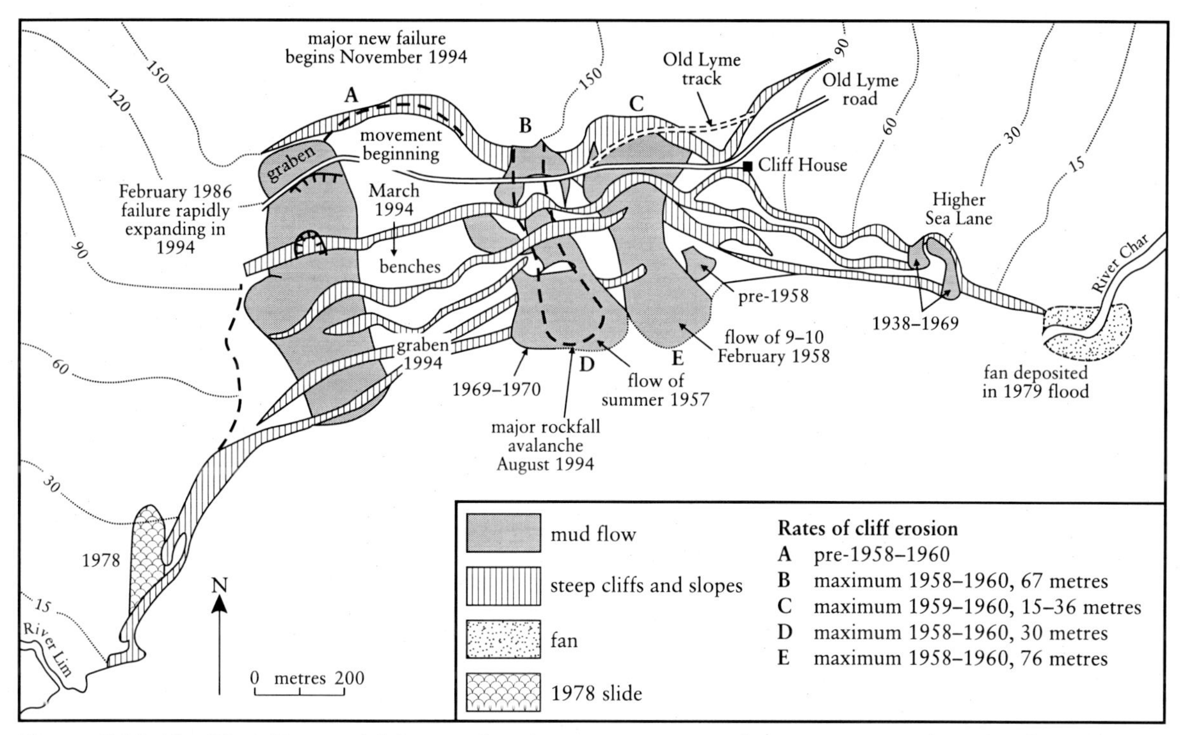

Figure 6.35 The Black Ven mudslide complex showing movements between 1958 and 1994. After Chandler and Brunsden (1995).

1974). These terraces or benches are caused by the presence of resistant horizons within the Lower Lias. A fourth resistant horizon gives rise to a less well-developed terrace feature at the top of the Blue Lias, but this is often obscured by landslip deposits. A number of minor terraces are developed above the other resistant horizons, but they are not very extensive horizontally.

In the Upper Greensand, brecciated Chert Beds consist of broken chert in a firm, coarse sandy clay matrix with some iron and manganese oxide concentration in the lower part. The beds are much harder than the underlying decalcified Foxmould sands and this has resulted in the development of a steep upper cliff, the height of which is sufficient to allow the generation of shear stresses far in excess of the resistance offered by the decalcified sands. This results in the propagation of single- and multiple-failure rotational slides that affect the full thickness of the Upper Greensand. Secondary iron oxides have been deposited at the base of the Chert Beds, impeding the downward movement of groundwater. This results in springs being thrown out at the cliff-face, which have cut deep gullies in the cliffs of Foxmould sand below. This process is greatly assisted by land drainage from the top of the cliffs. Many of the initial movements of slide blocks occur as a result of failure of the conical buttresses that develop between these gullies.

The base level to which the gullying and the rotational slides operate could be the top of the Gault Clay, the level of the highest terrace. Large accumulations of sand and chert debris build up on this terrace, and during the winter months are rapidly saturated. The water is discharged on the cliff-face at the junction of the debris with the Gault Clay. This results in extensive seepage erosion (Conway, 1974) and gullying, leading to failure, and the debris is carried down the gullies onto the next terrace below. The upper cliff sides are thus deprived of part of their toe areas and stress is again able to build up to the level required to regenerate the cycle of primary instability.

Although gullied by seepage erosion the major cause of the removal of material from the Belemnite Marl cliffs are joint-controlled rockfall caused by erosional unloading, and frost action. The material received by the second terrace, at the base of the Belemnite Marl cliffs, is again rapidly saturated in winter and loaded at its head by the cascade of mud from the terrace above. It discharges water and sediment onto

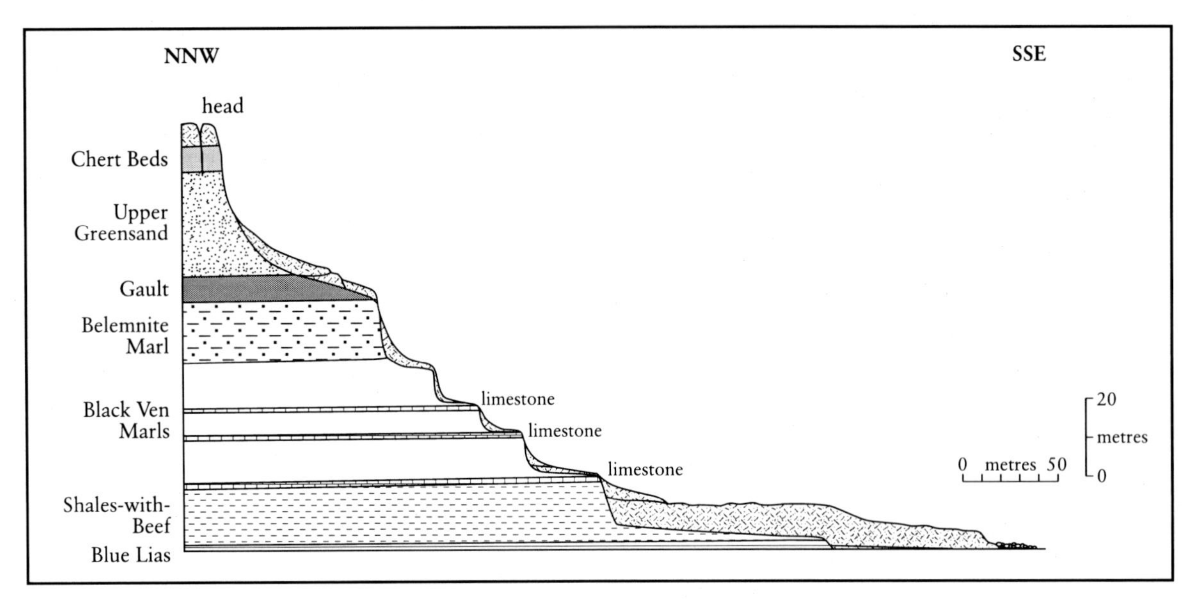

Figure 6.36 Geological cross-section of Black Ven showing the lobes of the 1958 mudslide. After Conway (1974).

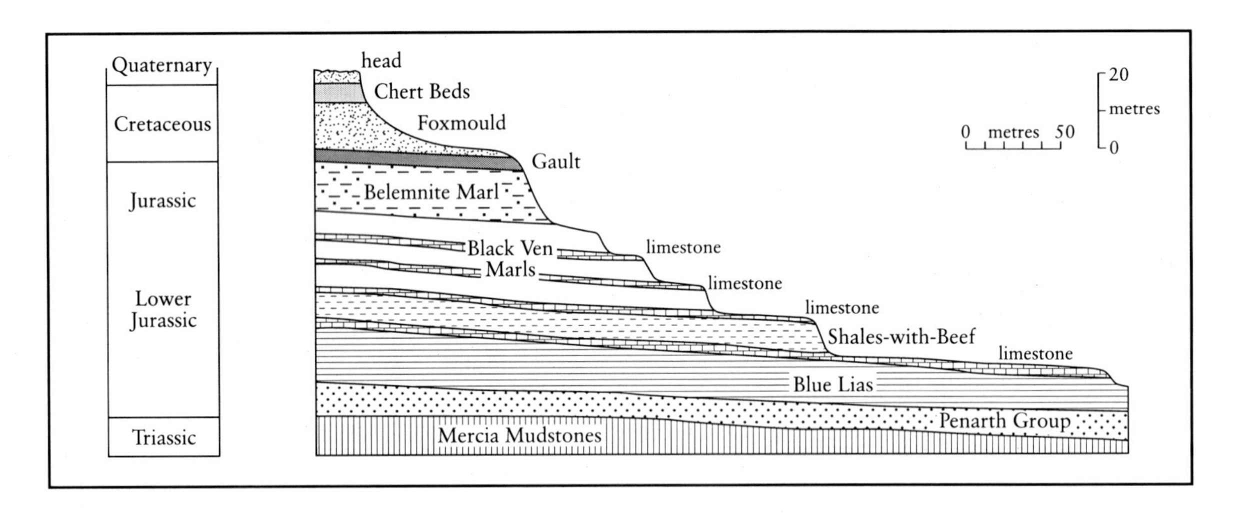

Figure 6.37 Section through cliffs to the west of Black Ven showing regional dip and benches under-scoured by landslide debris.

the next cliff-face from its lower boundary. The Black Ven Marls below is soft and fissured. Subjected to seepage erosion, it is rapidly gullied and small rotational failures occur. Water-charged debris, carried down gullies or pushed over the terrace edge by material accumulating behind, strips off the outer, weathered, layer of clay. Removal of this material from the terrace face results in steepening of the cliff and allows the generation of higher stresses in the clay, which in consequence approaches a failure condition. The clay, which is heavily over-consolidated, takes in water, swelling and weakening until failure takes place. On the benches major mudflows, mudslides and sheet-flows descend to the third and fourth terraces at the base of the Black Ven Marls.

The process is repeated from the fourth terrace over the soft Shales-with-Beef cliff down to the beach. Flows and slides coalesce to form large composite toe fans, which often (as in some recent years) completely envelop the fifth and lowest terrace, at the top of the Blue Lias, and much of the beach. This terrace is really a toe fan, a composite body formed by the accumulation of debris from many cycles of secondary instability, and built on an original which resulted from a single catastrophic event in 1958 (Lang, 1959). The upslope part of the fan, lying on the engulfed fifth terrace, displays transverse pressure ridges, while the downslope part shows extensive tension cracking and isolated pressure structures. Extensive sand-runs resulting from periodic flash-floods fill in and cover the irregular surface of the clay toe fan. At times of little or no mudslide activity the fine-grained material in the toe lobes is washed away to leave boulder arcs on the beach.

The above description is based on Arber (1941, 1973), Conway (1974), Brunsden and Allison (1990) and Koh (1990). More recent and much more detailed investigation has been made possible by the development of an archival, three-dimensional photogrammetric technique that is able to derive quantitative spatial information of known accuracy, from historical aerial photographs (Chandler and Cooper, 1988). The technique was itself developed using Black Ven as testbed and exemplar (Chandler and Cooper, 1988, 1989). These authors show how analytical photogrammetry can be applied to historical photographs, a hitherto untapped source of data for geomorphologists and other Earth scientists. They term their research the 'archival photo-grammetric technique'. They point out that, lacking camera calibration data and co-ordinated ground control points, conventional photogrammetry is impossible. To monitor the development of a feature a sequence of photographs is needed. The archival photogrammetric technique is based around computerized analytical techniques, mainly a *self calibrating bundle adjustment*. This establishes, digitally, the relationship between the photographs and a ground co-ordinate system. The replacement of the analogue stereoplotter with a digital mathematical model of this type is a well-established technique (Ghosh, 1979). The process involves photo acquisition, identification and derivation of control points, photo measurement, photogrammetric processing, data extraction, data processing/presentation, and interpretation.

This methodology was validated using Black Ven, selected because, being so active, it has shown marked changes. As pointed out by Chandler and Brunsden (1995), the site has been subject to several aerial photograph surveys constituting the 1946, 1958, 1969, 1976 and 1988 aerial photographic 'epochs' (see Figures 6.38–6.42). A further epoch, for 1995, is analysed by Brunsden and Chandler (1996).

Figure 6.38 Aerial photograph for the 1946 epoch. (Photo: English Heritage (NMR) RAF Photography.)

Figure 6.39 Oblique aerial photograph for the 1958 epoch. (Photo: Crown Copyright/MOD. Reproduced with the permission of the Controller of Her Majesty's Stationery Office.)

Figure 6.40 Aerial photograph for the 1969 epoch. (Photo: Copyright reserved Cambridge University Collection of Air Photographs.)

Figure 6.41 Aerial photograph for the 1976 epoch. (Photo: reproduced by permission of Ordnance Survey on behalf of HMSO © Crown Copyright (2006). All rights reserved. Ordnance Survey Licence number 100038718.)

Figure 6.42 Oblique aerial photograph for the 1988 epoch. (Photo: J. Chandler.)

The basic data units used for all photogrammetrically based methods are the three-dimensional co-ordinates that can be obtained with a density and efficiency which is unobtainable by other techniques. Data extraction is greater, denser and with subsequent data processing, more powerful and flexible. The co-ordinate data can be used to provide basic planimetry, slope profiles and contours (Chandler *et al.*, 1987) (see Figure 6.43), digital terrain models (DTMs) (shown as isometric views in Figure 6.44) and movement vectors. The technique can be regarded as an updating of previous methods of geomorphological mapping. As remarked by Chandler and Cooper (1988), precise definition and coding of morphological boundaries by rigorous photogrammetric techniques combines the benefits of geomorphological interpretation with positional relevance. Visual comparison between photographs at two widely differing times provides a basic tool which can be used to identify, quantify and interpret areas that display any degree of change.

Although contour plans provide a full description of site morphology at the different epochs it is difficult to identify areas of change by visual inspection. However, subtracting a grid surface produced at one epoch from the grid of a later or earlier epoch creates a grid surface that represents the change of form over the period defined by the photographs. This surface can be contoured, thus quantifying the spatial effects of processes: some areas will have lost material, others will have gained material, and some will have exhibited no change. Chandler and Cooper (1988) caution that the last-mentioned set of areas are not necessarily inactive areas. They can be areas where the input of material has equalled output over the defined period (see Figure 6.45a–f).

Chandler and Brunsden (1995) deal in more detail with the problems of applying photogrammetric methods to archive photographs, in particular the components of the self-calibrating bundle adjustment. At Black Ven the control points used at all epochs are derived from one Ordnance Survey plan, which therefore acts as a datum.

Koh (1990) set up an automated data-gathering system recording rainfall, porewater

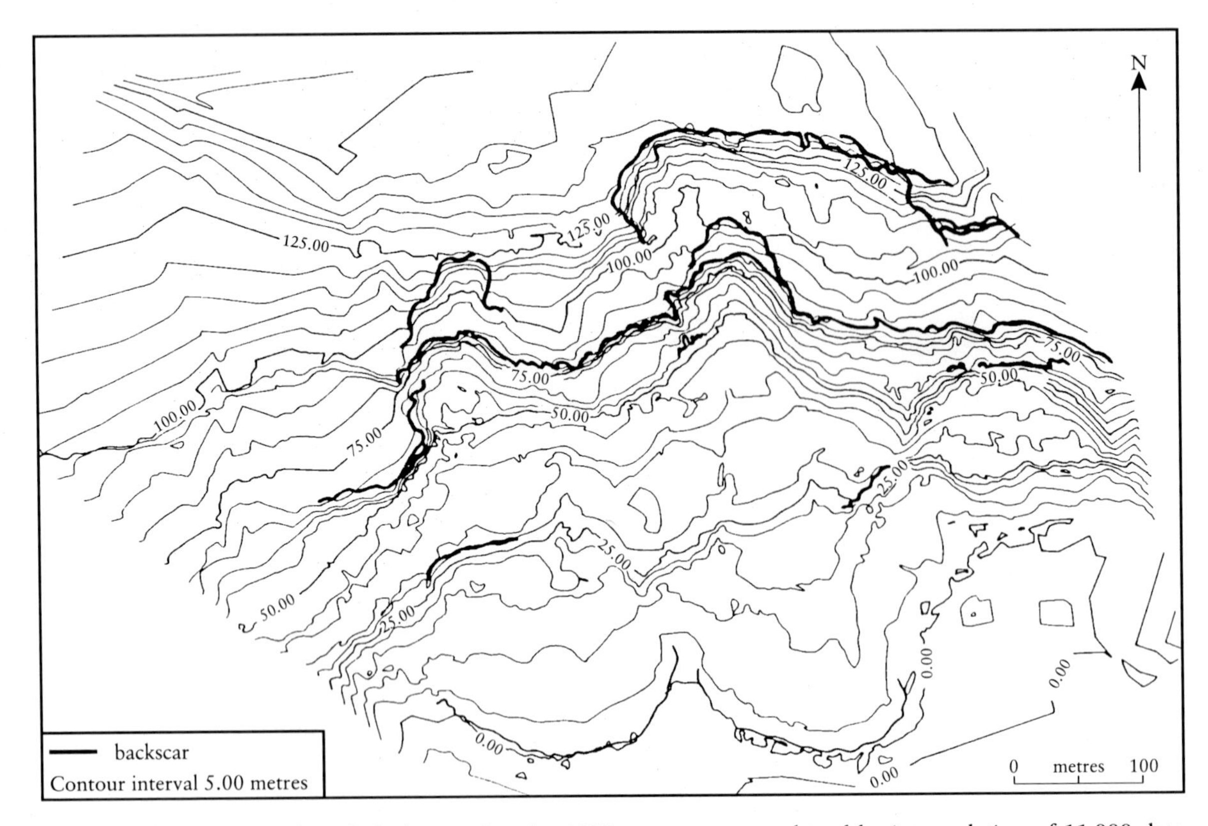

Figure 6.43 Contour plot of Black Ven after the 1958 movements produced by interpolation of 11 000 data points established by photogrammetry. After Chandler and Brunsden (1995).

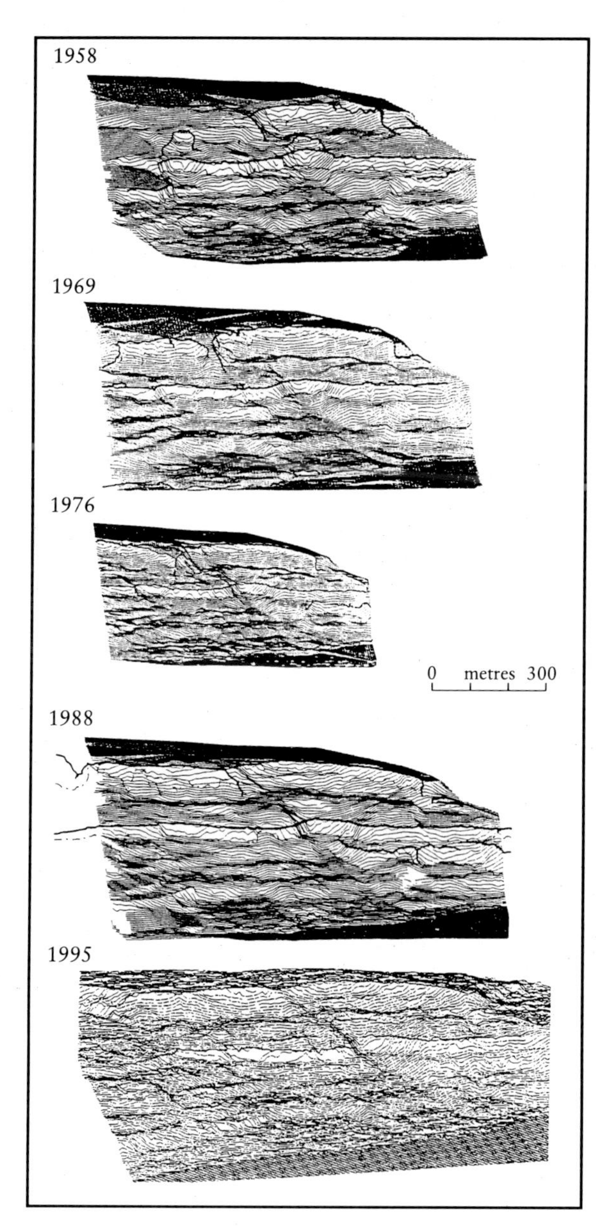

Figure 6.44 Digital terrain models (DTMs) shown as isometric views for the Black Ven mudslide at five epochs between 1958 and 1995. Note that the 1958–1988 epochs are based on analytical photogrammetry and an 11 000 point data set. The 1995 model is based on a larger data set aquired by digital photogrammetry. After Chandler and Brunsden (1995).

pressure, loading, surface movement and subsurface displacement, below the Belemnite Marl cliffs at Black Ven. His results are summarized in Figure 6.46 where the response of porewater pressure and cumulative displacement to monthly rainfall is very apparent. Koh suggests two alternative mechanisms: visco-plastic flow as described by the Bingham equation, and the release of dissolved solids, according to the groundwater chemistry shear-strength model of Moore (1988).

Interpretation

(a) 'The reservoir principle'

It has been suggested by Denness (1972) and Conway (1974) that the sequential process active at Black Ven may best be understood in terms of the presence, in intimate association with the instability, of bodies of material that behave as reservoirs of groundwater with effectively impermeable floors. Naturally occurring groundwater reservoirs may be seen as consisting of two kinds, primary and secondary, depending on whether the reservoir material itself is an in-situ rock body or an accumulation of rock debris resulting from slope degradation. Hence, the in-situ Upper Greensand is the primary reservoir at Black Ven The debris accumulations on the terraces below are secondary reservoirs, and it is the gradual release of accumulated water from these that leads to the unusually rapid degradation of the material on the terraces and the rapid transport of their material to the cliff-foot.

(b) The episodic landform change model of Chandler and Brunsden (1995)

Chandler and Brunsden (1995) include a 'Speculative Discussion' based on results from the archival photogrammetric technique. One view of morphological change is that landform change takes place when a state of process equilibrium and morphological stability is perturbed by an impulse of change of sufficient character to overcome the tolerance of the system (Brunsden and Thornes, 1979; Brunsden, 1985, 1990). This overcoming of tolerance may be divided into two phases: 'preparatory' impulses, which predispose a system to change, and 'triggering' impulses, which actually push the system over a threshold. In the case of Black Ven there is evidence that the system was prepared for a new phase of mudsliding by the erosion and steepening of the cliffs to a new average angle exceeding 19°. The 1958–1959 mudslides failed at about 19°, which can be regarded as a failure threshold. However, this is in part an

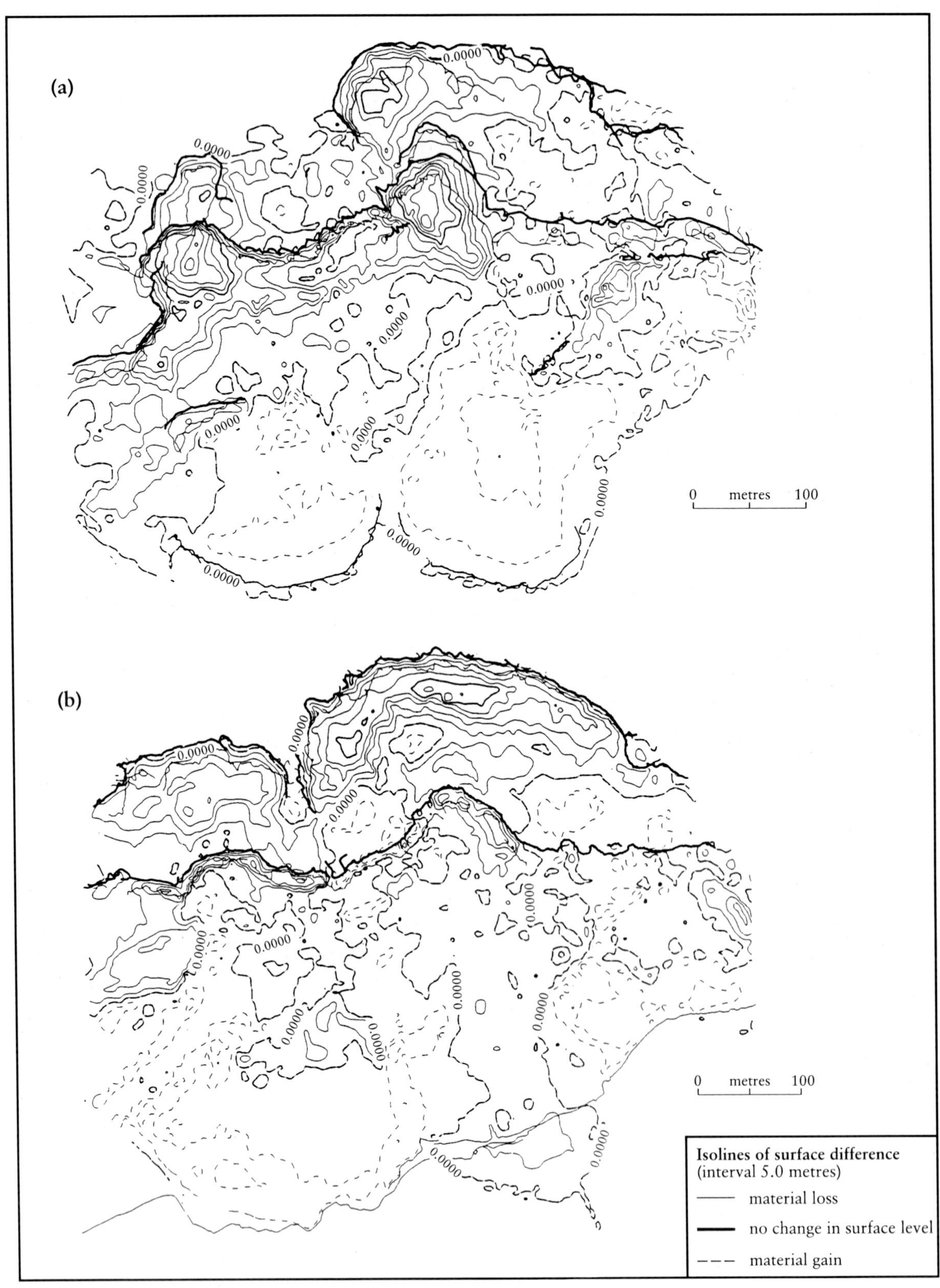

Figure 6.45 Contours of surface difference in elevation (i.e. erosion–deposition–no change) for the periods (a) 1958–1946, and (b) 1969–1958. Period (a) shows the location of the 1958 failures. This can be regarded as the formative event. Period (b) shows the diffusion of the wave erosion of the toe and continued input. The 'no change' along the main mudslide axis shows input = output and dynamic equilibrium over a decade interval. After Chandler and Brunsden (1995). *Continued opposite.*

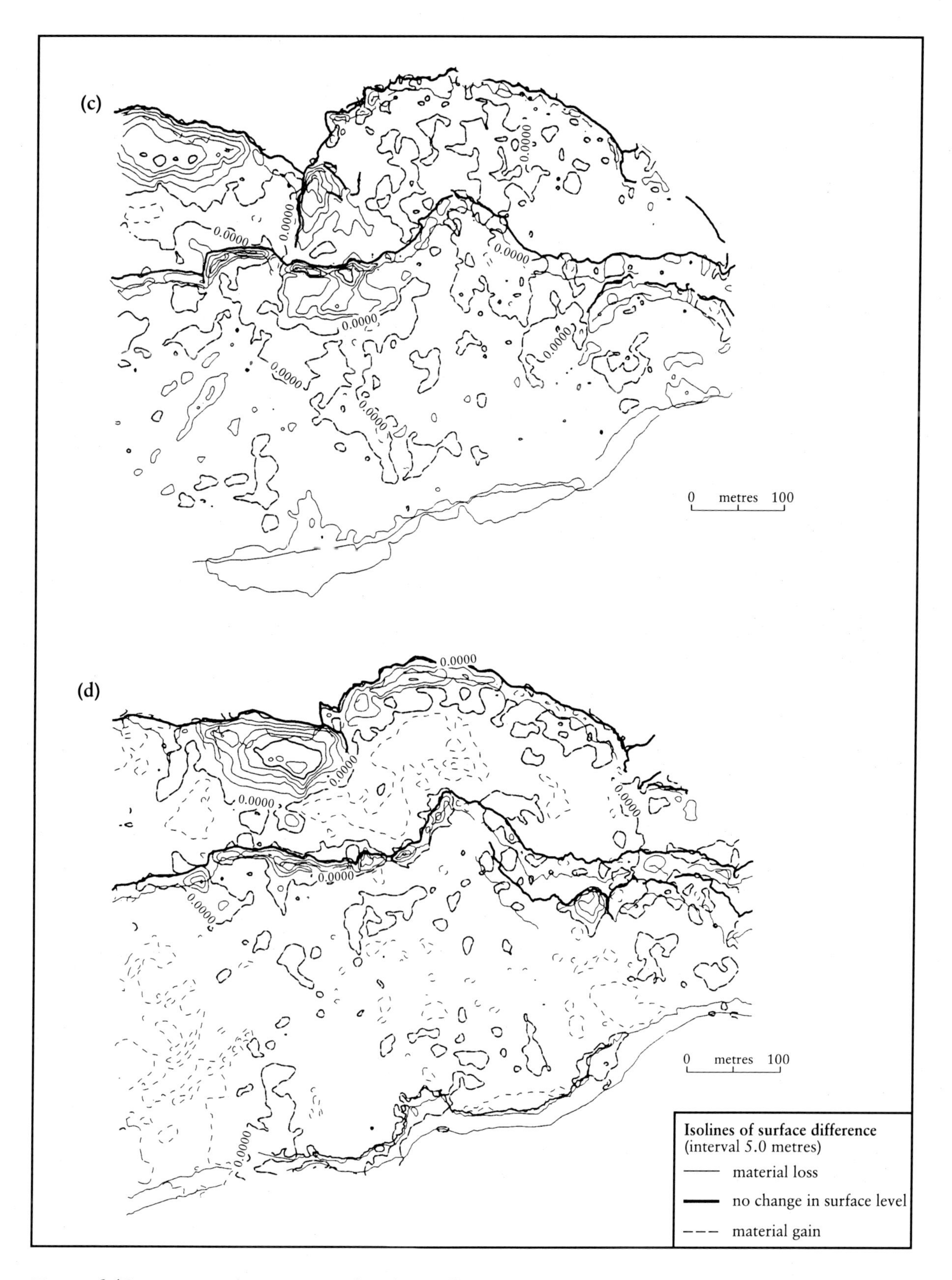

Figure 6.45 – *continued.* Contours of surface difference in elevation (i.e. erosion–deposition–no change) for the periods (c) 1976–1969, and (d) 1988–1976. The 'no change' along the main mudslide axis shows input = output and dynamic equilibrium over a decade interval. After Chandler and Brunsden (1995). *Contined overleaf.*

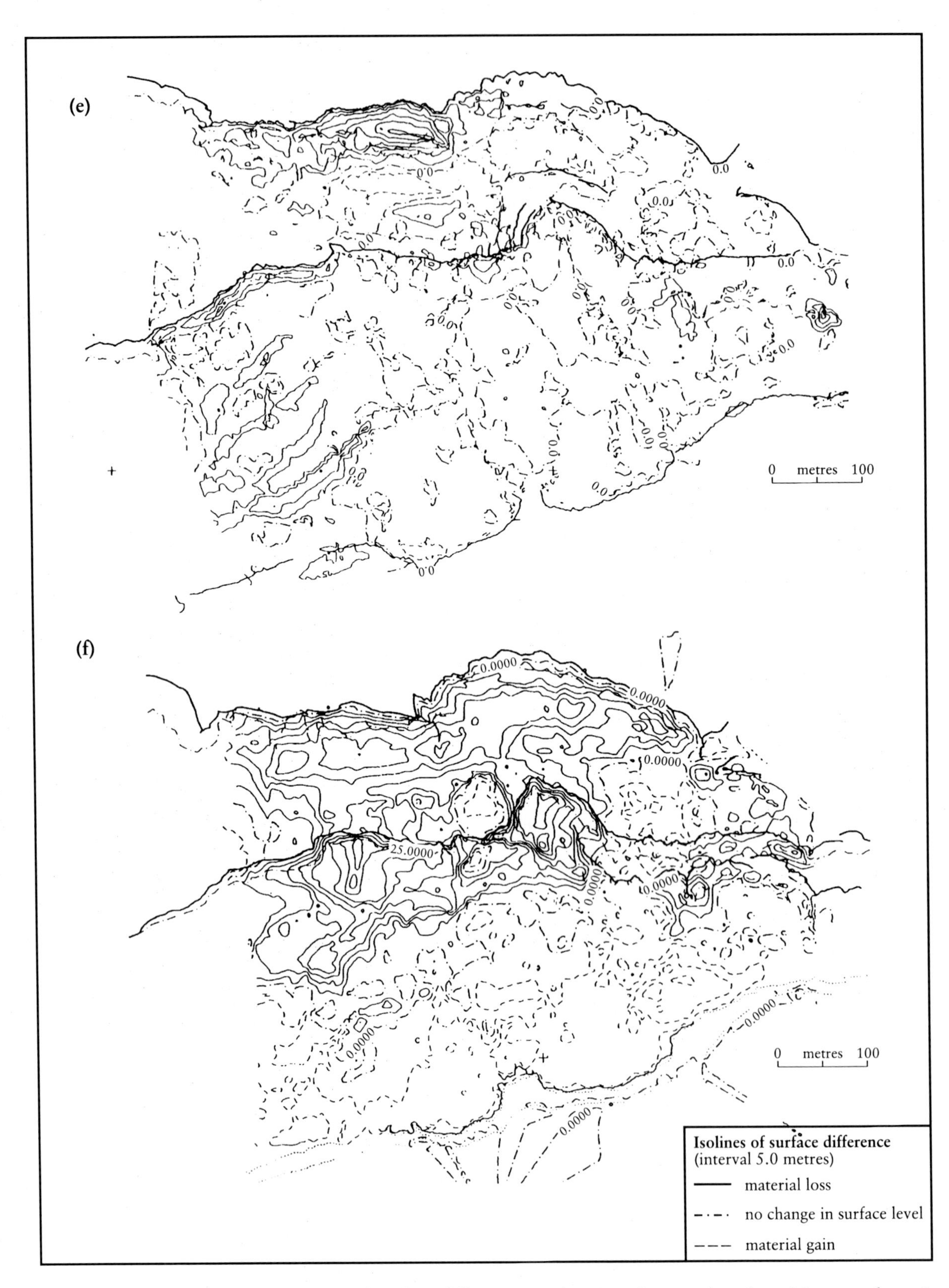

Figure 6.45 – *continued*. Contours of surface difference in elevation (i.e. erosion–deposition–no change) for the periods (e) 1995–1988, and (f) 1988–1946. The 'no change' along the main mudslide axis shows input = output and dynamic equilibrium over a decade interval. After Brunsden and Chandler (1996).

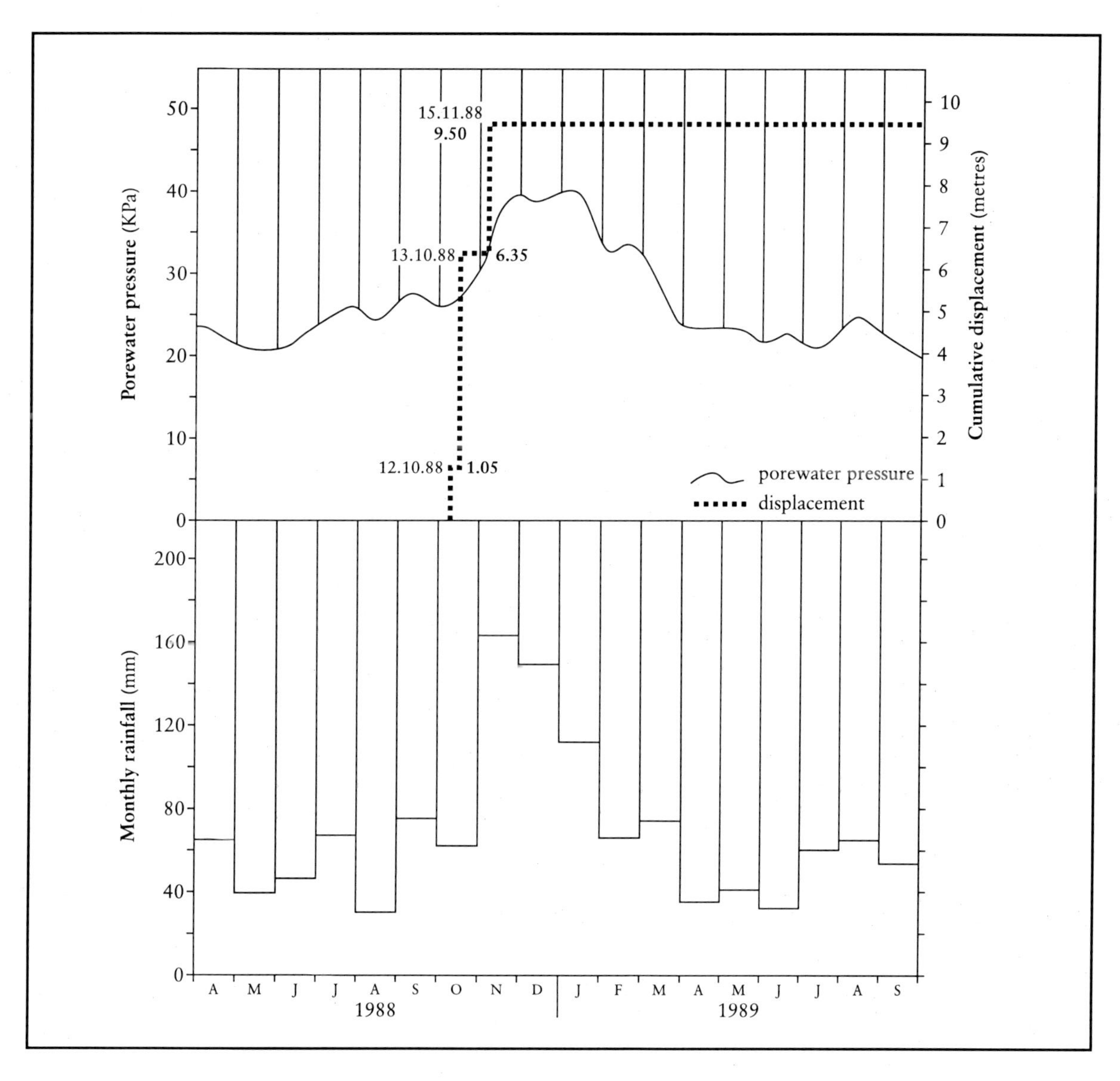

Figure 6.46 The seasonal behaviour of rainfall movement and porewater pressure at Black Ven. The observation that movement ceased on 5/11/88 suggests that there may have been a considerable strength gain following movement. After Koh (1990).

artefact of the fact that data for 1958 are available. The 1958 data therefore represent the first epoch, which happens to have been soon after failure, rather than failure activity at the threshold angle itself.

Following this initial rapid movement, the cliffs adjusted by building lobes of mud into the sea, and in the ensuing ten years the erosional wave diffused upslope to form low-angle slopes on the upper benches. In doing so, the form was maintained even though the whole complex moved inland by as much as 90 m. There was a change in the proportion of low-angle slopes between 1969 and 1988 because the accumulation lobes were being removed by the sea, but the degradation slopes maintained their form. The main processes involved were the cascades of material over the terraces, the parallel retreat of the undercliffs and the rapid transport of material away from the foot of the cliffs and across the benches by the mudslides. Chandler and Brunsden (1995) observe that this is a good example of a retreating but unchanging slope form being maintained by an efficient basal removal condition. Some impression of the rate of recession may be obtained from Figure 6.47.

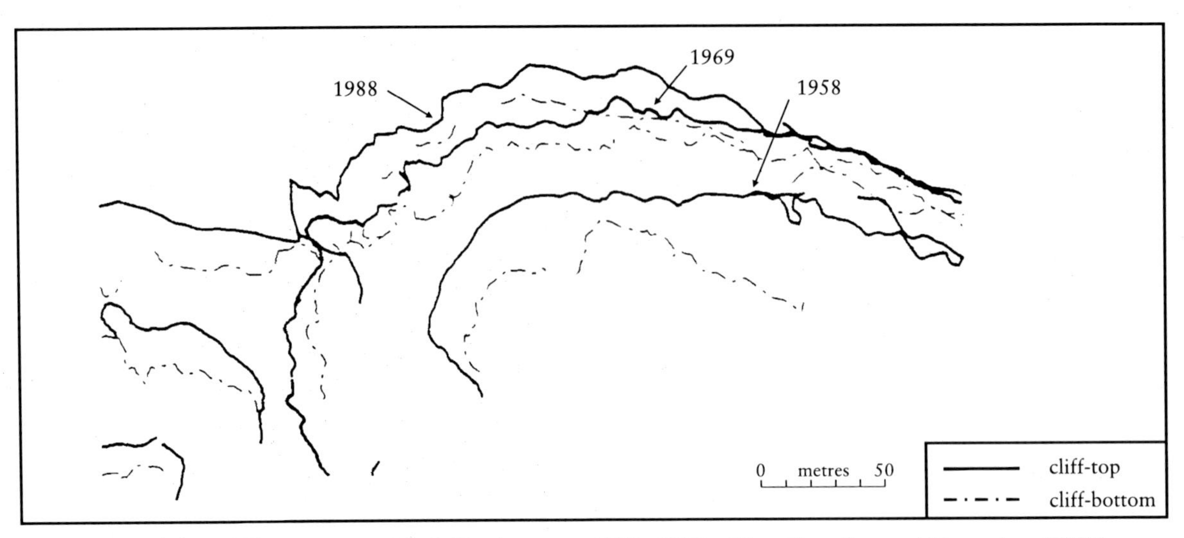

Figure 6.47 Cliff recession at Black Ven between 1958–1988. After Chandler and Brunsden (1995).

The data suggest that following an impulse of change, the Black Ven system adjusts dynamically and develops a new characteristic form. There is a rapid transmission of energy and material through the system and a remarkable interdependence of the slope-process components. The retreat of the cliffs and the total basal removal of each bench demonstrates almost perfect component coupling. Over a 30-year period the system fulfils most of the requirements of a system in steady state.

This opinion is supported by the extraordinary record of the DTMs of elevation difference between 1958 and 1988. These show that over 200 000 m^3 of sediment was transported from the cliff-top to the sea through one of the mudslide systems and yet the overall cliff form remained unchanged to a large extent.

The data may also be used to inform discussion of the timescales of landform change. By manipulating the mean slope-angles at all epochs in a Computer Assisted Design (CAD) system it is possible to produce projections of the mean slope-angle into the future. This can be used to set up a working hypothesis of the possible time adjustments required. The impulse of change or threshold activity crossing is followed by a rapid reduction of slope angles as the mudslides form and the seaward lobes accumulate. This was probably achieved early in the 1960s but the epoch interval (1958–1969) only permits a resolution of the reaction time to 11 years.

The system then relaxed over a further period of 7 years. Therefore, this model suggests that it takes about 20 years to achieve the current form, which by 1995 had been maintained for 16 years. The data suggest that during this period the elevations changed very little, that input was close to output, and that, overall, the erosion volumes were diminishing. Nevertheless, the mean slope-angle, based on 11 000 points shows a change from 17.7° to 18.1°. If this is significant it suggests that the characteristic form is a dynamic one of change at a constant rate. This would allow a linear projection and a prediction of the next major dynamic phase in about 2016 AD, a frequency of about 60 years. This may then be used as a basis of an episodic landform change model (Figure 6.48) based on the marine erosion rate that prepares the system by steepening the slope angle.

However, Chandler and Brunsden (1995) point out that this linear model is almost certainly incorrect. The 1969, 1976 and 1988 data points probably suggest an exponential decay towards the threshold activity angle. In this case, the characteristic form has not yet been achieved, the relaxation time is in progress and a long period of slow slope degradation can be expected. The length of time before the next active phase will then be determined by the rate at which the accumulation lobe, plus any input from upslope, is removed and the cliffs steepened towards the threshold. Chandler and Brunsden (1995) remark that because of the unknown input to the lobes during this basal removal phase, this will no doubt prove to be an example of complex response.

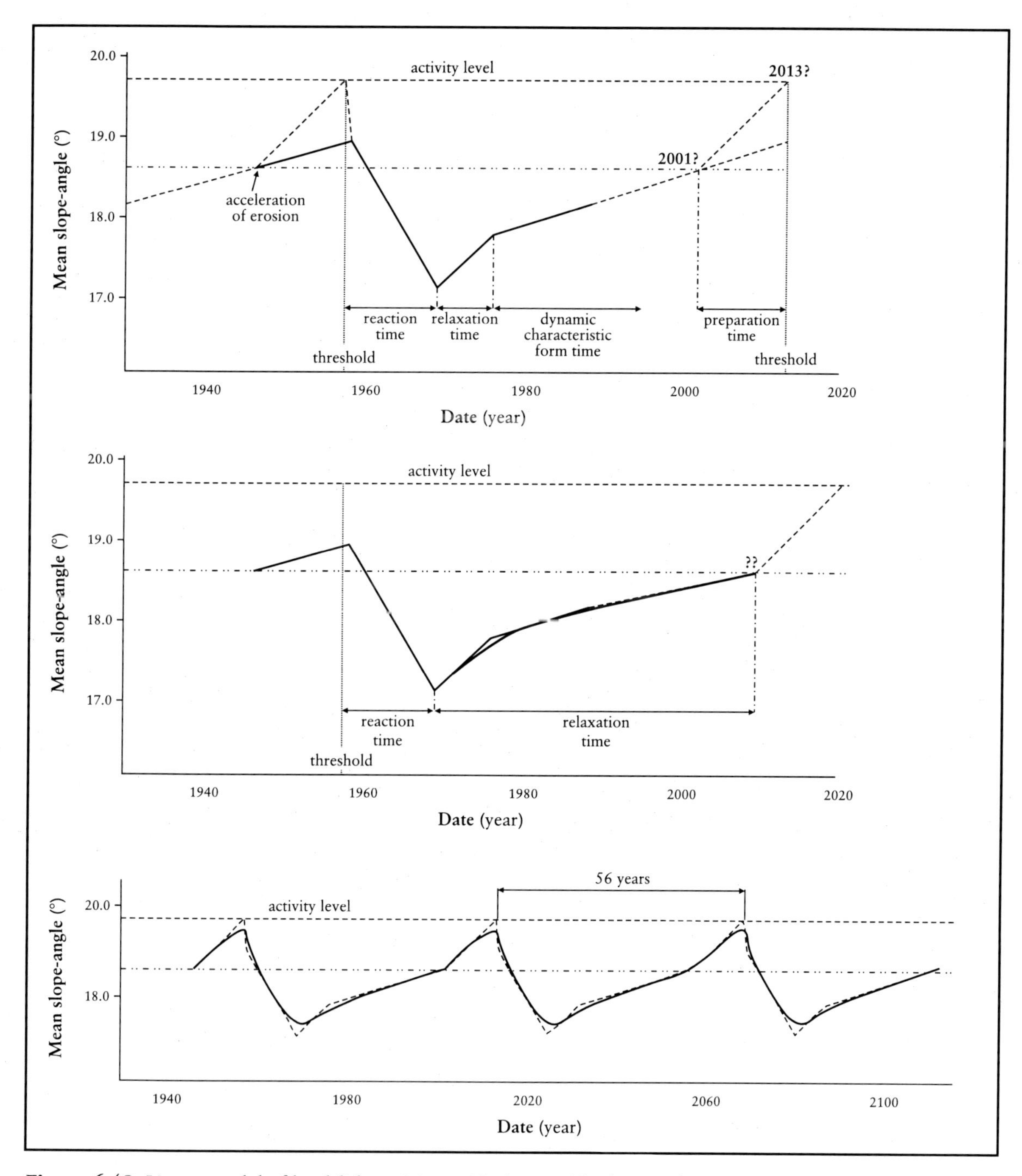

Figure 6.48 Linear model of landslide activity at Black Ven. The lowest diagram is a speculative cyclic model. After Chandler and Brunsden (1995).

(c) The episodic landform change model of Brunsden and Chandler (1996)

A year later, Brunsden and Chandler (1996) substantially revised the 1995 episodic landform change model, partly in response to a dramatic sequence of events that took place on 7 August 1994 at 7.35 in the evening. A large-scale, rare event took place at the western system of mudslides. The previous winter had been one of the wettest on record, with movement observed throughout the cascades. In particular, the dormant systems to the west developed large cracks at their head and very wet failures all along the top of the Belemnite Marl slope. Black Ven (west) began to develop three distinct feeder tracks on its western margin. In February a non-circular slide with a deep graben developed at the toe, and mudslides were re-activated on all terraces. The Spittles mudslide continued to extend in a headward direction with a graben developing along-side the abandoned Roman Road, where the road itself was split wide open.

During January and February 1994 at Black Ven (west), a tension crack 100 m long opened up across the edge of Lyme Regis golf course about 5 m from the cliff edge. The detached piece then settled very slowly so that, by the end of the winter, it had come to rest about 10 m down the cliff-face. A dry summer followed, but surprisingly in August the detached piece rapidly descended the cliff and suddenly loaded the accumulated debris on the uppermost bench, above the Belemnite Marl.

This debris, consisting of dry sand and gravel, was pushed forward between 40 m and 50 m so that approximately 60 000 m^3 of dry, fine-grained sand descended the vertical clay cliff. This mass appears to have fluidized (the exact mechanism is not known) because the material flowed in a few minutes, in a sheet form, to within 20 m of the sea. The flow track below the clay cliff descended 323 m horizontally and 90 m vertically. The dimensions of this landslide were: width 100–120 m, length 525 m, with a flow track of 0.3–1.0 m deep and an average angle of 13°. The deposit came to rest as a thin sheet of fine-grained sand with some mixture of clays. The deposit was laminated, had a clean margin to the mudslide surface below, with very sharp edges, shallow levées in places and a very abrupt termination of the frontal lobe. The surface was streamlined, boulders and gravel from the chert beds were strung out in lines and the overall surface was powdery.

Very shortly after the event all of the mudslides moved forwards, undoubtedly because of the rapid undrained loading of the terraces. On the edge of the Belemnite Marl cliff the loss of the toe of the upper terrace landslides caused a major rotational slide to develop, forming a very prominent scar in the undercliff.

The early autumn of 1994 occasioned significant rainfall. The loose sand, varying in depth between 0.3 m and 1.0 m, quickly became saturated and the surface was transformed into an inaccessible metastable sand. During the very wet winter of 1995 this landslide surface began to sort itself into distinct mud streams with pressure ridges and wet fans spread across the accumulation lobe. Unusually, a deep gully developed over the whole length of the track, which became a fully integrated stream system by the spring. Overall the event pushed a lobe of mud 10 m into the sea.

This event is one of the first dry sand-runs to have been observed. Brunsden and Chandler (1996) could find no other accounts in the literature. Certainly such events are unknown either on Black Ven or in West Dorset. It is known that the event occurred over a very short timescale because there were witnesses who could give approximate timings.

The effect of big event on the gross morphology was to flatten the whole landslide by blanketing everything in a thin layer of sand in just a few minutes. This reduced the average slope-angle by 2°, to 13°, and so delayed the return of the system, by undercutting and slope steepening, to a new unstable state.

As stated, this spectacular event contributed to the development of a revised episodic landform change model (Brunsden and Chandler, 1996). Other contributory factors were: further development of automated digital photogrammetry; new DTM software capabilities; an additional epoch of aerial photographs and derived spatial data, 1995; and new observations of mudslide activity in the period 1988–1995.

As pointed out by Brunsden and Chandler (1996), the episodic landform change model developed by Chandler and Brunsden (1995) could only be speculative, as it involved certain simplifications. For example, most of the functions available in the DTM processing package then available could only operate with a rectangular grid-based DTM. The consequent rectangular and imprecisely specified boundaries to the system resulted in probable distortion to the slope-angle histograms and the mean slope-angles derived from them. Inaccurate specification of boundaries also prevented separate processing and examination of what are in fact two independent mudslide systems. Also, the model was based exclusively on basal erosion; slope steepness triggering of mass-movement activity and other important controlling variables were omitted.

The 1995 epoch shows that the eastern mudslide remained effectively unchanged in form between 1988 and 1995, although very high rainfall in 1994 and 1995 caused some movement and tension cracking on the uppermost bench and slipping of some toe material over the cliff edge of the Belemnite Marl. The debris built up at the base of the cliffs and it must be assumed that the terraced mudslides have been loaded at their heads. Some gentle forward movement produced small slides in the main track, with lobes spread across the terraces, but the overall increase of slope angle from 18.1° to 19.6° must be destabilizing the system (Figure 6.49).

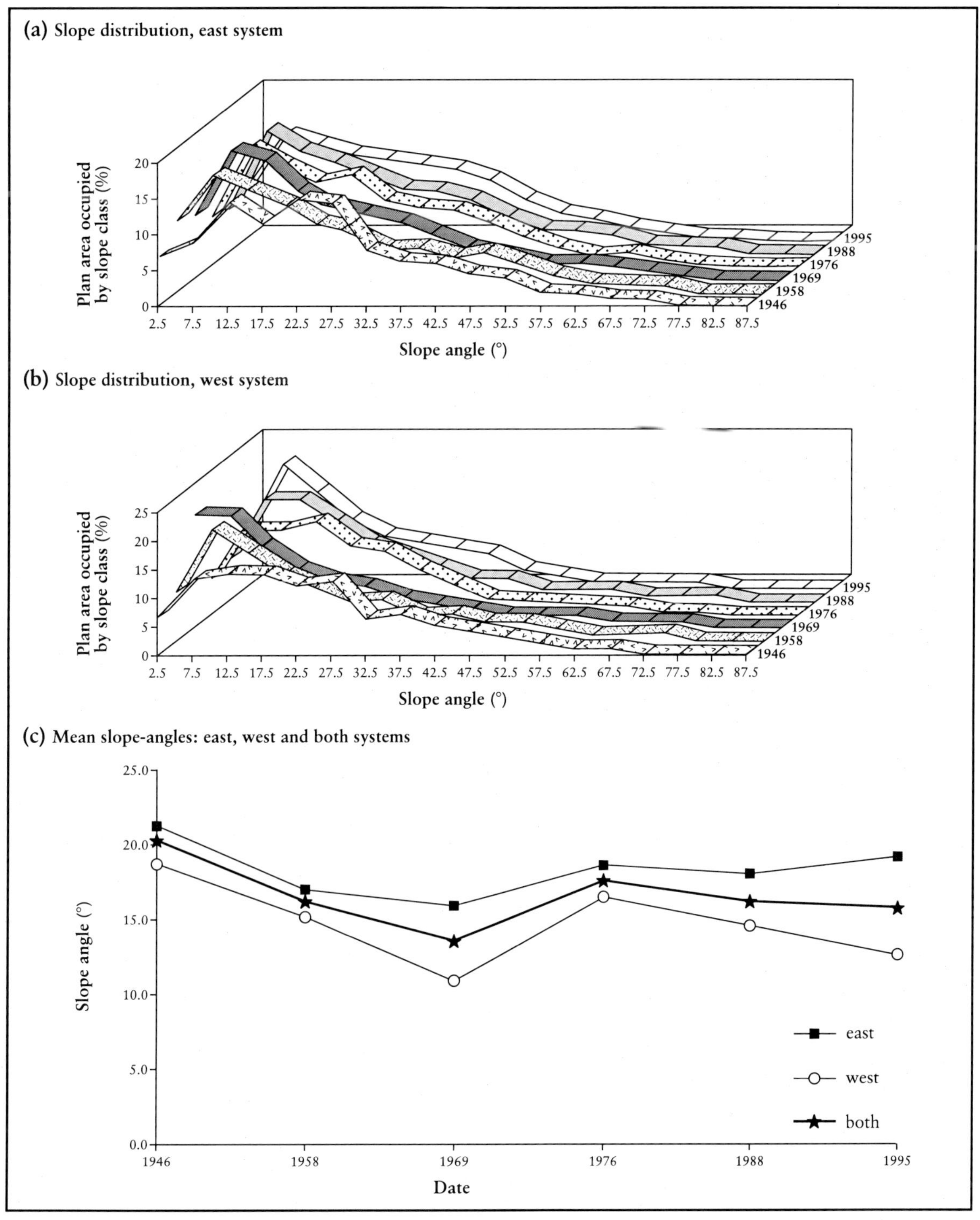

Figure 6.49 Slope-angle graphs for (a) the east system and (b) the west system of Black Ven. After Brunsden and Chandler (1996).

The western landslide continued to show erosion at the head, but with increasing evidence that the system was discharging sediment in a steady-state manner similar to the previous 20 years. There was no change in the morphology and storage of the track.

The most important discovery from Brunsden and Chandler's new (1996) analysis was that during the relatively drier and less-dynamic years of the mid-1970s the mudslides did not carry away all of the debris supplied at the back of the terraces that accumulated as medium-angle debris slopes. In consequence the slope-angle distributions began to move towards the 1946 bi-modal pattern and a higher mean value (Figure 6.49). The onset of years of greater rainfall in the 1980s reversed this trend to reveal the true nature of the relaxation variability.

Brunsden and Chandler (1996) were also able to make a more detailed analysis of climatic influences. Monitored data are not available for Black Ven; however, recent research on the occurrence of landslides and climatic change on the south coast of Britain (Brunsden and Ibsen, 1994a–c; Brunsden *et al.*, 1995; Ibsen and Brunsden, 1996) has provided data on landslide occurrence at a similar temporal scale to the Black Ven photographic evidence (Figure 6.50). This facilitates understanding of the cumulative data recorded on the photographs and enables refinements to the episodic landform change model to be made.

A recent investigation for the European EPOCH programme (Brunsden and Ibsen, 1994a–c) has shown that the European historical archive of landslide data, though incomplete, is very rich. It has been used to create a database (Brunsden *et al.*, 1995) on the occurrence of landslides on the south coast of England. It was found possible to derive a time-series that could be related to the broad diurnal series provided by such stations as Ventnor, Portland and Lyme Regis. The series for West Dorset displays a pattern similar to that for the south coast as a whole, and for Ventnor. The latter, which is a long, complete and homogeneous record, is used to determine the climatic landslide control for the south coast (Figure 6.50). Three points are helpful in the development of the episodic landform change model.

1. The Pleistocene history of Black Ven is not known, but the plateau top, the valley-side slopes to the Lym and Char rivers, the slopes of the neighbouring landslides at Stonebarrow and Golden Cap and the westernmost areas of the Black Ven complex are mantled with head up to 3 m thick. Late Pleistocene mudslides are known to underlie some of Charmouth and the lower valley-side slopes of the River Char. These can be up to 20 m deep (Brunsden and Jones, 1976). The Spittles area on the western side of Black Ven is a re-activation, in 1986, of a relict, very degraded landslide slope, which is mantled in solifluction debris and suggests that post-glacial erosion by the sea only reached the old system in historic times. The database has no records for the coast until 5500–3000 BP when it is thought that the rising sea first renewed its attack on the degraded, head-covered, pre-glacial cliffs.
2. The first records for Lyme Regis and the Axmouth to Lyme Regis Undercliffs Nature Reserve (specifically, Haven Cliffs) are from the 11th century, but details are scanty. There are more substantial records for the period of the Little Ice Age, with 13 records from West Dorset and the National Nature Reserve between 1592 and 1843, and a heavy concentration in the 16th and 17th centuries. The records are all historical narrative reports. The central, dormant, but only partly degraded, landslide of Black Ven may well date from this period since the oldest trees on the site are about 200 years old. However, these records need not relate to a period of greater rainfall. Their fortuitous recording may indicate movements due to marine erosion or weathering.
3. Records are far fuller for the modern period (the last 200 years), with annual and decadal data showing an apparent increase of events in the last century, and a sequence of troughs and peaks.

The obvious change in the nature and intensity of reporting leads to uncertainty as to whether all the changes shown are due to natural causes (Brunsden *et al.*, 1995). On the coast it is logical to ascribe some of the increase to sea-level rise and erosion of the sea cliffs. There appears to be

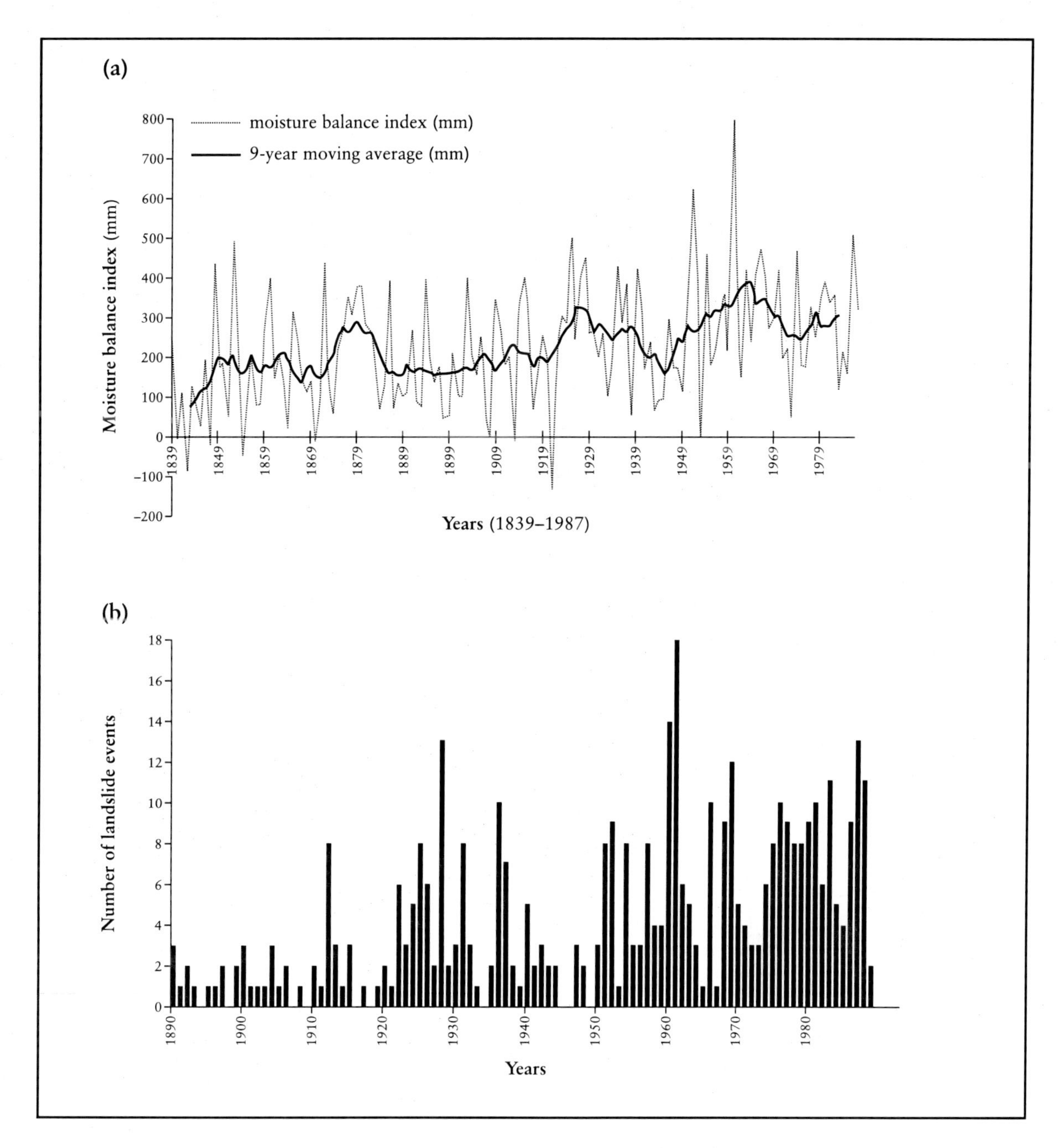

Figure 6.50 Climate and landslide series for the south coast of England. (a) Moisture balance (mm) for Ventnor (Isle of Wight) (1639–1987) plotted as a 9-year moving average. (b) The number of landslide events for the south coast. After Ibsen and Brunsden (1996). *Continued overleaf.*

an apparent concentration and periodicity, shortening towards the present-day, in the number of landslides this century (Ibsen, 1994). The periodicity that is most significant in that central and southern England has experienced a cyclical pattern in which the wet years gave rise to greater geomorphological activity. There is a concentration of landsliding during the periods 1912–1913, 1922–1932, 1936–1941, 1950–1970, 1975–1982 and 1993–1995. This is a frequency of 5–10 years, with the episodes of sediment transfer lasting several years. Wet-year sequences with three or more wet years in succession occurred during 1877–1882, 1913–1915, 1922–1932, 1936–1939, 1952–1954, 1963–1970 and 1978–1982, all co-inciding with landslide

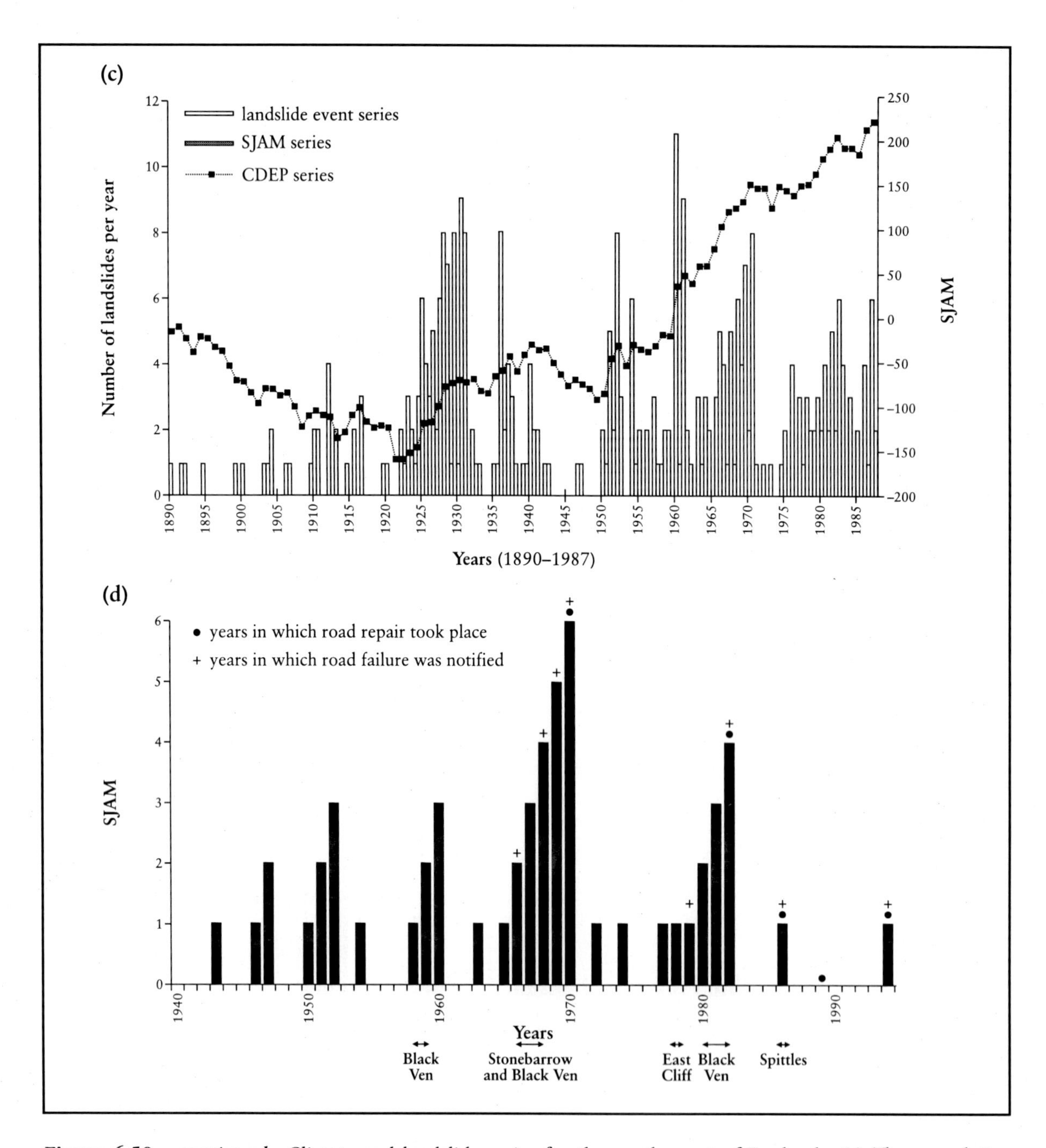

Figure 6.50 – *continued*. Climate and landslide series for the south coast of England. (c) The cumulative moisture balance departure from the mean (CDEP), the cumulative number of years with moisture balance greater than the mean (SJAM) and the landslide occurrence at Ventnor (after Ibsen and Brunsden, 1996). (d) The sequence of years of higher rainfall and landslide occurrences for west Dorset (after Brunsden and Chandler, 1996).

records. The precipitation record for the Axmouth to Lyme Regis Undercliffs National Nature Reserve broadly supports this pattern.

It therefore seems reasonable to accept, for the modelling of the episodic behaviour of the West Dorset landslides, a 'wet' year climatic control of 5–10 years lasting for 3–6 years which is superimposed on the trend for increasing wetness, sea-level rise, and slope steepening over the last 60 years.

These records allow a further speculative model for episodic landform change at Black

Ven to be developed (Brunsden and Chandler, 1996) (Figure 6.51), replacing that given by Chandler and Brunsden (1995). During the last glacial period the whole coastal slope was abandoned by the sea so that erosion was curtailed. The slope evolved under periglacial conditions, with solifluction and major landslides. During the post-glacial period the landslide scars and the structural benches degraded and became vegetated. After 5500 BP the rising sea began to remove the solifluction apron and eroded the abandoned pre-glacial sea cliffs. In the Black Ven complex, the uppermost slopes of the Spittles are still in a periglacial form perhaps because the toe is protected by a sea wall. The Spittles landslide itself was in a similar state until it became so undercut that it failed in 1986.

The morphology of Lyme Bay and the distribution of wave energy ensured that the cliffs to the east became unstable at an earlier date. The central part is now in a metastable, vegetated state and appears to have been through a phase of activity about 200 years ago. There are no solifluction deposits surviving on this slope except for the cappings on slump blocks from the backscar. The slides then degraded to an overall slope-angle of about 17°, with a bi-modal distribution of slope angles reflecting the vegetation of the rear scarp, the growth of talus and debris slopes at the base of the steep slopes, and the smoothing of the structural slopes. This state continued until 1994–1995 when movement began on all of the terraces in response to continued undercutting and the wettest year on record. Almost certainly, Black Ven (west and east) reacted in the same way, with re-activation rather earlier than 1957–1958, but no evidence survives (Brunsden and Chandler, 1996). Present understanding of the spatial relationships of its various parts is shown in the geomorphological map of Brunsden and Chandler (1996) (Figure 6.52).

Conclusions

Black Ven has a long history of landslide activity, but major movements have been episodic. It is remarkable for the detail in which it has been

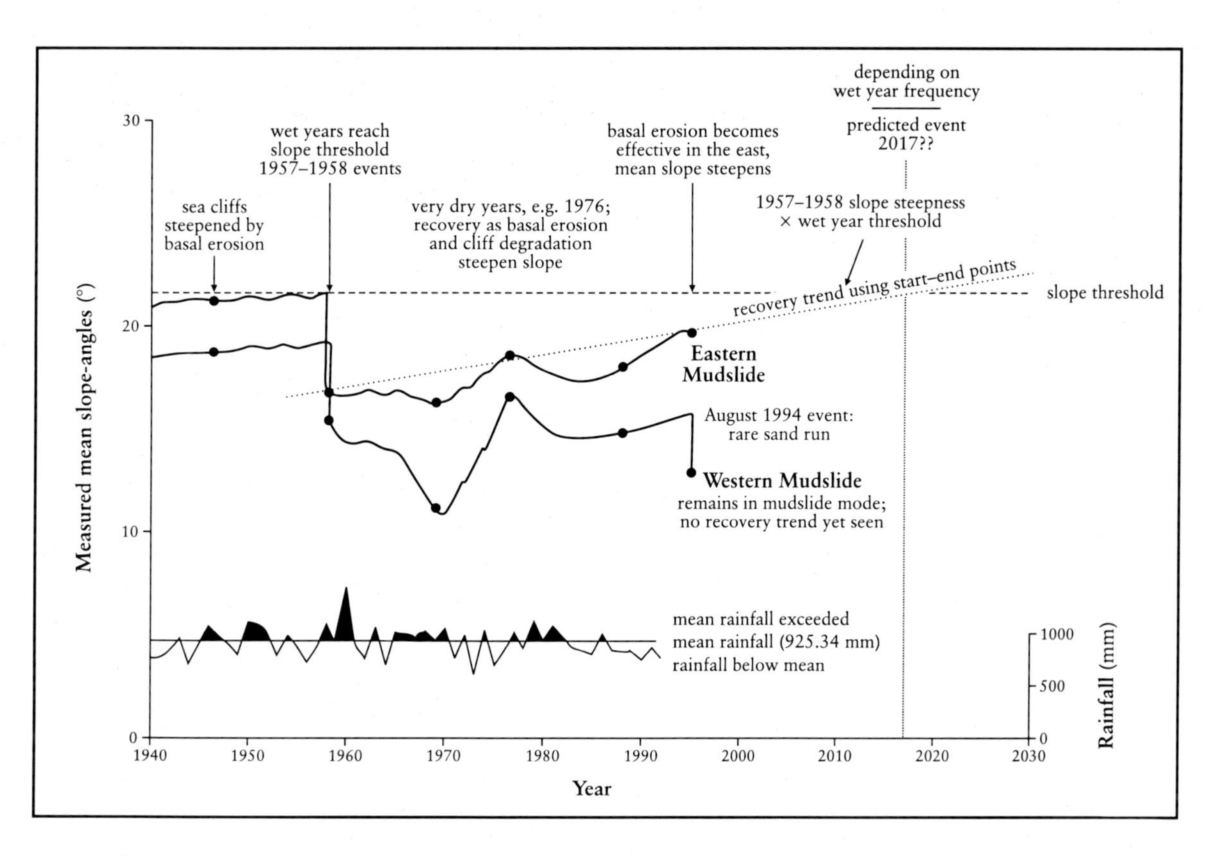

Figure 6.51 A temporal model of episodic landform change at Black Ven. After Brunsden and Chandler (1996).

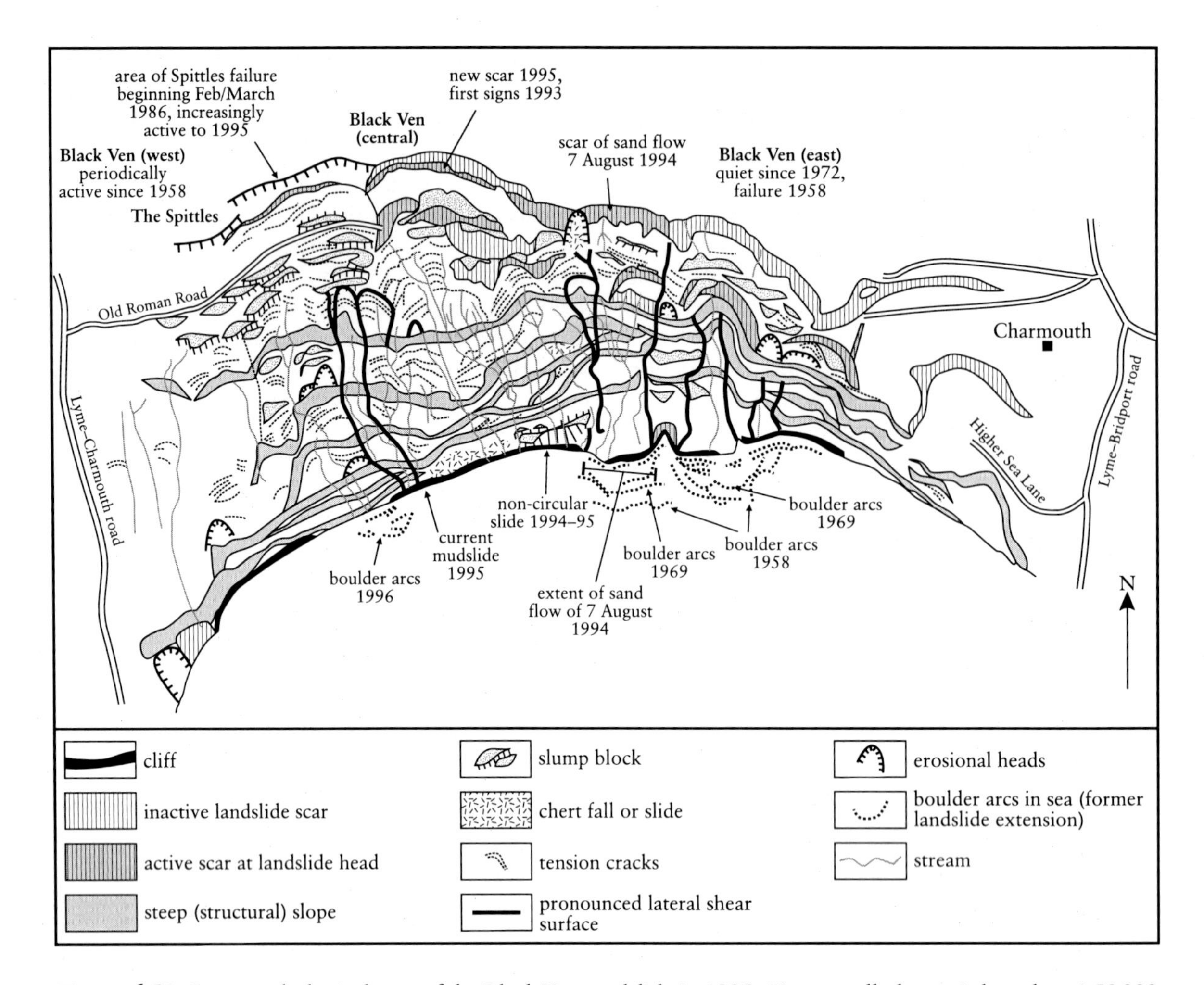

Figure 6.52 Geomorphological map of the Black Ven mudslide in 1995. Uncontrolled mosaic based on 1:50 000 scale aerial photographs, NERC 2/95, site no. 94/26, Charmouth, not to scale. After Brunsden and Chandler (1996).

studied, and the extent to which its mechanisms are understood, ranging from the notion that aquifers can act as reservoirs, the gradual release of groundwater from which has kept the downslope transport of materials at Black Ven in motion as far as the sea itself, to the sophisticated landform change models of Chandler and Brunsden (1995) and Brunsden and Chandler (1996). The rapid replacement, a year later and by the same authors, of the earlier of these two models is unusual but amply justified by the list of 'developments since 1989' given by Brunsden and Chandler (1996) in presenting the later model.

The sheer scale of current and recent activity on the Black Ven slope, and the close attention that has been accorded to it, strongly justify its inclusion as a Geological Conservation Review site.

MASS-MOVEMENT SITES IN UPPER JURASSIC STRATA

PEAK SCAR, NORTH YORKSHIRE (SE 530 883)

R.G. Cooper

Introduction

Peak Scar is a north-facing cliff 290 m long and about 30 m high, containing widened vertical joints. It is located at the head of the north-facing slope on the south side of Peak Scar Gill, which joins Gowerdale, a tributary valley of Ryedale in the Hambleton Hills of North Yorkshire. It is in the Lower Calcareous Grit Member of the Corallian (Upper Jurassic) strata. Between 9 m and 21 m downslope from the foot of the cliff is a row of massive detached blocks of rock, tilting away from the cliff and towards the valley, with an apparent valley-ward dip of 26°. The main mass is 130 m long and nearly 30 m broad. The crest of the ridge formed by these detached blocks rises 15 m above the floor of the trough that separates it from the cliff-face. The slope on the downslope side of the detached blocks is steeper than adjacent slopes to the west and east of Peak Scar (Figure 6.53).

Description

The sequence of features, on moving downslope, at Peak Scar is (Cooper, 1982):

- Vegetated plateau top/crest of slope
- Vertical cliff-face in bedrock
- Poorly vegetated trough or trench with large boulders; upper sides of detached blocks, in bare or moss-clad rock with pronounced downslope tilt
- Vegetated top and downslope sides of detached blocks
- Undulating slope to stream course at the slope foot

This sequence is remarkably similar to one of the sites described by de Freitas and Watters (1973) in their landmark paper on toppling failure. The site in Nant Gareg-lwyd, in the Rhondda Valley near Blaenrhondda, Powys, shows a joint-controlled vertical cliff-face in hard rock, at the foot of which lies a large toppled mass dipping downslope, and separated from it by a hillside trench. Peak Scar displays all of these features far more clearly, and with greater vertical development (see Figure 6.53).

The Corallian strata are a predominantly horizontally bedded, alternating series of calcareous sandstones and oolitic limestones (the Hambleton Oolite). The areal distribution of two kinds of features on the slopes of the Hambleton Hills suggest that a particular type of landsliding is taking place in this area. Firstly, enterable widened joints, locally termed 'windypits' are found on the slopes and on the surface of the plateau, close to its edges (Fitton and Mitchell, 1950; Cooper *et al.*, 1976). Secondly, ridge-and-trough features (Briggs and

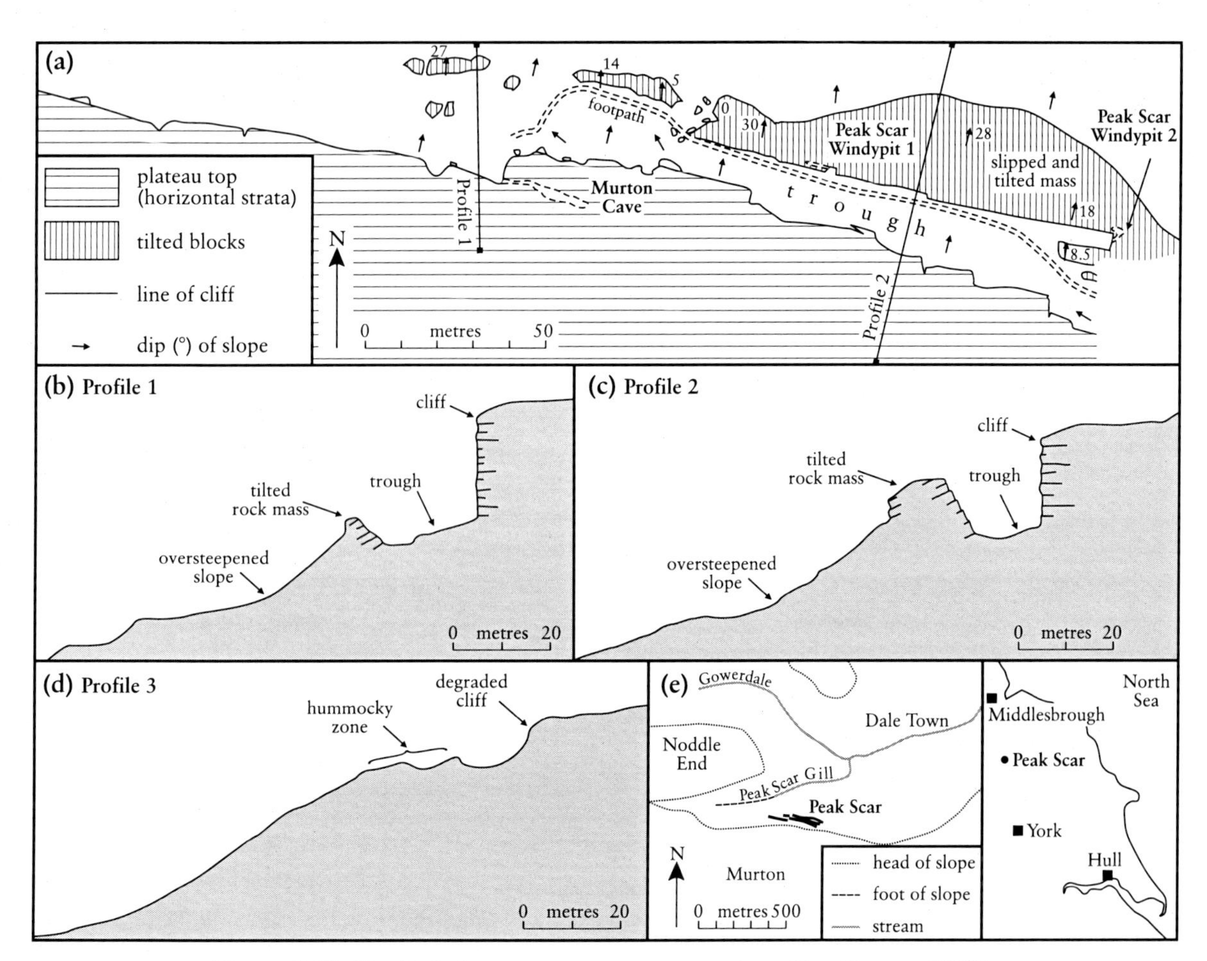

Figure 6.53 The Peak Scar mass-movement complex. After Cooper (1980).

Courtney, 1972), hillside trenches and uphill-facing scarps (Radbruch-Hall, 1978) are found on the slopes. Sites of enterable widened joints, and of ridge-and-trough features are shown in Figure 6.54.

At Peak Scar, the cliff-face is intersected by joints running sub-normal to it, which offset its line towards the valley at several places. These joints, and a series of pronounced horizontal bedding planes, divide the cliff-face into large

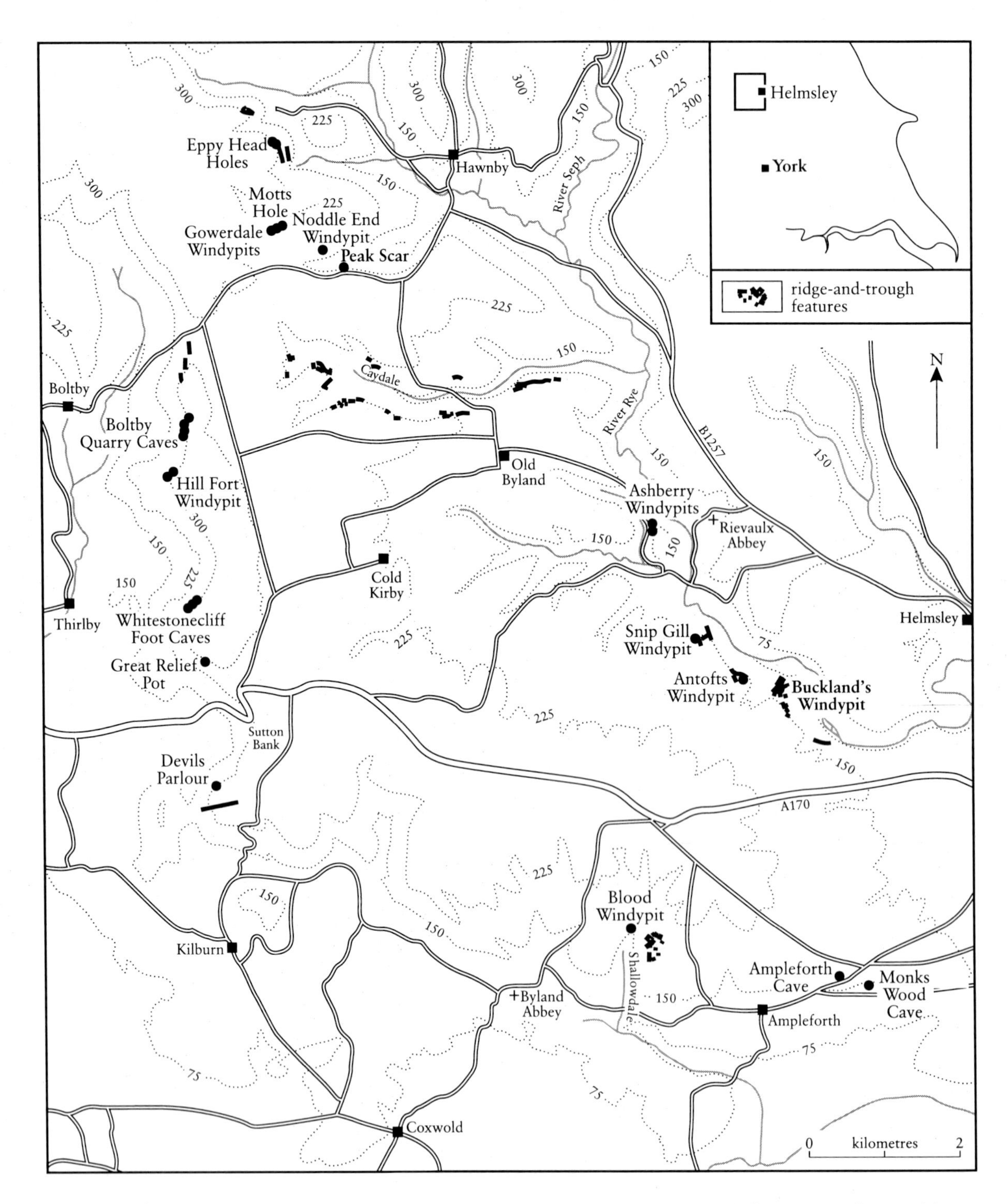

Figure 6.54 The distribution of widened joints ('windypits') and ridge-and-trough features on the Hambleton Hills between Hawnby and Ampleforth, including the locations of Peak Scar and **Buckland's Windypit** (described later in the present chapter). After Cooper (1980).

sub-cuboidal blocks. Many of the blocks appear to be loose as they are separated from the main mass of the rock by joints that run sub-parallel to the cliff-face, but behind it. About 70 m from its eastern end the line of the cliff-face is offset 8 m away from the valley by a joint. At the point of the inner corner produced by this offset is the entrance to Murton Cave, which is an enterable fissure running sub-parallel to the cliff-face.

Murton Cave (Cooper *et al.*, 1976) has been formed by the widening of a joint. Its upslope wall is continuous with the set-back face of the cliff. Between the fissure and the cliff-face in front of it (i.e. on its valley-ward side) is a tower of Corallian limestone 8 m thick and 30 m high, the upslope side of which forms the downslope wall of the fissure. The downslope wall of Murton Cave does not appear to have been displaced upwards or downwards, and appears to have undergone no tilting with respect to the upslope wall.

Some of the tilted rock-masses seem to be made up of several towers of rock, each tilted and leaning against the next one downslope (Figure 6.55). Thus different readings of apparent downslope dip were obtained from different parts of the tilted mass. This internal structure, and numerous bedding planes that have opened as a result of the tilting, are indicative of the shattered condition of the rock, which is broken up by many short fissures. Peak Scar Windypits 1 and 2 (Figure 6.53) (Cooper *et al.*, 1976) are merely two of these that happen to be wide enough to be entered.

The trench or trough between the cliff-face and the row of tilted rock-masses varies from 9 m to 21 m in width. Its floor slopes down from the cliff-foot towards the tilted rock-masses, with an overall fall varying from 2 m to 6 m, at angles of up to 20°, but more commonly about 10°. The surface is covered with hummocks and strewn with large angular boulders that appear to have fallen from the cliff and the tilted rock-masses. There is a clayey soil, with a vegetation of scrub and young deciduous trees. Neither the cliff-face nor the upslope-facing rampart of the tilted rock-masses bear any indication of solutional activity on the rock surfaces, or of fluvial erosion or deposition. The material in the intervening trough has not been deposited by water action.

The slope below the row of tilted rock-masses is afforested, and frequently stands at angles in excess of 30°. It incorporates short vertical exposures of rock up to 1.0 m in height. There are zones of loose scree material, and of scree that is grassed over or buried in leaf mould.

Figure 6.55 Peak Scar, showing a tower of rock just beginning to detach from the rockface owing to unloading and opening of the joints. (Photo: R.G. Cooper.)

Interpretation

It is clear that the rock tower between Murton Cave and the cliff-face has moved toward the valley by a distance at least equal to the width of the fissure. Since this has happened without vertical displacement or tilting, it is reasonable to suggest that the movement took place over some horizontal plane in the bedrock. Since the bedrock is effectively horizontal in its bedding, a bedding plane seems to be an obvious candidate for the failure surface.

It is possible that the tilted rock-masses were originally flush against what has now become the cliff, from which they were separated by a joint. A procedure developed by Caine (1982) for similar sites in Tasmania can be applied to profiles 1 and 2 in Figure 6.53. This involves projecting the line of the upslope rampart of the tilted mass and the line of the cliff until they cross. At this point rotation of the tilted mass may have taken place (Figure 6.56). It is worth noting that Caine's sites are about three times larger (in each dimension) than the features at Peak Scar, but the same considerations apply. Reconstruction in the manner of Caine (1982), assuming the later movement to have been purely rotational (Cooper, 1980), indicates a downslope rotation of 33° about a line parallel to and in line with the cliff, 40 m below the top of the cliff and 22 m below the trench floor. On reconstruction in this way, it is apparent that a considerable amount of rock has been lost from the tilted masses, as when set upright they do not reach the cliff-top. It is suggested that this material has slid off downslope over bedding planes, and is in part responsible for the steepness of the lower slope.

It has been suggested (Cooper, 1982) that three major types of mechanism acting in succession may be adduced as an explanation of the suite of features at Peak Scar.

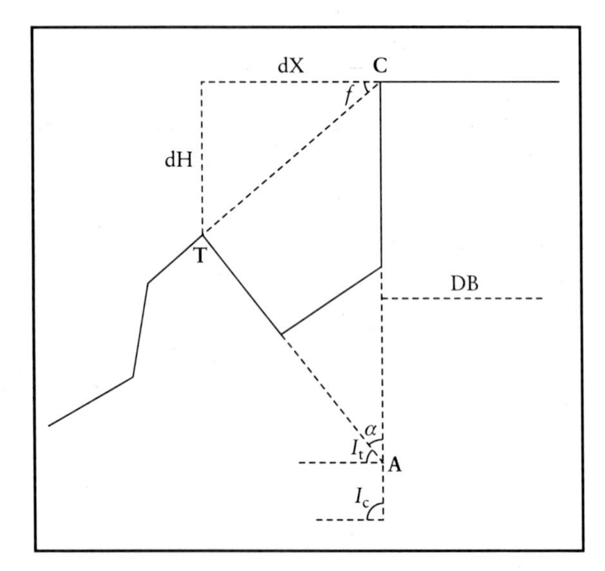

Figure 6.56 Terms and definitions used in modelling a hypothetical topple: (A: point of convergence based on extrapolation from joint surveys on the topple and the cliffs; C: cliff crest; T: crest of the topple; DB: elevation at the base of the dolerite sheet; LC: distance a–c; LT: distance a–t; I_c: joint inclination on the cliff; I_t: joint inclination on the topple; α: tilt angle of the topple (I_c–I_t); dH: vertical lowering of the topple crest; dX: lateral displacement of the topple crest; *f*: angle from c–t). After Caine (1982).

1. *Splitting mechanisms*: clearly, fissures must form in a massive rock before they can be widened. They could be tectonically initiated joints, and thus pre-date not only the slope features, but also the incision of the valley itself. Alternatively they could be 'valley joints', formed as a response to stress-release on valley incision.
2. *Sliding mechanisms*: as the initial translational movement seems to be effectively horizontal, it is likely that the plane of sliding is horizontal. It could be located at the junction with the Oxford Clay at the base of the Corallian. However, there is a broad lithological transition zone between the two. Alternatively, there could be thin 'mobile' bands of clay or shale interbedded in the Corallian strata, which would provide planes of sliding.
3. *Tilting mechanisms*: the tilting towards the valley of blocks that have slid forward could be due to undermining of blocks by erosional removal of weaker material beneath, and the settlement of the blocks. Alternatively, it is possible that the blocks move forward until their centres of gravity are unsupported due to the steepness of the slope, and then over-balance. Either way, the tilting seems to represent a form of toppling failure (de Freitas and Watters, 1973). The observation of uphill-facing scarps describes features widely regarded as being diagnostic of 'sagging failures'.

Conclusions

Peak Scar is the best British example of a de Freitas and Watters' intermediate-sized toppling failure, by virtue of its apparent 'freshness' (which may be deceptive) and its vertical range. The characteristic of the downslope wall of Murton Cave, that does not appear to have been displaced upwards or downwards, and has not rotated with respect to the upslope wall, is common to most of the other widened joints in the Hambleton Hills. The widespread distribution of both widened joints and hillside trenches in the Hambleton Hills suggests that the three-fold process evident at Peak Scar may have wider applicability than just the immediate site. Other interesting sites include the East Weare and Great Southwell slips on the Isle of Portland, Dorset, and Daddyshole Plain and Ansteys Cove, Torquay, Devon.

BLACKNOR CLIFFS, ISLE OF PORTLAND, DORSET (SY 677 715)

R.G. Cooper

Introduction

The cliffs on the western coast of the Isle of Portland (Figures 6.57 and 6.58) exhibit probably the best British examples of toppling failures. This type of failure involves the separation of a monolith or slab of hard rock from the surrounding rock, usually by jointing. At a cliff edge, the monolith or slab will be isolated between the cliff-face and the joint(s). Eventually the slab will fall. This site shows various stages of the process.

Toppling failure, which involves forward rotation of the mass, is classically preceded by a period of accelerating creep, which may last for several years. This may involve slow outward movement of the slab away from the surrounding rock, widening the joints, over some underlying incompetent horizon (Schumm and Chorley, 1964).

Portland is traversed by a series of NNE–SSW-trending major joints. These have been mapped by Coombe (1981). The joints are in the hard, often oolitic or shelly Portland Stone, running upwards into Purbeck Marls and downwards into Portland Sands (all Upper Jurassic).

Description

There are three distinct lines of evidence that this type of failure is occurring on the Isle of Portland. Firstly, the NNE–SSW joints are almost all widened to a greater or lesser degree. Some of them in the interior of Portland have been entered and explored, via the quarries on its plateau surface (Ford and Hooper, 1964; Churcher *et al.*, 1970; O'Conner and Graham, 1996). It is likely that the joints owe both their existence and their widening to Alpine Earth movements (Cooper, 1983a,b). Around the coast of Portland, the joints are wider than in the interior, presumably due to seaward movement of the huge slabs of limestone isolated between them and the cliff-line. On the western side of Portland, between Blacknor and Mutton Cove, these joints have been entered and extensively mapped (Cooper and Solman, 1983; Graham and Ryder, 1983; Cooper *et al.*, 1995) (Figures 6.59 and 6.60).

Figure 6.57 Blacknor Cliffs, Isle of Portland. (Photo: R. Edmonds, Dorset County Council.)

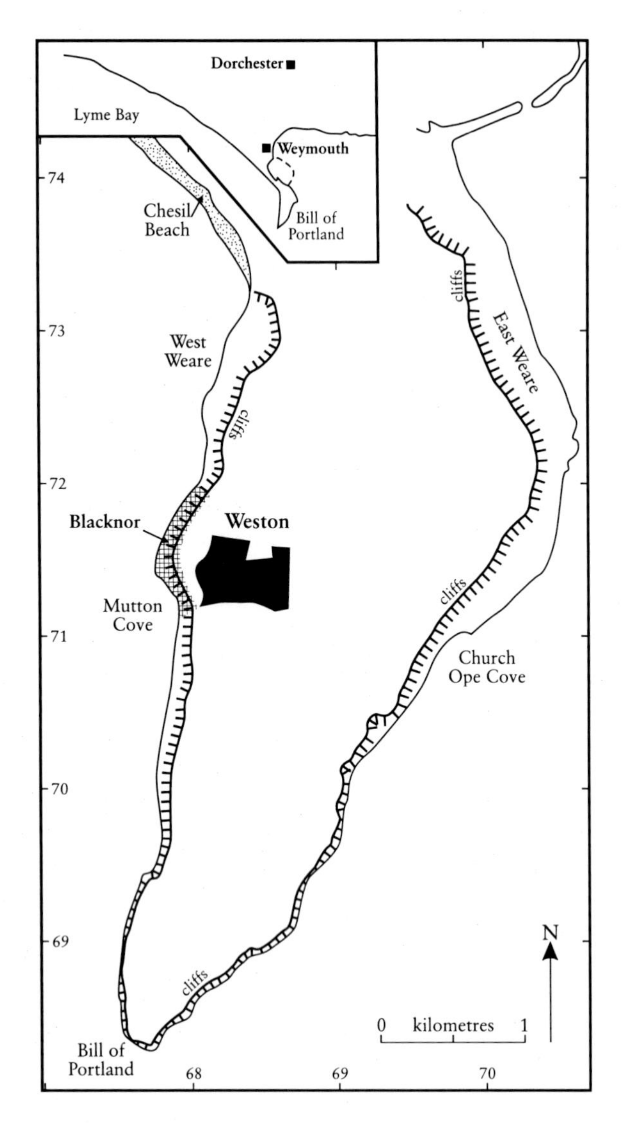

Figure 6.58 Location of the Blacknor Cliffs GCR site.

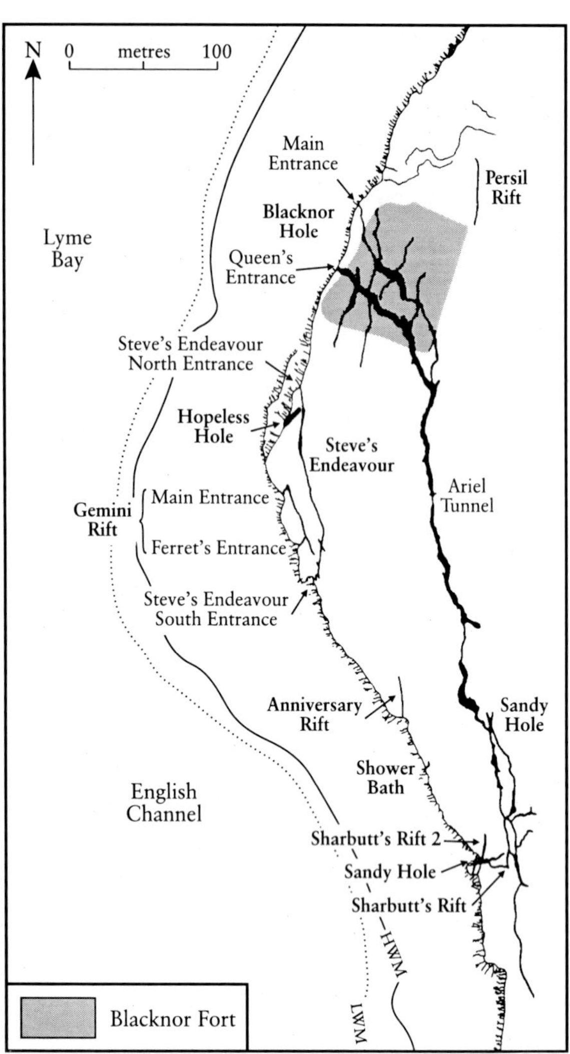

Figure 6.59 The fissures of Blacknor, Isle of Portland.

Secondly, at Blacknor, the 30 m-high cliff-face bears speleothems, principally calcite flowstone (travertine) coating the limestone faces, and often developed into 'organ-pipe' formations. Unsorted assemblages of clasts ranging from sands to boulders in size are cemented to the cliff-face by this calcite, at various heights above the cliff-foot (Cooper and Solman, 1983). It is difficult to imagine how a boulder, falling over the cliff, could be arrested part-way down by a thin film of deposited calcite on the cliff-face. However, it is easy to imagine how a boulder, falling into an open fisuure, could become wedged part-way down and then coated with calcite at the same time that the walls become coated, and so be 'cemented' in place.

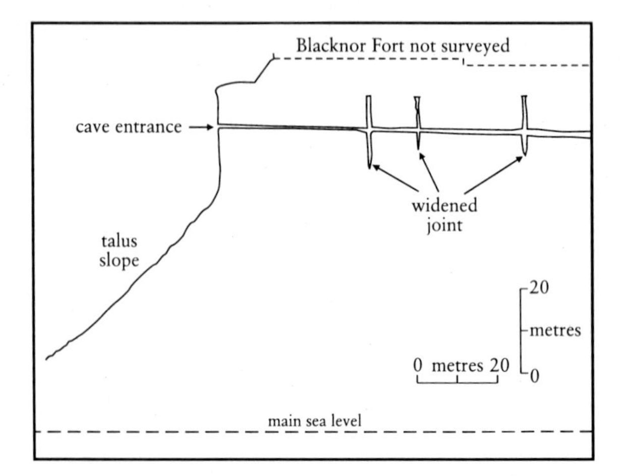

Figure 6.60 Cross-section through the cliff at Blacknor. After Cooper *et al.* (1995).

Thirdly, the 120 m-long debris slope below the cliffs, standing at angles of up to 40°, is littered at its lower end with large, joint-faced limestone blocks, chiefly Lower Purbeck and Portland Stone (Figure 6.61). Some of these bear patches of calcite flows. The blocks are up to 42 m long and 14 m broad, but the third dimension tends to be only a few metres, giving a marked tabular 'slab' shape. Individual units are up to 3500 m^3 in volume.

Interpretation

The wideness of the joints, particularly those near the cliff edge, is a clear indication that the blocks isolated by them have moved laterally, causing the present cliff-line itself to move seawards. Presumably the movement has taken place by bedding-plane shearing on an incompetent horizon in the 'parallel bands' of the Portland Sand, or even in the underlying Kimmeridge Clay.

The speleothem calcite deposits on subaerial cliff-faces suggest that major lengths of the present cliff-face were once the landward walls of widened joints, and that the rock masses which were once on the seaward sides of these joints have collapsed, leaving the present face with some of the joint deposits still adhering to it.

Presumably the large slabs in the talus below the present cliff-line have resulted from periodic collapse of masses of the cliff between the (then) cliff-face and joints parallel to and behind it (Cooper and Solman, 1983). It may be expected that the slabs currently isolated between the present cliff-face and the enterable widened joints behind it will eventually collapse.

Brunsden *et al.* (1996b) associate the joint widening with a large-scale process which accounts for the 2 km-long physiographic 'low' in part of the interior of Portland. Applying to Portland the ideas developed by Cancelli and Pellegrini (1987) in a study of the Northern Appenines in Italy, they propose differential settling of the 'prisms' of rock separated from each other by the major joints. This model 'has not been tested by sub-surface exploration' and needs to be checked against the observable features within the fissures (Brunsden, 1996b).

Conclusions

The site illustrates the sequence of events more clearly than any other site of toppling failure in Great Britain. The joint control of the process is very apparent. The intact collapsed slabs are much larger than those found elsewhere. The fortuitous preservation of unsorted debris up to boulder size, cemented to the once underground (but now subaerial) cliff-face by exhumed cave calcite formations, adds to the interest of the site, and provides clear evidence of the processes involved in cliff retreat at the site.

Figure 6.61 View of Blacknor Cliffs from the sea showing the build-up of limestone blocks at the base of the debris slopes. (Photo: R. Edmonds, Dorset County Council.)

BUCKLAND'S WINDYPIT, NORTH YORKSHIRE (SE 588 828)

R.G. Cooper

Introduction

Buckland's Windypit is located on the slope of Far Moor Park, on the right bank of the River Rye in the Duncombe Park Estate, just to the south of Castle Gill, 1 km north of Helmsley, North Yorkshire (see Figure 6.54). The slope at this point is crossed by a complex pattern of hillside trenches up to 2 m deep, leaving an irregular pattern of hummocks clearly visible on stereo-pairs of aerial photographs taken before the present plantation of conifers was begun in the late 1970s. Within this area, and doubtless related to the pattern of trenches, is a fissure in the bedrock, 5 m long and up to 1 m wide, divided into two parts by a large fallen tree trunk. This is the entrance to Buckland's Windypit. The bedrock is Lower Calcareous Grit (Corallian, Upper Jurassic), a fine-grained spicular sandstone conformably overlying Oxford Clay with a slight southerly and easterly dip. The major directions of jointing in this rock are 0°–5° and 95°–100° (Cooper, 1979).

The first recorded descent of the open widened joint was by the Rev. William Buckland, the Professor of Geology at Oxford, in 1822 (Buckland, 1823; Cooper, 1978). However, he penetrated no further than the chambers immediately below the entrance. At present, 366 m of passages have been explored and mapped, forming a labyrinth leading off the entrance chambers. The passages form a complex network on different levels, penetrating 40 m below the level of the entrance (Cooper *et al.*, 1982). Archaeological finds indicating occupation by Bronze Age man have been found in some of the passages (Hayes, 1962, 1987).

Description

The entrances to this underground labyrinth lie within a wire fence, in a plantation of conifers, a few metres upslope from a forestry track crossing Far Moor Park. Buckland (1823) described the entrance as a 'great irregular crack or chasm...about twenty feet long and three or four feet broad'. The tree trunk has fallen across this fissure, covering it in the middle section and leaving two entrances, one at either end. The labyrinth is large and complex. In the survey (Figure 6.62) and this description, the names given to passages and features are largely those of Fitton and Mitchell (1950) and Hayes (1962). The larger entrance, A, is distinguished by a holly bush growing above it. It requires a climb down of 7.6 m, which needs climbing equipment. However, using the other entrance, B, the entire labyrinth can be explored without climbing equipment. A drop of 2 m lands on a boulder wedged in the top of the fissure. A fixed chain on the right-hand wall can be used to traverse along a ledge leading past a boulder bridge before dropping into a large, light chamber at the foot of entrance A. To the south from this chamber, a short climb and squeeze emerges at the top of Fissure 'S'. Any descent here would involve using climbing equipment in a vertical descent of 22 m among very loose and potentially hazardous boulders wedged in the fissure. This route has been negotiated (Hayes, 1962) but is dangerous.

At the other end of the entrance fissure a descent through boulders leads after 7 m to a vertical drop of 2 m, with an overhang. A climb down to the right-hand side of this gains a ledge skirting under the overhang and following the left wall of the fissure to the head of a steep scree-slope. At this point a small fissure branches off north-west to Chamber 'R', a small chamber at the intersection of several fissures.

To the left in Chamber 'R' there is a low passage that can only be negotiated by crawling, which leads into Fissure 'S'. This is a lower level of the same fissure as that which forms the entrance. Fissure 'S' lies below the entrance fissure's floor of wedged boulders. It is about 20 m long, 13.7 m high and 1.2–2.1 m wide, and is blocked at the far end by boulders. It was the main location at which archaeological material was found in the labyrinth (Hayes, 1987). Two fissures are encountered part of the way along Fissure 'S'. Fissure 'T', on the left, becomes

Figure 6.62▸ Detailed surveys of Buckland's Windypit showing how the lateral spreading of the hillside opens up joint or fissure caves to form a typical labyrinth network. In this case the fissures are beneath the surface suggesting loss of support from below owing to ductile behaviour of the Oxford Clay.

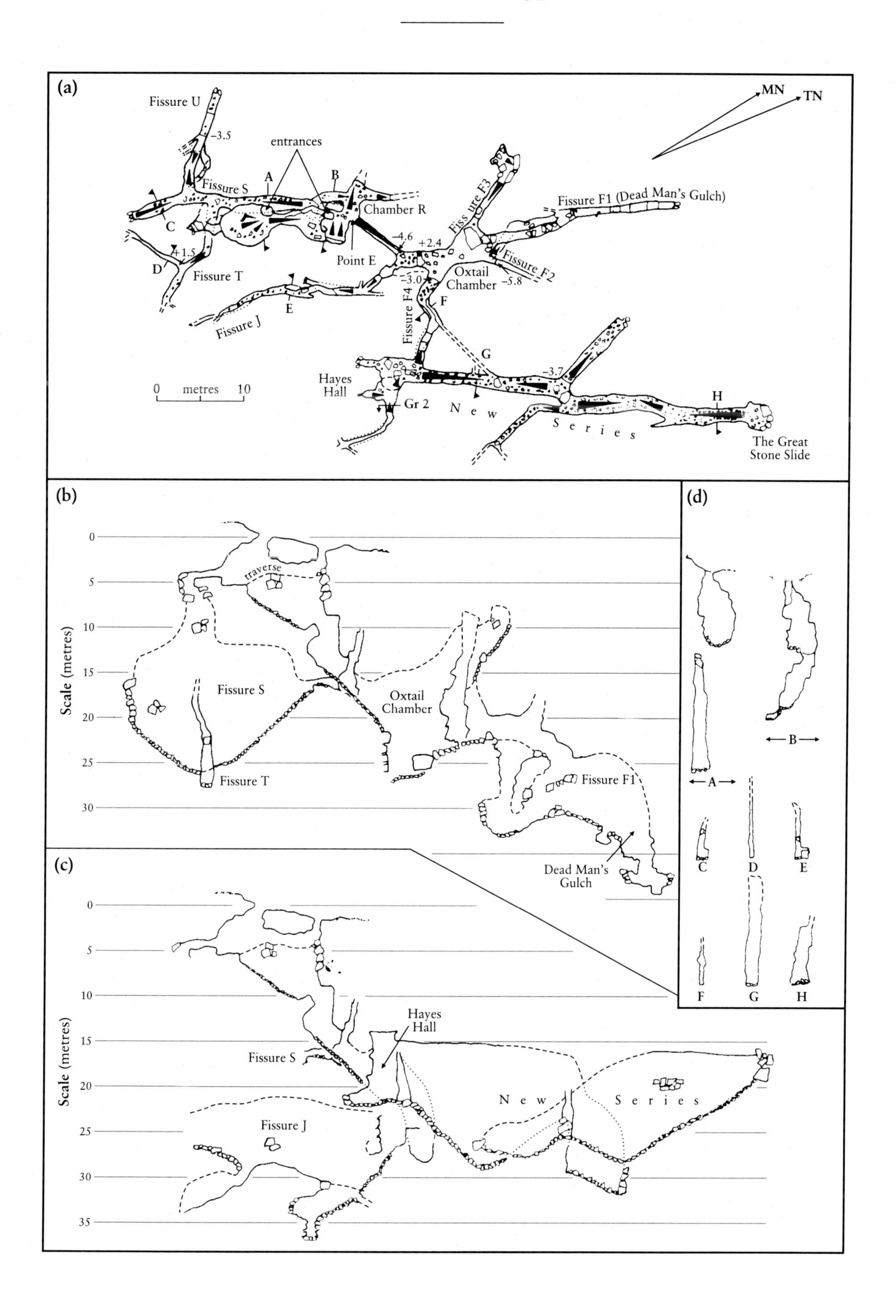
(a)
Fissure U
-3.5
entrances
MN
TN
Fissure S
A
B
Chamber R
Fissure F3
Fissure F1 (Dead Man's Gulch)
C
-4.6
+2.4
+1.5
Point E
Fissure F2
D
Fissure T
-3.0
Oxtail Chamber
-5.8
E
F
Fissure J
Fissure F4
G
0 metres 10
Hayes Hall
-3.7
H
Gr 2
New Series
The Great Stone Slide
(b)
Scale (metres)
traverse
Fissure S
Oxtail Chamber
Fissure T
Fissure F1
Dead Man's Gulch
(c)
Hayes Hall
Fissure S
New Series
Fissure J
(d)
A
B
C
D
E
F
G
H

impenetrable after 12.8 m, with a narrow branch fissure on the right which is 6.6 m long. Fissure 'U', running to the right from Fissure 'S', is blocked by boulders after 10.2 m. A drop of 3.5 m near its end leads to a narrow fissure 6.9 m long. At the foot of the 3.5 m drop, a former water level can be seen on the calcite-covered north wall. It is 27.7 m below the level of the entrance. From Chamber 'R' it is possible to climb up into the roof among very loose boulders, to a passage running north-west. Overhanging boulders make this dangerous to enter. At floor level a further passage runs NNE from Chamber 'R'. This would be difficult to explore as it is only 0.4 m wide. It is about 10 m high.

At the point where the small branch fissure runs from the main entrance route to Chamber 'R' (Point 'E'), a steep scree-slope, the Stony Corridor, descends for 9 m to the head of a 4.6 m drop, which is easy to climb down without climbing equipment. At its foot, Fissure 'J' runs off to the right. This is a descending fissure leading to a blockage of boulders. Six metres before this blockage a traverse leads above it into the farther reaches of the fissure, ending too narrow for further progress, but with a voice connection possible to the end of Fissure 'T'.

Returning to the main fissure, a climb behind an overhanging boulder leads into Oxtail Chamber (see Figure 6.63), an impressive chamber formed by the junction of five high fissures. These are:

1. The main fissure, by which Oxtail Chamber has been entered.
2. Fissure 'F1', or 'Dead Man's Gulch'. This is reached by a scramble between boulders. A traverse over a hole in the floor, followed by a series of climbable descents from false floors formed from boulders wedged in the fissure, reaches the floor (which could also be founded on wedged boulders). To the south the fissure is choked with boulders below Oxtail Chamber, while to the north a scramble down leads to the deepest point in the labyrinth, 40 m below the entrance.
3. Fissure 'F2'. This is very narrow, but may be descended to a depth of about 9 m with the aid of a ladder, before becoming too narrow.
4. Fissure 'F3'. This involves a scramble up among boulders and ends after 10 m in a loose pile of boulders.
5. Fissure 'F4'. This is a twisting fissure, about 13 m long. It connects with Fissure 'F3' in the roof of Oxtail Chamber.

In Fissure 'F4' a 3 m descent leads to a floor in the fissure, which at this level ends after 7.5 m. However, by continuing over the 3 m drop and following an ascending traverse and sections of false floor along the fissure, Hayes Hall is reached. This is a large collapse at the junction of several fissures. In this respect it resembles Oxtail Chamber, but it is not as large. To the south all ways but one are blocked by boulder piles. The exception is 9 m long, but low and narrow. The main fissure runs north-east into the New Series, and is 1.2 m wide and 12 m high. It continues for 17.5 m to a junction. Ahead is a 3.7 m descent into a roomy fissure, which, after 10 m, terminates in a boulder pile. At the junction, a crawl under, or climb over, a large boulder on the right leads into a parallel fissure, 2 m wide and 12 m high. To the north-east this fissure ends in a very large pile of

Figure 6.63 Buckland's Windypit showing part of Oxtail Chamber with animal bones. (Photo: © M. Roe.)

boulders after 23 m – the Great Stone Slide. Here the floor is of shattered angular debris noticeably smaller than the usual boulders that floor the fissures. To the south-east the passage continues narrower for 11 m, ending too narrow for further progress (Cooper *et al.*, 1982).

Interpretation

This type of feature, consisting of roofed passages between blocks of hard rock that have slipped valley-ward as part of a deep-seated translational slide, has been termed a 'mass-movement cave' (Cooper, 1983a,b). Such caves have certain characteristic features, well illustrated by Buckland's Windypit. These include the absence of any sign of ever having contained a stream, and the possession of high, parallel-sided passage shapes, with large blocks of rock wedged between the walls at irregular intervals and at various heights. Ledges on one wall are mirrored by overhangs on the opposite wall, offsetting the joint to one side and dividing it by means of this 'step' into passages on different levels. Protrusions on one wall can be matched to corresponding hollows on the opposite wall. The roofs are often formed of relatively undisturbed near-surface layers of the bedrock (Hawkins and Privett, 1981).

Mass-movement caves tend to be clustered in areas where the geological and physiographical conditions necessary for their formation are well developed. Buckland's Windypit is in such an area, the Hambleton Hills on the western border of the North York Moors, which contain 30 mass-movement caves, locally known as 'windypits (Fitton and Mitchell, 1950; Cooper *et al.*, 1976; Cooper, 1981).

Conclusions

Mass-movement caves are widespread in Great Britain. Buckland's Windypit is the longest and most complex so far discovered. Together with the ridge-and-trough features on the ground surface above, it testifies to a complex shifting of huge blocks of Lower Calcareous Grit towards the valley, most probably due to ductile movements of the underlying Oxford Clay, and possibly over bands within the Lower Calcareous Grit and/or Hambleton Oolite. This has involved various differential movements of the blocks relative to each other, opening up joints at high angles, as well as sub-parallel, to the valley-side.

Recently, it has been realized that the land-sliding resulting from lateral expansion has a very wide distribution and that the forms range from slight detachment and fissure opening to labyrinths, mass-movement caves, cambered structures, toppling, sagging and very large-scale mass movement.

Chapter 7

Mass-movement sites in Cretaceous strata

R.G. Cooper

INTRODUCTION

Considering their large outcrop (i.e. 4% greater in area than that of the Carboniferous strata) the Cretaceous strata of Britain have a relatively low density of landslides (38% of the number in Carboniferous strata). Of these, 58% are recorded in the national landslide survey as of unspecified type (Jones and Lee, 1994). Of those specified, 34% are complex and 19% are rockfalls. The Chalk, which is by far the largest formation considered areally (71% of the total Cretaceous outcrop) has a lower number of landslides (less than half) than each of the Upper Greensand and Gault, Upper Greensand, Weald Clay and Hastings Beds, each of which has less than one tenth of the area of the Chalk (Jones and Lee, 1994). Of these other formations, the Upper Greensand and Gault lead, with 31% of the landslips in the British Cretaceous outcrop. Unfortunately the survey does not provide statistics for these two formations separately. However, while 58% of the 273 slides identified in the Upper Greensand and Gault were of unspecified type, 49% were described as complex, and 23% were multiple rotational.

Two sites in Cretaceous strata have been selected (Figure 7.1). The first, **Folkestone Warren**, is one of the most intensively studied landslides in Great Britain. It is described in international reviews of landsliding and mass-movement processes (e.g. Zaruba and Mencl, 1969; Selby, 1982) and may therefore be claimed as a site of international significance for its mass-movement interest. The second site, **Stutfall Castle**, has two principal points of interest: it is on an abandoned marine cliff, and represents the types and sequence of mass movements characteristic of the degradation of such a cliff after removal of (marine) cliff-foot erosion. Secondly, it illustrates the way in which geotechnical understanding can, in certain circumstances, be enhanced by archaeological investigation.

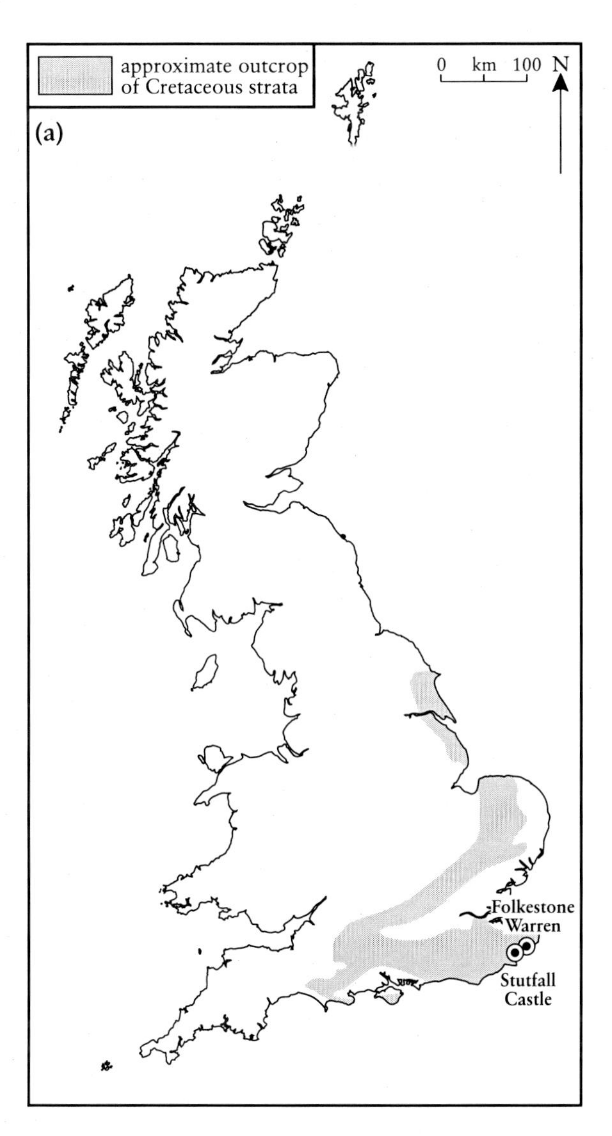

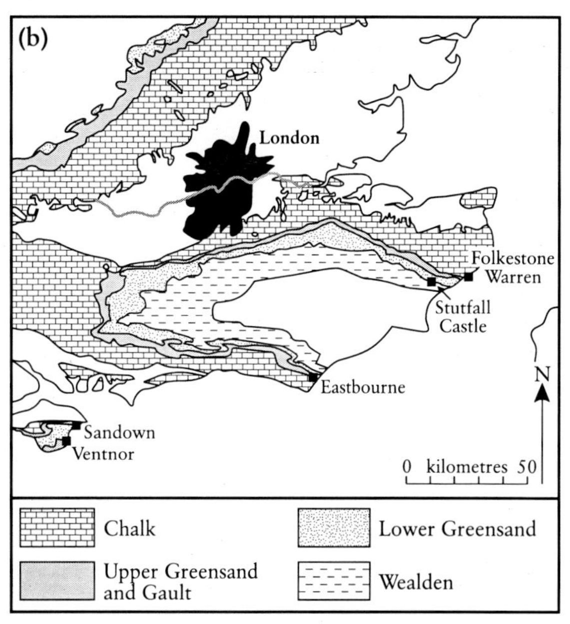

Figure 7.1 ◄(a) Areas of Cretaceous strata (shaded) and the locations of the GCR sites described in the present chapter. ▲(b) The Cretaceous strata of southern England showing the locations of the GCR sites described – **Folkestone Warren** and **Stutfall Castle**. After Hutchinson *et al.* (1980).

FOLKESTONE WARREN, KENT (TR 243 375–TR 268 385)

Introduction

(a) General

The principal reason for the detail in which Folkestone Warren has been studied has been that since 1844 the 3 km-long unstable area has been traversed by the main railway line from Folkestone to Dover (Figure 7.2). The photographs of the twisted railway tracks, one with a train on them, taken after the great landslip of December 1915, have become justly famous (Figure 7.3). Detailed investigations have been carried out by the Southern Railway and later by British Railways, and more recently by workers from a number of academic institutions and civil engineering consultancies.

The area of landslides is backed by the 'High Cliff', a 30 m-high Chalk cliff standing at about 55°, consisting of a succession of broad, irregularly spaced buttresses. In plan (Figures 7.4 and 7.5), the rear scarp of the Folkestone Warren has a generally arcuate form, concave to seaward, with the degree of concavity increasing towards the west. It is made up chiefly of three en-echelon sets of essentially joint-controlled faces (Hutchinson *et al.*, 1980), which trend at 67°,

Figure 7.2 The landslide complex at Folkestone Warren, Kent, showing the engineering structures. (Photo: Copyright reserved Cambridge University Collection of Air Photographs.)

Figure 7.3 The central part of Folkestone Warren, viewed eastwards from a point about 200 m east of the Warren Halt, shortly after the 1915 slip. (Photo: British Railways, Southern Region.)

47° and 109° (Figure 7.4). These give the rear scarp in detail a rather saw-toothed appearance. Its general form reflects the tendency of the faces at 47° to become increasingly predominant over those at 67° towards the west. The upper part of the cliff is vegetated, but much of its foot was freshened by the movements of 1915. The 300–400 m-wide area of slips between the High Cliff and the coastline is known as the 'Undercliff'. The multiple form of the Warren slips is best seen at the western end, where the rear scarps of slips of the 1930s and 1940s parallel both the associated rear scarp of the 1915 slip and the High Cliff above the ancient slide block of 'Little Switzerland' (Figure 7.6). Elsewhere the slide blocks are generally obscured by a mantle of Chalk debris derived from numerous falls from the High Cliff. There is a break in the talus at the foot of the High Cliff. This was produced by the 1915 slip. It can still be traced except where masked by later Chalk falls. The seaward edge of the Undercliff consists of sea cliffs up to 15 m high, exhibiting a great thickness of Chalk fall debris overlying the slip deposits.

(b) Stratigraphy

The arrangement of the undisturbed strata in the immediate vicinity of Folkestone Warren was described by Osman (1917). The cliffs are formed by the truncation of the scarp of the North Downs by the sea, and consist of Middle and Lower Chalk (Upper Cretaceous), overlying Gault Clay and the Folkestone Beds division of the Lower Greensand (Lower Cretaceous). All of the strata dip at about 1° in a direction between north-east and NNE. The Folkestone Beds formation is about 18 m thick and consists of coarse-grained yellowish greensands with bands of calcareous and glauconitic sandstone (Gallois, 1965). The junction between the Folkestone Beds and the overlying Gault Clay is at the top of the 'Sulphur Band', a bed of phosphatic nodules (Smart *et al.*, 1966).

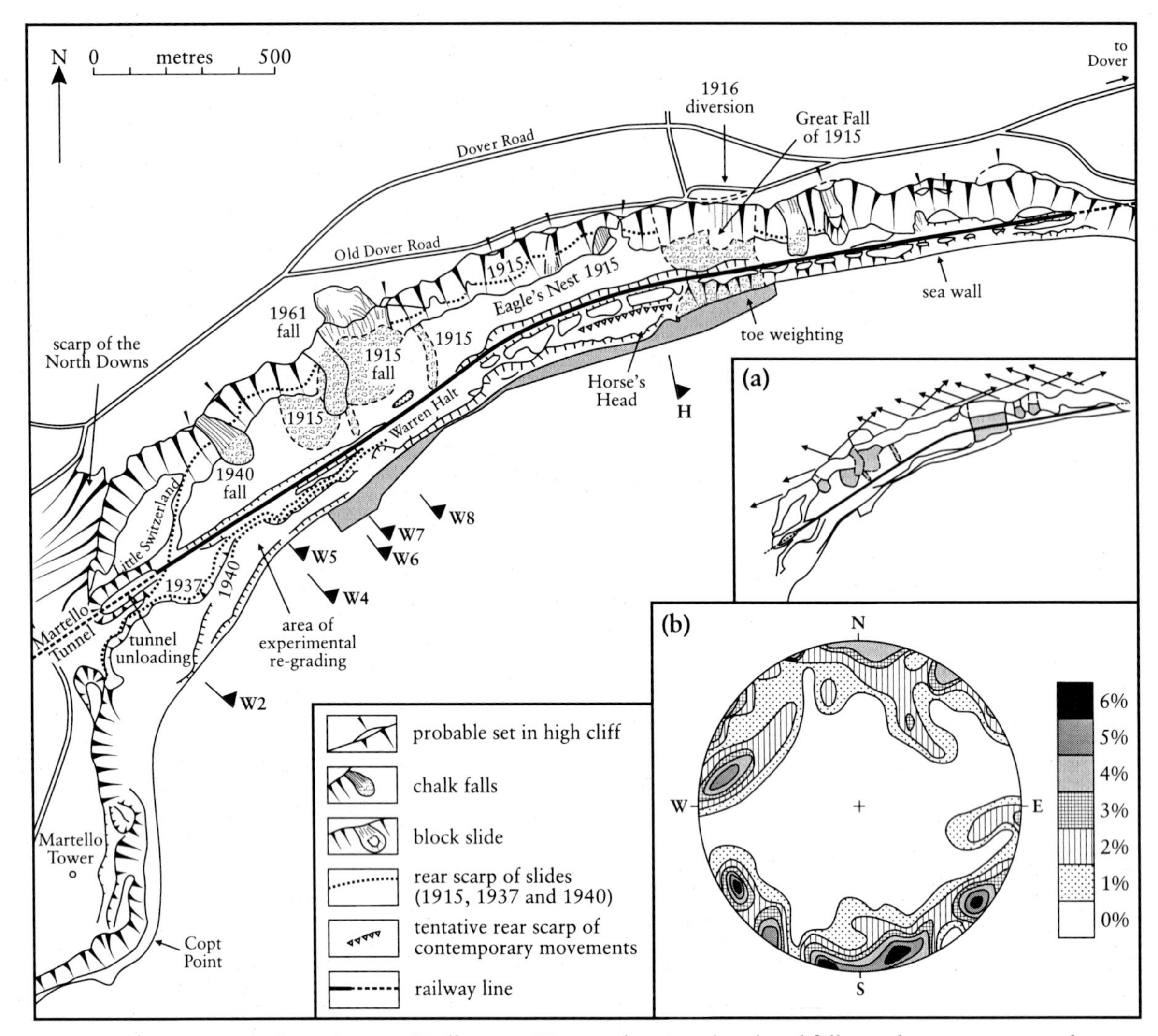

Figure 7.4 Geomorphological map of Folkestone Warren showing dated rockfalls on the near scarp and traces of some of the larger rotational slips. Position of cross-sections W2, W4–8 and G are shown (see Figure 7.5 for the postion of cross-sections W3–4, W6 and W8–9). Inset (a) shows the main scarp trend directions and inset (b) the predominant joint directions. After Hutchinson (1969).

The Gault Clay consists of hard, overconsolidated, fissured and jointed clays. It is between 40 m and 50 m thick. Thirteen lithological subdivisions have been recognized within it (Jukes-Brown, 1900). These possess large variations in physical properties, for example in the liquid limit – in the upper part of the Gault Clay, liquid limit values lie between 80% and 120%, falling to 60%–70% in the middle part, and rising again to 90%–110% in the lower part, with a

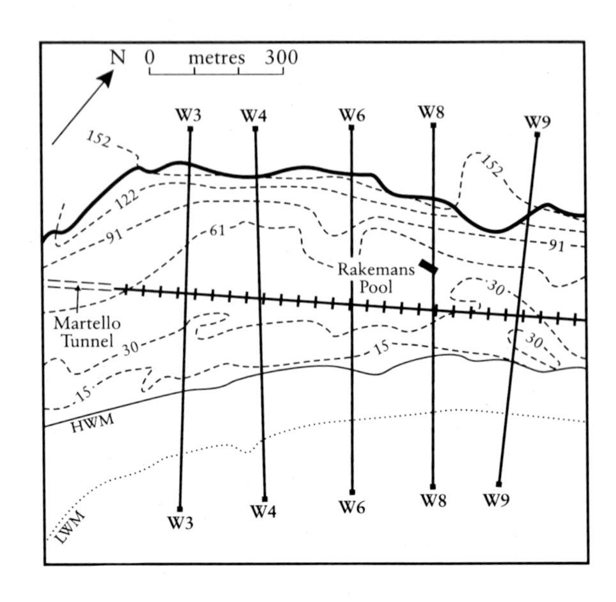

Figure 7.5▸ Part of the 2nd edition Ordnance Survey maps of 1899, showing the contours in the western part of Folkestone Warren and locations of cross-sections W3–4, W6 and W8–9. After Hutchinson (1969).

Figure 7.6 Folkestone Warren viewed from the west, with 'Little Switzerland' in the foreground. (Photo: R.G. Cooper.)

rapid fall to about 70% immediately above the Sulphur Band (Toms, 1953; Muir Wood, 1955a).

The lowest bed of the Lower Chalk is the Chloritic Marl, a relatively impermeable formation about 27 m thick, which can have as much as 50% argillaceous and arenaceous matter (Gallois, 1965). The overlying Grey Chalk, 24 m thick, becomes more pure, massive and blocky upwards into the overlying well-jointed White Chalk, 18 m thick and itself overlain by a 2 m-thick layer of *plenus* marls. The superincumbent Middle Chalk has at its base the Melbourn Rock, a band of nodular, gritty, yellowish-white Chalk about 12 m thick. This passes upward into a fine white Chalk in massive, well-jointed and highly pervious beds. At Folkestone Warren these beds extend up to the level of the presumed Pliocene or Plio–Pleistocene platform (Jones, D.K.C. 1980) which defines the top of the cliffs. Detailed descriptions of the Middle Chalk at Horse's Head, the Gault Clay west of Horse's Head, and Glauconitic Marl and Chalk Marl on the foreshore east of Horse's Head, are provided by Gale (1987).

(c) History of landsliding

The first recorded landslide at Folkestone Warren took place in 1716. Since then failures have recurred at frequent intervals. As noted by Muir Wood (1955b), slipping has been more frequent in the western part of the Warren. Since 1844 there have been 11 known deep-seated landslides. All have taken place within the December to March period, which is when groundwater levels are highest. Moreover, the three largest known failures have occurred in the first third of this period. This suggests that the landslides were brought about by particularly high seasonal peaks of groundwater pressure in the slipped masses. In the case of the 1915 slip at least, there is evidence that it took place, as might be expected, near the time of low tide.

Studies at The Roughs (Brunsden *et al.*, 1996a), 12 km to the west of Folkestone Warren, confirm this. Geotechnical data from investigations at the site have been plotted on every cell of a slope map derived from a digital terrain model (DTM) of the site, to produce a factor of safety map of the Hythe Beds escarpment. Combining this with 5% and 8% perturbations of the current climatic trend extrapolated to the years 2030 and 2050 respectively, shows that many currently dominant areas of the escarpment are vulnerable to a small rise in water levels. Statistical analysis of archive data confirms this conclusion. Since the Weald Clay escarpment is adjacent to Folkestone Warren (the next feature along the coast to the east), it is likely that this conclusion applies there too.

The incidence of Chalk falls appears to be little influenced by the seasonal variations in groundwater level. This no doubt reflects the fact that the falls involve chiefly the body of Chalk situated above the highest position of the groundwater table in the slipped masses and adjacent Chalk. The boreholes of Trenter and Warren (1996) reveal three en-echelon slips that commence north of the railway, trend eastwards and swing south-east to the beach, surfacing on the foreshore at the Warren Point, Horse's Head and Warren East End locations. Measurements made at the toes of each of these slips show that in each case the rate of movement increased eastward to a maximum where the slip crossed the foreshore. On the adjoining landward side, movements were markedly smaller.

There has also been sliding or falling of large masses of Chalk from the High Cliff at the rear of Folkestone Warren. Chalk falls from the rear of the Warren are commonly preceded by slight, chiefly downward, movements known as 'sets'. These affect the Chalk behind the High Cliff for distances of probably up to 20 m, and may also involve the underlying Gault Clay (Toms, 1953). A subsidence of as much as 1.5 m has been recorded, but movement is usually much less than this. It is noteworthy that the great majority of the three dozen recorded failures have consisted of renewals of movement in the slipped masses that form the Undercliff and have in no case involved a general recession of the rear scarp of the landslips. Thus, although locally scarred by Chalk falls, or slightly shifted by sets, the High Cliff is a feature of considerable age.

The largest slip about which detailed information has been collected took place in 1915. Other notable movements occurred in 1937 and 1940 (see Tables 7.1 and 7.2). The 1937 landslip (Toms, 1946; Hutchinson, 1969) was more than 900 m wide, and affected the whole of the slope seawards of the railway line (Figure 7.7). Upheaval of the foreshore took place. Seaward movement varied from about 27 m in the western part of the slip to zero at Warren Halt. The 1940 landslip (Toms, 1953; Muir Wood, 1955b; Hutchinson, 1969) took place in about 6 ha of the Warren, with a length along the coast approaching 700 m. Movements were slight and gradual, beginning in 1940 and continuing episodically to 1947, amounting to 1.5 m horizontally at most. Its essential features are shown on two cross-sections (Figure 7.8).

Since 1936–1940, apart from a slight renewal of movement of the 1940 slip in 1947, no major movements have occurred. The improved stability of the Warren since 1915 is probably the cumulative result of coast protection measures, drainage works in the slipped masses and the extensive weighting of the toe described by Viner-Brady (1955).

Description

(a) Hydrology

Investigations in 1948–1950 left little doubt that most of the groundwater in the slipped masses derives from the aquifer provided immediately inland by the Chalk. Using data from boreholes in the Warren, Muir Wood (1955b) has drawn

Table 7.1 Folkestone Warren: summary of the average values of ϕ_r' (°), σ_n' and s in the Gault Clay at failure in the 1940, 1937 and 1915 landslips. The original pre-metric data have been used. After Hutchinson (1969).

Landslip	ϕ_r' (°)		σ_n' (pounds per square foot)		s (pounds per square foot)	
	Max u	Min u	Max u	Min u	Max u	Min u
1940	15.1	14.0	4510	4950	1215	1235
1937	16.3	14.0	8340	9740	2440	2430
1915	16.6	13.9	13 170	15 620	3925	3865

σ_n' average effective normal stress on slip-surface in Gault Clay determined graphically using computed values of internal forces

u porewater pressure acting on slip-surface

s average shear-strength, $\sigma_n' \tan\phi_r'$, along slip-surface in Gault Clay

Table 7.2 Results of stability analyses. After Hutchinson (1969).

	Cross section	Pore pressure assumption on slip surface	Value of $\varnothing_r'$ Gault Clay required for $F = 1.0$		Remarks
			Janbu (°)	Morgenstern and Price (°)	
1915 landslip	W4 lower profile	Maximum Minimum	10.2 8.3		Slip surface entirely in Gault Clay
	W4 upper profile	Maximum Minimum	10.3 8.5		Slip surface entirely in Gault Clay
	W4 average profile	Maximum Minimum	10.25 8.4	9.7 7.7	Slip surface entirely in Gault Clay
	W6	Maximum Minimum	16.55 14.3	16.3 13.8	$\varnothing_r'$ for small length of slip surface in Chalk Marl at rear of slip = 20°
	W8	Maximum Minimum	23.85 20.1	22.2 18.7	$\varnothing_r'$ for small length of slip surface in Lower Chalk at rear of slip = 23°
	Weighted average of sections W4 (average), W6 and W8	Maximum Minimum	17.6 14.8	16.6 13.9	
1937 landslip	W1	Maximum Minimum	17.4 14.9		$\varnothing_r'$ for small length of slip surface in Lower Chalk at rear of slip = 23°
	W2	Maximum Minimum	18.3 15.9	16.2 14.0	$\varnothing_r'$ for small length of slip surface in Lower Chalk at rear of slip = 23°
1940 landslip	W5	Maximum Minimum	13.9 13.0	12.5 11.6	$\varnothing_r'$ for small length of slip surface in Chalk Marl at rear of slip = 20°
	W7	Maximum Minimum	19.2 17.6	17.6 16.4	$\varnothing_r'$ for small length of slip surface in Chalk Marl at rear of slip = 20°
	Average of (W5 + W7)/2	Maximum Minimum	16.55 15.3	15.05 14.0	$\varnothing_r'$ for small length of slip surface in Chalk Marl at rear of slip = 20°

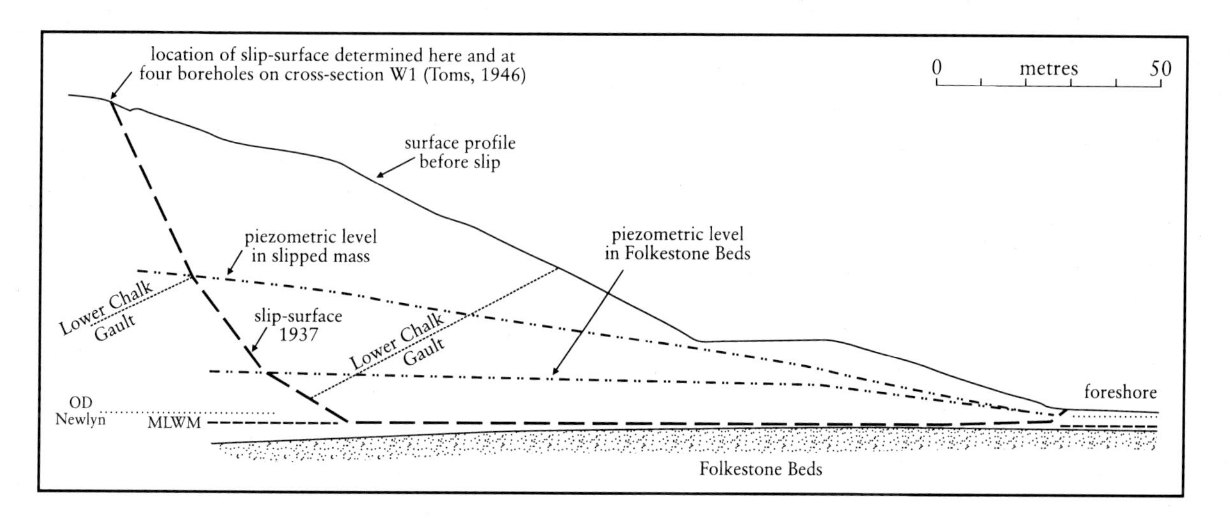

Figure 7.7 Section through the 1937 landslide of Folkestone Warren, transformed/transferred from cross-section W2. After Hutchinson (1969).

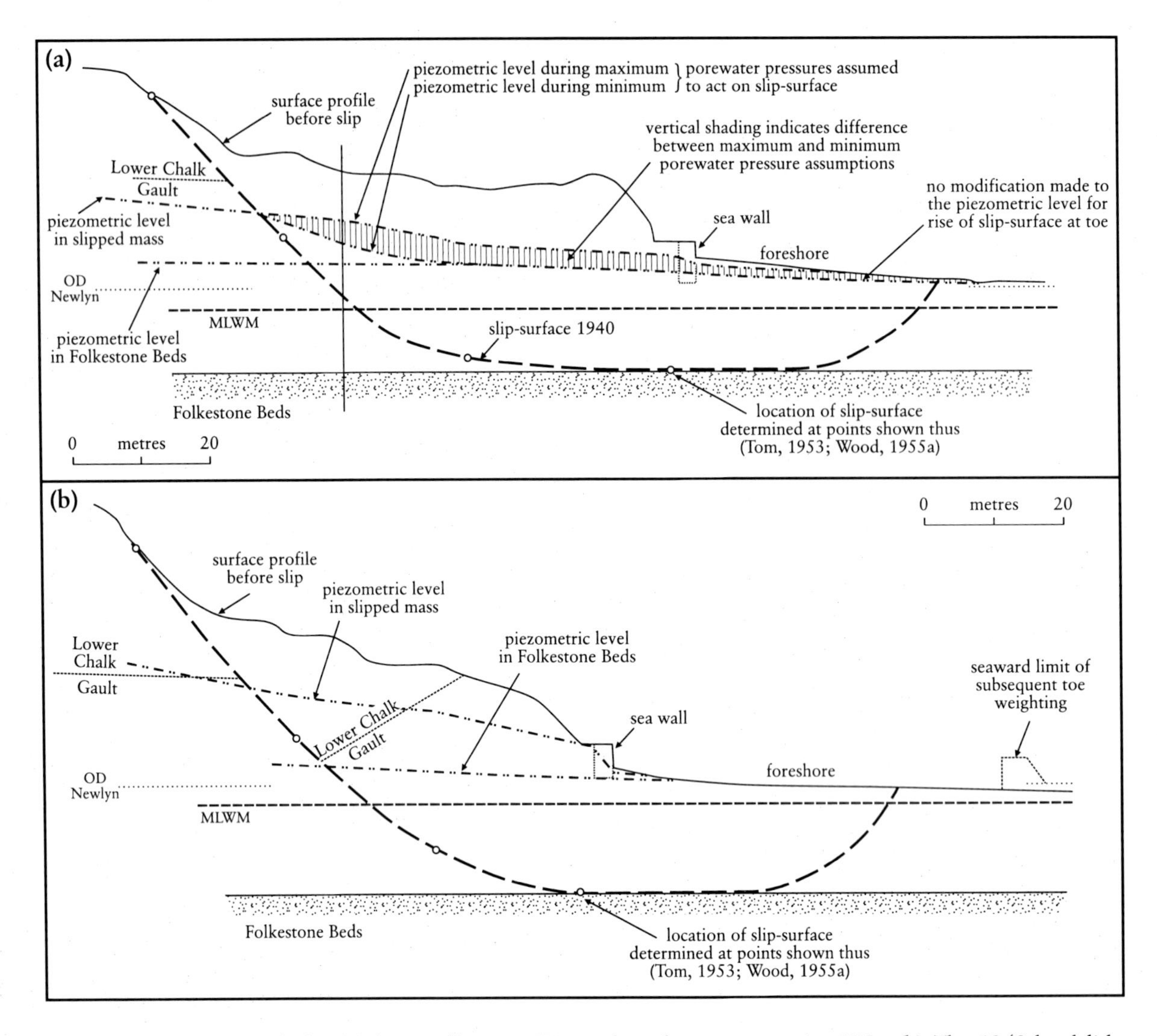

Figure 7.8 (a) The 1940 landslide at Folkestone Warren based on cross-section W5. (b) The 1940 landslide based on cross-section W7. After Hutchinson (1969).

contours on piezometric levels in the slipped masses. Monthly observations from 1953 onwards give a good indication of the seasonal fluctuation in piezometric levels, which is of the order of 3–9 m (Hutchinson, 1969) (Figure 7.9). Comparison of hydrological measurements within the Warren with records for a nearby Chalk well shows that an intimate connection exists between the groundwater bodies in the slipped masses and the adjacent in-situ Chalk (Muir Wood, 1955b).

At The Roughs, analysis using the 5% and 8% perturbations of the current climatic trend mentioned (Brunsden *et al.*, 1996a), showed that only a small rise in water levels would bring groundwater to ground surface level over large areas of the Hythe Beds escarpment. Again, since this is so close to Folkestone Warren, it is likely that the same is true there also.

Available evidence suggests that the seasonal variation in groundwater levels in the Folkestone Beds beneath the Warren is small. At the Warren it is likely that the levels in this confined aquifer fluctuate slightly in response to, but lag somewhat behind, the tidal variations in sea level (Hutchinson, 1969).

Figure 7.10▸ The influence of dominant wave energy and littoral drift on the stability of the landslide complex at Folkestone Warren. The asymmetrical, zeta-bay shape is a typical setting for such large landslides on the south coast of Britain. After Bromhead (1986), from Jones and Lee (1994).

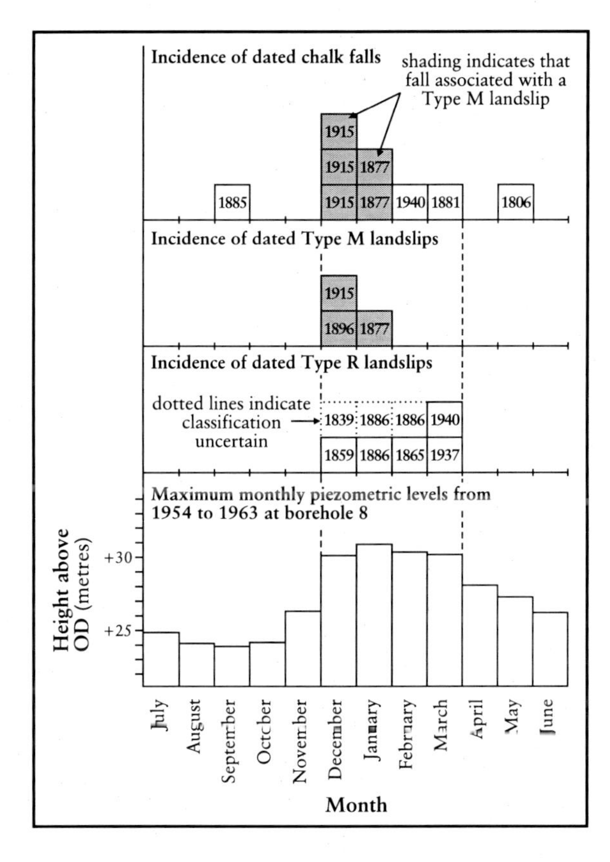

◂**Figure 7.9** The relationship between the incidence of landslides and seasonal variation in piezometric level in the slipped masses. After Hutchinson (1969).

(b) Coastal factors

Hutchinson *et al.* (1980) have discussed the predominant eastward littoral drift on this part of the coast (Figure 7.10), and the way that interference with this drift in the Folkestone area tends to deplete the beaches downdrift and hence stimulate landsliding in Folkestone Warren. Major interference dates from the construction of the masonry harbour at Folkestone in the first decade of the 19th century. Construction began in about 1807. The West Pier was completed by 1810 and the Haven by 1820. However, by 1830 the harbour mouth was completely choked by sand and shingle. From 1842 onwards, successive pier extensions were carried out in order to produce shingle-free, deep-water berths (Figure 7.11). In 1861–1863 the Promenade Pier was built, initially of timber but stone-cladded and

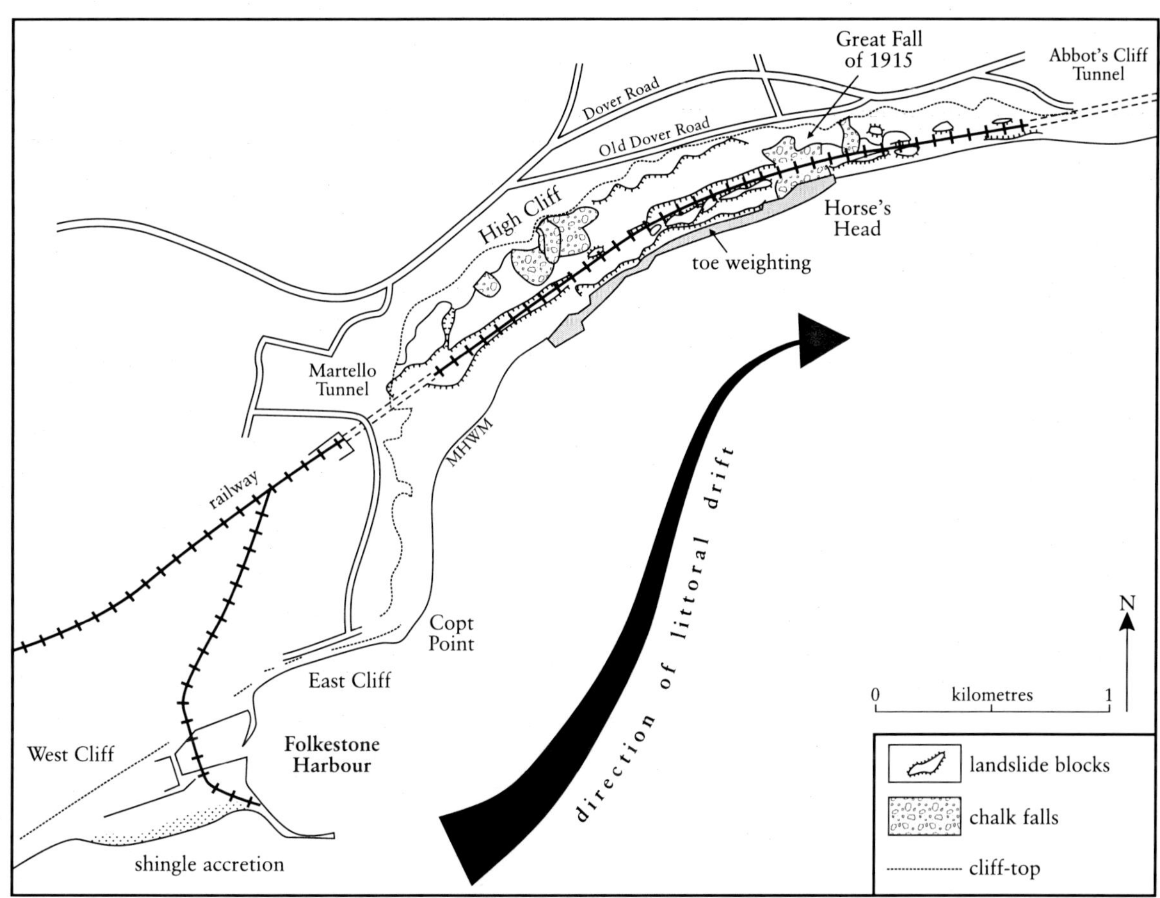

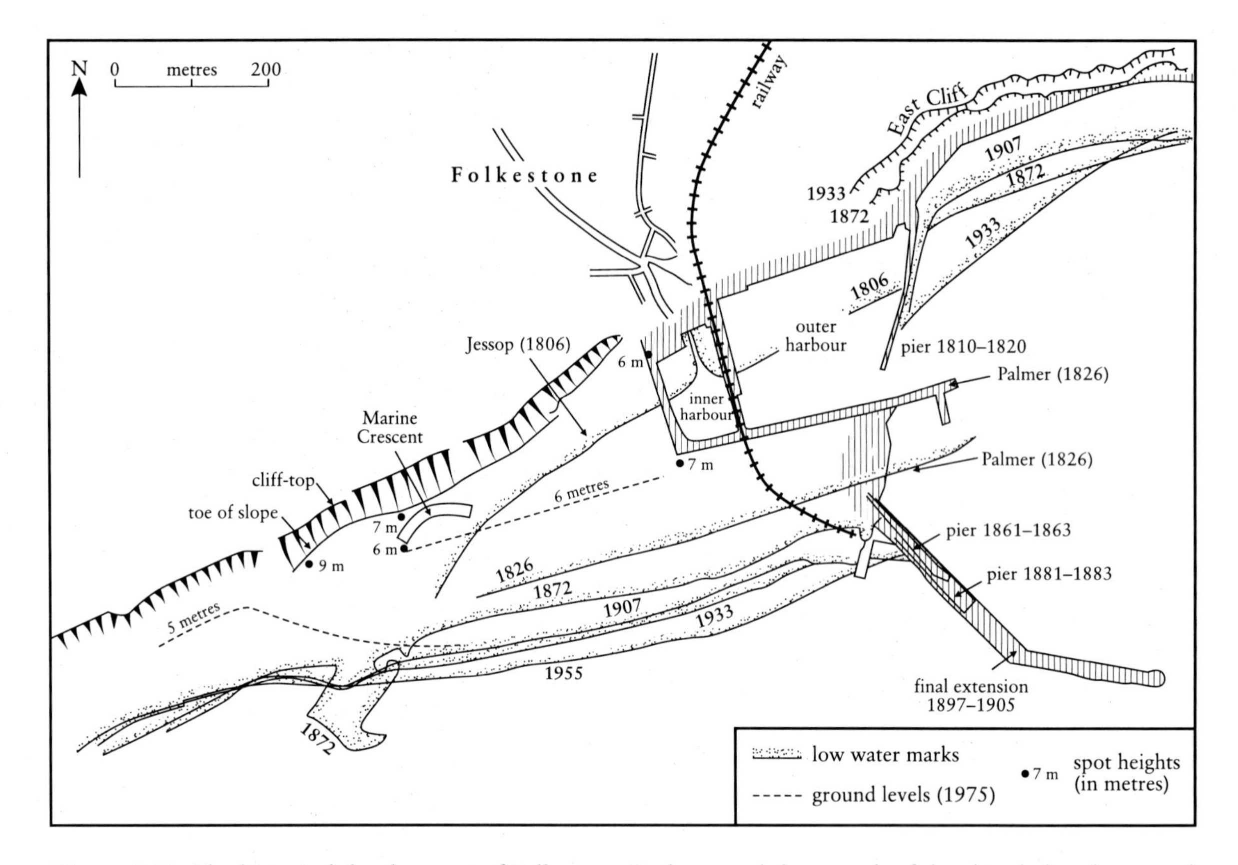

Figure 7.11 The historical development of Folkestone Harbour and the growth of the shingle beach trapped updrift. The effect of the harbour is two-fold. The prevention of lateral drift (Figure 7.10) removes beach protection below Folkestone Warren. The pier forms a headland that causes wave diffraction and the concentration of erosion at the mid-point of the Warren. Some counter-drift takes place to the west to form the beach below East Cliff. After Hutchinson *et al.* (1980).

extended in 1881–83 when it was renamed the New Pier. A further extension to this was made in 1897–1915. Accelerated coastal erosion at the Warren was commented upon by Drew (1864) and Osman (1917), as well as by many local authors. Hutchinson *et al.* (1980) conclude that these works led to a progressive increase of landsliding in the Warren, which culminated in the great slide of 1915.

(c) Geotechnical investigations

A few boreholes were put down in Folkestone Warren in 1916, shortly after the major movement, but they yielded little information. The first co-ordinated geotechnical investigation was made by the Southern Railway in 1938–1939, following the landslip movements in 1936 and 1937. The investigation was concentrated towards the western end of the unstable area, and was described by Toms (1946), who concluded that the slides are large-scale slumps of Chalk over underlying Gault Clay, on non-circular slip-surfaces. More extensive investigations made between 1948 and 1950, reported in detail by Toms (1953) and Muir Wood (1955b) led these authors to concur in this interpretation.

Muir Wood (1955b, 1971) noted that the rotational slips penetrate to the base of the Gault Clay and that failure is largely confined to a plastic sheet of clay immediately overlying the 'Sulphur Band'. Thus, testing of the strength of the materials involved in the failures has concentrated on the Gault Clay. Toms (1946) measured its unconfined drained shear strength, but as the landslips considered have all involved renewals of movement upon pre-existing slip-surfaces where shear displacements of tens of metres have taken place, the shear strength mobilized at failure can be taken to be residual. The residual strength of the Gault Clay, taking samples of both high and low liquid limit, has been determined by Hutchinson (1969) (Figure 7.12a) and, using more refined ring-shear techniques, by Hutchinson *et al.* (1980) (Figure 7.12b). A sample from the high liquid limit zone near the base of the stratum has a residual angle of shearing resistance, ϕ_r', of 12° (Hutchinson, 1969). For a sample from the lower liquid limit material forming the middle part of the stratum, he obtained a value of 19° for ϕ_r'. However, using ring-shear tests, Hutchinson *et al.* (1980) found 12° for the low liquid limit Gault Clay and 7° for the high liquid limit material (Figure 7.13). These values are up to 7° lower than those obtained by Hutchinson (1969) in cut-plane direct shear tests on similar material at lower normal effective stresses. In relation to the Chalk falls, Hutchinson also measured the residual strength of a sample of the Middle Chalk, obtaining $c_r' = 0$, $\phi_r' = 35°$.

The Folkestone Warren landslips have been the subject of a large number of stability analyses. The 1937 landslip was analysed in terms of total stresses and using a rotational landslide failure model, by Toms (1946) and Skempton (1946). Similar analysis of the 1915 and 1940 landslips was carried out by Muir Wood (1955b). The 1915, 1937 and 1940 landslips have been analysed in terms of effective stresses by Hutchinson (1969) and Hutchinson *et al.* (1980), an approach more appropriate for long-term problems. In contrast to the earlier work of Toms (1946, 1953) they employed a non-circular, multiple failure model. Hutchinson (1969) found that the average strengths mobilized on the non-circular failure surfaces in the Gault Clay approximate to the residual, and are bounded by the envelopes $c_r' = 0$, $\phi_r' = 13.9°$ and $c_r' = 0$, $\phi_r' = 16.6°$. However, Hutchinson *et al.* (1980) found that in the range of average normal effective stress levels in the Warren (about 200–800 kN m^{-3}), the values of ϕ_r' indicated as likely by stability analyses are more probably in the range 7.5° to 15°. Clearly, these values tend to be higher than those derived from ring-shear tests. Trenter and Warren (1996) made residual effective shear

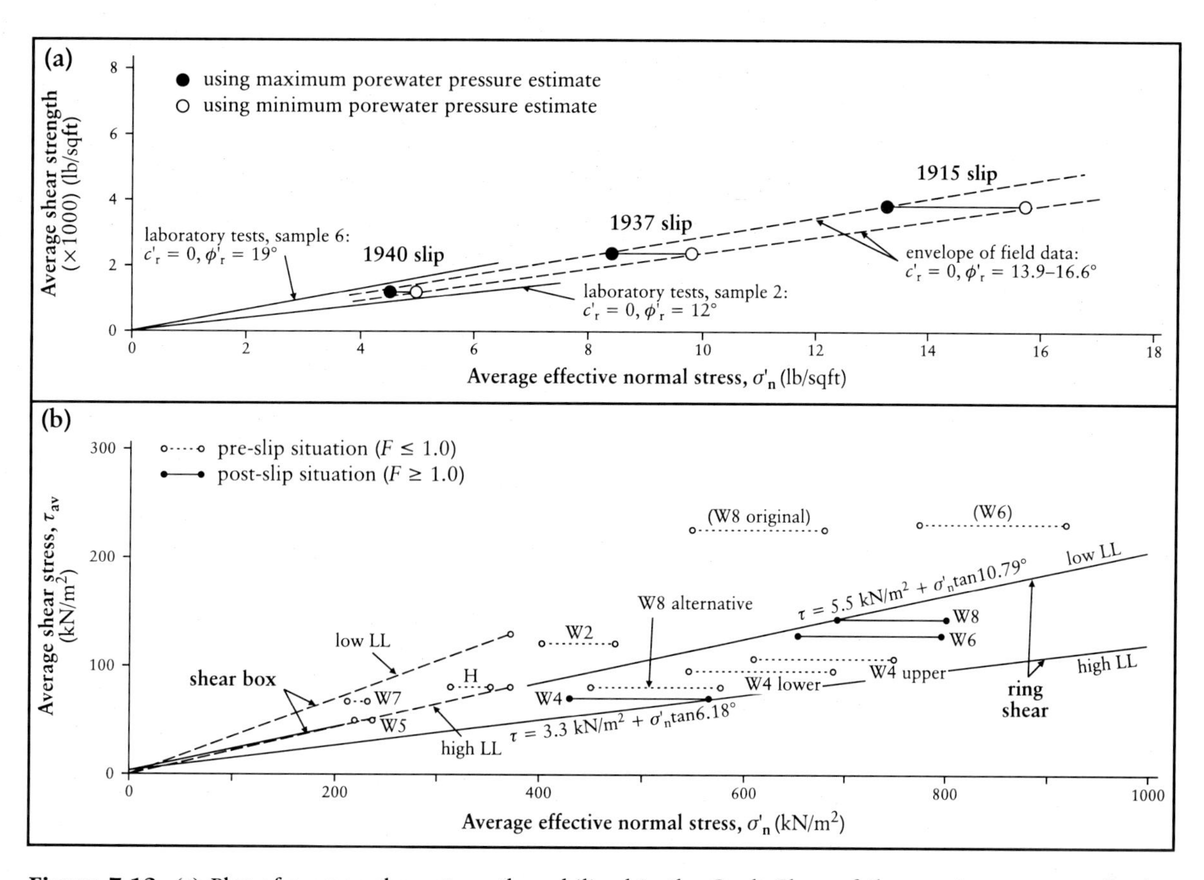

Figure 7.12 (a) Plot of average shear-strength mobilized in the Gault Clay at failure against average effective normal stress for the 1915, 1937 and 1940 Folkestone Warren landslips (after Hutchinson, 1969). (b) Comparison of residual strengths in the Gault Clay derived from back-analyses with the corresponding envelopes obtained in the laboratory (after Hutchinson *et al.*, 1980).

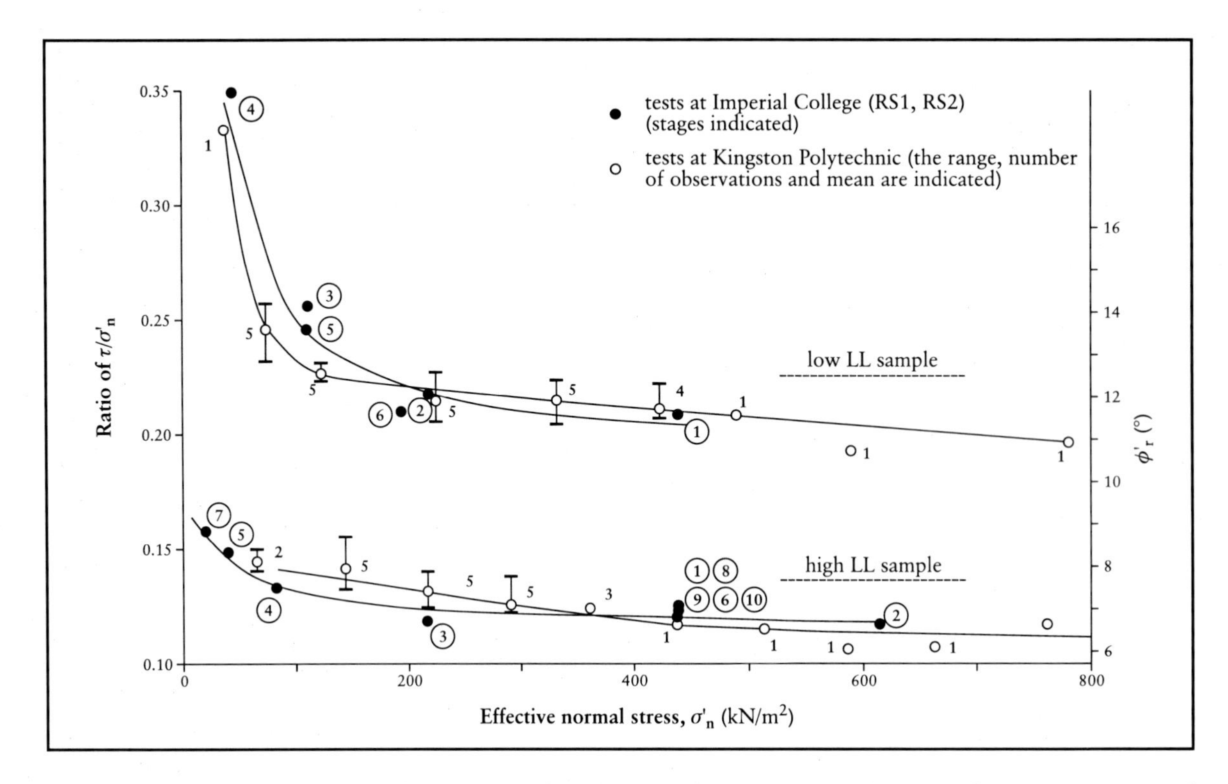

Figure 7.13 Summary of ring-shear test results, showing the comparison between the measurements of Imperial College and Kingston Polytechnic. After Hutchinson *et al.* (1980).

strength determinations on Gault Clay samples using the Bromhead ring-shear device and the reversal shear-box, obtaining results in good agreement, with an average ϕ_r' value of 9.5° (Figure 7.14). Back-analyses for two of their cross-sections (Warren Halt and Horse's Head) produced ϕ_r' values averaging 10.7°. They see the back-analysed results as being reasonably close to the average measured value. They are also close to the field residual strength line obtained by Skempton *et al.* (1989) at Mam Tor (Figure 7.13).

The value of 10.7° was obtained without including the result from the Horse's Head slip 2, which showed differences between back-analysed and measured ϕ_r' values. These are explainable in part as resulting from a proposed 'hinging' mechanism for these seaward landslides: 'as more movement occurs along pre-existing slip planes in the Gault Clay to the east, caused by a rise in the groundwater table in the chalk or chalk rubble (or, before the protection works, by erosion), some first-time movement will be provoked about the hinge further west' (Trenter and Warren, 1996, p. 618).

Interpretation

(a) Classifications of slide types

Two main views have emerged regarding the types of slides present at Folkestone Warren. Hutchinson (1968b, 1969; Hutchinson *et al.*, 1980) followed earlier workers in taking the view that most major movements at Folkestone Warren are rotational albeit controlled by a planar basal, bedding-plane, surface. He divided the recorded landslides into three main types. The largest of these ('Type M', for multiple rotational) involve a renewal of movement in virtually the whole of the landslips which form the Undercliff, and result in large seaward displacements of the railway lines. Smaller features of rotational character ('Type R', for single rotational) comprise a renewal of movement only in the slip masses in the vicinity of the sea cliff. The remaining type is the sliding or falling of large masses of Chalk from the High Cliff at the rear of the Warren (Figure 7.15).

The second and more recent view is that of Trenter and Warren (1996) who gave more emphasis to the effect of the planar bedding

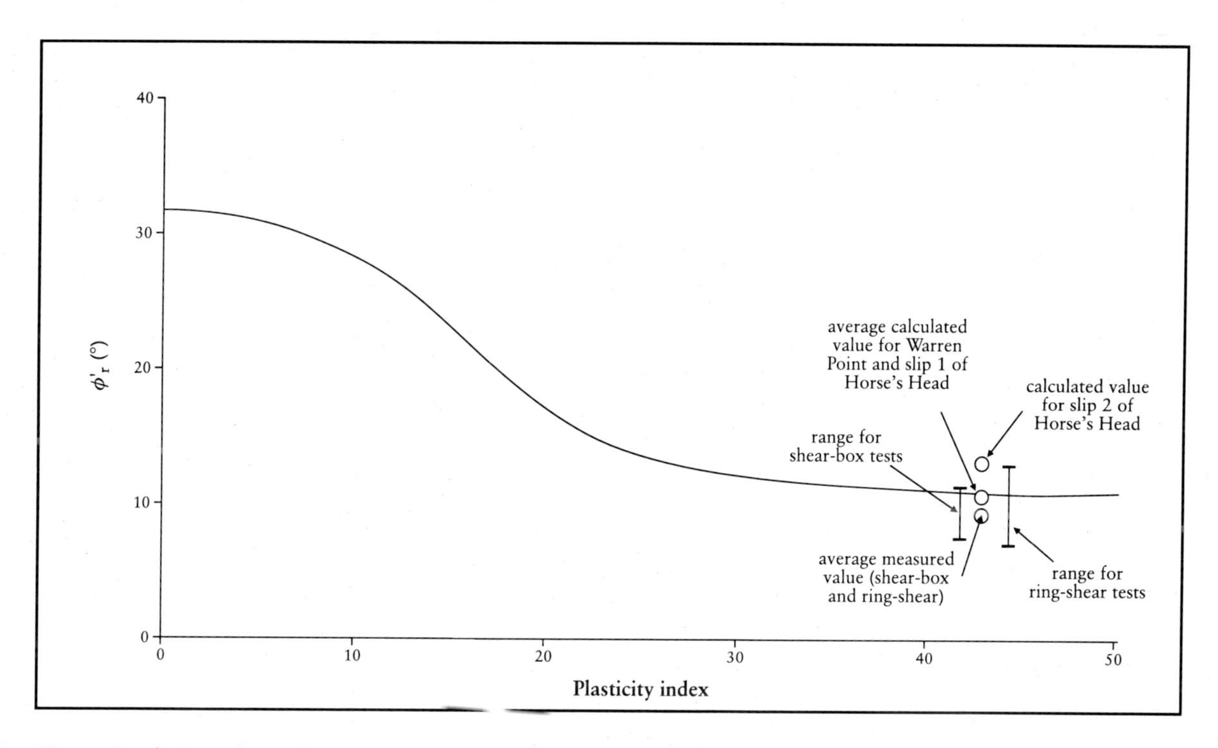

Figure 7.14 Correlation between residual strength and plastic index. From Skempton *et al.* (1989) with data from Trenter and Warren (1996).

control. Based on borehole observations, it provides an alternative two-fold classification of the mechanisms of the slips, which does not coincide with Hutchinson's (1969) Type M and Type R. Trenter and Warren's 'Slip 1' type consists of large translational slips extending from the High Cliff to the sea, with a failure surface passing through the basal Gault Clay, immediately above the Sulphur Band. Their 'Slip 2' type corresponds to smaller features, on circular failure surfaces at the east end of the Warren but compound at the west with a failure surface passing through substantial quantities of slipped and broken chalk.

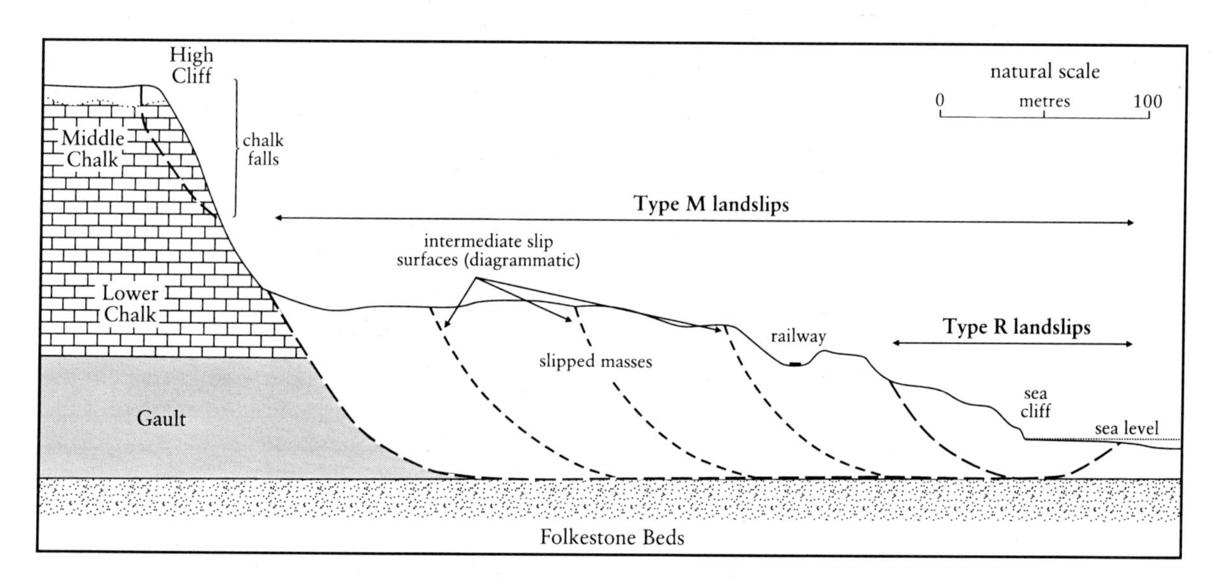

Figure 7.15 The main types of landslide occurring at Folkestone Warren as suggested by Hutchinson (1969).

(b) Retrogression mechanism

A possible mechanism for the retrogression of the rear scarp of the landslip has been suggested by Hutchinson (1969) (Figure 7.16). He recognized that the existence of large horizontal stresses in over-consolidated plastic clays is well attested, as is the lateral expansion that such deposits exhibit under reduction of their lateral support. At Folkestone Warren the over-consolidated Gault Clay lies between two more rigid strata, the Folkestone Beds below and the Chalk above. The available field evidence suggests that the lateral expansion of the seaward parts of the Gault Clay, resulting from the reduction of their side support by marine erosion and landsliding, will have been accompanied by the generation landwards of a shear surface of residual strength situated at or near the base of the Gault Clay. This in turn suggests that the failure of the 'sets' does not take place at peak strength. A possible mechanism for the progressive failure involved in the generation of such a shear surface is given by Bjerrum (1966). These 'first-time' failures take place at 'residual' strength and not at 'peak'. This is also the mechanism proposed for the Castle Hill landslide at the portal of the Channel Tunnel. A further result of the seaward expansion of the Gault Clay is that the Chalk caprock, acting as a sensitive movement indicator, will have been thrown into tension with the consequent opening-up of those vertical joints behind but close to the High Cliff. Widened joints locally known as 'vents' have been found from time-to-time behind the cliff-top. These are generally filled by superficial deposits, for example Pliocene sands. It seems likely, in the absence of anticlinal flexure and/or cambering, that they were initiated by lateral extension of the hill.

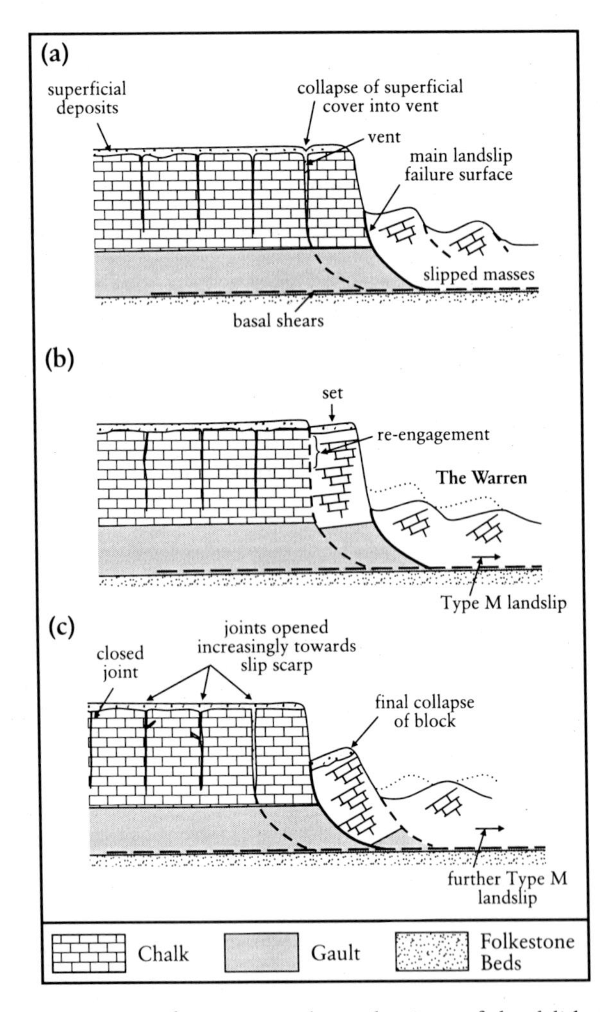

Figure 7.16 Suggested mechanism of landslide retrogression by Hutchinson (1969). The lateral expansion of the slide is dependent on the loss or the loss or extrusion of the underlying clay layer and the settlement ('sagging') of the more coherent beds above. This mechanism has become increasingly important in recent work.

The forces resisting the collapse of the Chalk mass isolated between the cliff-face and the most seaward vent will have been greatly diminished by these movements. If insufficient support is provided by the slipped masses to seaward, the block will fail and begin to subside. With increasing subsidence the curvature of the failure surface produces a back-tilt of the failing block (the apparent rotational failure) which will bring about a re-engagement of part of the irregular joint faces. Although the whole failure surface will now be at its residual strength, the stabilizing contribution from the re-engaged joint may be sufficient to effect a temporary cessation of movement. Such a mechanism is thought to be the explanation of the arrested failures in the High Cliff known as 'sets'. Final collapse of the block will generally co-incide with the next reduction of support by the slipped masses. The cycle of retrogression may then be repeated on the next block to landward.

The major part of the lateral expansion in the Gault Clay, and therefore of the formation of the associated vents and basal shears, seems likely to have been roughly contemporaneous with the initiation of the present landslides. The effect of this can be expected to have led to relatively rapid, successive cycles of retrogression of the rear scarp. These will have proceeded until a situation similar to that of the present day was reached in which the support provided by the slipped masses is generally sufficient to prevent the total collapse of the Chalk forming the current rear scarp.

These ideas are broadly supported by Trenter and Warren (1996), whose borehole information shows the form and nature of the slipped masses comprising the Undercliff varying along the Warren's length (Figures 7.17–7.21). At the western end the High Cliff reaches its highest point at about 165 m above OD while the Undercliff is at its lowest at about 70 m above OD. However, at the eastern end, the High Cliff

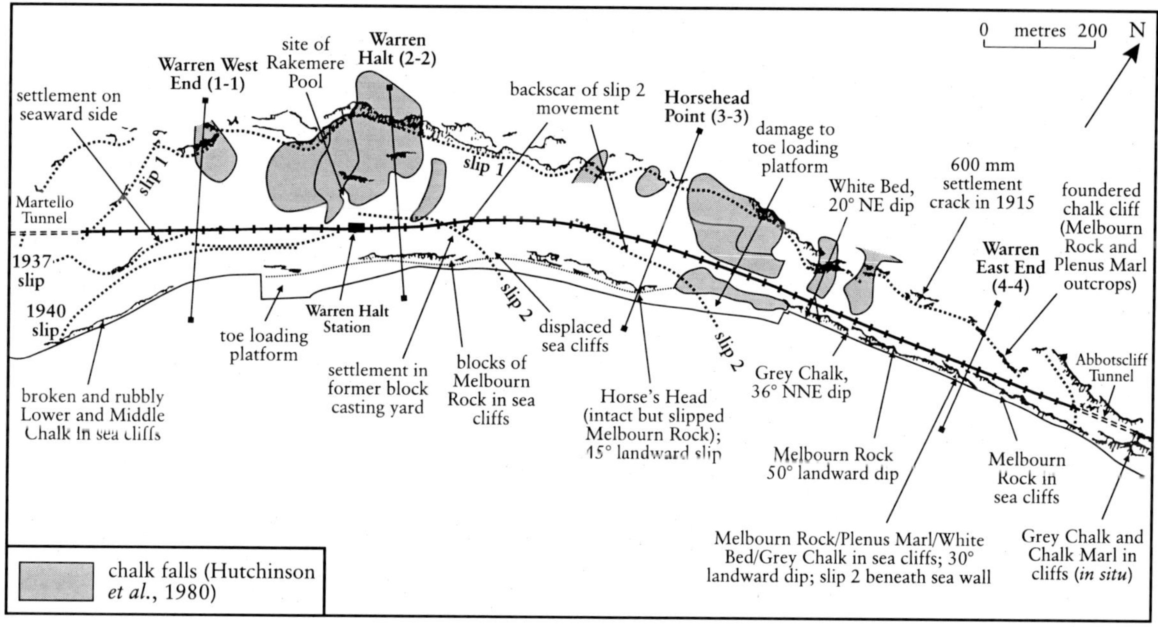

Figure 7.17 Site plan showing the location of boreholes and the cross-sections described by Trenter and Warren (1996).

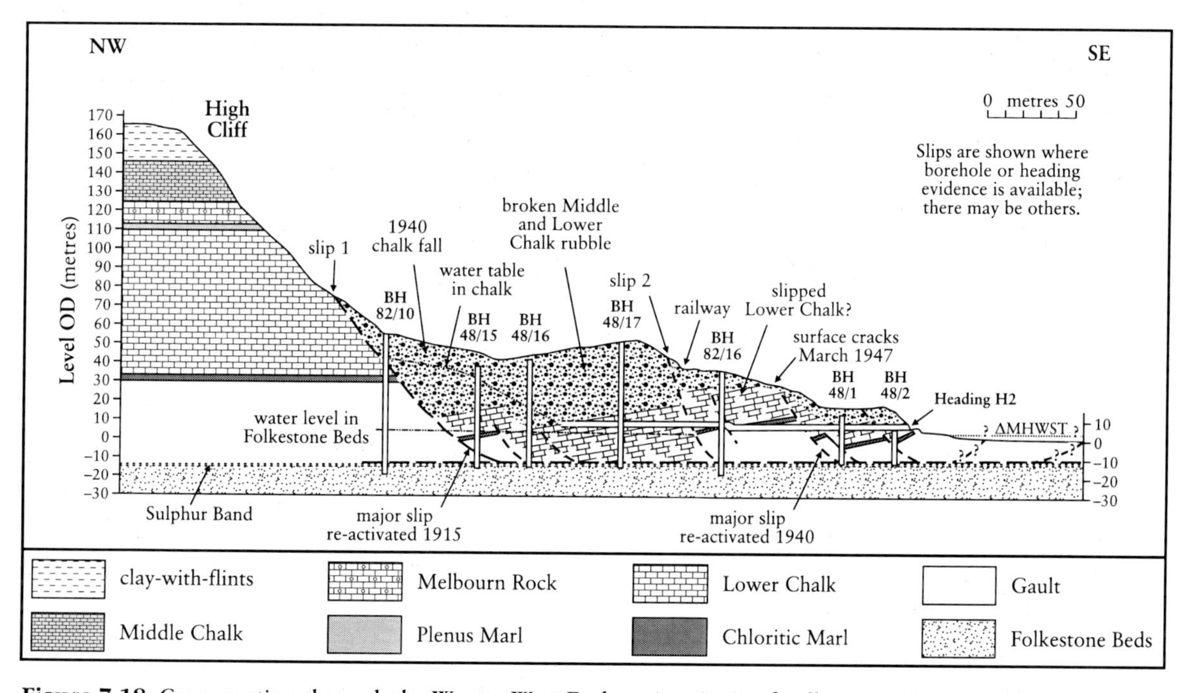

Figure 7.18 Cross-section through the Warren West End section (1–1) of Folkestone Warren. After Trenter and Warren (1996). Note the loss of thickness of the Gault Clay and Lower Chalk strata. The former suggests clay extrusion. The latter suggests that the original failure took place from a cliff that sloped towards the sea.

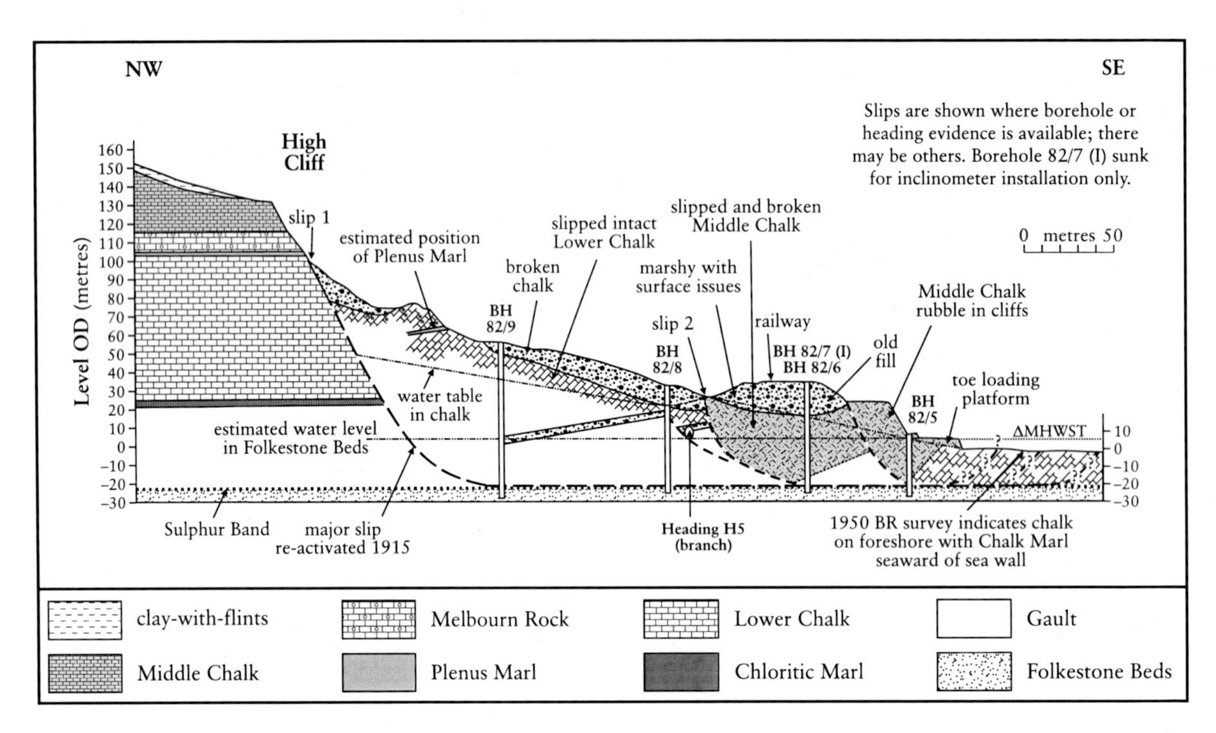

Figure 7.19 Cross-section through Warren Halt (2–2) at Folkestone Warren. After Trenter and Warren (1996).

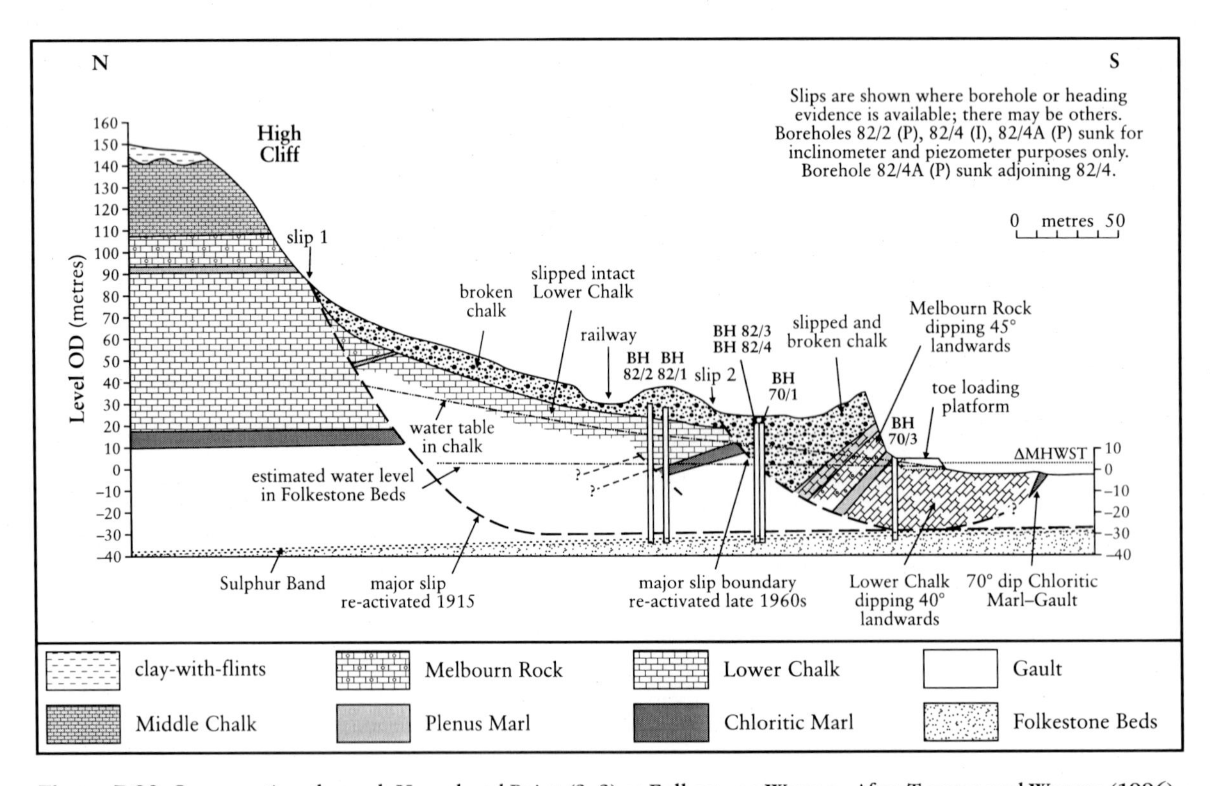

Figure 7.20 Cross-section through Horsehead Point (3–3) at Folkestone Warren. After Trenter and Warren (1996).

is lower at 135 m above OD, with the Undercliff at its highest, 90 m above OD. Trenter and Warren's (1996) boreholes show up to 45 m thickness of broken Chalk at the west end, while at the east end the mantle of broken Chalk is only 10–15 m thick. The boreholes also indicate larger amounts of slipped but intact Lower and Middle Chalk at the east end, the prime example

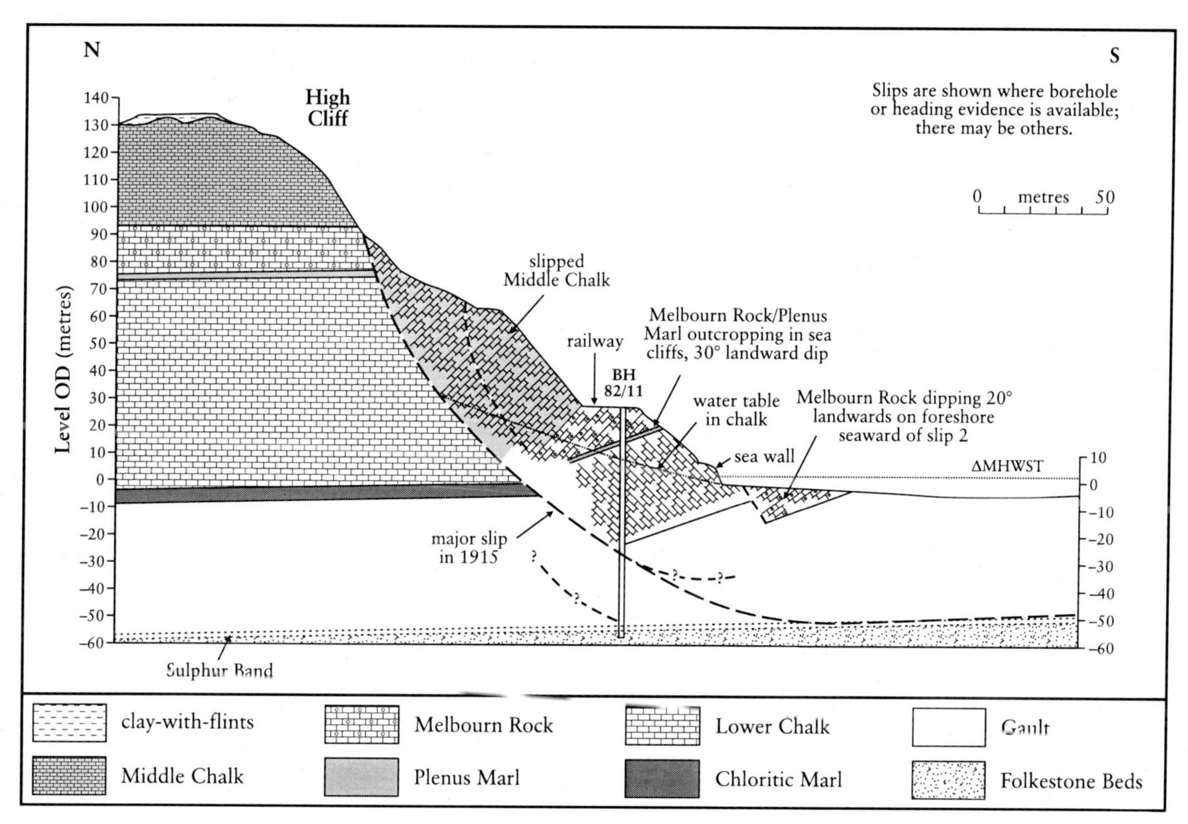

Figure 7.21 Cross-section through Warren East End (4–4) at Folkestone Warren. The thickness of the Middle Chalk is unexplained. After Trenter and Warren (1996).

being the Horse's Head, a prominent small hill, at Horsehead Point (Figures 7.18–7.21). They consider these results to indicate that the landslides at the west end of Folkestone Warren are of much greater size and mobility than the rest, and suggest that this is due partly to more massive Chalk falls from the High Cliff behind, and partly to the consequent undrained loading of the Undercliff landslides. They also point out, following Hutchinson (1969) and Muir Wood (1994), that most of the Folkestone Warren slides have been confined to the shoreward part of the Undercliff and the foreshore. Only the slides of 1877, 1896 and 1915 penetrated as far as the High Cliff.

(c) The 1915 landslip

The 1915 failure is complicated by the fact that several large Chalk falls from the High Cliff were associated with it (Figure 7.22). As noted by Hutchinson (1969) this introduces uncertainties into stability analyses. An unusually large number of eyewitness accounts of the slide were collected. These tend to show that the renewal of movement commenced at the western end of the Warren and spread eastwards. This disturbance was followed by three associated failures of the rear scarp in a west to east order, which were therefore triggered by the slips, rather than initiating them (Hutchinson *et al.*, 1980). However, the falls doubtless further stimulated the movements of the Undercliff. Field and historical evidence suggests that Chalk falls occur at the projecting corners of the individual 'saw-teeth' of the irregular rear scarp of the landslip area.

Hutchinson *et al.* (1980) also showed that in the year 1915 the rainfall recorded at Folkestone had the highest annual total since records had begun in 1868, 138% of the 1881–1915 average. The September–December 1915 total was the second highest since 1868 (158% of the 1881–1915 average), and the December 1915 total was the highest since 1868 (256% of the 1881–1915 average). They attribute the renewal of movement in the Undercliff to this unusually high rainfall, and to the erosion of the bay as intensified by pier construction. The low tide (two days before Spring tides) that preceded the

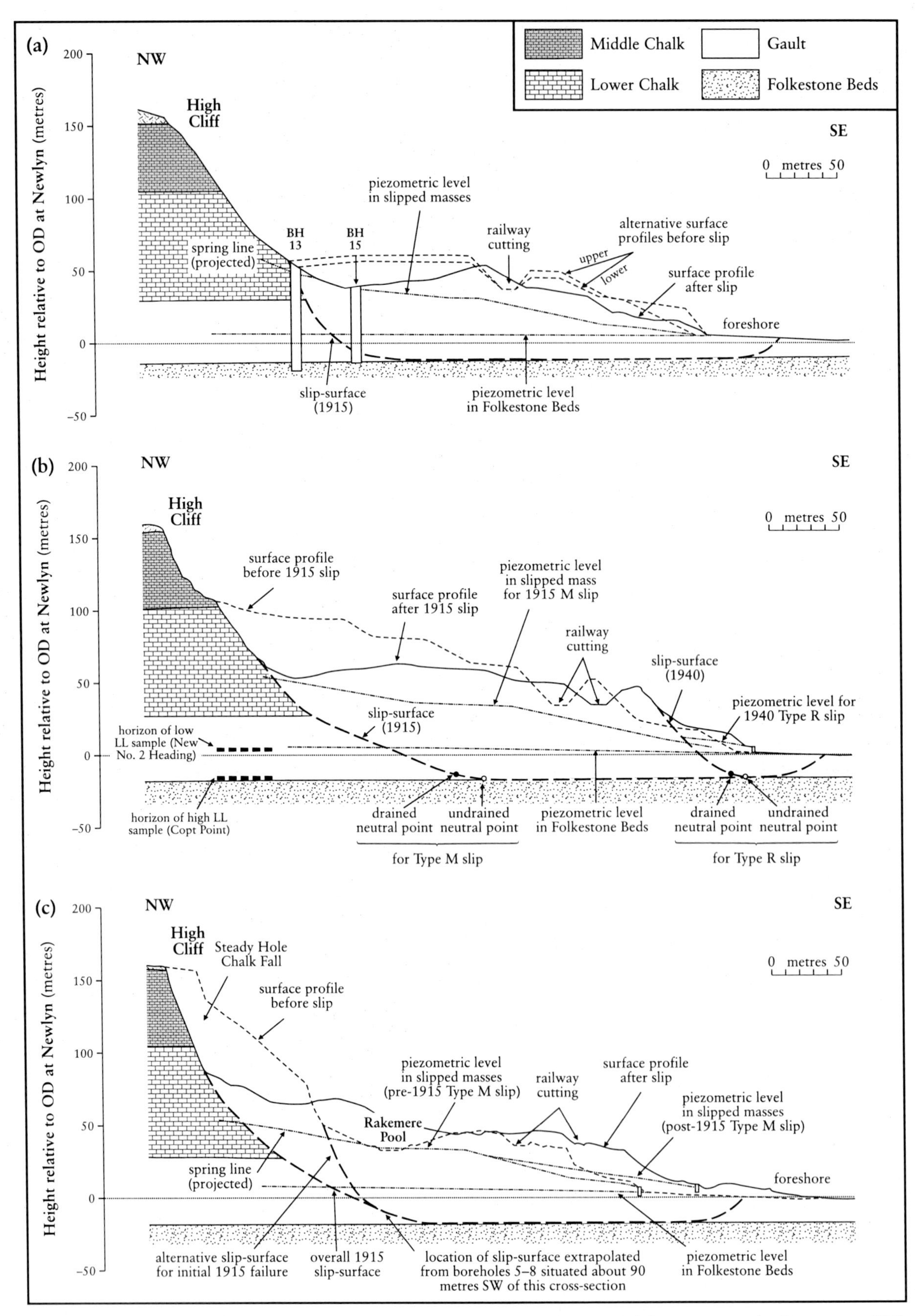

Figure 7.22 Cross-sections through Folkestone Warren. (a) 1915 landslide section W4; (b) section W6; (c) 1915, section W8. After Hutchinson (1969).

first reported movements by about 1.25 hours, was probably the final trigger. The study at The Roughs (Brunsden *et al.*, 1996a) indicates that, given only small perturbations to present climatic trends, similar conditions may become more common, and it may be presumed that this would be accompanied by similar slope movements.

In the main part of the 1915 slide, the displacement (as measured by the movement of the railway lines) was 10–20 m, increasing to a maximum of about 50 m at one point (Figure 7.22). In the central part of the Warren, the Undercliff is blanketed by flow slide debris from the 1915 fall. In a discussion of Hutchinson (1969), Muir Wood (1970) commented on the fact that the 1915 slip moved forward about 30 m, and asked how this could be reconciled with a renewal of movement on a pre-existing, residual slip-surface. Hutchinson *et al.* (1980) listed nine possible mechanisms for this, and in view of the large Chalk fall from the High Cliff that took place just after the start of the slow movement in 1915, concluded that sudden undrained loading (Hutchinson and Bhandari, 1971) of the rear of the slide by this rockfall, was the most likely explanation.

Casey (1955) described how the 1915 slide was accompanied by an upward heaving of the foreshore, involving a strip that extended between 140 m and 240 m out from the shoreline. In front of the central half of the main slip there was a second ridge of upheaved seabed farther seawards. The outer edge of this reached to between 260 m and 350 m from the former shoreline and it had a width of 80–90 m (Figure 7.23). The inner and outer zones of upheaved material were partly separated by a lagoon. Earlier large slides at Folkestone Warren were also accompanied by the development of long Gault Clay and Chalk ridges that formed on the foreshore. Trenter and Warren (1996) suggest that these ridges or reefs may have been due to seaward movement of their Slip 2 slides thrusting under the slipped masses lying beneath the foreshore, and so raising them. However, since Casey (1955) describes such ridge development as having taken place some time *before* the main movements of 1916 took place, they also consider it possible that prior to the onset of slipping there may have been some plastic flow of the Gault Clay, with the failed Gault Clay erupting at the foreshore.

Figure 7.23 View to the east of the sea cliff and heaved foreshore of the Warren, taken just after the 1915 slip. The partly eroded 'cape' of the debris of the Great Fall can be seen just beyond the Horse's Head. (Photo: British Railways, Southern Region.)

(d) Contemporary movements

Trenter and Warren's (1996) triangulation and observations within drainage headings, supplemented by piezometer readings, have shown that an area in the vicinity of the 'Horse's Head', which is the most prominent sea cliff in the Warren, in the central part of the Undercliff, has been moving seawards since at least 1938. The area of greatest movement co-incides with the location of the debris tongue of the 1915 fall, and is also the area in which coastal erosion was at a maximum until the extension of the present sea wall across it. Piezometer results suggest that at least in the vicinity of the seaward edge of the Warren and possibly elsewhere, the slipped masses consist of blocks of Gault Clay of various sizes among Chalk blocks and debris which in general have become unloaded as the slips have developed (Hutchinson *et al.*, 1980). The piezometric pressures within the Chalk and the smaller masses of Gault Clay had equalized with the long-term groundwater conditions, but negative excess porewater-pressures still existed within the larger blocks of slipped Gault Clay (Figure 7.24). As these swelled back to equilibrium the factor of safety on any slip-surfaces traversing these blocks steadily decreased.

(e) Age

It is unlikely that Folkestone Warren was initiated before the virtual completion of the Flandrian transgression. An immature soil profile situated

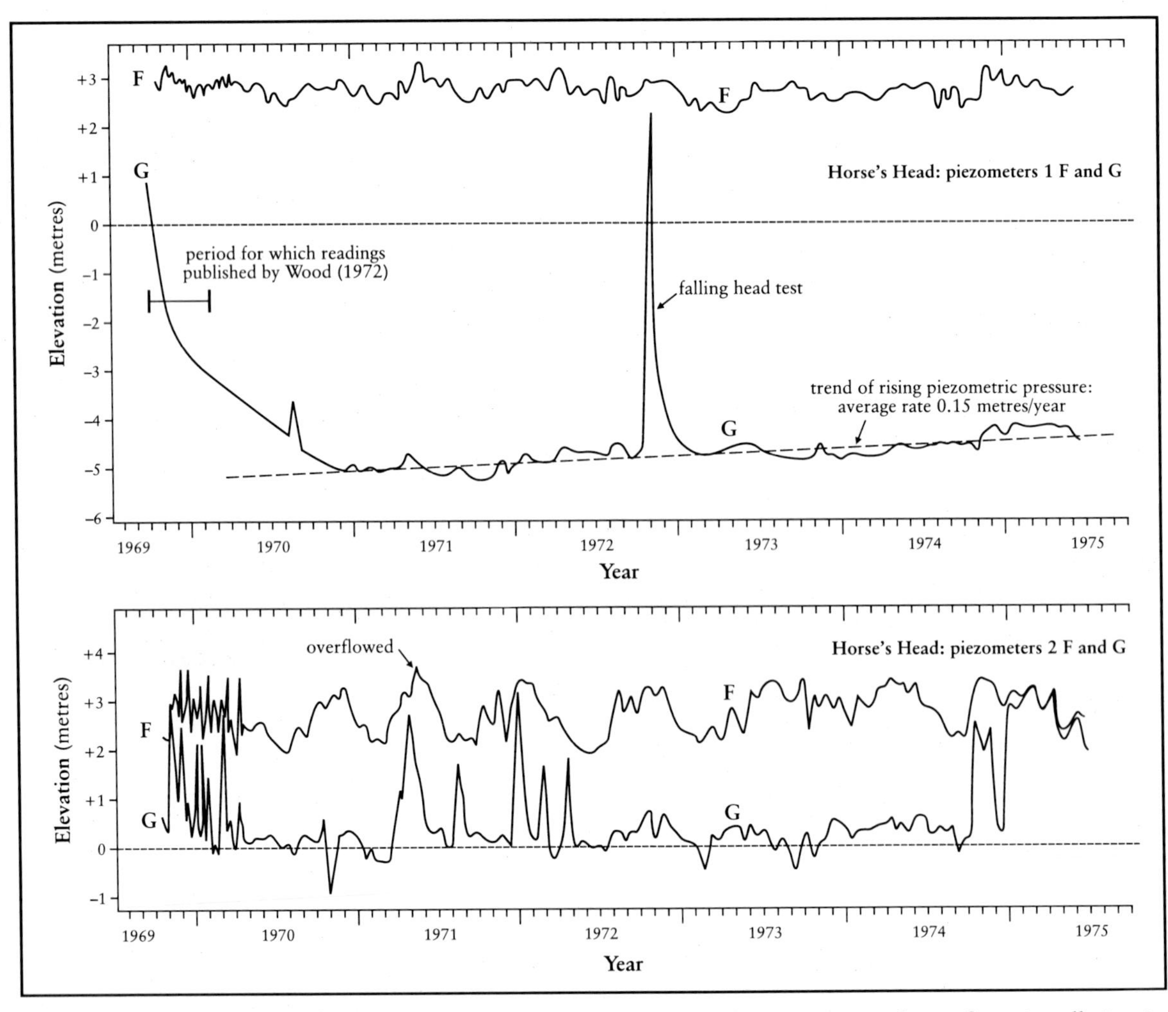

Figure 7.24 Record of piezometric levels in boreholes 1 and 2 in the Horse's Head area from installation in 1969 until readings ceased in 1975. The letters F, G refer respectively to piezometers with their tips in the Folkestone Beds and the slipped Gault Clay. After Hutchinson *et al.* (1980).

towards the base of Chalk debris overlying a bluff of steeply back-tilted Middle Chalk, at the Horse's Head, has been dated by its contained fauna to Atlantic or Sub-Boreal age. As these sea cliffs are the oldest remaining part of the present Warren landslides, and the soil layer is also strongly back-tilted, it can be inferred that the main slipping movements affecting it have occurred since its formation, suggesting a Sub-Boreal date (5500–2500 BC) for the initiation of the present landslides (Hutchinson, 1968b). The weathered and vegetated appearance of those parts of the present High Cliff that have not suffered relatively recent Chalk falls suggests that its age of at least two centuries indicated by the historical records is on the low side.

Conclusions

At least before the construction of the sea wall and toe weighting, Folkestone Warren was in a state of dynamic equilibrium under the combined influences of its topography, hydrology, geology and its exposure to marine attack. This is reflected chiefly in the widening of the Undercliff towards the west, which compensates for the gradual reduction in passive resistance at the toe as the level of the Gault Clay rises.

Folkestone Warren is a mass-movement site of great importance, particularly due to the great detail in which, and timescale over which, it has been studied. Other large-scale Chalk slips on the south coast, for example the Undercliff at Ventnor (Isle of Wight), and White Nothe in Dorset, seem to be broadly similar in nature. The results of the studies on Folkestone Warren are quoted in many studies of mass movements still further afield. The site is fundamental in the development of understanding of both translational and rotational slips, and the relationships between them.

The early recognition of the clay-extrusion model is of fundamental importance and has since become important in the interpretation of the major slides at Castle Hill on the Channel Tunnel portal, and is being increasingly recognized as a fundamental failure mechanism. New research on this topic is in progress in many countries.

STUTFALL CASTLE, LYMPNE, KENT (TR 117 345)

Introduction

Romney Marsh in Kent is backed by an abandoned marine cliff (Figure 7.25) (May, 2003). The cliff has been degrading by a process involving several types of mass movement since isolation from wave action by the growth of Romney Marsh. For example, The Roughs, a site in the Hythe Beds between Hythe and West Hythe, has 17 separate landslide scars, with lobate features in the accumulation zone down-slope. Stutfall Castle, at Lympne, is the remains of a Roman fort built in AD 280, which has broken up under the influence of these mass movements. The bastions which constitute the remaining fragments of the fort have moved relative to the crest and foot of the slope, and relative to each other, throwing light on the nature and effects of mass movements on the slope of the degraded cliff over the 1600+ years since the fort was abandoned by its last occupants.

Abandoned cliffs undergo successive changes following the cessation of erosion at their foot. Studies aimed at describing and explaining these changes have concentrated on locations where a cliff has been successively abandoned along its length, either by the growth of a coastal spit along a coastline (e.g. Savigear, 1952) or by the migration of a river meander (Brunsden and Kesel, 1973). The 'ergodic hypothesis' (substitution of space for time) is used to elucidate the probable sequence of events at any particular point along the cliff-line. At Stutfall Castle a different approach has been adopted: the abandoned cliff has been subject to investigation by a combined team of geotechnical engineers and archaeologists (Hutchinson *et al.*, 1985), in the expectation that the archaeology of the fort would throw light on the later stages of land-sliding and slope development (and that geotechnical investigation would illuminate some of the archaeological problems of the site). Previous excavations of the fort had been carried out by Roach Smith (1850, 1852), and by Cunliffe (1980a).

The crest of the slope is capped by the Hythe Beds, which form a plateau to the north. These are underlain thinly by Atherfield Clay, but most of the cliff slope is formed in Weald Clay, which

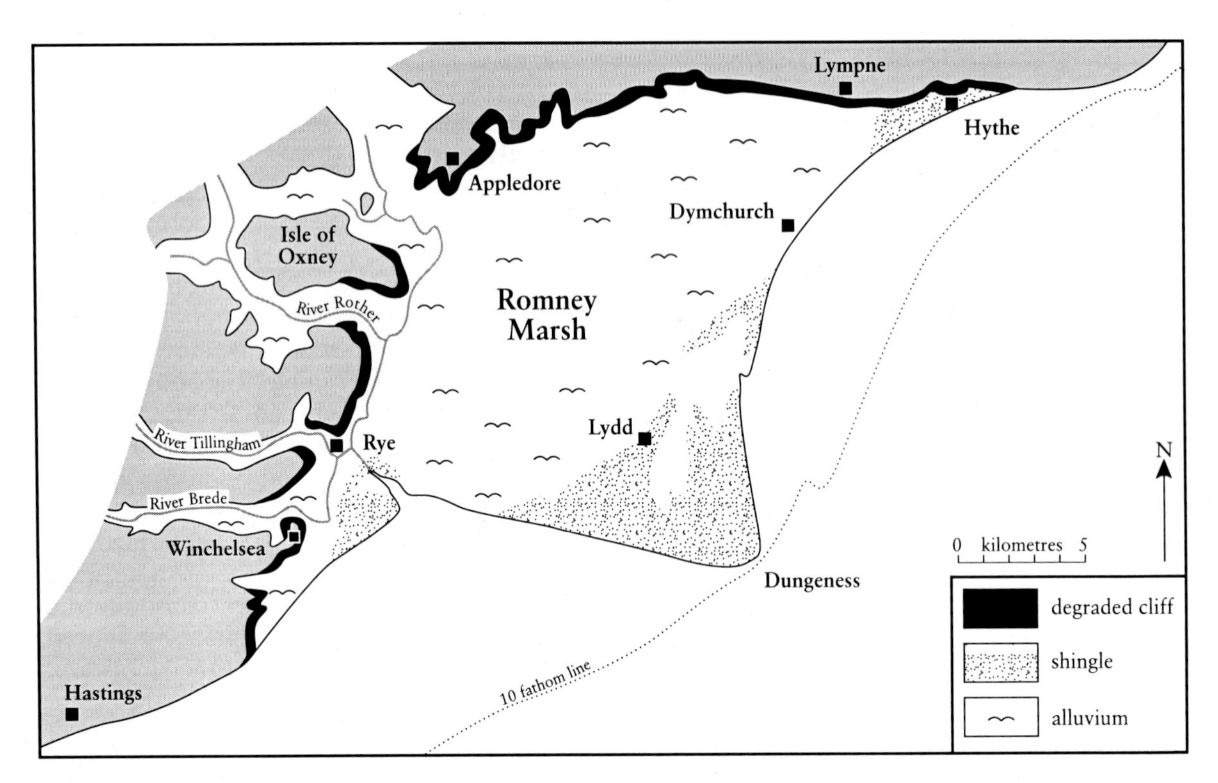

Figure 7.25 The former sea cliff which was abandoned as a result of the formation of Romney Marsh behind a major barrier of sand and shingle. Based on Jones, DKC (1981) and Jones and Lee (1994).

is estimated to extend to about 70 m below the slope foot (Smart *et al.*, 1966). The regional dip is 1°– 2° to the NNE or north-east.

Description

The slope at Lympne is part of an extensive abandoned marine cliff, protected from marine erosion by Romney Marsh and its associated shingle spits. The cliff extends about 8 km from Hythe, 3 km to the east of Lympne, continuously to Appledore, 18 km to the west, and then discontinuously southwards to the coast 5 km to the east of Hastings (Figure 7.25). It is of fairly constant height, at about 90–100 m above OD. There is a tendency for the average steepness and signs of instability to increase from west to east, which may result from geological changes, but suggests that active marine erosion ceased more recently in the eastern parts. For example, The Roughs, close to the eastern end of the abandoned cliff, shows signs of current instability (Brunsden *et al.*, 1996a).

At Lympne the abandoned cliff has a height about 100 m above the marsh at its foot. The present slope profile comprises two main elements: at its head is a 15 m-high scarp in the Hythe Beds, standing at 35°, below which a 550 m-long, slightly irregular slope, predominantly in Weald Clay, extends down to the marsh at about 9°. The 9° slope can be divided into an upper degradation zone and a lower accumulation zone. The degradation zone possesses marked cross-slope undulations, characteristic of landsliding. The accumulation zone is smoother. The fact that the average slope angles of the degradation and accumulation zones are virtually identical indicates that the slope has developed to a condition of long-term stability so that, under present conditions of climate and vegetation, further landsliding is unlikely (Hutchinson *et al.*, 1985). This is in contrast with The Roughs, 1 km to the east along the Hythe Beds escarpment, which is still showing signs of active degradation and has not yet reached the angle of ultimate stability (Brunsden *et al.*, 1996a).

In the accumulation zone a sheet of shallow landslide debris and hillwash obscures the traces of the earlier, more deep-seated landslides which dislocated the fort. Three significant features may be identified (see Figure 7.26), using the combined evidence of the slope morphology and the remains of the fort:

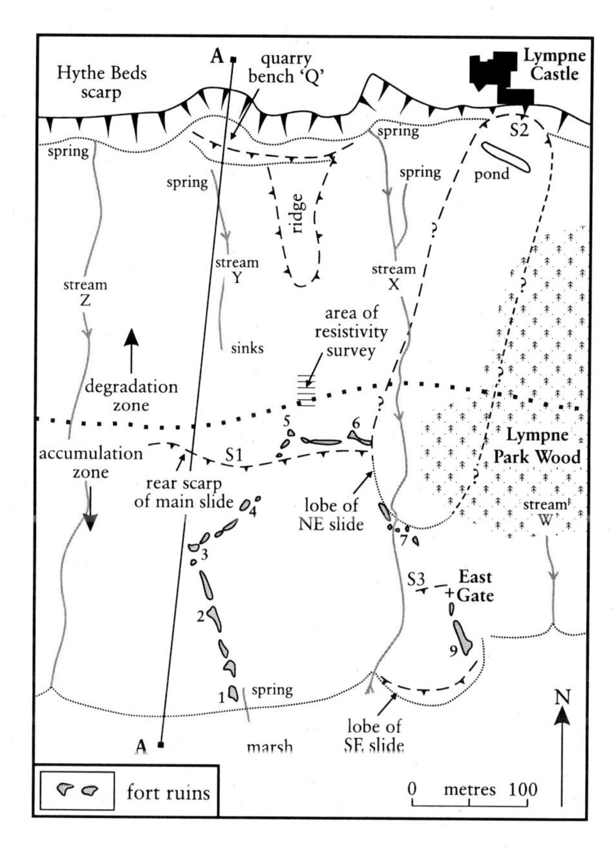

Figure 7.26 Geomorphology of the Stutfall Castle GCR site. The ruins of the fort are stippled. Numbers 1–7 and 9 are Bastion Numbers. After Hutchinson *et al.* (1985).

1. A scarp, S1, which runs beneath the north wall of the fort, cuts across the north-west wall and can be inferred to have transected the north-east wall. This represents the rear scarp of the slide termed the 'main landslide' (Hutchinson *et al.*, 1985), which chiefly disrupted the fort.
2. A lobe that projects well into the marsh between stream X and point K. This is associated with the toe of a 'south-east slide' which caused additional displacement and damage in much of the east wall of the fort and the eastern part of its south wall.
3. A pronounced lobe which encroaches on the north-east corner of the fort, forming the toe of a 'north-east slide'. This appears to cover the eastward continuation of scarp S1 and is therefore inferred to be the later event.

The toe of the slope is fronted by the extensive post-glacial, littoral and alluvial accumulations of Romney Marsh.

Descending the cliff slope at intervals along the slope of 100–150 m, is a succession of streams originating from the spring line near the crest. One of these, 'stream X' (Figure 7.26), runs through the eastern part of the fort ruins.

Interpretation

By analogy with other degraded cliffs (e.g. Hutchinson and Gostelow, 1976) it was expected that the area immediately below the crest of the escarpment would be occupied by the remains of rotational slips, consisting of back-tilted masses of Hythe Beds over Atherfield Clay, and possibly also Weald Clay. Geotechnical investigations (including boreholes) showed, however, that instead the area is occupied by a fairly level bench, Q (Figure 7.26), which a 2.5 m-deep pit showed to consist of 2.2 m of loose loam, with flecks of charcoal, over Hythe Beds debris. Hutchinson *et al.* (1985) concluded that this bench was produced in the course of quarrying of the slipped, and possibly also the in-situ, Hythe Beds in that vicinity for building materials. As Stutfall Castle is constructed predominantly of Hythe Beds material, for which this bench is the nearest accessible source, they consider it likely that the quarrying dates from the construction of the Roman fort.

From the quarry bench to about halfway down the main slope, the ground is mantled by a thin sheet of landslip debris, varying between 1.5 m and 3.5 m in thickness, which shows evidence of part successive-rotational, part translational slipping. This sheet thickens to around 8 m or 9 m from about 30 m downslope of the line of the north wall to just below Bastion 3.

In the main part of the accumulation zone, the landslide debris thickens further to nearly 20 m where it buries a former sea cliff, cut into the in-situ Weald Clay during the last phase of strong marine erosion. The sub-surface investigations define the position and form of this cliff and also show the extent to which the associated beach and alluvial deposits have subsequently been over-ridden by landsliding (Figure 7.27). The crest of the former sea cliff is situated just downslope of the present position of Bastion 3: its foot is a further 35 m downslope. Since erosion ceased there, the landslide debris has advanced approximately 130 m seawards of that point. The buried cliff is fronted by a shore platform, also formed in the Weald Clay. This is sub-planar and declines slightly towards the

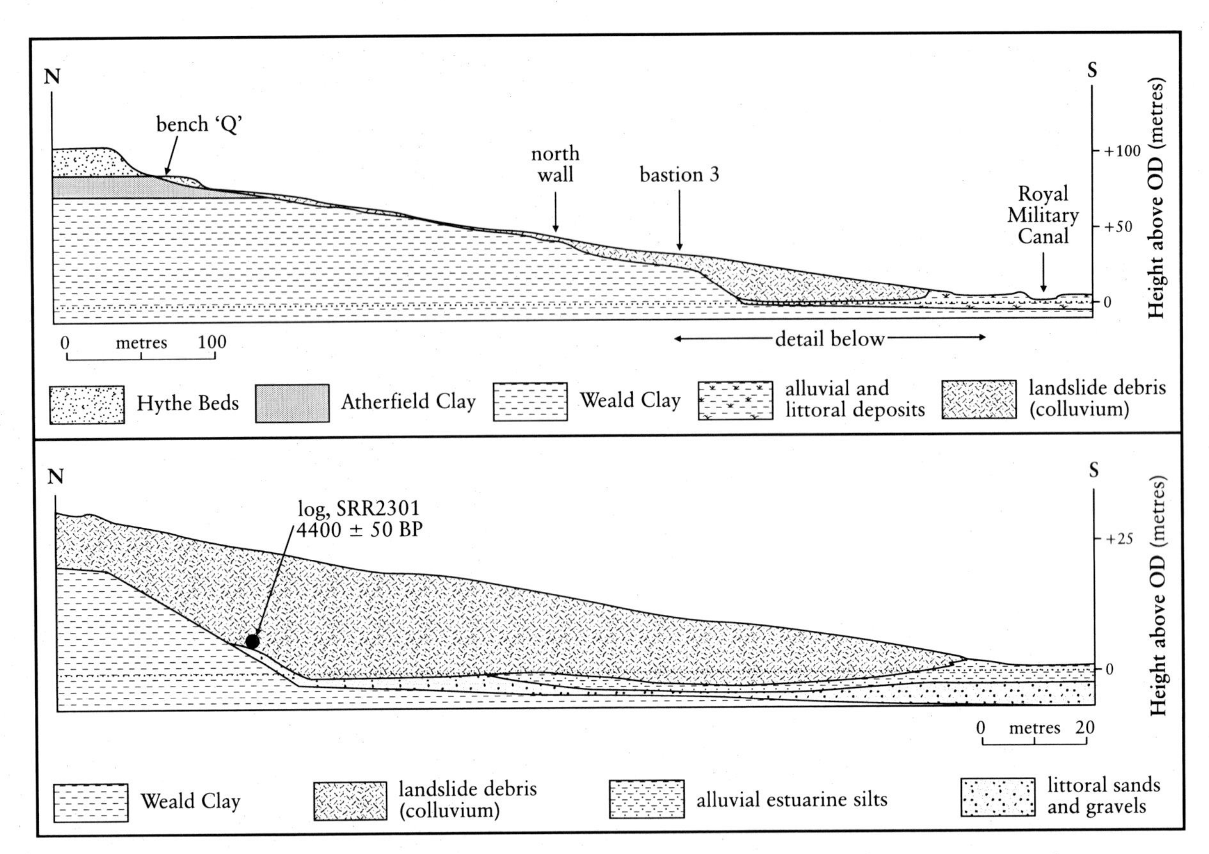

Figure 7.27 Cross-sections through the abandoned cliff at Stutfall Castle. Top – complete section; Bottom – detail of the base of the slope. After Hutchinson *et al.* (1985).

south. At its contact with the buried cliff it has an elevation of 1.0 m below OD. Overlying the shore platform is an irregular spread of alluvial silts, and littoral sands and gravels, often shelly, up to 3.8 m thick (Figure 7.27).

The absence of periglacial solifluction features on the Lympne slope, in contrast to their presence on the geologically analogous escarpment at Sevenoaks, Kent (Skempton and Weeks, 1976) indicates that marine erosion ceased at its foot after the Younger Dryas period of the Late-glacial (10 850–10 050 BP). A radiocarbon date of 4400 ± 50 years BP on a piece of waterlogged wood recovered from the landslip debris just above the foot of the buried cliff (Figure 7.27), suggests that the cliff was abandoned around 4000 BP. Since then the slope has developed chiefly through the burial of the cliff and the re-establishment of the accumulation zone by the supply of debris from the degradation zone upslope. This process must have been relatively far advanced when the Romans chose the site for the fort.

The previous archaeological excavations had not exposed the foundations of the fort. Hutchinson *et al.* (1985) used a tracked excavator to dig eight trenches, each about a metre wide and with one exception they were taken down into the weathered surface of the Weald Clay, 3–4 m below ground level. The northern walls of the fort enabled the investigators to establish the relationship of the Roman foundations to the landslide debris and the in-situ strata. Thirty timber piles were exposed in the floor of one of these trenches, in an area of about 0.9 × 5.0 m. Most of the piles were about 3 m in length. The northernmost of the piles, and those nearest to it, were close to vertical, but those to the south leant successively more and more downslope so that in the southern part of the trench, they occupied sub-horizontal positions. It was evident that the piles, each sharpened to a point at the lower end, had originally been driven and emplaced vertically. The increasing forward tilt of the piles towards the southern end of the trench was produced by the sliding downhill over them of the wall which they had supported (Figure 7.28). The wall slid downslope for a distance of about 5.5 m, tilting forward in the process, and leaving its piles

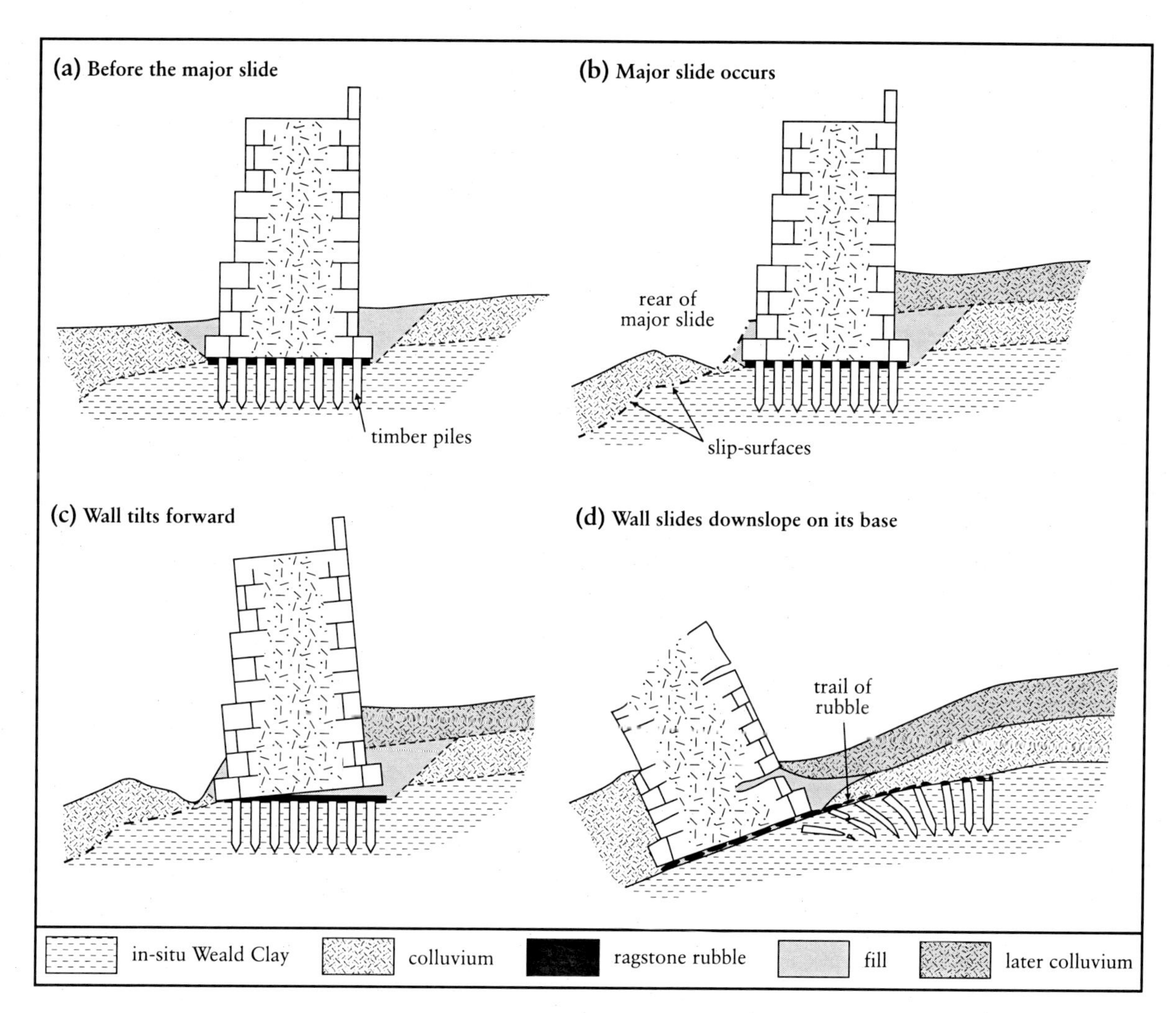

Figure 7.28▲ The inferred mode of failure of the north wall of the fort. After Hutchinson *et al.* (1985).

behind. Similar observations and measurements at other trenches enabled the original positions of the bastions and walls of the fort to be identified, and so the original shape and size of the fort was reconstructed (Figure 7.29).

The date (or dates) of the landslides which disrupted the fort are not easily established, but clearly they post-date the construction of the fort. According to Cunliffe (1980a) two classical texts refer to the fort, enabling it to be identified as *Portus Lemanis*: the Antonine Itinerary, compiled in the early third century, mentions Portus Lemanis, while the *Notitia Dignitarum*

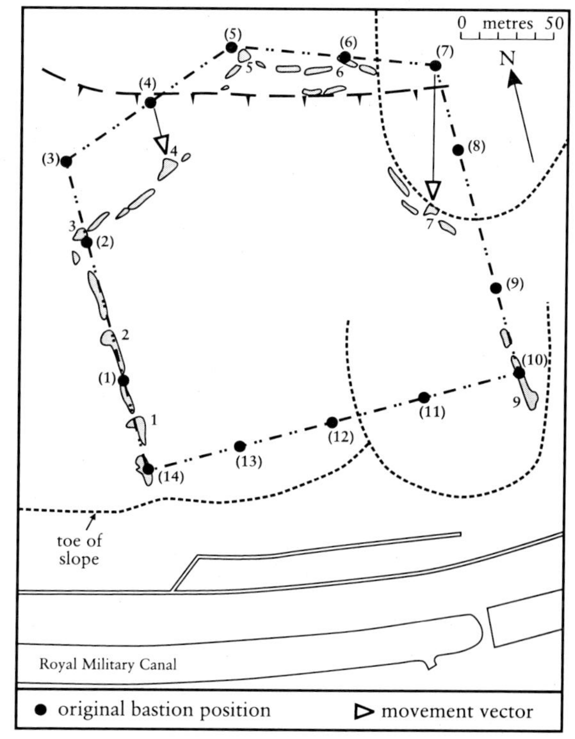

Figure 7.29▸ Reconstructed plan of the Roman Fort (Stutfall Castle) showing the original shape and absolute position of the central parts of the northern walls and the inferred outline of the remainder. After Hutchinson *et al.* (1985).

records that the *numerus Turnacensium*, a military detachment, was stationed at Lemanis in the fourth century. Cunliffe (1980b) has made a preliminary assessment of the state of Romney Marsh in the first millennium AD, arguing that in the early part of the Roman period the rivers Rother, Tillingham and Brede flowed into an extensive estuary that opened to the sea through a narrow outlet just to the east of the site of the fort (see Figure 7.30). The first historical account of the fort seems to be that of John Leland (published 1744), who observed, some time in the period 1535–1543, that the structure was then already ruined. The archaeological evidence concerning the period of the fort's abandonment is supported by what Cunliffe (1977, 1980a) describes as tenuous arguments. Based in part on the assemblage of pottery sherds found at the site (Young, 1980), it suggests that the fort was abandoned earlier than the Romans' other Saxon Shore forts, in *c*. AD 340–350 (Cunliffe, 1980a). Hutchinson *et al*. (1985) remark that it is tempting to ascribe this early abandonment to the commencement of landsliding in the fort. However, further evidence suggests that the slipping of the fort walls took place at a later date.

Hutchinson *et al*.'s (1985) investigations reveal that at Borehole 1 (see Figure 7.27) the marsh surface at 2.8 m above OD is underlain by 4.4 m of silty alluvium, which rests on a Roman beach at 1.6 m below OD. In section, the slide toe is shown to be double. The first phase of sliding, associated with the destruction of the fort, occurred when about 2.6 m of silt had accumulated above this beach. The second phase, involving movements in a shallower surface mantle, over-rode the marsh surface. Taking a

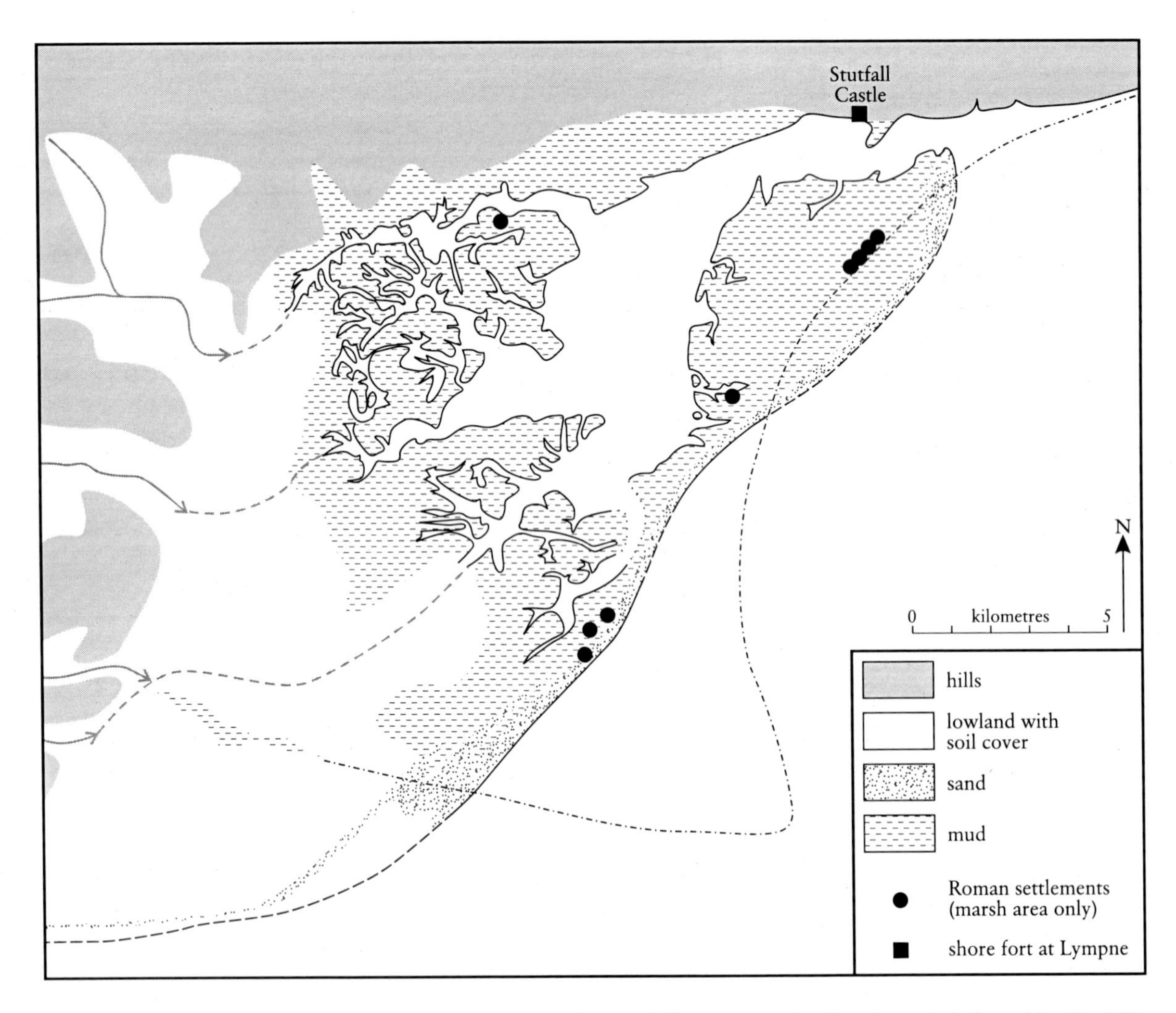

Figure 7.30 The Romney Marsh region *c*. AD 300. After Cunliffe (1980a); details of creeks inferred by Cunliffe from soils data in Green (1968).

nominal date of AD 300 for the commencement of siltation, and Cunliffe's (1980a) date of *c*. AD 700 for the cessation of sedimentation in this locality, Hutchinson *et al*. (1985) estimate the date of the first phase of sliding at around AD 540 (assuming a steady sedimentation rate). The second phase is, evidently, post AD 700. Bearing in mind the assumptions involved, Hutchinson *et al*. (1985) estimate tentatively that the landslides which disrupted the fort occurred in the sixth century AD, and that these were followed, after the end of the seventh century AD, by shallower movements of the surface mantle. Clearly the morphology of the currently visible toe of the landslides is a result only of these latter, second phase movements.

A definite trigger for the main sliding has yet to be identified. It is not clear, despite the intimate association of some of the sliding with the fort, whether its initiation was natural or the result of human activity.

Conclusions

The Stutfall Castle site may be compared with the site at Hadleigh Castle where there is also an abandoned marine cliff and a ruined castle. However, the quality of geological information derived from the archaeological excavation at Lympne is of greater significance in the understanding of the Lympne slope than is the case with Hadleigh Castle's role in the understanding of the Hadleigh slope. In particular, the fact that the development history of Romney Marsh is so well understood, and closely tied down to historically identifiable periods, increases the conservation value of the Lympne site. These two studies have, however, led to the development of an important model of slope degradation through time and a deepened understanding of slope evolution processes following the removal of basal erosion and debris removal.

Chapter 8

Mass-movement sites in London Clay (Eocene) strata

R.G. Cooper

INTRODUCTION

London Clay is an Eocene deposit of stiff blue clay that weathers brown. It is uniform in lithology, although it becomes sandier to the west. It is more argillaceous in its middle beds than in its upper and lower ones (Sherlock, 1960). It is up to 1800 m thick in the eastern part of the London Basin, thinning to less than 900 m westwards.

According to Jones and Lee (1994), of the 465 landslides recorded in Eocene strata, 44% were not classified into any particular type. Of those that were classified, single rotational slides were the most frequently recorded type (43%), and 84% of these were in London Clay. Of the 356 slides in London Clay, 40.7% were of unspecified type. Of those which were classified, 45% were identified as single rotational, 15% as compound and 12% as complex. Two GCR sites are located in Eocene strata, **High Halstow** and **Warden Point** (see Figure 8.1).

These figures are as would be expected from the detailed work that has been carried out on mass movements in the London Clay, particularly in two series of papers, one series by Hutchinson (1967, 1973; summarized in Hutchinson, 1979), and the other series by Bromhead and Dixon (Bromhead, 1979; Bromhead and Dixon, 1984, 1986; Dixon and Bromhead, 1986, 1991). Hutchinson's 1967 paper made a distinction between slopes that are undergoing 'free degradation' and those that are undergoing erosion. Free degradation is the development of a slope from which no material is being removed at the toe, by fluvial, marine, human or any other agent (Figure 8.2). His 1973 paper introduced a further distinction: the ratio of the rate of erosion to the rate of weathering. He distinguishes three situations, each of which gives rise to a characteristic type or combination of types of mass movement:

(1) the rate of erosion is broadly in balance with the rate of weathering;
(2) the rate of erosion exceeds the rate of weathering;
(3) the rate of erosion is zero, for example on an abandoned cliff.

Bromhead (1979) has strongly challenged this view. Disparaging Hutchinson's tripartite categorization as 'convenient', he points to the existence of an accumulating body of evidence that suggests that the correlation of slide processes with erosion rate is largely fortuitous. Any such subdivision, he continues, should rather distinguish between 'active' and 'inactive' types, with subcategories within each type based on other factors. Characterizing most of Hutchinson's evidence as observational rather than quantitative, he points out that it is difficult to come to firm conclusions using such evidence. Type of mass movement is most strongly controlled by the properties of the soil forming the head of the slope and by the groundwater hydrology in that material, the rate of erosion at the toe being a secondary

Figure 8.1 Areas of London Clay (Eocene) strata (shaded) and the locations of the GCR sites described in the present chapter.

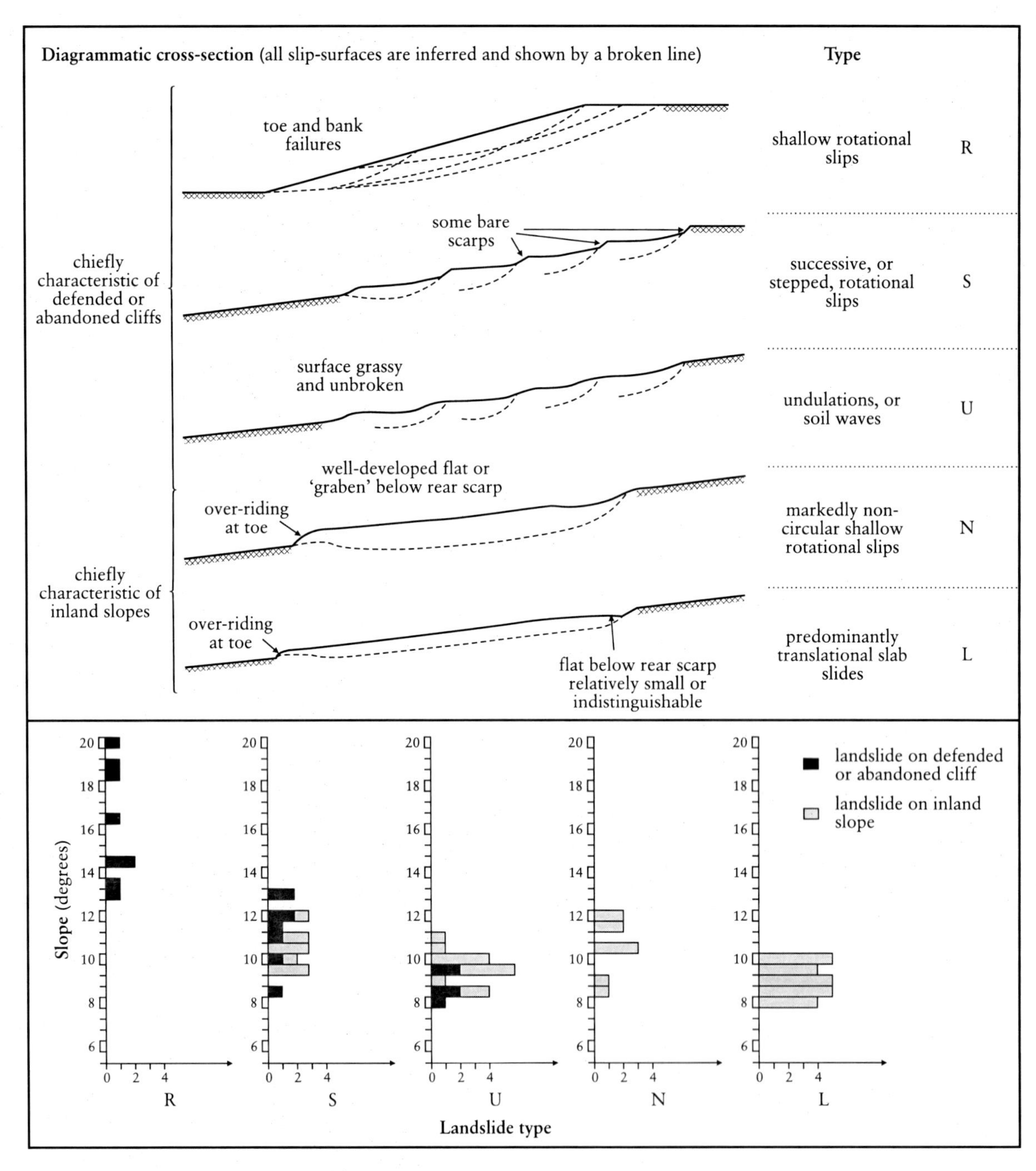

Figure 8.2 Behaviour of London Clay cliffs following cessation of erosion at the toe. Cross-sections and slope-angle histograms illustrate the types of landslide found on slopes in London Clay as described by Hutchinson (1979).

consideration. Hutchinson (1968b, 1973), he points out, ascribes control to the intensity of marine erosion. 'Unfortunately, [Hutchinson's] work is intended to apply only to cliffs of similar height, composed wholly of London Clay and subject to largely similar climatic conditions. This naturally leaves toe erosion intensity as the only remaining variable' (Bromhead, 1979).

Geotechnical characteristics of the London Clay deposit in the London Basin have been investigated by Burnett and Fookes (1974).

HIGH HALSTOW, KENT (TL 777 756)

Introduction

The area of north-facing hillslope immediately north of Cooling Road at High Halstow in north Kent (Figures 8.3 and 8.4) is one of very few remaining areas that demonstrate low-angle mass-movement phenomena on inland slopes of London Clay. Many London Clay slopes elsewhere have been stabilized by drainage works in connection with agricultural improvements and forestry, whereas this site remains in a relatively natural condition.

Description

High Halstow demonstrates failure on a slope of about 8°. The physical expression of this is a series of ridges and small scarps running across the slope.

Mass movement on the slope at High Halstow occurs at an extremely slow and irregular rate, responding only to exceptionally wet weather conditions. As landsliding proceeds, the slope naturally degrades to a gentler angle. The slope is undergoing 'free degradation', that is, degra-

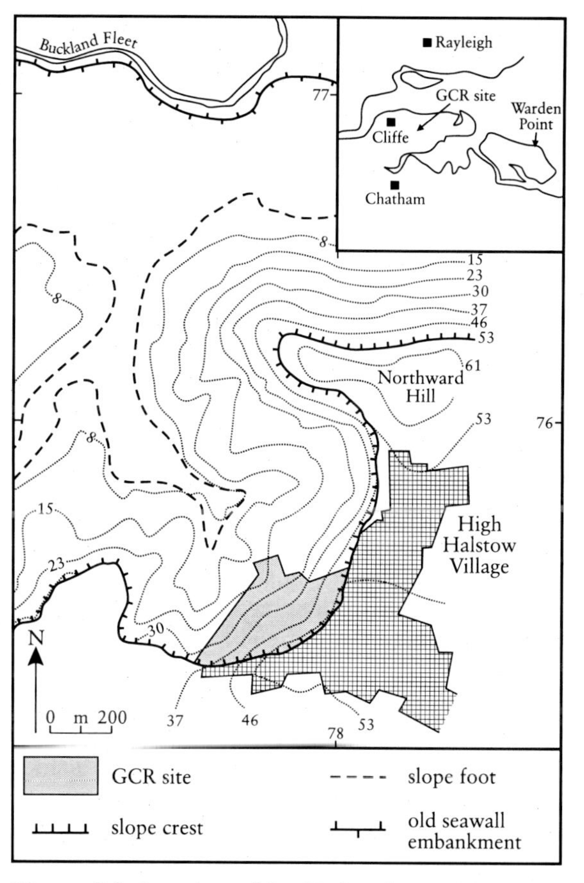

Figure 8.3 Location of the High Halstow GCR site.

Figure 8.4 High Halstow in 1938 showing the failure of the lower slopes of the landslide. (Photo: reproduced courtesy of British Geological Survey. IPR/88-06CGC.)

dation solely under the influence of weathering and climate, with no removal of material at the toe, a process examined in detail at Hadleigh Castle, Essex, by Hutchinson and Gostelow (1976). In time it will reach the angle of ultimate stability, that is, the angle at which future natural stability is assured (Skempton and DeLory, 1957). Hutchinson (1967, 1973) has shown through detailed study of many slopes that the angle of ultimate stability on inland London Clay slopes is about 8°, and that slipping does not take place on slopes of lower angle than this. The co-incidence of the angle of the slope at High Halstow with this limiting angle of 8° implies that the slope at the site is only just unstable, and that further degradation of the slope is unlikely. Hutchinson (1967) has described the slip as consisting of successive or stepped rotational slips, having some small bare scarps less than 1 m in height.

Interpretation

The scarp features are characteristic of hillsides that appear to represent a late stage in the free degradation of slopes once steepened by marine or fluvial action. The behaviour of London Clay cliffs following cessation of erosion at their toe has been described by Hutchinson (1967) and summarized by Skempton and Hutchinson (1969) and Hutchinson (1973) (Figure 8.2). The cessation of erosion may occur naturally through abandonment of a cliff, or be effected artificially through the building of defences against toe erosion. Immediately following abandonment, the cliff tends to fail by shallow rotational slips. No case is known of a base failure initiated on an abandoned cliff of London Clay. Initially, the shallow rotational slips may encroach upon the cliff-top, but with time their size tends to diminish. As no debris is removed from the foot of the cliff, all of the slipped material comes to rest there. A highly characteristic feature of many abandoned cliffs is thus a bi-linear profile composed of a steeper upper slope on which landsliding is initiated (the degradation zone) and a flatter lower slope (the accumulation zone) of colluvium.

Under the action of shallow rotational slides the slope of the degradation zone is reduced, with time, to about 13°. At about this inclination, the slips occupying the degradation zone and at slopes below 13° are found to be of successive rotational type. These may be either regular (stepped) as at High Halstow, or irregular (mosaic) in plan. The retrogressive succession is probably initiated by a shallow rotational slide at the lower end of the degradation zone. Successive rotational slipping then proceeds until the slope of the degradation zone has been brought down to its angle of ultimate stability against landsliding. The slips then become quiescent and are gradually converted first into undulations and finally into a smooth slope by hillwash and soil creep.

Since the site is remote from streams and rivers it is unlikely that it has been subjected to significant post-glacial erosion. It is probable, therefore, that it is mantled by solifluction deposits and hillwash, the thickness and properties of which largely determine the present slope form and behaviour. As both solifluction-mantled London Clay slopes and solifluction-free London Clay coastal cliffs appear to have the same 8° limiting angle, this points to the seat of failure lying beneath the mantle and in the London Clay. The 8° limiting inclination would suggest, using the assumption of Skempton (1964), that the residual angle of shearing resistance ϕ_r', lies between 11° and 15°.

Conclusions

The slope at High Halstow represents a late stage in the process of free degradation, and the slips, although still occasionally active, are liable to become quiescent. It is the lowest-angled slope on which failure is known to be current in Great Britain.

WARDEN POINT, ISLE OF SHEPPEY, KENT (TR 018 726)

Introduction

The cliffs in London Clay at Warden Point exhibit a series of deep-seated rotational slips (Figure 8.5) which are bench-shaped in plan, and generally extend along the coast for distances of between 3.1 and 7.1 times the cliff height, and average in width 5.0 times the cliff height (Figure 8.6).

Warden Point is the most exposed north-eastern extremity of the cliffs of the Isle of

Figure 8.5 The characteristic morphology of the Warden Point GCR site. (Photo: R.G. Cooper.)

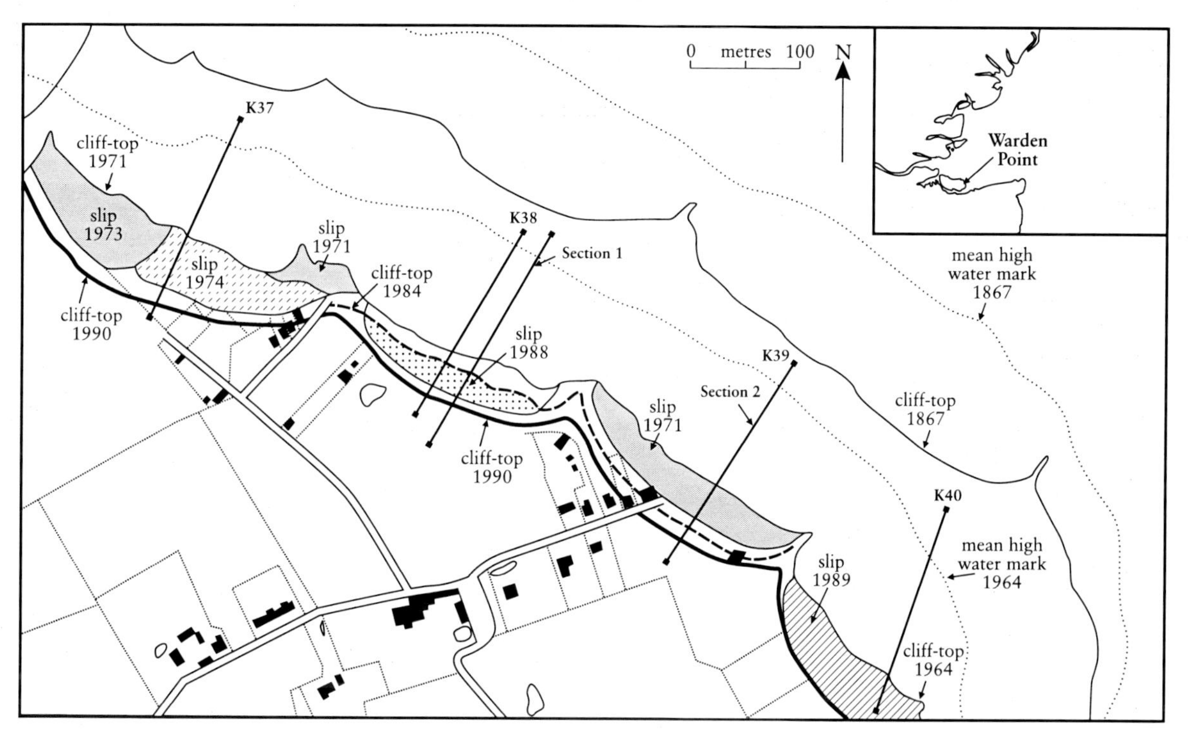

Figure 8.6 Landslides and location of cross-sections at Warden Point. Based on Hutchinson (1965) and Dixon and Bromhead (1991).

Sheppey, Kent, facing the Thames estuary (see Figure 8.6). It is formed in cliffs of London Clay (Eocene) (Figure 8.7) up to 40 m high, which stretch from Warden Point for 8 km to the east). According to Steers (1964), marine erosion is generally strong along the length of the cliffs giving a maximum recession rate of nearly 3 $m a^{-1}$ at their more exposed eastern end. Over the period 1865–1964 rates vary from 1.5 $m a^{-1}$ to 2.15 $m a^{-1}$ (Hutchinson, 1973).

Description

Throughout the last 100 years, the average length of time between consecutive first-time slides occurring at a particular length of the cliff has been 30–40 years. Hutchinson (1965), in conducting a survey of all the landslides on the Kent coast, labelled the landslides around Warden Point as K37 to K40; the positions of these are shown in Figure 8.6. Hutchinson intended these labels to refer to the individual slide masses, but, following Dixon and Bromhead (1991), they are used here to denote the sections (lengths) of cliff in which the slides occur.

(a) Section K40

A first-time slide took place at K40 during the night of Friday 13 January 1989. It involved a length of cliff-top about 150 m long, with a maximum land loss of 25–30 m. The landslide took the form of a large back-tilted block of relatively intact London Clay, which moved about 12 m down the cliff during the initial failure. The original cliff-top ground surface, recognizable by the grassed and partly wooded back-tilted area, remained predominantly undisturbed by internal shearing. However, immediately seaward of the original cliff edge a secondary scarp indicated some internal disturbance. The toe of the slide experienced 3 m of heave as a result of rotation of the slide mass. This rotation was considerable, and the absence of excessive internal distortion of the slide mass indicates that the slip-surface was close to being circular in shape (Dixon and Bromhead, 1991).

The previous slide at K40, which occurred in 1950, also involved a back-tilted slide mass with little associated disturbance (Hutchinson, 1965). In front of this length of cliff the foreshore comprises the planed-off remnants of previous

Figure 8.7 The cliffs at Warden Point looking westwards. Separate mudslide tongues are clearly shown. (Photo: J. Larwood, English Nature/Natural England.)

slides. These can be recognized by linear outcrops of septarian nodules (formed by originally sub-horizontal nodule beds having been rotated during the slide movements), and by the end shears, which clearly mark the boundary between the landslipped material and the in-situ clay. The in-situ clay marks the position of the former abutments between laterally adjacent slides. There are no lines of nodules cropping out in these areas of in-situ material, indicating that the slides at this location have been repeatedly deep-seated (Dixon and Bromhead, 1991).

(b) Section K39

The most recent failure in the K39 section of the cliff took place in November 1971, and was described by Hutchinson (1971). The initial slide movements involved a 200 m length of cliff-top with a maximum width of 30 m at its centre. From the evidence on the foreshore the slides at this location are also deep-seated.

Post-failure observations at K39 indicate that the shape of the slip-surface varies across the length of the slide (Dixon and Bromhead, 1991). Along the eastern and central section the landslide mass moved 18 m down the slope, back-tilted by 16°, and remained relatively intact. This suggests that at least the rear section of the slip-surface was close to being circular in shape. In the western part of the slide mass, which had an initial downward displacement of 15 m, not only was there a 7° angle of back-tilt, but also the landslide block was severely distorted by internal shearing; this included the formation of a graben-type feature with reverse shears and a zone of tension cracks which were up to 10 m deep. Thus it seems that in the western section of the cliff, failure took place involving a highly non-circular slip-surface (Dixon and Bromhead, 1991).

Figure 8.8a shows a cross-section through the central part of the slide mass which was surveyed by Gostelow (1974) shortly after the first-time failure. Also shown is the slope profile in 1985 and the position of the slip-surface as indicated by the above measurements and observations. The measured slip-surface is consistent with the observed mode of failure (Dixon and Bromhead, 1991).

(c) Section K38

A rotational slide occurred at K38 in 1945. A local resident who remembered the slide happening reported that a strip of cliff-top about 30 m wide was involved in the initial failure (Hutchinson, 1965). In addition, the slipped mass is reported to have exhibited considerable back-tilt. In front of this length of cliff the foreshore is formed of in-situ London Clay, and therefore it can be assumed that the slip-surfaces do not extend below about 0 m OD.

The section of cliff at K38 commenced falling around May 1988 while instrumentation was in place; deformations continued over a period of about 6 weeks. A detailed preliminary study was carried out by Koor (1989) who surveyed and mapped the slide area: it was clear that the failure was not a rotational slide. The following description and assessment of the failure mode are taken from his work.

Failure was initiated at the north-west end of the section and spread over a six-week period south-eastwards along the cliff. There are two distinct areas of landslipped material that are separated by a fault in the London Clay. The fault has a throw of about 1 m and should therefore be at residual shear-strength. The existence of the fault clearly controlled the two-phase nature of the movements. A cross-section surveyed through the eastern part of the slipped mass is shown in Figure 8.8b.

The following features were associated with the failure:

1. a zone of tension cracks extending up to 30 m landward of the main scarp, with associated settlement of the cliff crest;
2. the main slide mass was formed from forward rotating linear blocks of London Clay which had slickenslided counterscarps;
3. the toe of the slope was formed of scree-like material.

Koor (1989) proposed failure by flexural toppling as an explanation of the above features. This type of failure could occur as a result of vertical columns of London Clay being formed by stress-relief-induced vertical discontinuities. If these columns were not adequately supported by the toe of the slope, they would tend to topple seawards thus producing the slope profile observed. This failure mode would account for the retrogressive nature of the slope movements. Also shown in Figure 8.8b is the pre-failure slope profile. By comparing these two sections it is evident that substantial translational deformations took place.

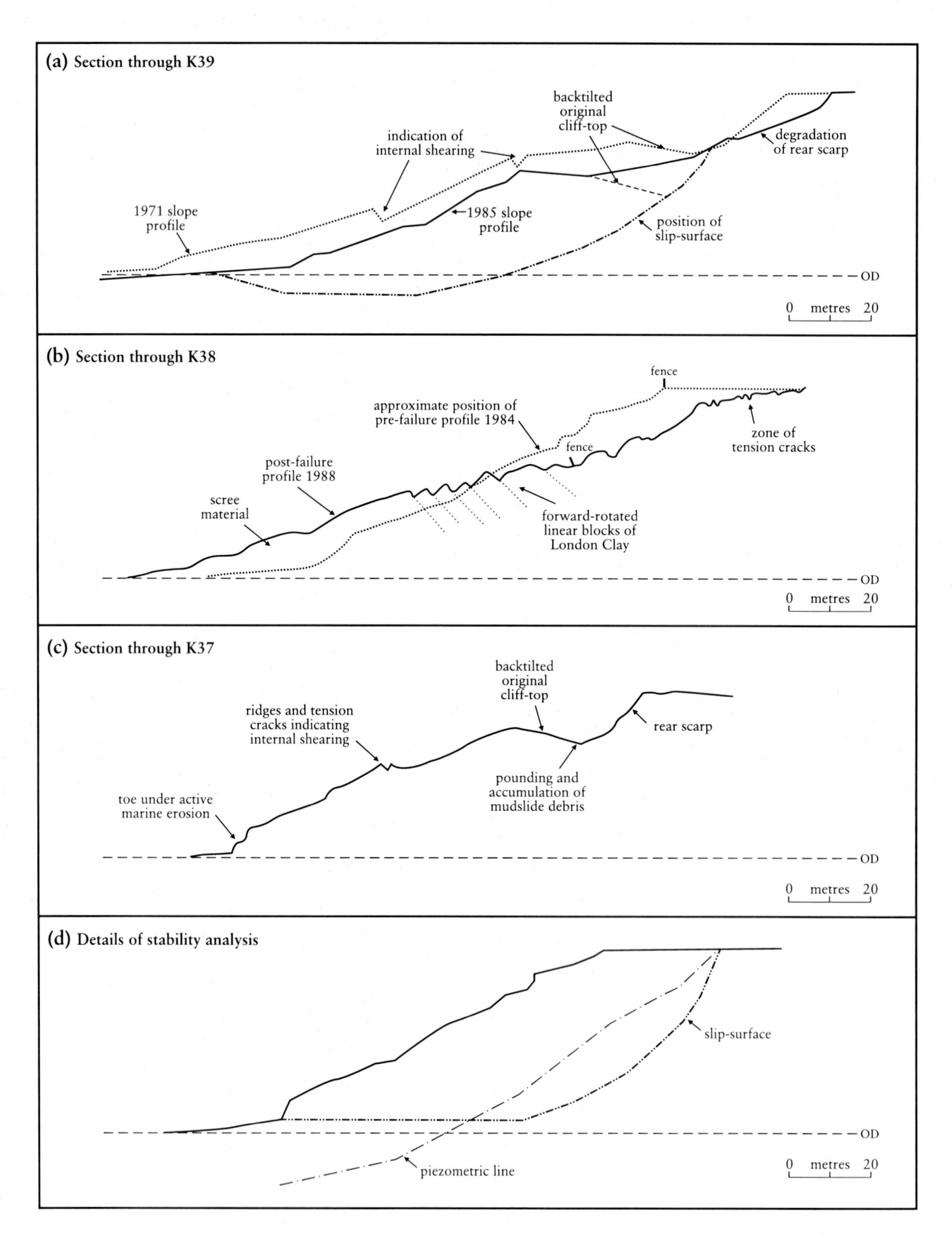

Figure 8.8 Sections K37–9 (a–c) at Warden Point (after Lee, 1976; Koor, 1989; Dixon and Bromhead, 1991), and (d) K38 'idealized' profile and slip-surface (after Bromhead and Dixon, 1984). See Figure 8.6 for locations of sections.

(d) Section K37

Two major rotational slides occurred at K37 during the 1970s. In 1973 a slip took place in the western half of the cliff and in 1974 the remaining length failed. Although both landslide blocks exhibited a considerable amount of back-tilt, this was accompanied by major internal shearing, as demonstrated by the presence of ridges on the surface of the slide mass and a large number of deep tension cracks. From this evidence it is clear that the slip-surface was non-circular. Inspection of the foreshore shows that previous slides at this location were not deep-seated. Figure 8.8c shows a cross-section through the 1974 landslide mass surveyed in 1975 by Lee (1976).

At the eastern end of K37 the base of the slip-surface outcrops in the cliff which forms the toe of the slope. The slip-surface follows the dip of the bedding, which is indicated by layers of septarian nodules exposed in the cliff-face, and is formed of approximately 50 mm of highly sheared re-moulded clay with the primary shear surface positioned at the base of the failure zone.

Interpretation

(a) Hutchinson

Hutchinson (1968b, 1973) described a form of cyclic mechanism of mass movement in London Clay cliffs. He illustrated this mechanism using examples from a range of sites as well as Warden Point, including The Lees and Beltinge at Herne Bay, and compared them with the mechanisms that operate on inland slopes. The stages are as follows (Figure 8.9):

(1) A deep-seated landslide is imminent. The virtually stable upper cliff is blanketed by thin deposits of exhausted mudslides, commonly grassed-over at inclinations of about 15°–20°. The lower cliff is bare, in-situ clay under active marine attack.

(2) The landslide has just taken place. It is characteristically a deep-seated rotational slip involving base failure. In failing, the slipped mass has moved downwards and seawards, rotating backwards in the process, until a position of equilibrium is again reached, commonly when the former cliff edge is roughly halfway down the cliff. Pools of water frequently collect in the hollow thus formed.

(3) This stage comprises two main processes: the commencement of marine erosion at the toe of the slipped mass and the rapid breakdown of the steep rear scarp of the slip, chiefly by soil-fall and shallow slides. The debris from these collects in the hollow at the back of the slipped mass. This loading, combined with the unloading resulting from the toe erosion, brings about a slight further rotation and sinking of the landslip.

(4) Erosion at the landslip toe has reached the line of the cliff-foot just before the slip occurred. The continuing degradation of the rear scarp is now affected predominantly by mudslides, chiefly during the wet season. The deposits of these gradually fill the hollow behind the slipped mass and begin to spill into the sea to form secondary mudslides around each extremity of the slipped mass and further slight rotation and sinking of the mass results.

(5) The gradual removal of the toe of the slipped mass, and the degradation of the rear scarp, continues. The lower cliff is now completely buried under mudslide deposits. The mudslides on the upper cliff are approaching a state of exhaustion, having almost destroyed the rear scarp from which their debris supply largely derived, and degraded to an inclination too low for an effective rate of debris transport to be maintained. The two secondary mudflows, however, being continuously stimulated by marine erosion, increase in power and enlarge their channels at the expense of the slipped mass and the sides of the landslip cavity. The sea continues to erode the face of the slipped mass and may cause it to sag forward and so reduce its degree of back-tilt. Vertical erosion of the slipped mass forming the foreshore occurs.

In the following stage, the sea, having removed the last remains of the slipped mass remains above sea-level, once again attacks the in-situ clay. The cycle has been completed and the cliff, displaced somewhat inland, again stands at stage 1 on the point of a deep-seated failure.

The time taken to complete this cycle appears to vary primarily with the intensity of the erosion and to a lesser extent with cliff height. At Warden Point the cycle length is estimated to

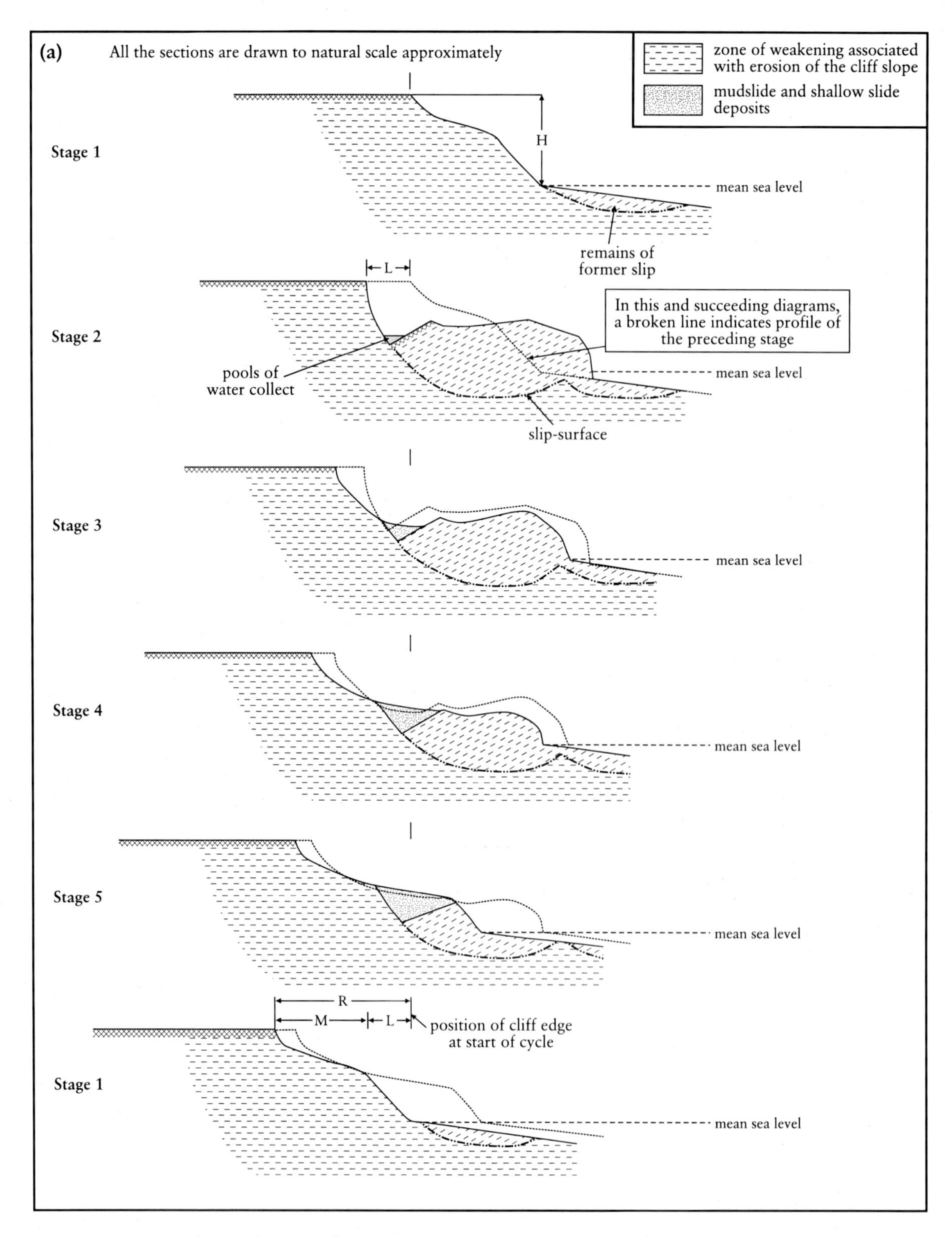

Figure 8.9 (a) Cyclic mechanism of mass movements in London Clay Cliffs, North Kent. After Hutchinson (1967, 1970, 1973). *Continued opposite.*

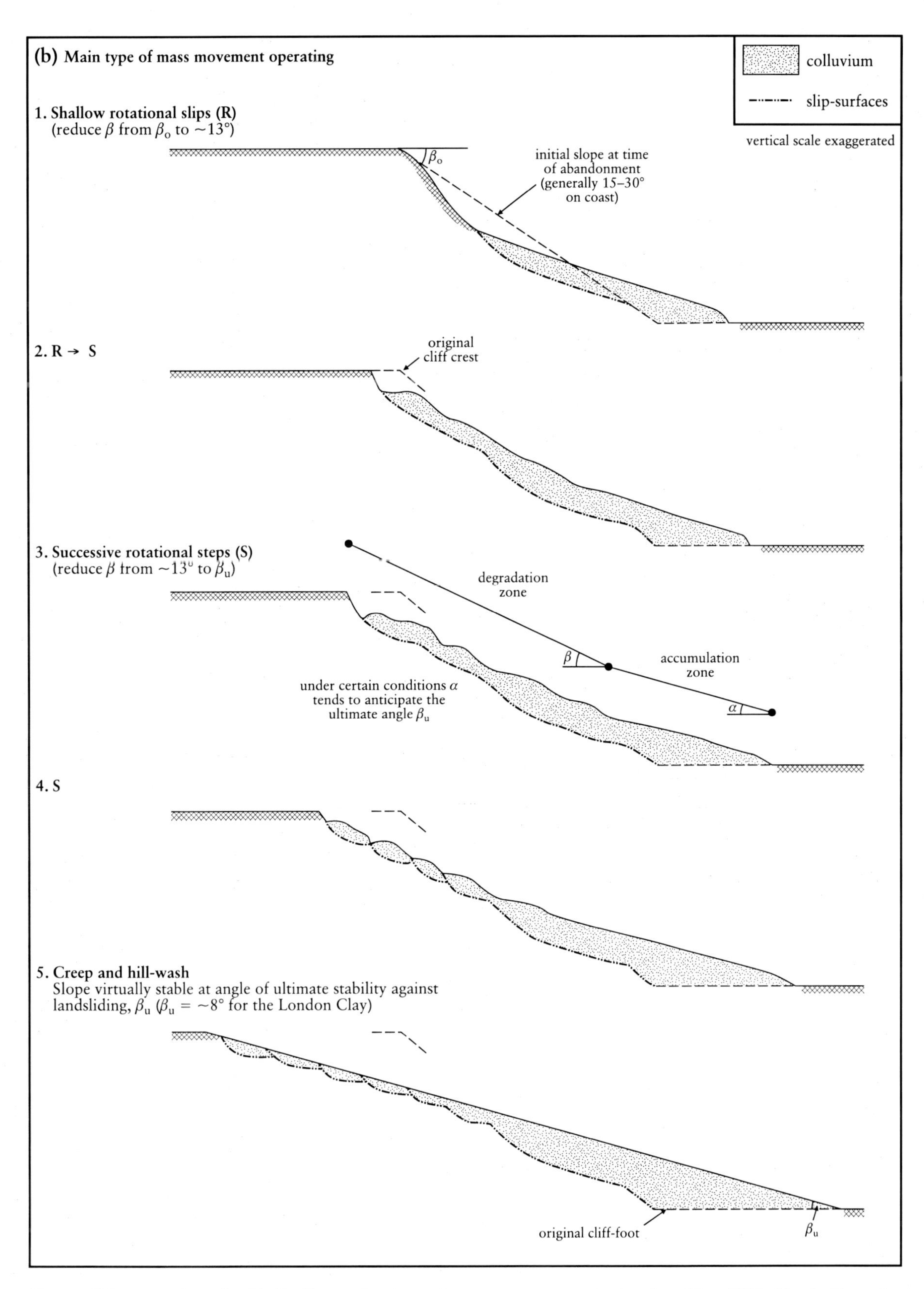

Figure 8.9 – *continued.* (b) Cyclic mechanism of mass movements in London Clay Cliffs, North Kent; for comparison with with Figure 8.9a. After Hutchinson (1967, 1970, 1973).

have been 30–40 years (Hutchinson, 1973). A cliff undergoing this type of cyclic degradation exhibits, not parallel retreat, but a mode of retreat which alternates in inclination between the steep-angled stage 1 and the shallow-angled stage 4. This results from the alternation of the predominantly erosional phase of cliff steepening with the succeeding phase dominated by degradation, initially by deep-seated landslip and subsequently by shallow slides and mudslides.

In plan view (see Figure 8.6) it is clear that the slips take place along the same axis and that the headlands between them maintain their positions. It is likely that this is due to the low porewater-pressure and because the debris is removed at the foot. This is in contrast to the landslides at Ventnor, Isle of Wight, where lack of removal at the foot of the slope results in offsetting of headlands in successive cycles of sliding.

(b) Bromhead and Dixon on depressed porewater-pressure and its implications

Bromhead and Dixon have adopted two main lines of investigation of the cliffs at Warden Point. Firstly, they have investigated porewater-pressure distribution in detail (Bromhead and Dixon, 1984; Dixon and Bromhead, 1986). This was done not so much to investigate the site's characteristics, as to use the site as a testbed for the idea that a condition often observed in unloaded artificial slopes (most commonly at excavations) might also exist in certain natural slopes. The condition is that stress-relief on excavation leads to reduced porewater-pressures around the excavation (Figure 8.10; Vaughan and Walbancke, 1973; Eigenbrod, 1975). Skempton (1977) attributed delayed failure of cutting slopes in brown London Clay to this effect. According to Dixon and Bromhead (1986), it can be expected that the effect will be modified in the case of a natural slope, by a mantle of shallow landslide deposits (e.g. mudslides; Hutchinson, 1967), with a localized, perched, and possibly artesian porewater-pressure regime (Hutchinson and Bhandari, 1971). If, in addition, the slope has been subject to deep-seated slide activity, it may be grossly modified by porewater-pressure distributions 'carried down' the slope, but otherwise unaltered (Dixon and Bromhead, 1986).

To measure the magnitude and distribution of a depressed porewater-pressure regime in a slope, the following criteria must be met:

(1) a large slope is needed to enable an adequate number of measurements to be made;
(2) the slope must be subject to rapid erosion so that the porewater-pressure changes are recent and not partly equilibrated (this is important owing to uncertainty about field equilibration rates (Chandler, 1984));
(3) the magnitude of porewater-pressure depression should be great so that inevitable inaccuracies in measurements are immaterial.

These criteria are met by the 46 m-high London Clay cliffs at Warden Point. Two sections were instrumented by Bromhead and Dixon (1984, 1986; Dixon and Bromhead, 1986) to allow differentiation between porewater-pressure changes resulting from stress-relief, and those 'carried down' the cliff when one of the characteristic large and deep-seated rotational slides takes place.

Dixon and Bromhead (1986; Bromhead and Dixon, 1984) installed 56 piezometers along two cross-sections (Figure 8.10a,b), which extend back behind the cliff-top, and onto the foreshore. The piezometric levels were read at about 10-day intervals. The two sections were chosen with the intention that Section 1 (Figure 8.10a) would yield important data on the initial failure of the cliff, providing peak field strength values from back-analysis. Section 2 (Figure 8.10b) would yield data for an existing landslide and provide a field residual strength. Each piezometer was installed in a PVC plastic tube located in a vertically augered hole (maximum available depth 45 m). Blockages in the tubing due to shearing or bending were used to identify the position of zones of movement. Most of the piezometers were far too deep for seasonal effects to be significant, and in fact no seasonal effects were recorded.

Dixon and Bromhead (1986) did not find it possible to explain the measured porewater-pressure distribution in terms of a steady seepage regime, which is the usual replenishment mechanism. At Warden Point the cliffs are composed of the uppermost part of the London Clay, which is 132 m thick at this location. The London Clay extends to a depth of 83 m below beach level at Warden Point, with a small amount of underdrainage taking place into the underlying sands of the Oldhaven Beds. This base boundary is not critical, the slope being isolated from the underdrainage by decreasing

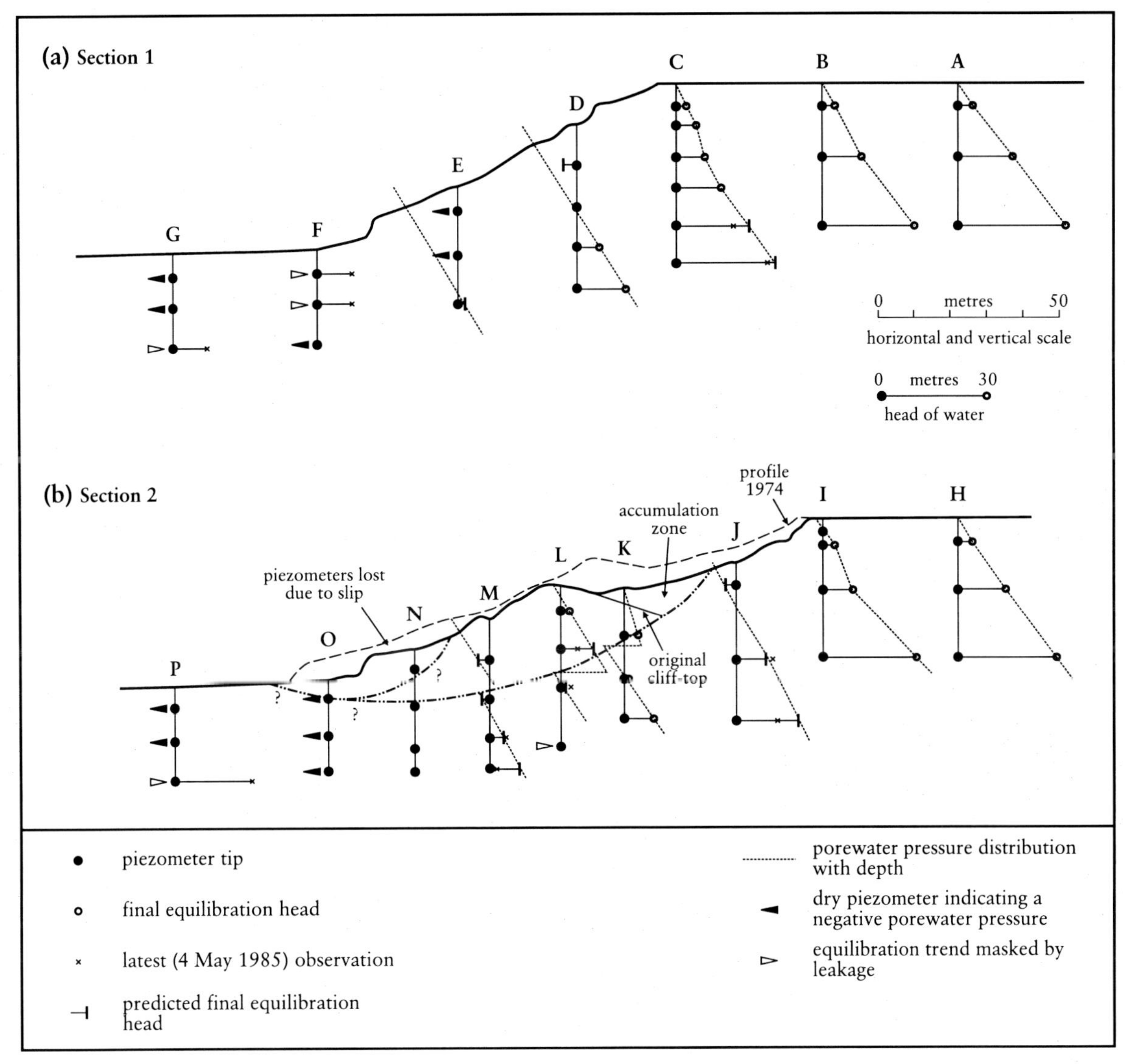

Figure 8.10 Porewater-pressure distributions at two sections at Warden Point: (a) pre-failure; and (b) post-failure. After Bromhead and Dixon (1984).

permeability with depth (Bromhead and Vaughan, 1980). The measured porewater pressures are consistently less than those predicted by seepage analyses.

Taking into consideration the depressed porewater-pressures found in artificial cutting slopes as discussed above, it seems likely that, as in such slopes, the low porewater-pressures at Warden Point are due to stress-relief effects. The difference between the measured and predicted porewater-pressures decreases as the distance inland from the cliff edge increases. It appears that the zone of depressed porewater-pressure is carried inland with the cliff as the coastline retreats. It is local to the slope, but may extend a comparatively large distance inland. The predicted equilibrium levels shown in Figure 8.10 were derived by Bromhead and Dixon (1984) from water-level/time plots using Gibson's (1963) theory and a curve-fitting technique. Equilibration rates in the London Clay are slow, as evidenced by the retention of suctions in the foreshore in areas first exposed more than 50 years ago. Permeability data from the initial response curves of the piezometers indicates a variation in permeability from 3.5×10^{-10} m s^{-1} in the weathered upper surface of the clay to 4×10^{-11} m s^{-1} at 36 m depth. This order of magnitude of permeability change has little effect on the porewater-pressure distribution inland of the slopes at Warden Point. The slow equilibration rate prevents loss of depressed porewater-pressure in the slope since soil is removed faster than moisture penetration takes place.

The cliff-line is retreating at about 1.6 m a^{-1}, taking place as shallow mudslides (Hutchinson, 1970; Hutchinson and Bhandari, 1971) or deep-seated rotational landslides (Hutchinson, 1965, 1973; Bromhead, 1979). The rate of removal is far too rapid for a steady seepage regime to develop in these slopes; instead, the original slightly underdrained, and hence downwards, seepage pattern is modified by the stress-relief-induced porewater effects. Where the slopes recede under the influence of shallow mudslides there is a perched water table in the mudslide debris (Hutchinson and Bhandari, 1971), often with porewater pressures significantly in excess of hydrostatic pressure due to undrained loading. Beneath these, in the in-situ clay, porewater pressures may be lowered substantially, possibly becoming negative in places. Such depressed porewater-pressures exist only in the vicinity of the cliff-face, as they are carried back inland with retreat of the cliff-face.

Porewater pressure/depth relationships are comparable for the two sections with the exception of the two cliff-edge installations. These are similar down to a depth of about 18 m, after which the porewater pressures are considerably greater for Section 2. The most probable explanation for this is that the existence of the large landslide block 20 m down the slope reduces the lateral stress-relief below 20 m and leaves higher porewater-pressures.

(c) Bromhead and Dixon on slope stability

First-time slides that occur along the north Sheppey coast are of particular interest because, unlike most failures in London Clay, they are located substantially in the unweathered (in-situ) material. This material therefore has a controlling influence on the mechanism of failure, and hence on the shear strength that is mobilized during the first-time sliding.

As shown, the failure mechanisms of the slopes at Warden Point are controlled by the magnitude and distribution of the depressed porewater-pressures. Slide behaviour in these cliffs is dominated by the effects of the stress-relief-induced suctions. If artificially stabilized, the long-term tendency of the porewater pressures to rise will result in destabilization. Using the above measurements Bromhead and Dixon (1986; Dixon and Bromhead, 1991) carried out an investigation based on back-analysis of the slide at K39, to provide information on the field residual shear-strength of London Clay that is mobilized during re-activated slide movements. Unfortunately the data obtained was still insufficient to enable a detailed back-analysis of one of the first-time failures to be carried out and hence to gain accurate information on the distribution of mobilized shear-strength. However, an analysis using a typical 'idealized' pre-failure slope profile, with an idealized porewater-pressure distribution and idealized slip-surface was attempted. The slope profile was based on the 1984 survey of K38, which was considered to be representative of the angle of the cliffs at Warden Point prior to failure. The porewater pressures were those measured on Section 1 at K38, as these related to the chosen slope profile and hence the magnitude of stress-relief experienced during the formation of the slope. A slip-surface with a curved rear section and planar basal portion (Figure 8.8d) was considered relevant as discussed above. Figure 8.11 shows details of the problem analysed.

Using the Morgenstern and Price (1965) slope stability method the initial analysis assumed a uniform distribution of shear strength around the slip-surface. For failure to occur an average shear-strength given by the parameters $c' = 0$ k Nm^{-2} (assumed), $\phi' = 16°$ is required. However, there is no field evidence to suggest that it is possible for the shear strength of London Clay to drop to this value in a first-time slide. Skempton (1970) proposed that the parameters $c' = 0$, $\phi' = 20°$ give the realistic limit for the drop in strength preceding failure. This conclusion is based on the fact that unrealistically large pre-failure deformations are required to reduce the shear-strength parameters below these values. It appears from this analysis that a non-uniform distribution of shear strength is needed around the slip-surface, and that a major part of the surface must have substantially reduced strength (Dixon and Bromhead, 1991).

Dixon and Bromhead (1991) carried out an additional analysis with the curved and planar sections of the slip-surface given different strengths. The shear strength mobilized along the curved part should be lower than the peak strength of London Clay, typically taken as $c' = 20$ k Nm^{-2}, $\phi' = 20°$, due to the progressive failure that will occur as stability decreases. Progressive failure was a factor in the 1971 slide at K39. This was demonstrated by the

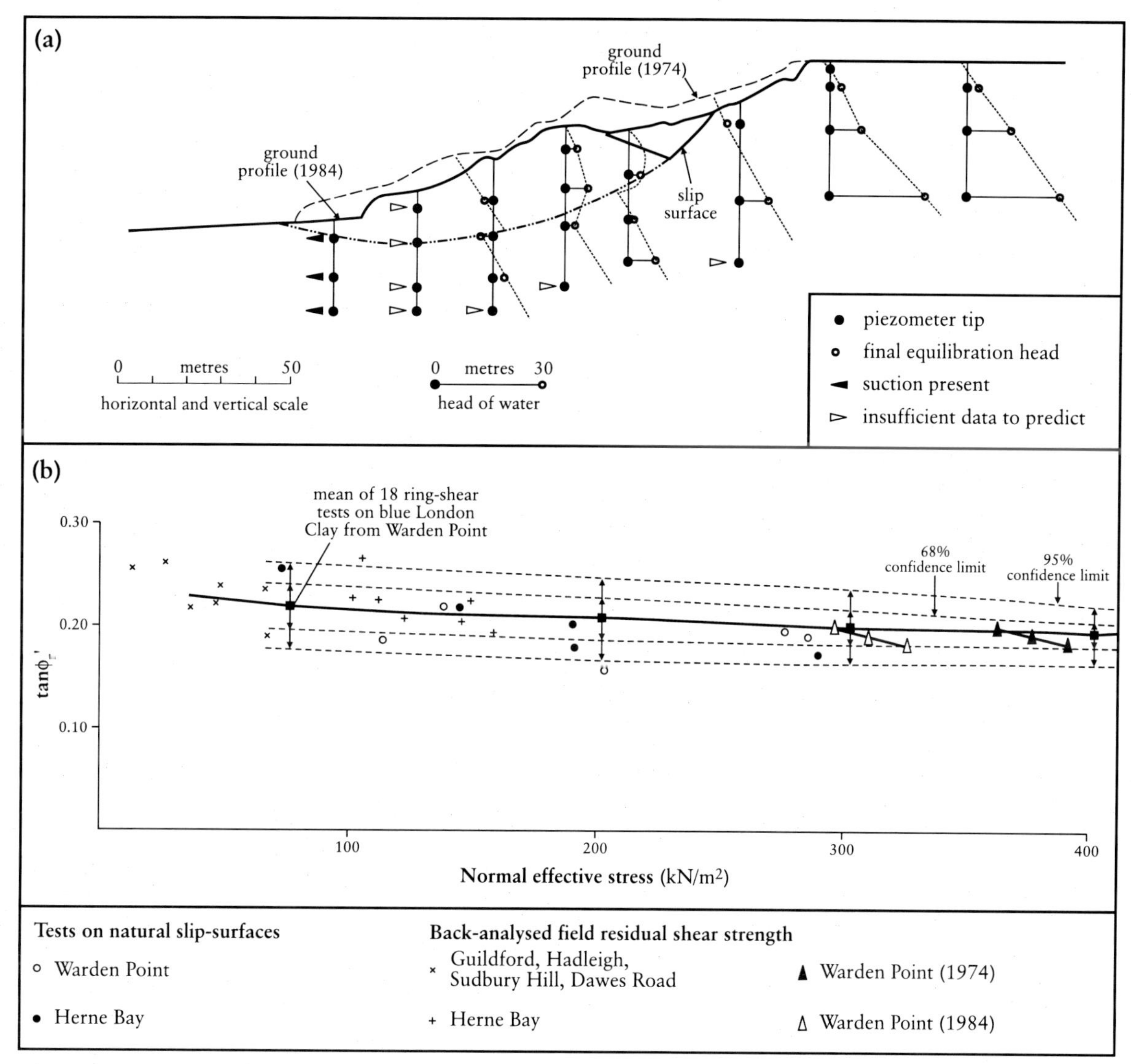

Figure 8.11 Results of back-analysis of the 1984 landslide at section K38, Warden Point. Based on Bromhead and Dixon (1984) and Dixon and Bromhead (1991.)

observation, made on the rear scarp shortly after failure, that the zone of softened clay contained numerous minor shear surfaces. The parameters $c' = 0$, $\phi' = 20°$ were used as they represented the lowest value of strength which could be sensibly expected. The shear-strength parameters required along the planar section of the slip-surface for limit equilibrium, were back-analysed as $c' = 0$ (assumed), $\phi' = 14°$ (i.e. only a few degrees above the residual value).

Deformations in the order of tens of centimetres are normally needed to substantially reduce the shear strength of intact London Clay. However, along discontinuities the shear strength can be lowered close to the residual value after only small displacements (i.e. in the order of tens of millimetres) as demonstrated by Skempton and Petley (1967). If it is assumed that the basal portion of the slip-surface is a bedding-plane-type feature, only small pre-failure movements would be required to reduce the shear strength to the back-analysed values of $c' = 0$, $\phi' = 14°$ (Dixon and Bromhead, 1991).

Despite the different failure conditions observed in the recent landslide at K38 the general mode of failure in the cliffs around Warden Point can still be classified as rotational sliding. However, considering the information on the individual first-time slides, it can be seen that the position and shape of the slip-surface changes from one section of the cliff to another. The most probable explanation for this is that the base of each slide is controlled by planes of weakness, the heights of which alter along the length of the coast. Field evidence from K37 indicates that the base of the slip-surface follows the dip of the bedding, and may in fact be a bedding plane. This demonstrates that the failure mechanism could be controlled by a bedding-related feature. The north-west dip of the London Clay in this area results in a 0.5° component of the dip parallel to the line of the coast. This means that the depth of a particular bed/plane will increase from east to west.

Field observations indicate that a bedding-related feature in the London Clay controls the depth and shape of the slip-surface. In order for the cliffs at Warden Point to fail at the slope angle observed, it is necessary for part of the slip-surface to obtain shear strength close to the residual value. Dixon and Bromhead (1991) have proposed a mechanism whereby the stress-relief that accompanies formation of the slope leads to deformations along the bedding feature, thus substantially reducing its shear strength prior to general slope failure.

Dixon and Bromhead (1991) point out that the above 'idealized' analysis can only give an indication of the magnitude and distribution of shear strength mobilized during a first-time slide. However, they considered that any inaccuracies in the specified slope profile, porewater pressures or slip-surface geometry, would be of insufficient magnitude to alter the general conclusions.

A mechanism for causing these strength reductions is readily available. The horizontal stress-relief that accompanies formation of the cliffs will result in general horizontal slope movements. The presence of a bedding plane would act as a strain concentrator, thus forming a plane within the slope that has a considerably reduced shear strength (Dixon and Bromhead, 1991).

Moving along the cliffs from K40 to K37, the basal slip-surface of the slides could be controlled by different bedding-related planes of weakness. This would explain the apparent change in failure mode from a deep-seated relatively circular slip at K40, through a non-circular deep-seated slide at K39, to a non-circular toe slip at K37. It is also worth noting that a planar sub-horizontal shear surface, which crops out at foreshore level, would help to account for the translational movements observed during the toppling type failure at K38.

It is of interest that the available histories on first-time slides in over-consolidated clays (see for example Barton, 1984) show that the large majority have the basal part of their slip-surface controlled by bedding planes. It is not unrealistic to suggest that this is also the case for slides in the unweathered London Clay. This idea is very close to the emerging idea of pre-failure displacement under the load of a cap-rock, as, for example, at Castle Hill, Kent, and at the famous landslide 'Le Chaos' on the Bessin Cliffs in Normandy, where the 'zone de décompression' is used as a fundamental planning boundary (Maquaire and Gigot, 1988).

Conclusions

Owing to the stabilization measures taken at other similar sites (at and near Herne Bay, Kent, and Walton, Essex) Warden Point is the only site where the cyclic mechanism described by Hutchinson (1973) is still in operation. Areas of cliff representing all the stages except Stage 5 can currently be seen there. Warden Point represents the condition of London Clay cliffs subject to active marine erosion, although there is disagreement about the significance of this. At Warden Point the porewater pressures do not equilibrate under the ambient hydraulic boundary conditions. Instead, they are held in dynamic equilibrium by the rapid retreat of the cliff-line.

Considering the recent landslides that have occurred in the cliffs at Warden Point, it can be seen that although their failure mechanisms are in most cases essentially the same, there are differences in the shape and position of the slip-surface which can be deduced from field observations. Assessing the landslides in turn, moving from east to west (K40 to K37) it becomes apparent that a change occurs in the mechanism of the first-time failures.

First-time slides occur in slopes that have been steepened by marine erosion. Slide behaviour is controlled by the stress-relief-modified porewater regime. Assessment of the field-mobilized shear strength and the failure mechanism require a detailed understanding of these depressed porewater-pressures.

The toppling mode of failure that occurred at K38 in 1988, while of considerable interest, is not common on slopes in London Clay. Further work is required to assess the special conditions that existed for this type of failure mechanism to take place preferentially to rotational sliding.

Chapter 9

A mass-movement site entirely in Pleistocene strata

R.G. Cooper

INTRODUCTION

In the landslide survey of Great Britain (Jones and Lee, 1994), 3042 landslides were recorded as either involving, or being entirely developed in, deposits of superficial origin. However, of the 346 at coastal locations, no classification was made, either of origin or type. For 19.7% of the 2696 at inland locations, origin was not specified further. Of the inland landslides for which the origin was specified, by far the largest numbers were in glacial deposits (74%) and periglacial deposits (18%). Of the remainder, landslides in fluvial deposits, fluvio-glacial deposits and deposits of contemporary processes each accounted for less than 2.7% of those of specified origin, while landslides in lacustrine, marine and aeolian deposits each accounted for less than 0.5%.

The one mass-movement GCR site entirely in Pleistocene deposits, at Trimingham in Norfolk (Figures 9.1 and 9.2), is a stretch of coastal cliffs undergoing rapid erosion. Such dynamic sites on coastal cliffs are frequently a focus for debate between those wishing to conserve the scientific features and those wishing to protect other interests (for example, where cliff retreat is consuming land that is valuable for some other purpose).

For mass-movement site-conservation purposes it is essential that the site is allowed to evolve naturally, unimpeded, in order for studies to be conducted into the natural evolution of the site. At Trimingham Cliffs, the cliff retreat is consuming agricultural land that is still valuable (specifically, for arable use). But this is not the case for the other mass-movement GCR sites that are located on the coast. For example, at the **Axmouth–Lyme Regis** GCR site, as at **Folkestone Warren**, retreat would be by means of collapses of the Chalk cliffs that back the slides. Specific prevention of this is not a practical possibility, but stabilization of the slipped masses in front of the Chalk cliffs would probably reduce the frequency and/or size of Chalk falls.

The landslips at Quiraing (see **Trotternish Escarpment** GCR site report, Chapter 6) on the Isle of Skye and **Hallaig** on the Isle of Raasay are considered far too large and deep-seated for stabilization to be contemplated. The land consumed by retreat of their backslopes would be, in any case, low-grade moorland, and this land is remote from human habitation. **Black Ven** in Dorset and **Folkestone Warren** in Kent are both subject to headslope retreat, but in both cases, the land consumed is low-grade pasture.

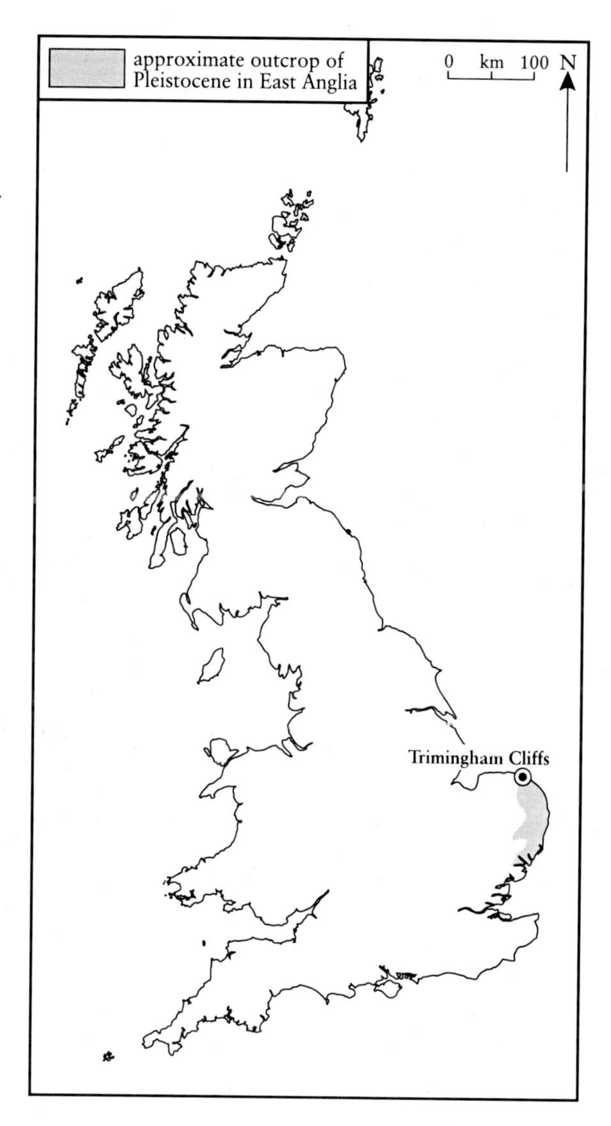

Figure 9.1 Areas of Pleistocene strata in East Anglia (shaded) and the location of the Trimingham Cliffs GCR site, described in the present chapter.

The implications of the fact that the one site entirely in Pleistocene deposits would, if allowed to continue receding, cut into valuable arable land, are discussed, along with the solutions advocated, in the account below.

TRIMINGHAM CLIFFS, NORFOLK (TG 280 390)

Introduction

(a) General

Along the northern coast of Norfolk there is a continuous line of cliffs from Weybourne in the west to Happisburgh in the east, a distance of 32 km (Figure 9.2). The cliffs are formed in materials of Pleistocene age, principally deposits of the Anglian (antepenultimate) glaciation. They are currently retreating by means of a variety of types of mass movements, but this retreat is potentially at risk from coast protection measures.

The site is one of considerable variety. Firstly, it includes an assortment of mass-movement types, at a wide range of scales. The mass movements represent types that are probably characteristic of sediments which are 'weak', if not actually unconsolidated. Secondly, it includes two of the three categories of coast protection described: in the west the cliff is unprotected, and acts as a feeder bluff for the beach system; in the east the cliff is protected by a revetment, which is intended to limit the rate of cliff recession, while maintaining sufficient input from the cliff to maintain an adequate beach (Figure 9.3).

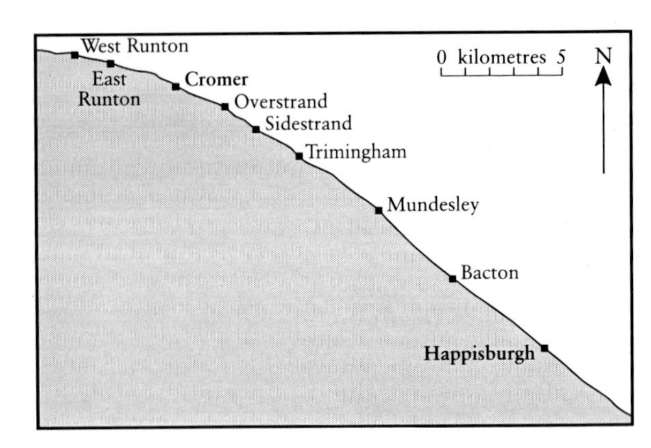

Figure 9.2 Locality map of the Happisburgh–Cromer area of the north Norfolk Coast. After Kazi and Knill (1969).

(b) Stratigraphy and lithology

The detailed stratigraphy of the Anglian deposits has been worked out by Banham (1968) (Table 9.1). In the area around Trimingham the lowest 3 m of the cliff is described by Banham (1968)

Figure 9.3 Trimingham Cliffs, showing mass movement around and over the revetment. (Photo: R.G. Cooper.)

Table 9.1 Geological succession in the cliffs of north Norfolk. After Banham (1968).

<table>
<tr><td>Chalky outwash sands and gravels</td><td></td><td>7.5 m</td></tr>
<tr><td>Chalky boulder clay</td><td></td><td>3.0 m</td></tr>
<tr><td>Brick Kiln Dale Gravels</td><td></td><td>12.5 m</td></tr>
<tr><td>Gimingham Sands</td><td rowspan="3">'Contorted Drift'</td><td>3.0 m</td></tr>
<tr><td>Third Till</td><td>12.0 m</td></tr>
<tr><td>Mundesley Sands</td><td>4.5 m</td></tr>
<tr><td>Second Till</td><td rowspan="2">'Cromer Till'</td><td rowspan="2">3.0 m</td></tr>
<tr><td>Intermediate Beds</td></tr>
<tr><td>First Till</td><td></td><td></td></tr>
</table>

as consisting of his Second Till. However, Hutchinson (1976) has pointed out that the cliffs to the east of Overstrand (i.e. close to Trimingham) are characterized by the presence of a sill consisting of all or part of the 'Cromer Till' sequence (Banham's First Till–Intermediate Beds–Banham's Second Till) in the cliff-foot. Accordingly, all formations from Banham's First Till upwards are described here. The Second Till is overlain by 4.5 m of Mundesley Sands, and 12 m of the Third Till. This is overlain successively by 3 m of Gimingham Sands, 12.5 m of Brick Kiln Dale Gravels and 3 m of a chalky boulder clay (Solomon, 1932). Above this is 6–9 m of chalky outwash sands and gravels.

West and Banham (1968) point out, following Reid (1882), that the succession can be broadly divided into a lower relatively undeformed and sub-horizontal zone, a middle zone of intense isoclinal deformation 30–35 m thick ('Contorted Drift'), and an upper zone of more open folding. Banham (1975) considers that the deformations originated through loading by the Gimingham Sands and the overlying gravels, with associated diapirism. The lower and middle zones are separated by a surface of décollement developed within the Mundesley Sands. Along most of the line of cliffs at Trimingham, the geology and structure are uncertain because of poor exposure: the cliffs are covered by mass-movement deposits.

Banham's First Till is a dense, fissured, grey or dark-grey sandy boulder clay resting on the Cromer Forest Bed Series, the Leda Myalis Sands or associated deposits. The Intermediate Beds (see Figures 9.4 and 9.5) lying above the First Till have been shown by Kazi and Knill (1969) to be laminated lake clays, anisotropic in their physical properties.

The Second Till occurs locally at Trimingham and again between Overstrand and Kirby Hill, east of Cromer. It is a dense, fissured, grey-blue sandy boulder clay which, depending upon the local structure, rests on any of the older formations. Commonly, more than 40% of the till is made up of Chalk pebbles ranging from a few millimetres to 5 cm in length. This till is readily distinguishable from the essentially Chalk-free First Till, particularly in areas where the Intermediate Beds are present. The structural arrangement of the two tills is different as the Second Till typically has both an irregular base and surface and, in addition, appears to have been laid down by an ice-sheet which ploughed into the older formations, thereby locally removing them completely (Kazi and Knill, 1969).

The Mundesley Sands are composed of uniformly textured, medium dense, dirty white silty sands resting on the hummocky surface of the Second Till. There is a local basal conglomerate of chalk pebbles. The sands are variously chalky or carbonaceous, and in the latter case a distinctive greyish tint is imparted to them. When weathered the sands are yellowish-brown in colour. This horizon can be traced at intervals along the cliffs at Trimingham, attaining a maximum thickness of about 13 m near Mundesley, the type locality for the deposit (Solomon, 1932).

The Third Till is characterized by a complex internal structure, frequent erratic masses of chalk and the presence of large-scale undulations. It, and possibly parts of the Mundesley Sands below and Gimingham Sands above, was

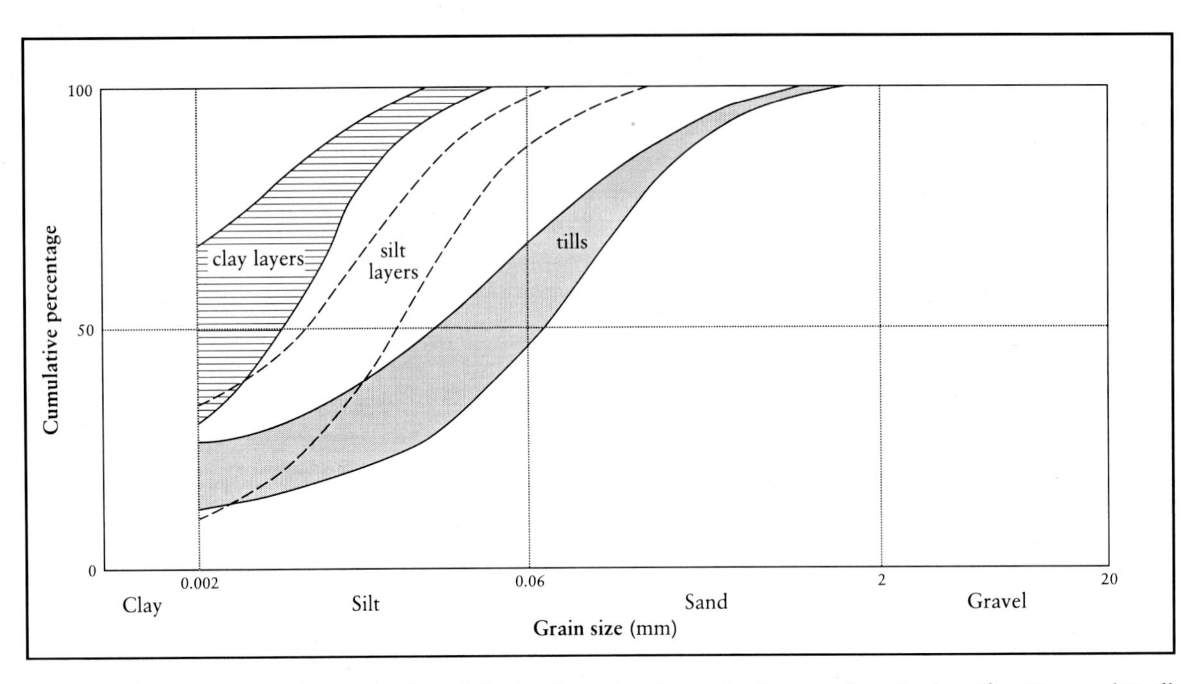

Figure 9.4 Grain-size analyses of tills and the laminated units of the Intermediate Beds. After Kazi and Knill (1969).

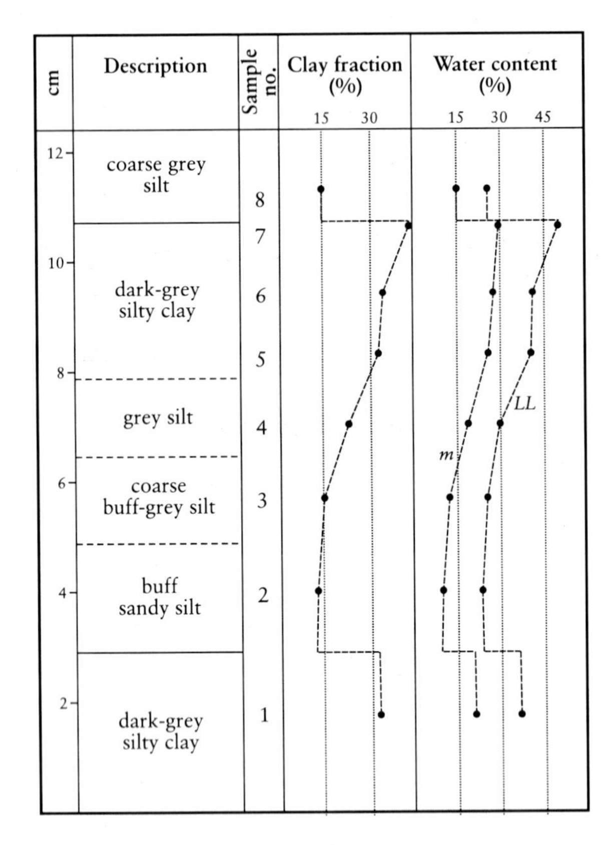

Figure 9.5 Variation of clay content, natural moisture content (*m*) and liquid limit (*LL*) through the individual graded bed in the Intermediate Beds. After Kazi and Knill (1969).

formerly known as the 'Contorted Drift'. The group is very variable in thickness and ranges to well in excess of 30 m near Cromer. The Third Till locally contains glacial lake sediments.

The Gimingham Sands comprise a unit of loose, stratified pale-yellow sands and gravels resting on the irregular top of Banham's Third Till, often occurring in large basin-shaped hollows in the till. The sands can be traced at intervals along the cliffs at Trimingham. Near Kirby Hill (east of Cromer) the group attains an apparent thickness of about 30 m.

The Brick Kiln Dale Gravels, in the one clear exposure near Trimingham, were shown to consist of 6 m of laminated stoneless and chalky clay, and 3.5 m of gravelly and chalky yellow sands (Solomon, 1932).

The sill described by Hutchinson (1976) is a common but discontinuous feature of the cliff morphology, forming a near-vertical, relatively resistant face up to several metres in height along the cliff-foot (Kazi and Knill, 1969). It is, however, only occasionally exposed at Trimingham; for most of the time it is buried in mass-movement deposits.

The cliffs are of considerable value in the study of the relationship between geological processes and engineering properties and behaviour, because of the variety of mass-movement processes that can be recognized between Trimingham and Overstrand, and new coastal works, which, if not restricted, could lead to stabilization of the cliffs, growth of vegetation and steady deterioration in the geological quality of the exposed section.

Many authors have described the cliffs of the Norfolk coast as consisting of 'unconsolidated' sediments, presumably meaning that there is little or no evidence of strengthening processes like cementation of the sand particles, and loading. However, there is evidence that in fact these sediments are over-consolidated, or at least normally consolidated (i.e. they have at each level the degree of consolidation which would be expected from the depth of the pile of sediments above them, from the ground surface down). Consolidation tests using a conventional oedometer showed that the Intermediate Beds at Cromer are markedly more over-consolidated than at Happisburgh. At both sites the sediments are over-consolidated with respect to the present height of the cliff and it has been suggested (Kazi and Knill, 1969) that the original surcharge included an ice-load. The topography of the cliff-top at Cromer is generally regarded as representing a moderately fresh glacial landscape. As a consequence the present cliff height is held to indicate the maximum loading provided by the glacial drift on the Intermediate Beds. The deficiency of pressure at Cromer is equivalent to an additional ice-load of 90 m (based on unit weights of 2024 kg m^{-3} for the drift, 923 kg m^{-3} for ice, and a groundwater level at the ground surface). At Happisburgh, an equivalent calculation indicates the pressure deficiency is equivalent to about 80 m of ice. Some erosion has occurred at Happisburgh, and a maximum ice-load of about 60 m is probably more reasonable. These thicknesses of ice appear to be generally in accord with knowledge of the thickness of modern glaciers and with the directions of ice movement during the Lowestoft stage (West, 1968). Kazi and Knill's (1969) view has, however, been challenged by Banham (1975), who has suggested that the Cromer landscape is not a fresh glacial surface but was once covered by superincumbent sediments the upper part of which have been removed by glacial meltwater erosion. This removal would have been sufficient to account for the degree of consolidation measured.

Description

(a) Cliff hydrology

During the winter, considerable quantities of groundwater discharge along the coast at the junction of the Mundesley Sands and Banham's Second Till beneath them. This is evidently a horizon at which permeability changes substantially, the Mundesley Sands being sandy and highly permeable, while the Second Till, at least in its uppermost part, is of low permeability. This junction forms the 'sill' in the cliff, noted by Hutchinson (1976). The junction varies significantly in level, typically from about 5–10 m above OD, and naturally the main discharges are concentrated at the depressions in the undulating 'sill' surface. As the Mundesley Sands in this area are largely composed of fine-grained sands, the discharge zones at the base of the formation are marked by active seepage erosion and the resultant formation of outwash fans at the cliff-foot. The seepage erosion is accompanied by back-sapping. These processes are generally absent from the areas between the depressions in the 'sill'. Hutchinson (1976) reported that the 'sill' is not the only water-table control in the cliffs. There are groundwater tables perched on the Third Till and on till inclusions and/or erratics in the Mundesley Sands, the Third Till and the Gimingham Sands.

At the depressions in the sill the back-sapping, combined with the effects of porewater pressures in the lower parts of the sand cliff, leads to a series of relatively shallow slides in the sands and consequent degradation of the sand cliff. Between the depressions, however, the sands are well drained, seepage erosion is absent and the sand cliffs stand at much steeper angles. With the progress of coastal erosion, these steeper slopes eventually suffer rotational slips of which the failure surfaces generally descend to about the level of the top of the sill (Hutchinson, 1976). These slips are larger than those associated with the back-sapping, but much smaller than the deep-seated failures that occur from time-to-time in the adjacent cliffs to the west.

At Section I (Figure 9.6) (Hutchinson, 1976), which was located at a depression in the surface of the sill, seepage erosion was very active and had led to the formation of a mudslide of mixed

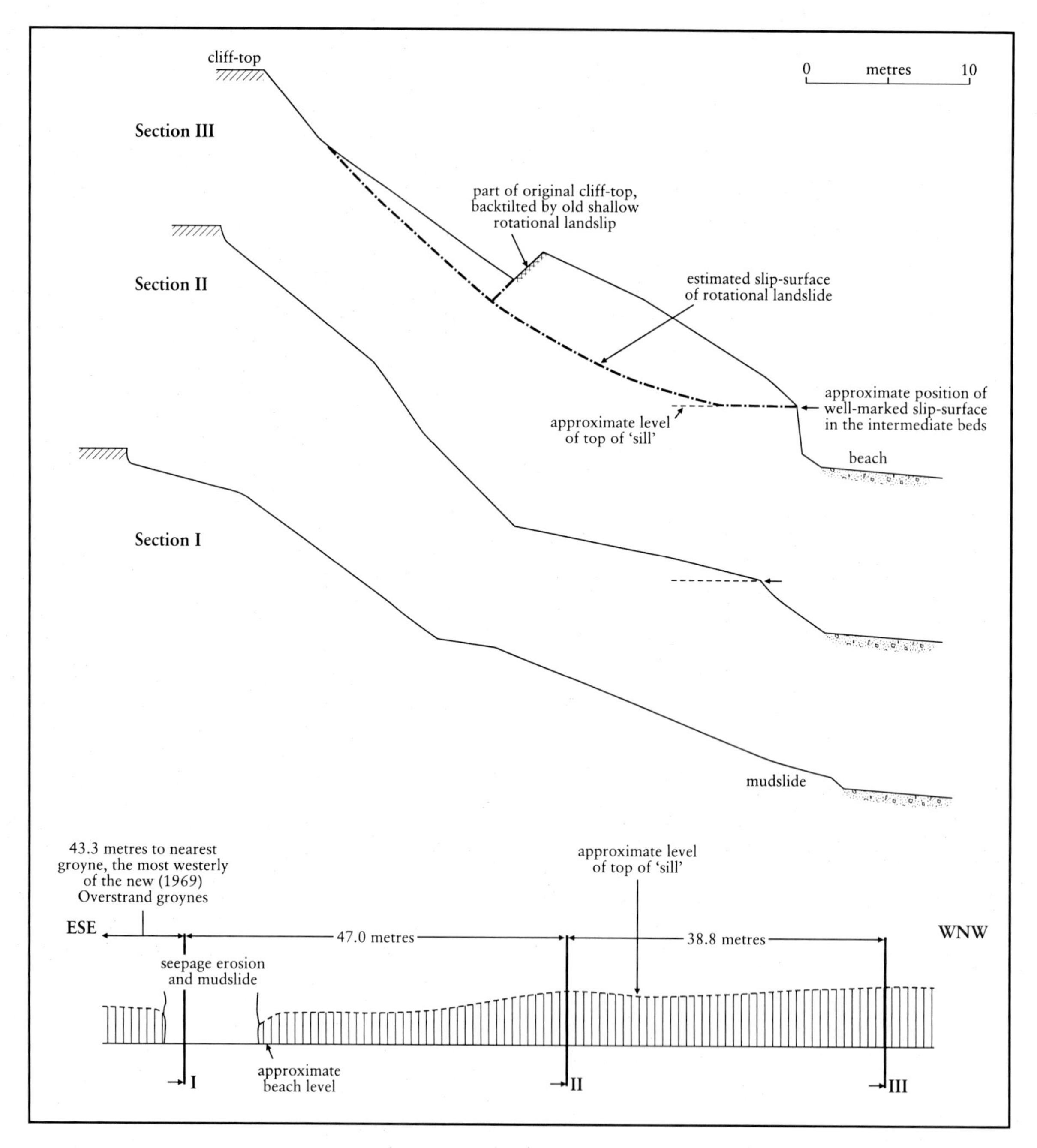

Figure 9.6 Section through the eastern cliffs at Trimingham. After Hutchinson (1976).

sand and clay (Hutchinson, 1976). As shown, this had eroded down some distance into the sill. As a result of the back-sapping and resulting shallow slides in the sands, the overall inclination of the cliff was only 24°. In the lower part of the sand cliff the average inclination was about 21°. Such a slope is just stable for an average porewater-pressure ratio, r_u, of 0.37, if the effective shear parameters obtaining are $c' = 0$, $ø' = 34°$ (Bishop and Morgenstern, 1960). If a small value of c' exists, the value of r_u required to cause failure would, of course, be increased. A similar situation is treated by Henkel (1967).

Section III (Figure 9.6) (Hutchinson, 1976) was located on the crest of an undulation in the sill. Seepage erosion and back-sapping were absent and the sand cliff stood at a much steeper angle, averaging about 31° overall. A rotational slip, involving a slice of the cliff-top, had recently occurred in the sand cliff. Prior to this slip the average inclination of the cliff was probably about 35°. At this location one would expect the porewater pressure in the base of the sand to be low or even zero. Even so, a small *c*' value, or negative porewater-pressures, would be required for such a slope to be stable if ø' were again 34° everywhere on a potential slip-surface.

The cliff profile at Section II (Figure 9.6) (Hutchinson, 1976) is similar to that at Section III in being located at a crest in the undulating sill. Section II represents a later stage of development, however, in which a rotational landslip in the sand, as before exploiting the slip-surface in the Intermediate Beds at its toe, had moved farther down the cliff and spilled over the sill (Hutchinson, 1976). This had left the upper cliff over-steepened, at an average inclination of nearly 45°, and probably soon to be involved in a further rotational slip. Assuming that it was not held up by included masses of till, the steep angle of the sand cliff on Section II provides further evidence for the existence of a cohesion intercept in these sands. This may well be made up from a combination of slight cementation with some capillary porewater tensions. Taking the sand cliff at Section II to have an average inclination of 45° and a height of 35 m, with ø' = 34° and zero porewater-pressures, an average *c*' value of about 10 kN m^{-2} can be inferred to be necessary just to maintain the stability of the cliff (Hutchinson, 1976).

(b) Mass movements

The cliffs at Trimingham expose a variety of Pleistocene sediments, and are subject to active coastal erosion (Figure 9.7). This has resulted in extensive slope instability and the development of a wide range of mass-movement features. Kazi and Knill (1969) have observed blockfalls, seepage failures, mudflows, 'sand glaciers' (their term) and deep-seated non-circular slips along this length of coast, and have carried out detailed analysis of the geotechnical properties of the 'Cromer Till' (Figures 9.8–9.10). This rather complex group of inter-related mass movements is responsible for a rate of coastline recession of up to 1.1 m a^{-1} (Hutchinson, 1976). The cliff-top and the cliff-foot are receding at about the same rate, maintaining the overall slope angle.

Figure 9.7 Erosion, undercutting, and, in the background, toe erosion at the Trimingham Cliffs GCR site. (Photo: R.G. Cooper.)

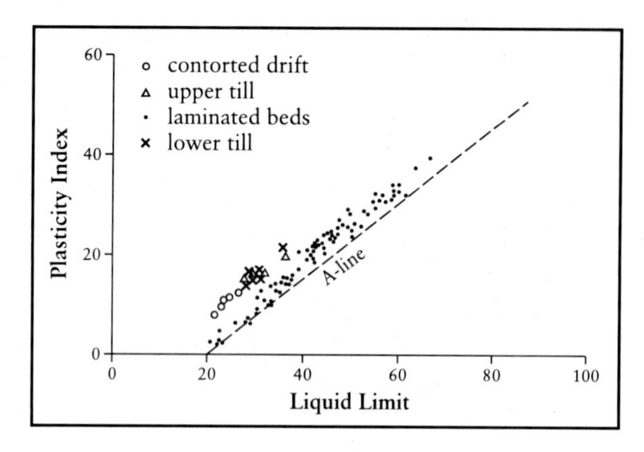

Figure 9.8 Plasticity charts for tills and Intermediate Beds. After Kazi and Knill (1969).

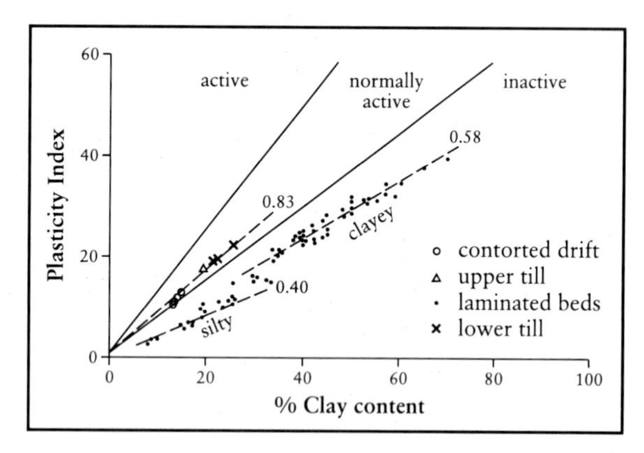

Figure 9.9 Activity chart of tills and Intermediate Beds. After Kazi and Knill (1969).

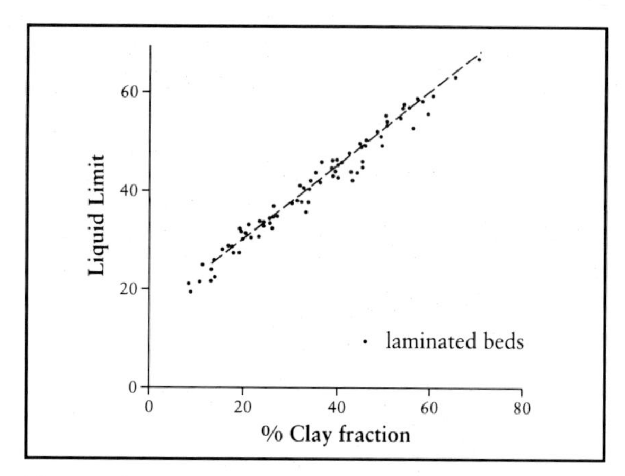

Figure 9.10 The relationship between the clay fraction and liquid limit for the Intermediate Beds. After Kazi and Knill (1969).

Another important feature of at least some parts of the cliffs, however, is the presence of the well-marked slip-surface within the Intermediate Beds near the top of the sill (Hutchinson, 1976). The slip-surface, although largely situated within the sands, follows for some distance at its toe a slip-surface in the Intermediate Beds. As the value of ø' on this latter surface was 19°, the average ø' mobilized in the rotational slip will have been less than 34° and the necessity for some *c*' component of strength, or negative porewater-pressure, to exist in the sand mass will have correspondingly increased.

Where the cliffs are highest, deep-seated rotational slips occasionally take place. Examples between Cromer and Overstrand, where they are also most frequent, were examined in detail by Hutchinson (1976). An example immediately west of Trimingham was noted by Ward (1962). However, between Cromer and Overstrand the main features on the cliffs are a series of rotational slips which generally toe out at about the level of the top of the 'sill', and large mudslides which from place to place erode down into the 'sill' and run down onto the beach.

(c) Recession rates

Taking measurements from published maps, Cambers (1973, 1976) found that the average rate of retreat of the cliffs from 1880 to 1967 was 0.9 m a^{-1}. Records of former villages recorded in the Domesday Book (1086) and now missing through erosion, as well as other historical accounts, suggest that a similar average rate of erosion has persisted for at least the past 900 years. Clayton (1989) showed that the cliffs at Trimingham had the greatest amount of retreat on the Norfolk coast over the 100 years to 1985 (Figure 9.11). The waves incident on the

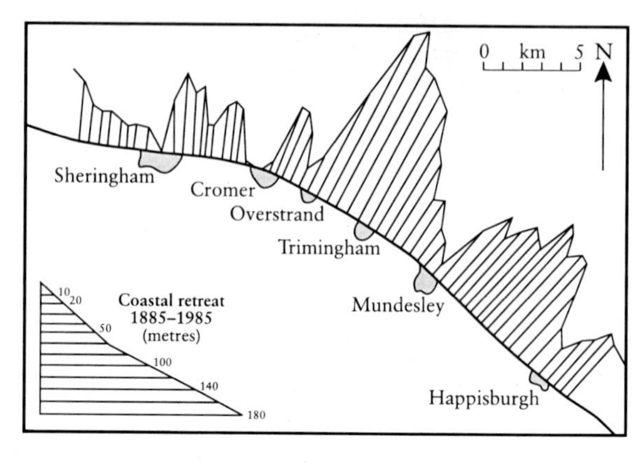

Figure 9.11 Total coastal retreat between 1885 and 1985 based on the First Ordnance Survey at 1:10 560 scale and field survey in 1985. After Clayton (1989).

coastline have cut back the cliffs 1–2 km over the past 900 years. Field sampling on the cliffs by Cambers (1973, 1976) established that the erosion of the Norfolk cliffs provides well over 500 000 $m^3 a^{-1}$ of sediment, and that up to two-thirds of this is sand and gravel which may remain in the beach system. Littoral drift transports this sediment: a small part moves westwards along the north Norfolk coast, but most moves southwards towards Lowestoft (the overall sand budget was calculated by Clayton *et al.*, 1983) (see Figure 9.12). Thus the beaches south of the Trimingham Cliffs for the 42 km to Lowestoft are largely, if not entirely, dependent on the cliffs for their throughput of sand. The cliffs act as 'feeder bluffs' for the beaches.

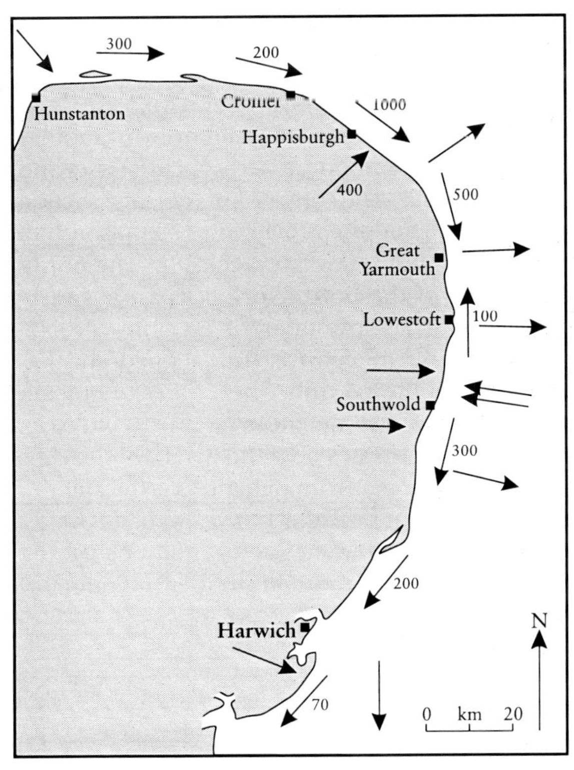

Figure 9.12 Net longshore transport values (m^3) for April 1974–March 1975 computed from wave observer data on wave height and direction. Based on Cambers (1976) and Clayton (1980).

(d) Cliff protection

A total length of more than 14 km of the cliffed Norfolk coast is defended by inclined permeable timber revetments, usually fronted by groynes. The purpose of the revetments is to reduce the energy of the waves reaching the cliff-foot, while the purpose of the groynes is to inhibit down-drift movement of the beach sand. The groynes are of two types: impermeable, which are effective until the beach builds up to a level on the updrift side at which it overtops the groyne, and permeable, which are much less effective for their purpose. Prior to the local government re-organization of April 1974, the coastal defence authority along the Trimingham stretch of coast was Erpingham Rural District Council, which installed timber revetments and permeable timber groynes, the revetments standing on concrete sheet piling on the seaward side, and timber piles on the landward side (Figure 9.13a). The revetments are designed to stand a short distance down the beach and far enough in front of the cliff to dissipate as much as possible of the energy of waves breaking at the revetment, before they reach the cliff-foot. This distance is usually between 16 m and 20 m. The revetments have planks, which may or may not have spaces between them; the planks can either run up the face of the revetment, or be placed horizontally. Where each plank is flush against the next, the revetment is essentially 'impermeable', although some waves may overtop it. Where the planks have been fitted with spaces in between them (Figure 9.13b), the revetment is 'permeable'. The slope of the face is generally about 45°. The design life is considered to be 40 years (with some repair). The revetments on the Norfolk coast are installed on a coast undergoing erosion where the beach is gradually losing volume; they will therefore be noticeably farther down the beach after 20–30 years (Clayton and Coventry, 1986). By this stage the sheet piling will be exposed when the beach is low, acting as 'hard' engineering: the waves will be reflected rather than having their power absorbed, and increased scour of the beach will result.

At Trimingham the cliffs are protected by a timber revetment the face of which consists of horizontal slats with gaps in between. This is sufficient to reduce substantially the power of incoming waves, while still allowing waves to reach the cliff-foot. The result is that the rate of cliff recession is lowered, but cliff recession is not halted. It allows the cliff-foot to be eroded by the waves, and a range of mass movements are taking place on the cliff as a result. Although this does not include deep-seated slips, these can be seen in the unprotected stretch of cliffs between

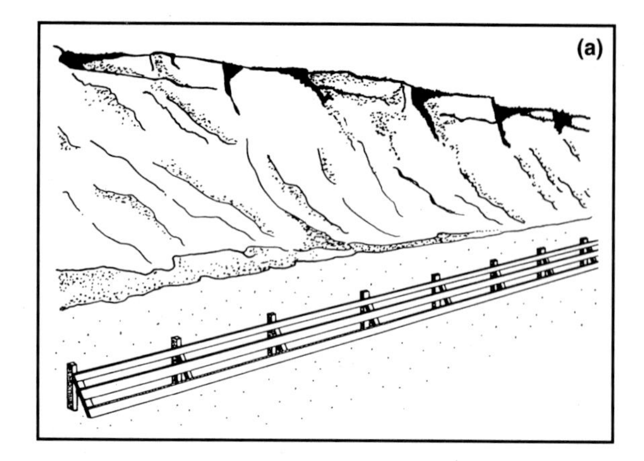

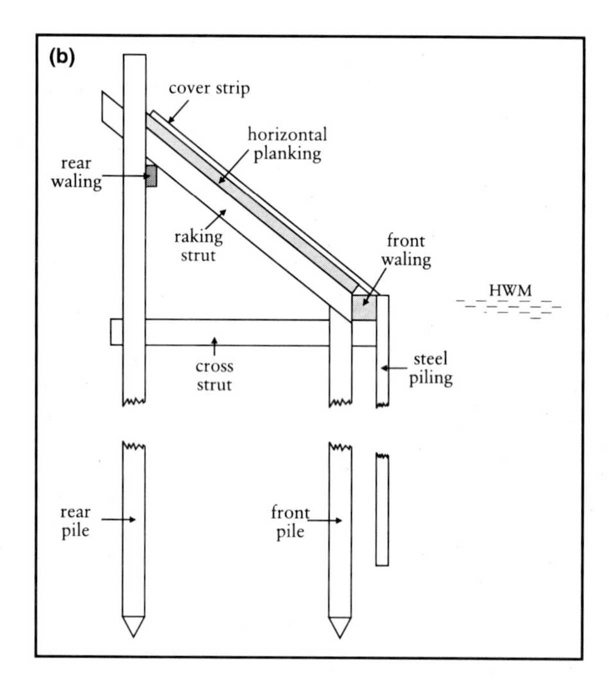

Figure 9.13 (a) Design of the timber revetments used on the Norfolk Coast (after McKirdy, 1990). (b) Design of the timber revetments used at West Runton (after Clayton and Coventry, 1986).

Trimingham and Overstrand, part of which is also within the GCR site.

Revetments are seen as 'softer' in character than walls, i.e. less reflective. An experiment was designed by Clayton and Coventry (1986) to measure the effects of reducing the number of planks on a revetment at West Runton. Some stretches of revetment were left with the original maximum of 10 planks, while other lengths had 4 planks; 7-plank stretches provided a physical intermediary between the two. On all of the sections measured, the beach level fell over the three-year period of observations, both behind the revetment and in front of it. Measurements showed that, by reflecting the waves less efficiently, the four-plank stretches of revetment reduce the amount of beach loss in front of the revetment. So a revetment with a smaller number of planks will reduce beach lowering in front of the revetment by almost half the amount where the full number of planks is in position. It appears, therefore, that the standard 10-plank design is too effective as a reflector of waves and causes rather rapid beach loss in front of the revetment.

(e) Cliff aspect

Clayton (1989) introduces a further distinction concerning the cliffs, based upon aspect with respect to the dominant wave direction. The section of coast to the west of West Runton is almost straight for 7.5 km, and faces on average 4° east of north. There is a gradually increasing curvature through to Overstrand (6.5 km) and then a fairly straight alignment for another 19 km to the end of the cliffs beyond Happisburgh. The first 7 km of this section averages 31°, and the remaining 12 km, south of Marl Point, Mundesley, averages 38°. The north-facing part of the north Norfolk coast is swash-aligned and has low rates of erosion; the NE-facing part of the coast is drift-aligned and has high rates of littoral drift where rates of erosion are high and sea defences are less effective. Figure 9.14 illustrates how the height of the cliffs is at its maximum at Trimingham, as is the amount of coastal retreat over the 100 years to 1985.

The most important factor influencing the rate of retreat is thought to be retention of

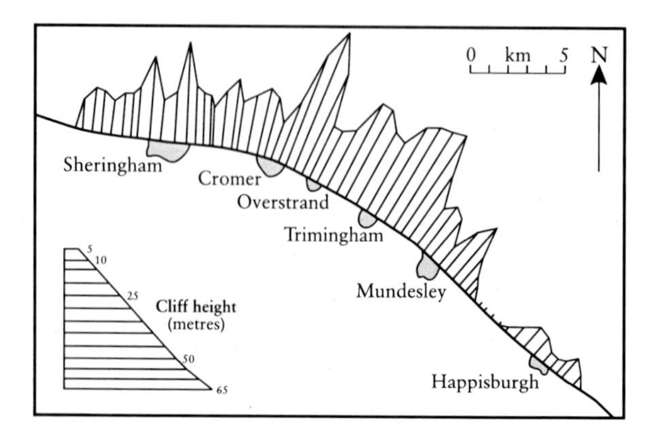

Figure 9.14 The height of the cliff of north-east Norfolk plotted for each measurement cell. After Clayton (1989).

material (which includes many large flints) on the swash-aligned coast and rapid removal by longshore drift on the drift-aligned coast. The sediment volumes from cliff erosion may be divided into material from the swash-aligned coast and material from the drift-aligned sector. Prior to the construction of defences, 33% of the sediment volume came from the swash-aligned coast. By 1983 this had been reduced to less than 19%, a clear indication that coast protection structures had been more successful on the swash-aligned coast than on the drift-aligned coast. By 1985 the swash-aligned coast was producing only 40% of the 1885–1905 volume (68% of the 1906–1946 volume). The drift-aligned coast, however, was producing 92% of the 1885–1905 volume (83% of the 1906–1946 volume) (Clayton *et al.*, 1983). Accordingly, it seems that the defences (all types of structure) had reduced sediment output by about 50% on the north-facing coast, but only by about 10–15% on the SE-facing coast.

Interpretation

The area of the Trimingham Cliffs mass-movement GCR site overlaps an SSSI which was already in existence at the time of designation. This site, the Sidestrand and Trimingham Cliffs SSSI, comprises a length of cliffs which have been left free of coastal defence works. Referring to the SSSI, McQuhae (1977) remarked that clear geological exposures are maintained so long as mass movements are allowed to occur. Also, investigation of the relationships between the engineering properties of the drift and mass movement is made possible. Coast defences in many places, however, have led to stabilization of the cliffs, and loss of these areas for observation of the different types of failure. She therefore concluded that it is essential that areas like this remain available for study, and that erosion be allowed to continue. The particular variety of types of instability and the value of their study warrants this stretch of cliffs for SSSI status for its mass-movement features.

There is also a practical reason why some cliffs should be allowed to erode. Clayton (1980) pointed out that beach-sand drifts long distances and is highly dependent on eroding cliffs which act as feeders. While these sand feeds ('feeder bluffs') survive, most coastal defence work is reasonably successful and can maintain a good beach. However, if the cliffs were to be fully protected, the entire system would face irreversible decline.

> *'There have been signs in recent years that the process of extension and elaboration of coastal defences in East Anglia has reached the point where the return on additional expenditure is small or zero. Indeed estimates of the coastal sand budget suggest that any extension of defences that successfully reduced the retreat of the Norfolk cliffs would actually threaten beach stability over a length of coast that eventually could extend to 50 km downdrift. The benefit/cost advantage of accepting continued retreat of feeder bluffs is very considerable: the only alternative would be beach nourishment on a very large scale.'*

Here, as in a later article (1991), Clayton (1980) sets out the main points of the argument about leaving feeder bluffs undefended from the sea, so that beaches can form a defence against the sea. If all feeder bluffs are proofed against marine erosion, for example by the construction of concrete walls, the long-term effect on the beach which they feed will be that it will be washed away, not only at the site of the former feeder bluffs, but potentially for many kilometres downdrift. The effects on any holiday resorts and on coastal recreation generally would be catastrophic. Further, unprotected coastline downdrift *beyond* the extent of the cliffs would be subject to much stronger marine attack, since it would lack the protection provided by a beach.

The argument that some of the cliffs must be left unprotected to act as feeder bluffs for the beach is now largely accepted. Shoreline Management Plans for the Norfolk coast include:

- leaving some cliffs unprotected, so that they can act as feeder bluffs;
- cliff protection of the 'hard engineering' type (e.g. concrete walls), to protect coastal towns, for example at Cromer;
- cliff protection of the 'soft engineering' type (e.g. timber revetments) in more rural areas, for example at Trimingham.

It is clear from a document produced in 1990 that the then Nature Conservancy Council (NCC) accepted the argument about the need to leave feeder bluffs (McKirdy, 1990). The document (a report to the NCC from HR Wallingford Ltd) made it clear that a coastal mass-movement site would have to be maintained not only in terms of the marine action upon it, but also in terms of groundwater entering it.

An NCC report of 1991 applies the suggestions in McKirdy (1990) to a selection of coastal SSSIs, including the Sidestrand and Trimingham Cliffs SSSI, which 'includes both Sidestrand and Trimingham Cliffs', the latter being 'primarily a mass-movement site'. The preferred option selected is 'the construction of a series of off-shore breakwaters', with sufficient space between the breakwaters to allow enough wave action at the cliff-foot for mass movement to continue, but at a reduced rate.

The breakwaters were never constructed, and the relevant local authority, Norfolk County Council, have allowed the wooden revetment on the beach at Trimingham to fall into disrepair, so its effectiveness as a moderator of wave power has diminished. There is evidence that it had never been effective: the construction of the revetment in 1974 was not without problems. Hutchinson (1983) shows an oblique aerial photograph of the sea defences under construction at Trimingham in 1974. Slope stabilization was not then attempted, on the basis of high cost. The revetment and groynes were installed specifically to check toe erosion. The photograph showed that the line of the revetment works was located too close to the cliff-foot: the excavations needed to permit construction of the revetment stimulated further slides by the concomitant unloading (Hutchinson, 1983). Further consequences of this error in positioning persist to the present: from time-to-time mass movements in the lower part of the cliff give rise to accumulations of cliff materials at the cliff-foot that burst through the revetment (causing damage) or even overtop it (Figure 9.3).

Norfolk County Council have adopted a strategy involving the sinking of deep (deeper than usual) drains in the cliff below the village of Trimingham. The result is two-fold: the improved drainage has reduced mass movement on the upper parts of the cliff. The best evidence of this is that the cliff has, in its upper parts, become vegetated, in contrast to the cliffs both to the east and west of Trimingham. The result of the ineffective protection of the cliff-foot by the revetment is that material is being removed from the cliff-foot by wave action, giving rise to mass movements on the lower part of the cliff. In some places these extend far up the cliff. Of the types described by Kazi and Knill (1969), the deep drains seem, so far, to have prevented conditions from arising which would precipitate deep-seated movements of the type described by Hutchinson (1976).

The Norfolk coast between Happisburgh and Cromer has been critical in two related debates about how the coastline is to be best conserved. The first debate concerns coast protection to prevent loss of land, which is important because of the relatively rapid rate of cliff retreat in these cliffs, which has led to considerable loss of land over the last 900 years, and indeed over the 100 years to 1985. This characteristic is shared with the till cliffs of Holderness in east Yorkshire. It appears necessary to protect the cliffs from wave action, i.e. basal removal, in order to reduce or actually halt cliff recession and so protect the cliff-top and the land behind it.

The second area of debate is concerned with the conservation of the cliffs and their characteristics for their own sake, particularly for their scientific and/or pedagogic value, be it biological or geological. There is one area of general agreement: grading of the cliffs, with protection of the cliff-foot by a hard structure like a concrete wall, would be inimical to such interests. The reason for this is obvious in the case of the conservation interest of the cliff: such interest would be destroyed.

Clayton (1995) points out that the Royal Commission on Afforestation and Coastal Erosion stated in 1911 that it is only possible to protect parts of the coast if other parts are left to erode and so supply sediment to adjacent beaches. He continues:

> *'Yet our ambitious coastal engineers have protected very high proportions of our eroding coastline and ignored this long-standing wise advice. The history of coastal engineering shows that at most it wins two or three decades of stability, but by the end of that period the problems are increasingly intractable; indeed, after storm damage, most rebuilt coastal defences are set back by the very same amount that nature would have achieved year by year had things been left alone.'*

Acceptance of the need for 'soft' engineering and sustainable coastal defences has in recent years become recognized by both government and the coastal engineering profession. In part this has been brought about by the institution of Shoreline Management Plans (SMPs) in England and Wales (MAFF, 1995), drawn up by 'responsible authorities' for a set of 'sub-cells' of littoral cells defined by Motyka and Brampton (1993). The process involves public consultation. It is described in some detail by Hooke and Bray (1995). English Nature [now Natural England], which was one of the sponsors of the guidelines for Shoreline Management Plans (MAFF, 1995), recognized that such guidance could include an obligation for local authorities to take cliff conservation (and hence 'soft' engineering in coast protection works) into account (Leafe and Radley, 1994; Swash *et al.*, 1995). By 1998 it was apparent that wide acceptance of this had been achieved (Leafe, 1998). Also, Richardson (1996) explains how the UK policy for sustainable development applies to coast protection, and how in many situations it necessitates consideration of 'soft' engineering solutions.

Therefore the Trimingham Cliffs mass-movement site, the western half of which is in a section of unprotected cliff while the eastern half is protected by an unmaintained revetment, should provide an interesting opportunity for monitoring the development or maintenance of mass-movement types and forms, and the effect or success of the conservation/protection policy.

An independent description on the situation at Trimingham is given by Younger (1990):

> '...*attention has now turned to the area between Sidestrand and Trimingham. Here is the last undefended stretch of north-east Norfolk, together with an area of decaying and derelict revetments. The land at the cliff-top is high quality agricultural land, together with the village of Trimingham, clustered around a mediaeval parish church, with a substantial minority of new four-bedroomed houses and bungalows.*'

The view that measures to reduce or eliminate cliff recession would be unwise is perhaps counter-intuitive, and depends upon the argument detailed above. As Clayton (1980) remarks, such arguments are unlikely to find favour with those who live in the vicinity of the eroding cliffs, and Hutchison and Leafe (1996) extend the point: 'An operating authority [responsible for the production of an SMP] is currently unlikely to consider the relocation of property in order to allow continued erosion as an input to the sediment budget.'

However, Lee (1998) has sounded a warning note, pointing out that coast protection schemes such as those now advocated, which slow down rather than stop marine erosion, can result in changes in the rates and types of processes acting on a cliff. This can lead to the development of new cliff forms, and, in the context of mass movements, could change or destroy the mass-movement interest.

Conclusions

In summary, the cliffs at Trimingham are worthy of conservation as an example of soft rock degradation, with a characteristic set of different types of mass movement. To this may be added a second conservation imperative, that the site should continue to be allowed to contribute sediment to the beach, in order that the beach itself, and the coast which it fronts for many kilometres to the south, is protected from accelerated erosion. The risk noted by Lee (1998) is probably worth taking in this case, as each length of cliff is different, and the mass movements on the cliffs at Trimingham are dynamic features. Any changes in the style of cliff retreat in the protected part of the site will become apparent by comparison with the unprotected cliff, which will act as a control for comparison.

The decision to leave a length of cliff in an unprotected state so as to act as a feeder bluff for a long length of beach is secondary but opportune in the present context. As a result of this pragmatic decision, which has no geological conservation intention, conservation of a suite of mass-movement types is served.

References

In this reference list the arrangement is alphabetical by author surname for works by sole authors and dual authors. Where there are references that include the first-named author with others, the sole-author works are listed chronologically first, followed by the dual author references (alphabetically) followed by the references with three or more authors listed *chronologically*. Chronological order is used within each group of identical authors.

Abele, G. (1974) Bergstürze in den Alpen – ihre Verbreitung, Morphologie und Folgeerscheinungen. *Wissenshaftliche Alpenvereinshefte*, **25**, 230.

Addison, K. (1987) Debris flow during intense rainfall in Snowdonia, North Wales – a preliminary survey. *Earth Surface Processes and Landforms*, **12**, 561–6.

Aitkenhead, N., Chisholm, J.I. and Stevenson, I.P. (1985) *Geology of the Country Around Buxton, Leek and Bakewell*, Memoir of the British Geological Survey, Sheet 111 (England and Wales), HMSO, London, 168 pp.

Al-Dabbagh, T. (1985) A study of residual shear strength of Namurian shale in respect of slopes in North Derbyshire. Unpublished PhD thesis, University of Sheffield.

Anderson, F.W. and Dunham, K.C. (1966) *The Geology of Northern Skye*, Memoir of the Geological Survey of Great Britain, Sheet 80 and parts of 90, 81 and 91 (Scotland), HMSO, Edinburgh, 216 pp.

Anon. [Cooper, R.G.] (1982) Mass movement phenomena. In Further Progress of the Geological Conservation Review (ed. G.P. Black). *Earth Science Conservation*, **19**, pp. 27–9.

Arber, M.A. (1939) The great landslip in 1839. *Country Life*, **23 December 1939**, 658–9.

Arber, M.A. (1940) The coastal landslips of South-East Devon. *Proceedings of the Geologists' Association*, **51**, 257–71.

Arber, M.A. (1941) The coastal landslips of West Dorset. *Proceedings of the Geologists' Association*, **52**, 273–83.

Arber, M.A. (1962) Discussion. In Coastal cliffs: report of a symposium (J.A. Steers). *Geographical Journal*, **124**, pp. 303–20.

Arber, M.A. (1971) The plane of landslipping on the coast of South-East Devon. In *Applied Coastal Geomorphology* (ed. J.A. Steers), Macmillan, London, Appendix.

Arber, M.A. (1973) Landslips near Lyme Regis. *Proceedings of the Geologists' Association*, **84**, 121–33.

Ashford, S.A. and Sitar, N. (1995) Topographic amplification in steep slopes. In *Proceedings of IS-Tokyo '95: The First International Conference on Earthquake Geotechnical Engineering, Tokyo, 14–16 November 1995* (ed. K. Ishihara), Balkema, Rotterdam, pp. 1153–8.

Bailey, E.B. and Anderson, E.M. (1925) *The Geology of Staffa, Iona and Western Mull*, Memoir of the Geological Survey of Great Britain, Sheet 43 (Scotland), HMSO, Edinburgh, 107 pp.

References

Bailey, E.B. and Maufe, H.B. (1916) *The Geology of Ben Nevis and Glen Coe, and the Surrounding Country*, Memoir of the Geological Survey of Great Britain, Sheet 53 (Scotland), HMSO, Edinburgh, 247 pp.

Bailey, E.B. and Maufe, H.B. (1960) *The Geology of Ben Nevis and Glen Coe, and the Surrounding Country*, 2nd edn, Memoir of the Geological Survey of Great Britain, Sheet 53 (Scotland), HMSO, London, 307 pp.

Baird, P.D. and Lewis, W.V. (1957) The Cairngorm floods 1956: summer solifluction and distributary flormation. *Scottish Geographical Magazine*, **73**, 91–100.

Ballantyne, C.K. (1981) Periglacial landforms and environments on mountains in the northern highlands of Scotland. Unpublished PhD thesis, University of Edinburgh.

Ballantyne, C.K. (1986a) Landslides and slope failures in Scotland: a review. *Scottish Geographical Magazine*, **102**, 134–50.

Ballantyne, C.K. (1986b) Protalus rampart development and the limits of former glaciers in the vicinity of Baosbheinn, Wester Ross. *Scottish Journal of Geology*, **22**, 13–25.

Ballantyne, C.K. (1987a) The Beinn Alligin 'rock glacier'. In *Wester Ross: Field Guide* (eds C.K. Ballantyne and D.G. Sutherland), Quaternary Research Association, Cambridge, pp. 134–7.

Ballantyne, C.K. (1987b) Some observations on the morphology and sedimentology of two active protalus ramparts, Lyngen, northern Norway. *Arctic and Alpine Research*, **19**, 167–74.

Ballantyne, C.K. (1990) The late Quaternary glacial history of the Trotternish Escarpment, Isle of Skye, Scotland, and its implications for ice-sheet reconstruction. *Proceedings of the Geologists' Association*, **101**, 171–86.

Ballantyne, C.K. (1991a) The landslips of Skye. In *The Quaternary of the Isle of Skye: Field Guide* (eds C.K. Ballantyne, D.I. Benn, J.J. Lowe and M.J.C. Walker), Quaternary Research Association, Cambridge, pp. 82–9.

Ballantyne, C.K. (1991b) Scottish landform examples – 2. The landslides of Trotternish, Isle of Skye. *Scotish Geographical Magazine*, **107**(2), 130–5.

Ballantyne, C.K. (1992) Rock-slope failure and debris flow, Gleann na Guiserein, Knoydart: comment. *Scottish Journal of Geology*, **28**, 77–80.

Ballantyne, C.K. (1994) Gibbsitic soils on former nunataks: implications for ice sheet reconstruction. *Journal of Quaternary Science*, **9**, 73–80.

Ballantyne, CK (1997) Holocene rock-slope failures in the Scottish Highlands. In *Rapid Mass Movement as a Source of Climatic Evidence for the Holocene* (eds J.A. Matthews, D. Brunsden, B. Frenzel, B. Gläser and M.M. Weiss), *Palaoklimaforschung*, No. **19**, Gustav Fischer, Stuttgart, pp. 197–205.

Ballantyne, C.K. (1998) Aeolian deposits on a Scottish mountain summit: characteristics, provenance, history and significance. *Earth Surface Processes and Landforms*, **23**, 625–41.

Ballantyne, C.K. (2002a) Paraglacial geomorphology. *Quaternary Science Reviews*, **21**, 1935–2017.

Ballantyne, C.K. (2002b) Debris flow activity in the Scottish Highlands: temporal trends and wider implications for dating. *Studia Geomorphologica Carpatho-Balcanica*, **36**, 7–27.

Ballantyne, C.K. (2002c) Geomorphological changes and trends in Scotland: debris flows. *Scottish Natural Heritage Commissioned Report*, **F00AC107A**, 27 pp.

Ballantyne, C.K. (2003) A Scottish sturzstrom: the Beinn Alligin rock avalanche, Wester Ross. *Scottish Geographical Journal*, **119**, 159–67.

Ballantyne, C.K. and Eckford, J.D. (1984) Characteristics of evolution of two relict talus slopes in Scotland. *Scottish Geographical Magazine*, **100**, 20–33.

Ballantyne, C.K. and Harris, C. (1994) *The Periglaciation of Great Britain*, Cambridge University Press, Cambridge, 330 pp.

Ballantyne, C.K. and Stone, J.O. (2004) The Beinn Alligin rock avalanche, NW Scotland: cosmogenic ^{10}Be dating, interpretation and significance. *The Holocene*, **14**, 448–53.

Ballantyne, C.K., McCarroll, D., Nesje, A., Dahl, S.O. and Stone J.O. (1998a) The last ice sheet in North-West Scotland: reconstruction and implications. *Quaternary Science Reviews*, **17**, 1149–84.

Ballantyne, C.K., Stone, J.O. and Fifield, L.K. (1998b) Cosmogenic Cl-36 dating of postglacial landsliding at the Storr, Isle of Skye, Scotland. *The Holocene*, **8**, 347–51.

References

Banham, P.H. (1968) A preliminary note on the Pleistocene stratigraphy of north-east Norfolk. *Proceedings of the Geologists' Association*, **79**, 507–12.

Banham, P.H. (1975) Glaci-tectonic structures: a general discussion with particular reference to the contorted drift of Norfolk. In *Ice Ages: Ancient and Modern: Proceedings of the 21st Inter-University Geological Congress Held at the University of Birmingham, 2–4 January 1974* (eds A.E. Wright and F. Moseley) *Geological Journal Special Issue*, No. **6**, Seel House Press, Liverpool, pp. 69–94.

Barton, M.E. (1984) The preferred path of landslide shear surfaces in over-consolidated clays and soft rocks. In *IV International Symposium on Landslides*, Canadian Geotechnical Society, Toronto, Vol. **3**, pp. 75–9.

Bass, M.A. (1954) A study of the characteristics and origin of some major areas of landsliding in the eastern Pennines and the Isle of Wight. Unpublished MSc thesis, University of Sheffield.

Bateman, T. (1861) *Ten Years' Diggings in Celtic and Saxon Grave Hills, in the Counties of Derby, Stafford and York, from 1848 to 1858; with Notices of some Former Discoveries, Hitherto Unpublished, and Remarks on the Crania and Pottery from the Mounds*, Smith, London, and Bemrose, Derby, 309 pp.

Beck, A.C. (1968) Gravity faulting as a mechanism of topographic adjustment. *New Zealand Journal of Geology and Geophysics*, **11**, 191–9.

Beckinsale, R.P. (1970) Physical problems of Cotswold rivers and valleys. *Proceedings of the Cotteswold Naturalists Field Club*, **35**, 194–205.

Bell, B.R. and Harris, J.W. (1986) *An Excursion Guide to the Geology of the Isle of Skye*, Geological Society of Glasgow, Glasgow, 317 pp.

Bell, B.R. and Williamson, I.T. (2002) Tertiary igneous activity. In *The Geology of Scotland*, 4th edn (ed. N.H. Trewin), The Geological Society, London, pp. 371–407.

Bennett, M.R. and Langridge, A.J. (1990) A two stage rockslide in Gleann na Guiserein, Knoydart. *Scottish Journal of Geology*, **97**, 653–65.

Bentley, M. and Dugmore, A.J. (1998) Landslides and the rate of glacial trough formation in Iceland. In *Mountain Glaciation* (ed. L.A. Owen), *Quaternary Proceedings*, No. **6**, Wiley, Chichester, pp. 11–15.

Bisci, C., Dramis, F. and Sorriso-Valvo, M. (1996) Rock flow (sackung). In *Landslide Recognition: Identification, Movement and Causes* (eds R. Dikau, D. Brunsden, L. Schrott and M.-L. Ibsen), *International Association of Geomorphologists Publication*, No. **5**, John Wiley and Sons, Chichester, pp. 150–60.

Bishop, A.W. and Morgenstern, N. (1960) Stability coefficient for earth slopes. *Géotechnique*, **10**, 129–50.

Bjerrum, L. (1954) Geotechnical properties of Norwegian marine clays. *Géotechnique*, **4**, 49–60.

Bjerrum, L. (1966) Mechanism of progressive failure in slopes of overconsolidated plastic clays and clay shales. *Journal of the Soil Mechanics and Foundations Division: Proceedings of the American Society of Civil Engineers*, **93**(SM5), 1–49.

Bjerrum, L. and Jørstad, F.A. (1968) Stability of rock slopes in Norway. *Norwegian Geotechnical Institute Publication*, **79**, 1–11.

Bjerrum, L. and others (1969) A field study of factors responsible for quick clay slides. In *Proceedings of the Seventh International Conference on Soil Mechanics and Foundation Engineering, Mexico City, 1969*, Sociedad Mexicana de Mecánica de Suelos, Mexico City, Vol. **2**, pp. B31–B40.

Blikra, LH. and Nemec, W. (1998) Postglacial colluvium in western Norway: depositional processes, facies and palaeoclimatic record. *Sedimentology*, **45**, 909–59.

Boggett, A.D. (1989) The Bretton Clough landslides, Derbyshire. *Mercian Geologist*, **11**, 223–35.

Botch, S.G. (1946) Les névés et l'érosion par la neige dans la partie Nord de l'Oural. *Bulletin of the USSR Geographical Society*, **78**, 207–34 (translated from Russian by C.E.D.P., Paris).

Boulton, G.S., Peacock, J.D. and Sutherland, D.G. (1991) Quaternary. In *Geology of Scotland*, 3rd edn (ed. G.Y. Craig), The Geological Society, London, pp. 503–43.

Bovis, M.J. and Dagg, B.R. (1992) Debris flow triggering by impulsive loading: mechanical modelling and case studies. *Canadian Geotechnical Journal*, **29**, 345–52.

Bovis, M.J. and Evans, S.G. (1996) Extensive deformations of rock slopes in southern Coast Mountains, southwest British Columbia, Canada. *Engineering Geology*, **22**, 163–82.

References

Brabb, E.E. and Harrod, B.L. (eds) (1989) *Landslides: Extent and Economic significance: Proceedings of the 28th International Geological Congress: Symposium on Landslides, Washington D.C., 17 July 1989*, A.A. Balkema, Rotterdam, 385 pp.

Brazier, V. and Ballantyne, C.K. (1989) Late Holocene debris cone evolution in Glen Feshie, western Cairngorm Mountains, Scotland. *Transactions of the Royal Society of Edinburgh: Earth Sciences*, **80**, 17–24.

Briggs, D. and Courtney, F. (1972) Ridge and trough topography in the North Cotswolds. *Proceedings of the Cotteswold Naturalists Field Club*, 94–103.

Bromehead, C.E.N., Edwards, W., Wray, D.A. and Stevens, J.V. (1933) *The Geology of the Country Around Holmfirth and Glossop*, Memoir of the Geological Survey of Great Britain, Sheet 86 (England and Wales), HMSO, London, 209 pp.

Bromhead, E.N. (1979) Factors affecting the transition between the various types of mass movement in coastal cliffs consisting largely of overconsolidated clay with special reference to Southern England. *Quarterly Journal of Engineering Geology*, **12**, 291–300.

Bromhead, E.N. (1986) *The Stability of Slopes*, Surrey University Press, Glasgow, 373 pp.

Bromhead, E.N. and Dixon, N. (1984) Pore water pressure observations in the coastal clay cliffs at the Isle of Sheppey, England. In *IV International Symposium on Landslides*, Canadian Geotechnical Society, Toronto, Vol. **1**, pp. 385–90.

Bromhead, E.N. and Dixon, N. (1986) Technical note: The field residual strength of London Clay and its correlation with laboratory measurements, especially ring shear tests. *Géotechnique*, **36**, 449–52.

Bromhead, E.N. and Vaughan, P.R. (1980) Solutions for seepage in soils with an effective stress dependent permeability. In *Numerical Methods for Non-linear Problems: Proceedings of the International Conference Held at University College, Swansea, 2–5 September 1980* (eds C. Taylor, E. Hinton and D.R.J. Owen), Pineridge Press, Swansea, pp. 567–78.

Brown, E. (2003) Interpreting Quaternary landscapes in the Loch Lomond and Trossachs National Park. In *The Quaternary of the Western Highland Boundary: Field Guide* (ed. D.J.A. Evans), Quaternary Research Association, London, pp. 69–74.

Brown, M.C. (1973) Limestones on the eastern side of the Derbyshire outcrop of the Carboniferous Limestone. Unpublished PhD thesis, University of Reading.

Brown, R.D. (1966) Recent landslips at Mam Tor. *Don*, **10**, 13–18.

Brown, R.D. (1977) *Excursion Itineraries for the 6th British–Polish Seminar, Sheffield*, Department of Geography, University of Sheffield, Sheffield.

Brückl, E. and Parotidis, M. (2005) Prediction of slope instabilities due to deep-seated gravitational creep. *Natural Hazards and Earth System Sciences*, **5**, 155–72.

Brunsden, D. (1969) The moving cliffs of Black Ven. *Geographical Magazine*, **41**(5), 372–4.

Brunsden, D. (1979) Mass movements. In *Process in Geomorphology* (eds C.E. Embleton and J.B. Thornes), Edward Arnold, London, pp. 130–86.

Brunsden, D. (1984) Mudslides. In *Slope Instability* (eds D. Brunsden and D.B. Prior), Wiley, Chichester, pp. 363–418.

Brunsden, D. (1985) The revolution in geomorphology: a prospect for the future. In *Geographical Futures* (ed. R. King), The Geographical Association, Sheffield, pp. 30–55.

Brunsden, D. (1990) Tablets of stone: towards the ten commandments of geomorphology. *Zeitschrift für Geomorphologie, Supplementband*, **79**, 1–17.

Brunsden, D. (1996a) The case of the missing ductile layer. *Journal of the Geological Society of China*, **39**(4), 535–56.

Brunsden, D. (1996b) Landslides of the Dorset coast: some unresolved questions: The Scott Simpson lecture. *Proceedings of the Ussher Society*, **9**, 1–11.

Brunsden, D. and Allison, R.J. (1990) Geomorphology. In *Landslides of the Dorset Coast* (ed. R.J. Allison), British Geomorphological Research Group Field Guide, British Geomorphological Research Group, Sheffield, pp. 37–50.

Brunsden, D. and Chandler, J.H. (1996) Development of an episodic landform change model based upon the Black Ven mudslide, 1946–1995. In *Advances in Hillslope Processes* (eds M.G. Anderson and S.M. Brooks), British Geomorphological Research Group Symposia Series, Wiley, Chichester, pp. 869–96.

References

Brunsden, D. and Goudie, A.S. (1981) *Classic Coastal Landforms of Dorset*, Landform Guides, No. **1**, The Geographical Association and British Geomorphological Research Group, Sheffield, 42 pp.

Brunsden, D. and Ibsen, M.-L. (1994a) The nature of the European archive of historical landslide data, with specific reference to the United Kingdom. In *Temporal Occurrence and Forecasting of Landslides in the European Community. Final Report, Part 1. Methodology (Reviews) for the Temporal Study of Landslides* (ed. J.C. Flageollet), Contract 90 0025, Prog. EPOCH, European Community, pp. 21–70.

Brunsden, D. and Ibsen, M.-L. (1994b) The temporal causes of landslides on the south coast of Great Britain. In *Temporal Occurrence and Forecasting of Landslides in the European Community. Final Report, Part 2, Vol. 1. Case Studies of the Temporal Occurrence of Landslides in the European Community* (ed. J.C. Flageollet), Contract 90 0025, Prog. EPOCH, European Community, pp. 339–83.

Brunsden, D. and Ibsen, M.-L. (1994c) The spatial and temporal distribution of landslides on the south coast of Great Britain. In *Temporal Occurrence and Forecasting of Landslides in the European Community. Final Report, Part 2, Vol. 1. Case studies of the Temporal Occurrence of Landslides in the European Community* (ed. J.C. Flageollet), Contract 90 0025, Prog. EPOCH, European Community, pp. 385–423.

Brunsden, D. and Ibsen, M.-L. (1997) The temporal occurrence and forecasting of landslides in the European Community: summary of relevant results of the European Community EPOCH programme, Contract No. EPOC-CT-90.0025 (DTTE). In *Rapid Mass Movement as a Source of Climatic Evidence for the Holocene* (eds J.A. Matthews, D. Brunsden, B. Frenzel, B. Gläser and M.M. Weiß), *Paläoklimaforschung – Paleoclimatic Research*, No. **19**, Gustav Fischer, Stuttgart, pp. 401–7.

Brunsden, D. and Jones, D.K.C. (1976) The evolution of landslide slopes in Dorset. *Philosophical Transactions of the Royal Society of London, Series A*, **283**, 605–31.

Brunsden, D. and Kesel, R.H. (1973) Slope development on a Mississippi river bluff in historic time. *Journal of Geology*, **81**, 576–97.

Brunsden, D. and Thornes, J.B. (1979) Landscape sensitivity and change. *Transactions of the Institute of British Geographers*, **4**, 463–84.

Brunsden, D., Ibsen, M.-L., Lee, M. and Moore, R. (1995) The validity of temporal archive records for geomorphological processes. *Quaestiones Geographicae*, Special Issue No. **4**, 1–13.

Brunsden, D., Ibsen, M.-L., Bromhead, E. and Collison, A. (1996a) *Final national report (June 1996). King's College London, United Kingdom. The Temporal Stability and Activity of Landslides in Europe with Respect to Climatic Change (TESLEC)*, European Commission Environment Programme, Contract No. EV5V-CT94-0454.

Brunsden, D., Coombe, K., Goudie, A.S. and Parker, A.G. (1996b) The structural geomorphology of the Isle of Portland, southern England. *Proceedings of the Geologists' Association*, **107**, 209–30.

Buckland, W. (1823) *Reliquiae Diluvianae: or Observations on the Organic Remains Contained in Caves, Fissures and Diluvial Gravel, and on other Geological Phenomena Attesting the Action of an Universal Deluge*, John Murray, London, 303 pp.

Buckland, W. (1840) On the landslipping near Axmouth. *Proceedings of the Ashmolean Society*, **1**, 1832–42.

Burnett, A.D. and Fookes, P.G. (1974) A regional engineering geological study of the London Clay in the London and Hampshire Basins. *Quarterly Journal of Engineering Geology*, **7**, 257–95.

Cailleux, A. and Tricart, J. (1950) Un type de solifluction: les coulées boueuses. *Révue de Géomorphologie Dynamique*, **1**, 4–46.

Caine, N. (1982) Toppling failures from Alpine cliffs on Ben Lomond, Tasmania. *Earth Surface Processes and Landforms*, **7**, 133–52.

Cambers, G. (1973) The retreat of unconsolidated Quaternary cliffs. Unpublished PhD thesis, University of East Anglia.

Cambers, G. (1976) Temporal scales in coastal erosion systems. *Transactions of the Institute of British Geographers*, **1**(2), 246–56.

Campbell, S. and Bowen, D.Q. (1989) *Quaternary of Wales*, Geological Conservation Review Series, No. **2**, Nature Conservancy Council, Peterborough.

Cancelli, A. and Pellegrini, M. (1987) Deep-seated gravitational deformation in the Northern Appenines, Italy. In *Proceedings of the 5th International Conference and Field Workshop on Landslides (ICFL), Christchurch, New Zealand, 10–12 August 1987*, pp. 1–8.

References

Casey, R. (1955) Contribution to 'Discussion on "Folkestone Warren landslips: investigations, 1948–50"'. *Proceedings of the Institution of Civil Engineers, Railway Paper*, No. **56**, 429–60.

Chandler, J.H. and Brunsden, D. (1995) Steady state behaviour of the Black Ven mudslide: the application of archival analytical photogrammetry to studies of landform change. *Earth Surface Processes and Landforms*, **20**, 255–75.

Chandler, J.H. and Cooper, M.A.R. (1988) Monitoring the development of landslides using archival photography and analytical photogrammetry. *Land and Minerals Surveying*, **6**, 576–84.

Chandler, J.H. and Cooper, M.A.R. (1989) The extraction of positional data from historical photographs and their application to geomorphology. *Photogrammetric Record*, **13**(73), 69–78.

Chandler, J.H., Kellaway, G.A., Skempton, A.W. and Wyatt, R.J. (1976) Valley slope sections in Jurassic strata near Bath, Somerset. *Philosophical Transactions of the Royal Society of London, Series A*, **283**, 527–56.

Chandler, J.H., Clark, J.S., Cooper, M.A.R. and Stirling, D.M. (1987) Analytical photogrammetry applied to Nepalese slope morphology, *Photogrammetric Record*, **12**(70), 443–58.

Chandler, R.J. (1984) Recent European experience of landslides in over-consolidated clays and soft rocks. In *IV International Symposium on Landslides*, Canadian Geotechnical Society, Toronto, Vol. **1**, pp. 61–81.

Chesher, J.A., Smythe, D.K. and Bishop, P. (1983) The geology of the Minches, Inner Sound and Sound of Raasay. *Report of the Institute of Geological Sciences*, **83/6**, 30 pp.

Chigira, M. (1992) Long-term gravitational deformation of rocks by mass rock creep. *Engineering Geology*, **32**, 157–84.

Churcher, R.A., Butler, B. and Bartlett, P.D. (1970) A further report on the caves of the Isle of Portland. *Transactions of the Cave Research Group of Great Britain*, **12**, 291–8.

Clark, A., Lee, M., and Moore, R. (1996) *Landslide Investigation and Management in Great Britain: a Guide for Planners and Developers*, Department of the Environment, HMSO, London, 120 pp.

Clark, R. and Wilson, P. (2004) A rock avalanche deposit in Burtness Combe, Lake District, northwest England. *Geological Journal*, **39**, 419–30.

Clayton, K.M. (1974) Zones of glacial erosion. In *Progress in Geomorphology* (eds E.H. Brown and R.S. Waters), *Institute of British Geographers Special Publication*, No. **7**, Institute of British Geographers, London, pp. 163–76.

Clayton, K.M. (1980) Coastal protection along the East Anglian coast, U.K. *Zeitschrift für Geomorphologie, Supplementband*, **32**, 165–72.

Clayton, K.M. (1989) Sediment input from the Norfolk cliffs, Eastern England – A century of coast protection and its effect. *Journal of Coastal Research*, **5**(3), 422–33.

Clayton, K.M. (1995) The coast. In *The National Trust: the Next Hundred Years* (ed. H. Newby), The National Trust, London, pp. 70–86.

Clayton, K.M. and Coventry, F. (1986) An Assessment of the Conservation Effectiveness of the Modified Coast Protection Works at West Runton SSSI, Norfolk. *Nature Conservancy Council Report*, **CSD 675**.

Clayton, K.M., McCave, I.N. and Vincent, C.E. (1983) The establishment of a sand budget for the East Anglian coast and its implications for coastal stability. In *Shoreline Protection: Proceedings of a Conference Organised by the Institution of Civil Engineers and Held at the University of Southampton on 14–15 September 1982*, Thomas Telford, London, pp. 91–6.

Clough, C.T. (1897) Landslips. In *The Geology of Cowal* (eds W. Gunn, C.T. Clough and J.B. Hill), Memoir of the Geological Survey of Scotland, Sheet 29 and parts of 37 and 38 (Scotland), HMSO, Edinburgh, 333 pp.

Conway, B.W. (1974) The Black Ven landslip, Charmouth, Dorset. *Report of the Institute of Geological Sciences*, **74/3**, 16 pp.

Conybeare, W.D. (1840) Extraordinary land-slip and great convulsion of the coast of Culverhole Point, near Axmouth. *Edinburgh New Philosophical Journal*, **29**, 160–4.

Conybeare, W.D., Buckland, W. and Dawson, W. (1840) *Ten Plates Comprising a Plan, Section and Views Representing the Changes produced on the Coast of East Devon between Axmouth and Lyme Regis by the Subsidence of the Land and the Elevation of the Bottom of the Sea on 26th December, 1839 and 3rd February 1840*, John Murray, London.

References

Coombe, E.D.K. (1981) Some aspects of coastal landslips and cliff-falls at Portland. Unpublished BA Dissertation, University of Oxford.

Cooper, R.G. (1978) The discovery and exploration of the North Yorkshire windypits. *Ryedale Historian*, **9**, 10–21.

Cooper, R.G. (1979) Geomorphological studies in the Hambleton Hills, North Yorkshire. Unpublished PhD thesis, University of Hull.

Cooper, R.G. (1980) A sequence of landsliding mechanisms in the Hambleton Hills, Northern England, illustrated by features at Peak Scar, Hawnby. *Geografiska Annaler*, **62A**(3–4), 149–56.

Cooper, R.G. (1981) Four new windypits. *Caves and Caving*, **14**, 3–4.

Cooper, R.G. (1982) Further progress of the Geological Conservation Review. Mass movement phenomena. *Earth Science Conservation*, **19**, 27–9.

Cooper, R.G. (1983a) Fissures in the interior of the Isle of Portland. *William Pengelly Cave Studies Trust Newsletter*, **42**, 1–3.

Cooper, R.G. (1983b) Mass movement in caves in Great Britain. *Studies in Speleology*, **4**, 37–44.

Cooper, R.G. (1985) Conservation of geological features in the USA. *Earth Science Conservation*, **2**, 9–12.

Cooper, R.G. (1997) *John Wesley at Whitestone Cliff, North Yorkshire, 1755*, Borthwick Papers, No. **91**, University of York, York, 25 pp.

Cooper, R.G. and Solman, K.R. (1983) Fissures at Westcliffe, Isle of Portland. *Caves and Caving*, **22**, 22–3.

Cooper, R.G., Ryder, P.F. and Solman, K.R. (1976) The North Yorkshire windypits: a review. *Transactions of the British Cave Research Association*, **3**, 77–94.

Cooper, R.G., Ryder, P.F. and Solman, K.R. (1977) Caves in Lud's Church, North Staffordshire. *Bulletin of the British Cave Research Association*, **16**, 7.

Cooper, R.G., Ryder, P.F. and Solman, K.R (1982) The windypits in Duncombe Park, Helmsley, North Yorkshire. *Cave Science*, **9**, 1–14.

Cooper, R.G., Graham, N. and Read, M. (1995) Solutional cave passages intersected by mass movement rifts in the Isle of Portland, Dorset. *Studies in Speleology*, **10**, 29–35.

Cornish, R. (1981) Glaciers of the Loch Lomond Stadial in the western Southern Uplands of Scotland. *Proceedings of the Geologists' Association*, **92**, 105–14.

Corominas, J., Remond, J., Farias, P., Estevao, M., Zézere, J., Díaz de Terán, R., Dikau, R., Schrott, L., Moya, J. and González, A. (1996) Debris flow. In *Landslide Recognition, Identification, Movement and Causes* (eds R. Dikau, D. Brunsden, L. Schrott, and M.-L. Ibsen), John Wiley and Sons, Chichester, pp. 161–80.

Cotton, C. (1685) *The Wonders of the Peake*, 2nd edn, 86 pp.

Coussot, P. and Meunier, M. (1995) Recognition, classification and mechanical description of debris flows. *Earth Science Reviews*, **40**, 209–27.

Cripps, J.C. and Hird, C.C. (1992) A guide to the landslide at Mam Tor. *Geoscientist*, **2**(3), 22–7.

Cross, M. (1987) An engineering geomorphological investigation of hillslope stability in the Peak District of Derbyshire. Unpublished PhD thesis, University of Nottingham.

Crosta, G. (1996) Landslide, spreading, deep-seated gravitational deformation: analysis, examples, problems and proposals. *Geografia Fisica e Dinamica Quaternaria*, **19**, 297–313.

Cruden, D.M., Krauter, E., Beltran, L., Lefebvre, G., Ter-Stepanian, G.I. and Zhang, Z.Y. (1994) Describing landslides in several languages: The Multilingual Landslide Glossary. In *Proceedings 7th International Congress International Association of Engineering Geology, 5–9 September 1994, Lisboa Portugal*, (eds R. Oliveira and others), Balkema, Rotterdam, pp. 1325–33.

Cullum-Kenyon, S. (1991) A post-glacial debris flow near the Spittal of Glenshee, Perthshire. Unpublished BSc Honours Dissertation, University of St Andrews.

Cunliffe, B.W. (1977) The Saxon Shore – some problems and inconsistencies. In *The Saxon Shore* (ed. D.E. Johnston), Council for British Archaeology, London, pp. 1–6.

Cunliffe, B.W. (1980a) Excavations at the Roman Fort at Lympne, Kent 1976–1978. *Britannia*, **11**, 227–88.

Cunliffe, B.W. (1980b) The evolution of Romney Marsh: a preliminary survey. In *Archaeology and Coastal Change* (ed. F.H. Thompson), *Society of Antiquaries of London Occasional Papers*, No. **1**, Society of Antiquaries, London, pp. 37–53.

Curry, A.M., Walden, J. and Cheshire, D.A. (2001) The Nant Ffrancon 'protalus rampart': evidence for Late Pleistocene paraglacial landsliding in Snowdonia, Wales. *Proceedings of the Geologists' Association*, **112**, 317–330.

References

Dade, W.B. and Huppert, H.E. (1998) Long-runout rockfalls. *Geology*, **26**, 803–6.

Darton, D.M., Dingwall, R.G. and McCann, D.M. (1981) Geological and geophysical investigations in Lyme Bay. *Report of the Institute of Geological Sciences*, **79/10**, 24 pp.

Davenport, C.A., Ringrose, P.S., Becker, A., Hancock, P. and Fenton, C. (1989) Geological investigation of late and post glacial earthquake activity in Scotland. In *Earthquakes at North Atlantic Passive Margins: Neotectonics and Postglacial Rebound* (eds S. Gregersen and P. Basham), Kluwer Academic Publishers, Dordrecht, pp. 175–94.

Dawson, A.G., Matthews, J.A. and Shakesby, R.A. (1986) A catastrophic landslide (sturzstrom) in Verkilsdalen, Rondane National Park, southern Norway. *Geografiska Annaler*, **68A**, 77–87.

De Freitas, M.H. and Watters, R.J. (1973) Some field examples of toppling failure. *Géotechnique*, **23**, 495–514.

Delderfield, E.R. (1976) *The Lynmouth Flood Disaster*, 8th edn, E.R.D. Publications, Exmouth, 160 pp.

Denness, B. (1972) The reservoir principle of mass movement. *Report of the Institute of Geological Sciences*, **72/7**, 13 pp.

Dikau, R., Brunsden, D., Schrott, L. and Ibsen, M.-L. (eds) (1996) *Landslide Recognition*, Wiley, Chichester, 251 pp.

Dixon, N. and Bromhead, E.N. (1986) Groundwater conditions in the coastal landslides of the Isle of Sheppey. In *Groundwater in Engineering Geology: Proceedings of the 21st Annual Conference of the Engineering Group of the Geological Society* (eds J.C. Cripps, F.G. Bell and M.G. Culshaw), *Engineering Geology Special Publication*, No. **3**, The Geological Society, London, pp. 51–8.

Dixon, N. and Bromhead, E.N. (1991) The mechanics of first-time slides in the London Clay cliff at the Isle of Sheppey, England. In *Slope Stability Engineering: Developments and Applications: Proceedings of the International Conference on Slope Stability organized by the Institution of Civil Engineers and Held on the Isle of Wight on 15–18 April 1991* (ed. R.J. Chandler), Thomas Telford, London, pp. 277–82.

Doornkamp, J.C. (ed.) (1988) *Planning and Development: Applied Earth Science Background: Torbay*, Geomorphological Services for the Department of the Environment, Newport Pagnall, 109 pp.

Doornkamp, J.C. (1990) Landslides in Derbyshire. *East Midland Geographer*, **13**(2), 22–62.

Drayton, M. (1622) *The Second Part or Continuance of Poly-Olbion from the Eighteenth Song. Containing all the Tracts, Rivers, Mountaines, and Forests Intermixed with the most Remarkable Stories, Antiquities, Wonders, Rarities, Pleasures, and Commodities of the East, and Northerne parts of this Isle, Lying Betwixt the Two Famous Rivers of Thames, and Tweed*, Printed by Augustine Mathewes for John Marriott, John Grismond and Thomas Dewe, London.

Drew, F. (1864) *The Geology of the Country Between Folkestone and Rye, Including the Whole of Romney Marsh*, Memoir of the Geological Survey of Great Britain, Sheet 4 (England and Wales), HMSO, London, 27 pp.

Eigenbrod, K.D. (1975) Analysis of the pore pressure changes following the excavation of a slope. *Canadian Geotechnical Journal*, **12**, 424–40.

Elliott, R.V. (1979) The Upper Carboniferous rocks of the Ewden Valley, South Yorkshire. *Mercian Geologist*, **7**, 43–9.

Elliott, R.W.V. (1977) Staffordshire and Cheshire landscapes in *Sir Gawain and the Green Knight*. *North Staffordshire Journal of Field Studies*, **17**, 20–49.

Ellis, N.V., Bowen, D.Q., Campbell, S., Knill, J.L., McKirdy, A.P., Prosser, C.D., Vincent, M.A. and Wilson, R.C.L. (1996) *An Introduction to the Geological Conservation Review*, Geological Conservation Review Series, No. **1**. Joint Nature Conservation Committee, Peterborough, 131 pp.

Ellis-Gruffydd, I.D. (1972) The glacial geomorphology of the upper Esk basin (South Wales) and its right bank tributaries. Unpublished PhD thesis, University of London.

Emeleus, C.H. and Bell, B.R. (2005) *British Regional Geology: the Palaeogene volcanic districts of Scotland*, 4th edn, British Geological Survey, Nottingham, 212 pp.

Emeleus, C.H. and Gyopari, M.C. (1992) *British Tertiary Volcanic Province*, Geological Conservation Review Series, No. **4**, Chapman and Hall, London, 259 pp.

Evans, D.J.A. and Hansom, J.D. (1998) The Whangie and the landslides of the Campsie Fells. *Scottish Geographical Magazine*, **114**, 192–6.

References

Evans, D.J.A. and Hansom, J.D. (2003) The Whangie and paraglacial landslides of the Campsie Fells. In *The Quaternary of the Western Highland Boundary: Field Guide* (ed. D.J.A. Evans), Quaternary Research Association, London, pp. 76–80.

Evans, I.S. (1997) Process and form in the erosion of glaciated mountains. In *Process and Form in Geomorphology* (ed. D.R. Stoddart), Routledge, London.

Evans, S.G. (1984) Landslides in Tertiary basaltic successions. In *IV International Symposium on Landslides,* Canadian Geotechnical Society, Toronto, pp. 503–10.

Farmer, I.S. (1968) *Engineering Properties of Rocks*, E. and F.N. Spon Ltd, London, 180 pp.

Fearnsides, W.G., Bisat, W.S., Edwards, W., Lewis, H.P. and Wilcockson, W.H. (1932) The geology of the eastern part of the Peak District. *Proceedings of the Geologists' Association*, **43**, 152–91.

Fenton, C.H. (1991) Neotectonics and palaeoseismicity in North West Scotland. Unpublished PhD thesis, University of Glasgow.

Fenton, C.H. (ed.) (1992) *Neotectonics in NW Scotland: a Field Guide*, University of Glasgow, Glasgow.

Fiennes, C. (1947) *The Journeys of Celia Fiennes* (ed. C. Morris), The Cresset Press, London, 376 pp.

Firth, C.R. and Stewart, I.S. (2000) Postglacial tectonics of the Scottish glacio-isostatic uplift centre. *Quaternary Science Reviews*, **19**, 1469–93.

Fitton, E.P. and Mitchell, D. (1950) The Ryedale windypits. *Cave Science*, **2**(12), 162–84.

Flageollet, J.C. (1989) Landslides in France: a risk reduced by recent legal provisions. In *Landslides: Extent and Economic Significance* (eds E.E. Brabb and B.L. Harrod), Balkema, Rotterdam, pp. 157–67.

Fleming, S. (1978) *Reports and Documents in Reference to the Location of the Line and a Western Terminal Harbour*, Maclean Roger & Co., Ottawa.

Flint, R.F. (1957) *Glacial and Pleistocene Geology*, Wiley, New York, 553 pp.

Folk, R.L. (1959) Practical petrographic classification of limestones. *Bulletin of the American Association of Petroleum Geologists*, **79**, 1–38.

Ford, T.D. (1977) Carboniferous volcanic activity. In *Limestones and Caves of the Peak District* (ed. T.D. Ford), Geo Abstracts, Norwich, pp. 61–9.

Ford, T.D. and Hooper, M.J. (1964) The caves of the Isle of Portland. *Transactions of the Cave Research Group of Great Britain*, **7**, 13–37.

Franks, J.W. and Johnson, R.H. (1964) Pollen analytical dating of a Derbyshire landslip: the Crown Edge landslides, Charlesworth. *New Phytologist*, **63**, 209–16.

Gale, A.S. (1987) Field meeting at Folkestone Warren, 29th November, 1987. *Proceedings of the Geologists' Association*, **100**, 73–82.

Gallois, R.W. (1965) *British Regional Geology: The Wealden District*, 4th edn, HMSO, London, 101 pp.

Geomorphological Services Ltd (1988) Applied Earth Science mapping for planning and development: Torbay, Devon. *Report to the Department of the Environment.*

Gerber, E. and Scheidegger, A.E. (1969) Stress-induced weathering of rock masses. *Eclogae Geologicae Helveticae*, **62**, 401–15.

Ghosh, S.K. (1979) *Analytical Photogrammetry*, Pergamon, New York, 203 pp.

Gibson, R.E. (1963) An analysis of system flexibility and its effect on time lag in pore-water pressure measurements. *Géotechnique*, **13**, 1–11.

Gifford, J. (1953) Landslides on Exmoor caused by the storm of 15th August, 1952. *Geography*, **38**, 9–17.

Godard, A. (1965) *Recherches de Géomorphologie en Écosse du Nord-Ouest, Publications de la Faculté des lettres de l'Université de Strasbourg*, No. **1**, les Belles Lettres, Paris, 703 pp.

Golledge, N.R. and Hubbard, A. (2005) Evaluating Younger Dryas glacier reconstructions in part of the western Scottish Highlands: a combined empirical and theoretical approach. *Boreas*, **34**, 274–86.

Gordon, J.E. (1977) Morphometry of cirques in the Kintail–Affric–Cannich area of northwest Scotland. *Geografiska Annaler*, **59A**, 177–94.

Gordon, J.E. (1992) Conservation of geomorphology and Quaternary sites in Great Britain: an overview of site assessment. In *Conserving Our Landscape: Proceedings of the Conference Conserving our Landscape: Evolving Landforms and Ice-Age Heritage, Crewe UK, May 1992* (eds C. Stevens, J.E. Gordon, C.P. Green and M.G. Macklin), Nature Conservancy Council, Peterborough, pp. 11–21.

Gordon, J.E. (1993) Beinn Alligin. In *Quaternary of Scotland* (eds J.E. Gordon and D.G. Sutherland), Geological Conservation Review Series, No. **6**, Chapman and Hall, London, pp. 118–22.

Gordon and Sutherland (1993) *Quaternary of Scotland*, GCR Volume No. 6, Chapman and Hall, London, pp. 118–22.

Gordon, J.E. and Mactaggart, F. (1997) The Quaternary of Islay and Jura – report of short Field Meeting. *Quaternary News*, **83**, 31–4.

Gostelow, T.P. (1974) Slope development in stiff overconsolidated clays. Unpublished PhD thesis, University of London.

Goudie, A.S. and Hart, J. (1976) The Cleeve Hill slip-troughs. In *Field Guide to the Oxford Region* (ed. D. Roe), University of Reading for the Quaternary Research Association, Reading, pp. 55–60.

Graham, N. and Ryder, P.F. (1983) Sandy Hole, Isle of Portland. *Cave Science*, **10**, 171–80.

Grainger, P. and Kalaugher, P.G. (1995) Renewed landslide activity at Pinhay, Lyme Regis. *Proceedings of the Ussher Society*, **8**, 421–5.

Grainger, P., Tubb, C.D.N. and Neilson, A.P.M. (1985) Landslide activity at the Pinhay water source, Lyme Regis. *Proceedings of the Ussher Society*, **6**, 246–52.

Green, A.H., Le Neve Foster, C. and Dakyns, J.R. (1887) *The Geology of the Carboniferous Limestone, Yoredale Rocks and Millstone Grit of North Derbyshire*, Memoir of the Geological Survey of Great Britain, parts of sheets 88SE, 81NE, 81SE, 72NE, 82NW, 82SW and 71NW (England and Wales), HMSO, London, 212 pp.

Green, R.D. (1968) *The Soils of Romney Marsh, Bulletin of the Soil Survey of Great Britain (England and Wales)*, No. **4**, Agricultural Research Council, Harpenden, 158 pp.

Gregersen, O. and Sandersen, F. (1989) Landslide: extent and economic significance in Norway. In *Landslides: Extent and Economic Significance* (eds E.E. Brabb and B.L. Harrod), Balkema, Rotterdam, pp. 133–9.

Hall, A.M. (1991). Pre-Quaternary landscape evolution in the Scottish Highlands. *Transactions of the Royal Society of Edinburgh: Earth Sciences*, **82**, 1–26.

Hall, A.M. (ed.) (2003) Cairngorms landscape website: http://www.fettes.com/Cairngorms/index.htm

Hall, A.M. and Jarman, D. (2004). Quaternary landscape evolution – plateau dissection by glacial breaching. In *The Quaternary of the Central Grampian Highlands* (eds S. Lukas, J.W. Merritt and W. Mitchell), Quaternary Research Association, London, pp. 26–40.

Hallam, A. (1991) Jurassic, Cretaceous and Tertiary sediments. In *Geology of Scotland*, 3rd edn (ed. G.Y. Craig), The Geological Society, London, pp. 439–53.

Hawkins, A.B. (1977) Jurassic rocks of the Bath area. In *Geological Excursions in the Bristol District* (ed. R.J.G. Savage), University of Bristol, Bristol, pp. 119–32.

Hawkins, A.B. and Kellaway, G.A. (1971) Field Meeting at Bristol and Bath with special reference to new evidence of glaciation. *Proceedings of the Geologists' Association*, **82**, 267–92.

Hawkins, A.B. and Privett, K.D. (1979) Engineering geomorphological mapping as a technique to elucidate areas of superficial structures; with examples from the Bath area of the south Cotswolds. *Quarterly Journal of Engineering Geology*, **12**, 221–33.

Hawkins, A.B. and Privett, K.D. (1981) A building site on cambered ground at Radstock, Avon. *Quarterly Journal of Engineering Geology and Hydrogeology*, **14**, 151–67.

Hawkins, A.B. and Privett, K.D. (1985) Residual strength of cohesive soils. *Ground Engineering*, **18**, 22–9.

Hayes, R.H. (1962) Buckland's Windypit: an account of the excavations. *Peakland Archaeological Society Newsletter*, **9**, 23–9

Hayes, R.H. (1987) Archaeological finds in the Ryedale Windypits. *Studies in Speleology*, **7**, 31–74.

Haynes, V.M. (1977a) The modification of valley patterns by ice-sheet activity. *Geografiska Annaler*, **59A**, 195–207.

Haynes, V.M. (1977b) Landslip associated with glacier ice [Ben Hee]. *Scottish Journal of Geology*, **13**, 337–8.

Haynes, V.M. (1995) Scotland's landforms: a review. *Scottish Association of Geography Teachers Journal*, **24**, 18–37.

Henkel, D.J. (1967) Local geology and the stability of natural slopes. *Proceedings of the American Society of Civil Engineeers: Journal of the Soil Mechanics and Foundations Division*, **93**(SM4), 437–46.

Hepworth, O. (1954) *Some Historical Notes on Stocksbridge and District*, Stocksbridge Urban District Council, 38 pp.

Higgs, G. (1997) Black Monutain Scarp, Carmarthenshire. In *Fluvial Geomorphology of Great Britain* (ed. K.J. Gregory), Geological Conservation Review Series, No. **13**, Chapman and Hall, London, pp. 167–171.

References

Hill, H.P. (1949) The Ladybower Reservoir. *Journal of the Institution of Water Engineers*, **3**, 414–33.

Hinchliffe, S. (1998) The structure and evolution of relict talus accumulations in the Scottish Highlands. Unpublished PhD thesis, University of St Andrews.

Hinchliffe, S. (1999) Timing and significance of talus slope reworking, Trotternish, Isle of Skye, Scotland. *The Holocene*, **9**, 483–94.

Hinchliffe, S. and Ballantyne, C.K. (1999) Talus accumulation and rockwall retreat, Trotternish, Isle of Skye, Scotland. *Scottish Geographical Journal*, **115**, 53–70.

Hinchliffe, S., Ballantyne, C.K. and Walden, J. (1998) The structure and sedimentology of relict talus, Trotternish, northern Skye, Scotland. *Earth Surface Processes and Landforms*, **23**, 545–60.

Hoek, E. and Bray, J.W. (1977) *Rock Slope Engineering*, revised 2nd edn, Institution of Mining and Metallurgy, London, 402 pp.

Hoek, E. and Bray, J.W. (1981) *Rock Slope Engineering*, 3rd edn, Institution of Mining and Metallurgy, London.

Hollingworth, S.E. and Taylor, J.H. (1951) *The Northampton Sand Ironstone: Stratigraphy, Structure and Reserves*, Memoir of the Geological Survey of Great Britain, HMSO, London, 211 pp.

Hollingworth, S.E., Taylor, J.H. and Kellaway, G.A. (1944) Large-scale superficial structures in the Northampton ironstone field. *Quarterly Journal of the Geological Society of London*, **100**, 1–44.

Holmes, G. (1984) Rock-slope failure in parts of the Scottish highlands. Unpublished PhD thesis, University of Edinburgh.

Holmes, G. and Jarvis, J.J. (1985) Large-scale toppling within a Sackung type deformation at Ben Attow, Scotland. *Quarterly Journal of Engineering Geology*, **18**, 287–9.

Hooke, J.M. and Bray, M.J. (1995) Coastal groups, littoral cells, policies and plans in the UK. *Area*, **27.4**, 358–68.

Hsü, K. (1975) Catastrophic debris sreams (sturzstroms) generated by rockfalls. *Bulletin of the Geological Society of America*, **86**, 129–40.

Huddart, D. and Glasser, N.E. (2002) *Quaternary of Northern England*, Geological Conservation Review Series, No. **25**, Joint Nature Conservation Committee, Peterborough, 745 pp.

Hudson, J.D. and Trewin, N.H. (2002) Jurassic. In *The Geology of Scotland*, 4th edn (ed. N.H. Trewin), The Geological Society, London, pp. 323–50.

Hull, E. and Green, A.H. (1866) *The Geology of the Country Around Stockport, Macclesfield, Congleton and Leek*, Memoir of the Geological Survey of Great Britain, sheets 81NW and 81SW, HMSO, London, 102 pp.

Hunter, J. (1869) *Hallamshire, the History and Topography of the Parish of Sheffield*, revised by Alfred Gatty, Sheffield, from an 1819 original.

Hutchison, J. and Leafe, R. (1996) Shoreline management: a view of the way ahead. In *Coastal Management: Putting Policy into Practice* (ed. C.A. Fleming), Thomas Telford, London, pp. 352–60.

Hutchinson, J.N. (1965) Survey of the coastal landslides of Kent. *Building Research Station Note*, **EN35/65**, 9–26.

Hutchinson, J.N. (1967) The free degradation of London Clay cliffs. In *Proceedings of the Geotechnical Conference, Oslo 1967: on Shear Strength Properties of Natural Soils and Rocks*, Norwegian Geotechnical Institute, Oslo, Vol. **1**, pp. 113–18.

Hutchinson, J.N. (1968a) Mass movement. In *The Encyclopedia of Geomorphology* (ed. R.W. Fairbridge), Encyclopedia of Earth Sciences Series, Vol. **3**, Reinhold Book Corp., New York, pp. 688–96.

Hutchinson, J.N. (1968b) Field meeting on the coastal landslides of Kent, 1–3 July 1966. *Proceedings of the Geologists' Association*, **79**, 227–37.

Hutchinson, J.N. (1969) A reconsideration of the coastal landslides at Folkestone Warren, Kent. *Géotechnique*, **19**, 6–38.

Hutchinson, J.N. (1970) A coastal mudflow on the London Clay cliffs at Beltinge, north Kent. *Géotechnique*, **20**, 412–38.

Hutchinson, J.N. (1971) Warden Point landslide. *Annual Slide Report for IAEG/UNESCO*, 1971.

Hutchinson, J.N. (1973) The response of London Clay cliffs to differing rates of toe erosion. *Geologia Applicata e Idrogeologia*, **7**, 222–39.

Hutchinson, J.N. (1976) Coastal landslides in cliffs of Pleistocene deposits between Cromer and Overstrand, Norfolk, England. In *Laurits Bjerrum Memorial Volume: Contributions to Soil Mechanics* (eds N. Janbu, F. Jørstad and B. Kjaernsli), Norges Geotekniske Institutt, Oslo, pp. 155–82.

Hutchinson, J.N. (1979) Various forms of cliff instability arising from coast erosion in south-east England. *Fjellsprengningsteknikk Bergmekanikk Geoteknikk*, Vol. 19, pp. 1–32.

Hutchinson, J.N. (1983) The geotechnics of cliff stabilisation. In *Shoreline Protection: Proceedings of a Conference Organized by the Institution of Civil Engineers and Held at the University of Southampton on 14–15 September 1982*, Thomas Telford, London, pp. 215–22.

Hutchinson, J.N. (1984) Landslides in Britain and their countermeasures. *Journal of the Japan Landslide Society*, **21**, 1–21.

Hutchinson, J.N. (1988) General report: morphological and geotechnical parameters of landslides in relation to geology and hydrogeology. In *Landslides: Proceedings of the Fifth International Symposium on Landslides, 10–15 July 1988, Lausanne* (ed. C. Bonnard), Balkema, Rotterdam, Vol. **1**, pp. 3–35.

Hutchinson, J.N. (1991) Periglacial and slope processes. In *Quaternary Engineering Geology* (eds A. Forster, M.G. Culshaw, J.C. Cripps, J.A. Little and C.F. Moon), *Geological Society Special Publication*, No. **7**, Geological Society Publishing House, London, pp. 283–331.

Hutchinson, J.N. and Bhandari, R.K. (1971) Undrained loading, a fundamental mechanism of mudflows and other mass movements. *Géotechnique*, **21**, 353–8.

Hutchinson, J.N. and Gostelow, T.P. (1976) The development of an abandoned cliff in London Clay at Hadleigh, Essex. *Philosophical Transactions of the Royal Society of London, Series A*, **283**, 557–604.

Hutchinson, J.N. and Millar, D.L. (2001) The Graig Goch landslide dam, Meirionydd, mid-Wales. In *The Quaternary of West Wales: Field Guide* (eds M.J.C. Walker and D. McCarroll), Quaternary Research Association, London, pp. 113–25.

Hutchinson, J.N., Bromhead, E.N. and Lupini, J.F. (1980) Additional observations on the Folkestone Warren landslides. *Quarterly Journal of Engineering Geology*, **13**, 1–31.

Hutchinson, J.N, Poole, C., Lambert, N. and Bromhead, E.N. (1985) Combined archaeological and geotechnical investigations of the Roman fort at Lympne, Kent. *Britannia*, **16**, 209–36.

Hutchinson, P.O. (1840) *A Guide to the Landslip Near Axmouth, Devonshire, Together with a Geological and Philosophical Enquiry into its Nature and Causes, and a Topographic Description of the District*, John Harvey, Sidmouth.

Ibsen, M.-L. (1994) Evaluation of the temporal distribution of landslide events along the south coast of Britain, between Straight Point and St Margaret's Bay. Unpublished MPhil thesis, Kings College, University of London.

Ibsen, M.-L. and Brunsden, D. (1996) The nature, uses and problems of historical archives for the temporal occurrences of landslides, with specific reference to the south coast of Britain, Ventnor, Isle of Wight. *Geomorphology*, **15**, 241–58.

Ibsen, M.-L. and Brunsden, D. (1997) Mass movement and climatic variation on the south coast of Great Britain. In *Rapid Mass Movement as a Source of Climatic Evidence for the Holocene* (eds J.A. Matthews, D. Brunsden, B. Frenzel, B. Gläser and M.M. Weiß), *Paläoklimaforschung – Paleoclimatic Research*, No. **19**, Gustav Fischer, Stuttgart, pp. 171–82.

Innes, J.L. (1983) Lichenometric dating of debris-flow deposits in the Scottish Highlands. *Earth Surface Processes and Landforms*, **23**, 545–60.

Innes, J.L. (1985) Magnitude-frequency relations of debris flows in northwest Europe. *Geografiska Annaler*, **67A**, 23–32.

Iversen, R.M. and Major, J.J. (1986) Groundwater seepage vectors and the potential for hillslope failure and debris flow mobilization. *Water Resources Research*, **22**, 1543–48.

Janbu, N. (1973) Slope stability computations. In *Embankment Dam Engineering* (eds R.C. Hirschfield and S.J. Poulos), Wiley, New York, pp. 47–86.

Jarman, D. (2002) Rock slope failure and landscape evolution in the Caledonian Mountains, as exemplified in the Abisko area, northern Sweden. *Geografiska Annaler*, **84A**, 213–24.

Jarman, D. (2003a) Paraglacial landscape evolution – the significance of rock slope failure. In *The Quaternary of the Western Highland Boundary: Field Guide* (ed. D.J.A. Evans), Quaternary Research Association, London, pp. 50–68.

Jarman, D. (2003b) The Glen Shiel rock slope failure cluster. In *The Quaternary of Glen Affric and Kintail: Field Guide* (ed. R. Tipping), Quaternary Research Association, London, pp. 165–83.

Jarman, D. (2003c): The An Sornach rock slope failure. In *The Quaternary of Glen Affric and Kintail: Field Guide* (ed. R. Tipping), Quaternary Research Association, London, pp. 63–74.

Jarman, D. (2003d) Tullich Hill rock slope failures, Glen Douglas. In *The Quaternary of the Western Highland Boundary: Field Guide* (ed. D.J.A. Evans), Quaternary Research Association, London, pp. 200–208.

Jarman, D. (2003e) Beinn Fhada – the context for large-scale slope deformation. In *The Quaternary of Glen Affric and Kintail: Field Guide* (ed. R. Tipping), Quaternary Research Association, London, pp. 149–56.

Jarman, D. (2004a) Rock slope failures of the Gaick Pass. In *The Quaternary of the Central Grampian Highlands: Field Guide* (eds S. Lukas, J. Merritt and W. Mitchell), Quaternary Research Association, London, pp. 103–111.

Jarman, D. (2004b) The Cobbler – a mountain shaped by rock slope failure. *Scottish Geographical Journal*, **120**, 227–40.

Jarman, D. (2006) Large rock slope failures in the Scottish Highlands: characterisation, causes and spatial distribution. *Engineering Geology*, **83**(1–3), 161–82.

Jarman, D. and Ballantyne, C.K. (2002) Beinn Fhada, Kintail: a classic example of large-scale paraglacial rock slope deformation. *Scottish Geographical Journal*, **118**, 59–68.

Jarman, D. and Reid, E. (2003) Postglacial rock slope failures and slope features in Strathfarrar. In *The Quaternary of Glen Affric and Kintail: Field Guide* (ed. R. Tipping), Quaternary Research Association, London, pp. 119–121.

Jarman, D. and Stewart, I.S. (2004) What role have palaeoseismic shocks played in triggering large paraglacial slope instabilities in the mountains of Western Europe? Abstract, European Geophysical Union, Nice, NH 3.06.

Johnson, A.M. and Rahn, P.H. (1970) Mobilization of debris flows. *Zeitschrift für Geomorphologie, Supplementband*, **9**, 168–86.

Johnson, A.M. and Rodine, J.R. (1984) Debris flow. In *Slope Instability* (eds D. Brunsden and D.B. Prior), John Wiley and Sons, Chichester, pp. 257–361.

Johnson, R.H. (1965) A study of the Charlesworth landslides near Glossop, north Derbyshire. *Transactions of the Institute of British Geographers*, **37**, 111–26.

Johnson, R.H. and Vaughan, R.D. (1983) The Alport Castles, Derbyshire: a south Pennine slope and its geomorphic history. *East Midland Geographer*, **8**(3), 79–83.

Johnson, R.H. and Walthall, S. (1979) The Longdendale landslides. *Geological Journal*, **14**, 135–58.

Johnstone, G.S. and Mykura, W. (1989) *British Regional Geology: the Northern Highlands of Scotland*, 4th edn, HMSO, London, 219 pp.

Jones, C.M. (1980) Deltaic sedimentation in the Roaches Grit and associated sediments (Namurian R_2b) in the south-west Pennines. *Proceedings of the Yorkshire Geological Society*, **43**, 39–67.

Jones, D.K.C. (1980) The Tertiary evolution of south-east England with particular reference to the Weald. In *The Shaping of Southern England* (ed. D.K.C. Jones), *Institute of British Geographers Special Publication*, No. **11**, Academic Press, London, pp. 13–47.

Jones, D.K.C. (1981) *Southeast and Southern England*, The Geomorphology of the British Isles, Methuen, London, 332 pp.

Jones, D.K.C. and Lee, E.M. (1994) *Landsliding in Great Britain*, HMSO, London, 351 pp.

Jukes-Browne, A.J. (1900) *The Cretaceous rocks of Britain, Volume 1: The Gault and Upper Greensand of England*, Memoir of the Geological Survey of the United Kingdom, HMSO, London, 499 pp.

Kazi, A. and Knill, J.L. (1969) The sedimentation and geotechnical properties of the Cromer Till between Happisburgh and Cromer, Norfolk. *Quarterly Journal of Engineering Geology*, **2**, 63–86.

Keeping, W. (1882) The glacial geology of central Wales. *Geological Magazine*, **9**, 251–7.

Kellaway, G.A. (1972) Development of non-diastrophic Pleistocene structures in relation to climate and physical relief in Britain. In *Proceedings of the 24th International Geological Congress, Section 12: Quaternary Geology* (ed. J.E. Gill), International Geological Congress, Ottawa, pp. 136–46.

Kellaway, G.A. and Taylor, J.H. (1968) The influence of land-slipping on the development of the City of Bath, England. In *Report of the 23rd Session of the International Geological Congress, Czechslovakia, 1968* (ed. M. Malkovský), Academia, Prague, Vol. **12**, pp. 65–76.

Kellaway, G.A., Redding, J.H., Shephard-Thorne, E.R. and Destombes, J.P. (1975) The Quaternary history of the English Channel. *Philosophical Transactions of the Royal Society of London, Series A*, **279**, 189–218.

Kilburn, C.R.J. and Sørensen, S-A. (1998) Runout lengths of sturzstroms: the control of initial conditions and of fragment dynamics. *Journal of Geophysical Research*, **103**, 877–84.

Koh, A. (1990) Black Ven. In *Landslides of the Dorset Coast* (ed. R.J. Allison), British Geomorphological Research Group Field Guide, British Geomorphological Research Group, London, pp. 95–105.

Koor, N. (1989) A slope failure in the London Clay cliffs at Warden Point, Isle of Sheppey. Unpublished MSc Thesis, University of London.

Lang, W.D. (1928) Landslips in Dorset. *Natural History Magazine*, **1**, 201–9.

Lang, W.D. (1959) Report on Dorset natural history for 1958 – Geology. *Proceedings of the Dorset Natural History and Archaeological Society*, **80**, 22.

Lant, C. (1973) The Mam Tor Landslip. Unpublished MSc Thesis, Imperial College, University of London.

Lapworth, H. (1911) The geology of dam trenches. *Transactions of the Institution of Water Engineers*, **16**, 25.

Leafe, R. (1998) Conserving our coastal heritage – a conflict resolved? In *Coastal Defence and Earth Science Conservation* (ed. J.M. Hooke), Geological Society of London, Bath, pp. 10–19.

Leafe, R. and Radley, G. (1994) Environmental benefits of soft cliff erosion. *Proceedings of the 29th MAFF Conference of River and Coastal Engineers, 4–6 July, 1994*, MAFF, London, pp. 3.1.1–3.1.13.

Lee, E.M. (1998) Problems associated with the prediction of cliff recession rates for coastal defence and conservation. In *Coastal Defence and Earth Science Conservation* (ed. J.M. Hooke), Geological Society of London, Bath, pp. 46–57.

Lee, G.W. (1920) *The Mesozoic Rocks of Applecross, Raasay and North-East Skye*, Memoir of the Geological Survey of Great Britain (Scotland), HMSO, Edinburgh, 93 pp.

Lee, S.J. (1976) Aspects of coastal landslides on the Isle of Sheppey, Kent. Unpublished BSc project, Kingston Polytechnic.

Leland, J. (1744) *The Itinerary of John Leland the Antiquary*, 2nd edn, J. Fletcher, Oxford.

Linton, D.L. (1940) Some aspects of the evolution of the Rivers Earn and Tay. *Scottish Geographical Magazine*, **56**, 1–11, 61–79.

Linton, D.L. (1949) Watershed breaching by ice in Scotland. *Transactions of the Institute of British Geographers*, **15**, 1–16.

Linton, D.L. (1957) Radiating valleys in glaciated lands. *Tidjschrift van het Nederlandsch Aardrijkskundig. Geootschap*, **74**, 297–312.

Linton, D.L. (1967). Divide elimination by glacial erosion. In *Arctic and Alpine Environments* (eds H.E. Wright and W.H. Osburn), Indiana University Press, pp. 241–8.

Linton, D.L. and Moisley, H.A. (1960) The origin of Loch Lomond. *Scottish Geographical Magazine*, **76**, 26–37.

Lounsbury, R.W. (1962) Landslips in the Ashop valley, Derbyshire, England. In *Abstracts for 1961, Geological Society of America Special Paper*, No. **68**, Geological Society of America, New York, p. 219.

Lukas, S. (2005) Younger Dryas moraines in the NW Highlands of Scotland: genesis, significance, and potential modern analogues. Unpublished PhD thesis, University of St Andrews.

Lukas, S. and Lukas, T. (2006) A glacial geology and geomorphology map of the far NW Highlands, Scotland, parts 1 and 2. *Journal of Maps*, **2006**, 43–58

Lupini, J.F. (1980) The residual strength of soils. Unpublished PhD thesis, Imperial College, University of London.

Macfadyen, W.A. (1970) *Geological Highlights of the West Country*, Butterworth, London, 296 pp.

MAFF and others (1995) *Shoreline Management Plans: a Guide for Coastal Defence Authorities*, Ministry of Agriculture, Fisheries and Food, London, 24 pp.

Mahr, T. (1977) Deep-reaching gravitational deformations of high mountain slopes. *Bulletin of the International Association of Engineering Geologists*, **16**, 121–7.

Maquaire, O. and Gigot, P. (1988) Reconnaissance par sismique réfraction de la décompression et de l'instabilité des falaises vives du Bessin (Normandie, France). *Geodinamica Acta (Paris)*, **2**(3), 151–9.

May, F., Peacock, J.D., Smith, D.I. and Baker, A.J. (1993) *Geology of the Kintail District*, Memoir of the British Geological Survey, Sheet 72W and part of 71E (Scotland), HMSO, London, 74 pp.

May, V.J. (2003) Dungeness and Rye Harbour, Kent and East Sussex. In *Coastal Geomorphology of Great Britain* (V.J. May and J.D. Hansom), Geological Conservation Review Series, No. **28**, Joint Nature Conservation Committee, Peterborough, pp. 310–26.

May, V.J. and Hansom, J.D. (2003) *Coastal Geomorphology of Great Britain*, Geological Conservation Review Series, No. **28**, Joint Nature Conservation Committee, 737 pp.

McKirdy, A.P. (1990) A handbook of earth science conservation techniques. In *Earth Science Conservation in Great Britain: a Strategy* (ed. Nature Conservancy Council), Nature Conservancy Council, Peterborough, Appendices.

McQuhae, E.A. (1977) In *East Anglia: Localities of Geomorphological Importance* (ed. Nature Conservancy Council, Geology and Physiography Section), Nature Conservancy Council, Newbury, pp. 44–5.

Millward, R. and Robinson, A. (1975) *The Peak District*, The Regions of Britain, Eyre Methuen, London, 301 pp.

Moore, I. and Kokelaar, P. (1998) Tectonically-controlled piecemeal caldera collapse: a case study of Glencoe volcano, Scotland. *Bulletin of the Geological Society of America*, **110**, 1448 66.

Moore, R.M. (1988) The clay mineralogy, weathering and mudslide behaviour of coastal cliffs. Unpublished PhD thesis, University of London.

Morgenstern, N.R. and Price, V.E. (1965) The analysis of the stability of general slip surfaces. *Géotechnique*, **15**, 79–93.

Morton, N. (1969) Lower and middle Jurassic of Raasay. In *International Field Symposium on the British Jurassic. Excursion No. 4: Guide for Western Scotland* (eds J.D. Hudson and N. Morton), University of Keele, Keele, pp. D10–D16.

Motyka, J.M. and Brampton, A.H. (1993) Coastal Management. Mapping of littoral cells. *Hydraulics Research Wallingford Report*, **SR 328**, 102 pp.

Muir Wood, A.M. (1955a) Reply to discussion on "Folkestone Warren landslips: investigations, 1948–50". *Proceedings of the Institution of Civil Engineers, Railway Paper*, No. **56**, 460–4.

Muir Wood, A.M. (1955b) Folkestone Warren landslips: investigations, 1948–50. *Proceedings of the Institution of Civil Engineers, Railway Paper*, No. **56**, 410–28.

Muir Wood, A.M. (1970) Correspondence on Hutchinson 1969. *Géotechnique*, **20**, 110–13.

Muir Wood, A.M. (1971) Engineering aspects of coastal landslides. *Proceedings of the Institution of Civil Engineers*, **50**, 257–76.

Muir Wood, A.M. (1994) Geology and geometry: period return to Folkestone Warren. In *Proceedings of the Thirteenth International Conference on Soil Mechanics and Foundation Engineering, New Delhi, 5–10 January 1994*, Oxford and IBH Publishing Co., New Delhi, pp. 23–30.

Musson, R.M.W (1989) Accuracy of historical earthquake locations in Britain. *Geological Magazine*, **126**, 685–9.

Musson, R.M.W., Neilson, G. and Burton, P.W. (1984) Macroseismic Reports on Historical British Earthquakes: III – Central Scotland and Western Highland. *Global Seismology Research Group Reports*, **209**, 202 pp.

NCC (1991) *A Guide to the Selection of Appropriate Coast Protection Works for Geological Sites of Special Scientific Interest*, Nature Conservancy Council, Peterborough.

Neilson, G. and Burton, P.W. (1985) Instrumental magnitudes of British earthquakes. *In Earthquake Engineering in Britain: Proceedings of a Conference Organized by the Institution of Civil Engineers and the Society of Earthquake and Civil Engineering Dynamics, held at the University of East Anglia, 18–19 April 1985*, Thomas Telford Ltd, London, pp. 41–3.

Nichol, D. (2002) Slope instability and landslide research in North Wales. In *Landslides and Landslide Management in North Wales* (eds D. Nichol, M.G. Bassett, and V.K. Deisler), National Museum of Wales Geological Series, No. **22**, Cardiff, pp. 9–13.

O'Connor, M. and Graham, N. (1996) *The caves of the Isle of Portland, Wessex Cave Club Occasional Publication, Series 3*, No. **3**, Wessex Cave Club, Pangbourne, 104 pp.

Ordnance Survey (1964) *Geological Survey of Great Britain (Scotland): Northern Skye*, 1:63 360 scale map, drift edition, Ordnance Survey, Chessington.

Osman, C.W. (1917) The landslips of Folkestone Warren and thickness of the Lower Chalk and Gault near Dover. *Proceedings of the Geologists' Association*, **28**, 59–84.

Palmer, J. and Radley, J. (1961) Gritstone tors of the English Pennines. *Zeitschrift für Geomorphologie, Neue Folge, Supplementband*, **5**, 37–52.

Parks, C.D. (1991) A review of the possible mechanisms of cambering and valley bulging. In *Quaternary Engineering Geology* (eds A. Forster, M.G. Culshaw, J.C. Cripps, J.A. Little and C.F. Moon), *Geological Society Special Publication*, No. **7**, Geological Society Publishing House, London, pp. 373–80.

References

Peacock, J.D. and May, F. (1993) Pre-Flandrian slope deformation in the Scottish Highlands: examples from Glen Roy and Glen Gloy. *Scottish Journal of Geology*, **29**, 183–9.

Peacock, J.D., Mendum, J.R. and Fettes, D.J. (1992) *Geology of the Glen Affric District*, Memoir of the British Geological Survey, Sheet 72E (Scotland), HMSO, London, 81 pp.

Pearson, G.W. and Stuiver, M. (1986) High-precision calibration of the radiocarbon time scale, 500–2500 B.C. *Radiocarbon*, **B28**, 839–62.

Pitts, J. (1974) The Bindon landslip of 1839. *Proceedings of the Dorset Natural History and Archaeological Society*, **95**, 18–29.

Pitts, J. (1979) Morphological mapping in the Axmouth–Lyme Regis Undercliffs, Devon. *Quarterly Journal of Engineering Geology*, **12**, 205–17.

Pitts, J. (1982) An historical survey of the landslips of the Axmouth–Lyme Regis undercliffs, Devon. *Proceedings of the Dorset Natural History and Archaeological Society*, **103**, 101–6.

Pitts, J. (1983a) The temporal and spatial development of landslides in the Axmouth–Lyme Regis Undercliffs National Nature Reserve, Devon. *Earth Surface Processes and Landforms*, **8**, 584–603.

Pitts, J. (1983b) The recent evolution of landsliding in the Axmouth–Lyme Regis Undercliffs National Nature Reserve. *Proceedings of the Dorset Natural History and Archaeological Society*, **105**, 119–25.

Pitts, J. (1986) The form and stability of a double undercliff: an example from South-West England. *Engineering Geology*, **22**, 209–16.

Pitts, J. and Brunsden, D. (1987) A reconsideration of the Bindon landslide of 1839. *Proceedings of the Geologists' Association*, **98**, 1–18.

Pitty, A.F. (1966) A simple device for the field measurement of hillslopes. *Journal of Geology*, **76**, 717–20.

Prior, D.B., Stephens, N. and Douglas, G.R. (1970) Some examples of modern debris flows in N.E. Ireland. *Zeitschrift für Geomorphologie*, **14**(3), 275–88.

Radbruch-Hall, D.H. (1978) Gravitational creep of rock masses on slopes. In *Rockslides and Avalanches, 1: Natural phenomena* (ed. B. Voight), Developments in Geotechnical Engineering, 14A, Elsevier, Amsterdam, pp. 607–57.

Radbruch-Hall, D.H., Varnes, D.J. and Savage, W.Z. (1976) Gravitational spreading of steep-sided ridges ("sackung") in western United States. *Bulletin of the International Association of Engineering Geology*, **14**, 23–35.

Rapp, A. (1960) Recent development of mountain slopes in Karkevagge and surroundings, northern Scandinavia. *Geografiska Annaler*, **42**, 65–200.

Reid, C. (1882) *The Geology of the Country Around Cromer*, Memoir of the Geological Survey of Great Britain, Sheet 68E (England and Wales), HMSO, London, 103 pp.

Rib, H.T. and Liang, T. (1978) Recognition and identification. In *Landslides: Analysis and Control* (eds R.L. Schuster and R.J. Krizek), *National Research Council (U.S.) Transportation Research Board Special Report*, No. **176**, National Academy of Sciences, Washington D.C., pp. 34–80.

Richards, A. (1971) The evolution of marine cliffs and related landforms in the Inner Hebrides. Unpublished PhD thesis, University of Wales, Aberystwyth.

Richardson, B.D. (1996) Soft engineering on the coast: where to now? In *Coastal Management: Putting Policy into Practice* (ed. C.A. Fleming), Thomas Telford, London, pp. 219–28.

Ringrose, P.S. (1989) Recent fault movement and palaeoseismicity in western Scotland. *Tectonophysics*, **163**, 305–14.

Rizzo, V. and Leggeri, M. (2004) Slope instability and sagging reactivation at Maratea (Potenza, Basilicata, Italy). *Engineering Geology*, **71**, 181–98.

Roach Smith, C. (1850) *The Antiquities of Richborough, Reculver and Lymne, in Kent*, J.R. Smith, London, 272 pp.

Roach Smith, C. (1852) *Report on Excavations Made on the Site of the Roman Castrum at Lymne, in Kent, in 1850*, London.

Roberts, G. (1840) *An Account and Guide to the Mighty landslip of Dowlands and Bindon, Near Lyme Regis, December 25, 1839*, Daniel Dunster, Lyme Regis.

Robinson, A.H.W. (1949) Deep clefts in the Inner Sound of Raasay. *Scottish Geographical Magazine*, **65**, 20–5.

Rose, J (2001) Moelwyn Mawr. In *The Quaternary of West Wales: Field Guide* (eds M.J.C. Walker and D. McCarroll), Quaternary Research Association, London, pp. 153–63.

Russell, W.A. (1985) Investigation into andslides around Hallaig, East Raasay. Unpublished BSc (Hons) thesis, University of Strathclyde.

Sandeman, E. (1918) The Derwent Valley Waterworks. *Minutes of Proceedings of the Institution of Civil Engineers*, **206**, 152.

Savage, W.Z. and Varnes, D.J. (1987) Mechanics of gravitational spreading of steep-sided ridges ('Sackung'). *Bulletin of the International Association of Engineering Geologists*, **35**, 31–6.

Savigear, R.A.G. (1952) Some observations on slope development in South Wales. *Transactions of the Institute of British Geographers*, **18**, 31–51.

Schumm, S.A. and Chorley, R.J. (1964) The fall of Threatening Rock. *American Journal of Science*, **262**, 1041–54.

Selby, M.J. (1982) *Hillslope Materials and Processes*, Oxford University Press, Oxford, 264 pp.

Selby, M.J. (1993) *Hillslope Materials and Processes*, 2nd edn, Oxford University Press, Oxford, 451 pp.

Sellier, D. and Lawson, T.J. (1998) A complex slope failure on Beinn nan Cnaimhseag, Assynt, Sutherland. *Scottish Geographical Magazine*, **114**, 85–93.

Sharp, R.P. (1942) Mudflow levées. *Journal of Geomorphology*, **5**, 222–7.

Sharpe, C.F.S. (1938) *Landslides and Related Phenomena*, Columbia University Press, New York, 137 pp.

Sherlock, R.L. (1960) *British Regional Geology: London and Thames Valley*, 3rd edn, HMSO, London, 62 pp.

Sissons, J.B. (1967) *The Evolution of Scotland's Scenery*, Oliver & Boyd, Edinburgh, 259 pp.

Sissons, J.B. (1975) A fossil rock glacier in Wester Ross. *Scottish Journal of Geology*, **11**, 83–6.

Sissons, J.B. (1976) A fossil rock glacier in Wester Ross. Reply to W.B. Whalley. *Scottish Journal of Geology*, **12**, 178–9.

Sissons, J.B. (1977) The Loch Lomond Readvance in the northern mainland of Scotland. In *Studies in the Scottish Lateglacial Environment* (eds J.M. Gray and J.J. Lowe), Pergamon Press, Oxford, pp. 45–59.

Sissons, J.B. and Cornish, R. (1982) Differential isostatic uplift of crustal blocks at Glen Roy, Scotland. *Quaternary Research*, **18**, 268–88.

Skempton, A.W. (1946) Discussion of Folkestone Warren landslips: research carried out in 1939 by the Southern Railway. *Proceedings of the Institution of Civil Engineers, Railway Paper*, No. **19**, 29–33.

Skempton, A.W. (1964) Long-term stability of clay slopes. *Géotechnique*, **14**, 77–101.

Skempton, A.W. (1970) First time slides in over-consolidated clays. *Géotechnique*, **20**, 320–4.

Skempton, A.W. (1977) Slope stability of cuttings in brown London Clay. In *Proceedings of the ninth International Conference on Soil Mechanics and Foundation Engineering, Tokyo, 1977*, Japanese Society of Soil Mechanics and Foundation Engineering, Tokyo, Vol. **3**, pp. 261–70.

Skempton, A.W. and DeLory, F.A. (1957) Stability of natural slopes in London Clay. In *Proceedings of the Fourth International Conference on Soil Mechanics and Foundation Engineering, London, 12–24 August 1957*, Butterworths Scientific Publications, London, Vol. **2**, pp. 378–81.

Skempton, A.W. and Hutchinson, J.N. (1969) Stability of natural slopes and embankment foundations. In *Proceedings of the Seventh International Conference on Soil Mechanics and Foundation Engineering, Mexico City, 1969*, Sociedad Mexicana de Mecánica de Suelos, Mexico City, pp. 291–340.

Skempton, A.W. and Petley, D.J. (1967) The strength along structural discontinuities in stiff clays. In *Proceedings of the Geotechnical Conference, Oslo, 1967: on Shear Strength Properties of Natural Soils and Rocks*, Norwegian Geotechnical Institute, Oslo, Vol. **2**, pp. 29–46.

Skempton, A.W. and Weeks, A.G. (1976) The Quaternary history of the Lower Greensand escarpment and Weald Clay vale near Sevenoaks, Kent. *Philosophical Transactions of the Royal Society of London, Series A*, **238**, 493–526.

Skempton, A.W., Leadbeater, A.D. and Chandler, R.J. (1989) The Mam Tor landslide, north Derbyshire. *Philosophical Transactions of the Royal Society of London, Series A*, **329**, 503–47.

Smart, J.G.O., Bisson, G. and Worssam, B.C. (1966) *Geology of the Country Around Canterbury and Folkestone*, Memoir of the Geological Survey of Great Britain, sheets 289, 305 and 306 (England and Wales), HMSO, London, 337 pp.

References

Soldati, M (2004) Deep-seated gravitational slope deformation. In *Encyclopedia of Geomorphology* (ed. A. Goudie), Routledge, London, pp. 226–8.

Solomon, J.D. (1932) The glacial succession on the north Norfolk coast. *Proceedings of the Geologists' Association*, **43**, 241–71.

Spears, D.A. and Amin, M.A. (1981) A mineralogical and geochemical study of turbidite sandstones and interbedded shales, Mam Tor, Derbyshire, UK. *Clay Minerals*, **16**, 333–45.

Statham, I. (1976) Debris flows on vegetated screes in the Black Mountain, Carmarthenshire. *Earth Surface Processes*, **1**, 173–80.

Steers, J.A. (1964) *The Coastline of England and Wales*, 2nd edn, Cambridge University Press, Cambridge, 750 pp.

Stephenson, D., Bevins, R.E., Millward, D., Highton, A.J., Parsons, I., Stone, P. and Wadsworth, W.J. (1999) *Caledonian Igneous rocks of Great Britain*, Geological Conservation Review Series, No. **17**, Joint Nature Conservation Committee, Peterborough, 648 pp.

Stevenson, I.P. and Gaunt, G.D. (1971) *The Geology of the Country Around Chapel-en-le-Frith*, Memoir of the Geological Survey of Great Britain, Sheet 99 (England and Wales), HMSO, London, 444 pp.

Steward, H.E. and Cripps, J.C. (1983) Some engineering implications of chemical weathering of pyritic shale. *Quarterly Journal of Engineering Geology*, **16**, 281–9.

Stewart, I.S., Sauber, S. and Rose, J. (2000) Glacio-seismotectonics: ice sheets, crustal deformation and seismicity. *Quaternary Science Reviews*, **19**, 1367–89.

Stone, B. (ed.) (1974) *Sir Gawain and the Green Knight*, 2nd edn, Penguin Books, Harmondsworth, 185 pp.

Stone, J.O., Ballantyne, C.K. and Fifield, L.K. (1998) Exposure dating and validation of periglacial weathering limits, northwest Scotland. *Geology*, **26**, 587–90.

Sutherland, D.G. (1984) The Quaternary deposits and landforms of Scotland and the neighbouring shelves: a review. *Quaternary Science Reviews*, **3**, 157–254.

Swash, A.R.H., Leafe, R.N. and Radley, G.P. (1995) Shoreline Management Plans and environmental considerations. In *Directions in European Coastal Management* (eds R. Healy and P. Doody), Samara Publishing, Cardigan, pp. 161–7.

Takahashi, T. (1981) Debris flow. *Annual Review of Fluid Mechanics*, **13**, 57–77.

Tallis, J.H. and Johnson, R.H. (1980) The dating of landslides in Longdendale, north Derbyshire, using pollen-analytical techniques. In *Timescales in Geomorphology* (eds R.A. Cullingford, D.A. Davidson and J. Lewin), Wiley, Chichester, pp. 189–205.

Tate, C.J. (1995) Late Quaternary glacial history and environmental change in southern Ross-shire, Scotland. Unpublished PhD thesis, University of St Andrews.

Taylor, D.W. (1948) *Fundamentals of Soil Mechanics*, John Wiley and sons, New York, 700 pp.

Thompson, R.W.S. (1949) Discussion. In The Ladybower Reservoir (ed. H.P. Hill). *Journal of the Institution of Water Engineers*, **3**, pp. 414–33, 427–8.

Thorn, C.E. (1976) Quantitative evaluation of nivation in the Colorado Front Range. *Bulletin of the Geological Society of America*, **87**, 1169–78.

Thorn, C.E. (1979) Ground temperatures and surficial transport in colluvium during snow-patch meltout: Colorado Front Range. *Arctic and Alpine Research*, **11**, 41–52.

Thorn, C.E. (1988) Nivation: a geomorphic chimera. In *Advances in Periglacial Geomorphology* (ed. M.J. Clark), Wiley, Chichester, pp. 3–31.

Thorn, C.E. and Hall, K. (1980) Nivation: an arctic–Alpine comparison and reappraisal. *Journal of Glaciology*, **25**, 109–24.

Thorp, P.W. (1981) A trimline method for defining the upper limit of the Loch Lomond Advance glaciers: examples from the Loch Leven and Glen Coe areas. *Scottish Journal of Geology*, **17**, 49–64.

Thorp, P.W. (1987) Late Devensian ice sheet in the western Grampians, Scotland. *Journal of Quaternary Science*, **2**, 103–12.

Tinkler, K.J. (1966) Slope profiles and scree in the Eglwyseg valley, North Wales. *Geographical Journal*, **132**, 379–85.

Tolkein, J.R.R. and Gordon, E.V. (eds) (1967) *Sir Gawain and the Green Knight*, 2nd edn, revised by N. Davis, Clarendon Press, Oxford, 232 pp.

Toms, A.H. (1946) Folkestone Warren landslips: research carried out in 1939 by the Southern Railway. *Proceedings of the Institution of Civil Engineers, Railway Paper*, No. **19**, 3–25.

Toms, A.H. (1953) Recent research into the coastal landslides at Folkestone Warren, Kent, England. In *Proceedings of the Third International Conference on Soil Mechanics and Foundation Engineering, Mexico City, Switzerland, 16th–17th August 1953*, Organising Committee ICOSOMEF, Zurich, Vol. **2**, pp. 288–93.

Treagus, J.E. (2003) The Loch Tay Fault: type section geometry and kinematics. *Scottish Journal of Geology*, **39**, 135–44.

Trenter, N.A. and Warren, C.D. (1996) Further investigations at the Folkestone Warren landslide. *Géotechnique*, **46**(4), 589–620.

Turnbull, J.M. and Davies, T.R.H. (2006) A mass movement origin for cirques. *Earth Surface Processes and Landforms,* **31**, 1129–48.

Van Steijn, H., de Ruig, J. and Hoozemans, F. (1988) Mophological and mechanical aspects of debris flows in parts of the French Alps. *Zeitschrift für Geomorphologie*, **32**, 143–61.

Varnes, D.J. (1978) Slope movement types and processes. In *Landslides: Analysis and Control* (eds R.L. Schuster and R.J. Krizek), *National Research Council (U.S.) Transportation Research Board Special Report*, No. **176**, National Academy of Sciences, Washington D.C., pp. 11–33.

Vaughan, P.R. and Walbancke, H.J. (1973) Pore pressure changes and the delayed failure of cutting slopes in overconsolidated clay. *Géotechnique*, **23**, 531–9.

Vear, A. and Curtis, C. (1981) A quantitative evaluation of pyrite weathering. *Earth Surface Processes and Landforms*, **6**, 191–8.

Viner-Brady, N.E.V. (1955) Folkestone Warren landslips: remedial measures, 1948–1954. *Proceedings of the Institution of Civil Engineers, Railway Paper*, No. **57**, 429–41.

Wanklyn, C. (1927) *Lyme Regis: a Retrospect*, 2nd edn, Hatchards, London, 283 pp.

Ward, W.H. (1945) The stability of natural slopes. *Geographical Journal*, **105**, 170–91.

Ward, W.H. (1962) Discussion in 'Coastal cliffs: report of a symposium'. *Geographical Journal*, **128**, 309–13.

Watson, E. (1966) Two nivation cirques near Aberystwyth, Wales. *Biuletyn Peryglacjalny*, **15**, 79–101.

Watson, E. (1968) The periglacial landscape of the Aberystwyth region. In *Geography at Aberystwyth* (eds E.G. Bowen, H. Carter and A.J. Taylor), University of Wales Press, Cardiff, pp. 35–49.

Watson, E. (1970) The Cardigan Bay area. In *The Glaciations of Wales and Adjoining Regions* (ed. C.A. Lewis), Geographies for Advanced Studies, Longman, Harlow, pp. 125–45.

Watson, E. (1976) Field excursions in the Aberystwyth region. *Biuletyn Peryglacjalny*, **26**, 79–112.

Watson, E. and Watson, S. (1977) Nivation forms and deposits in Cwm Ystwyth. In *Guidebook for Excursion C9: Mid and North Wales, 10th INQUA Congress* (eds D.F. Ball and E. Watson), Geo Abstracts, Norwich, pp. 24–7.

Watters, R.J. (1972) Slope stability in the metamorphic rocks of the Scottish Highlands. Unpublished PhD thesis, Imperial College, University of London.

Werritty, A. (1997) Allt Coire Gabhail, Highland (NN 164553). In *Fluvial Geomorphology of Great Britain* (ed. K.J. Gregory), Geological Conservation Review Series, No. **13**, Chapman and Hall, London, pp. 81–3

West, R.G. (1968) *Pleistocene Geology and Biology*, Longman, London, 377 pp.

West, R.G. and Banham, P.H. (1968) Short field meeting on the north Norfolk coast. *Proceedings of the Geologists' Association*, **79**, 493–507.

Whalley, W.B. (1976) A fossil rock glacier in Wester Ross. *Scottish Journal of Geology*, **12**, 175–9.

Whittow, J.B. (1977) *Geology and Scenery in Scotland*, Penguin, Harmondsworth.

Wilson, P. (2005) Paraglacial rock-slope failures in Wasdale, western Lake District, England: morphology, styles and significance. *Proceedings of the Geologists' Association*, **116**, pp. 349–61.

Wilson, P. and Smith, A. (2006) Geomorphological characteristics and significance of Late Quaternary paraglacial rock-slope failures on Skiddaw Group terrain, Lake District, NW England. *Geograkiska Annaler*, **88A**, 237–52.

Wilson, P., Clark, R. and Smith, A. (2004) Rock-slope failures in the Lake District: a preliminary report. *Proceedings, Cumberland Geological Society*, **7**, 13–36.

Wilson, V., Welch, F.B.A., Robbie, J.A. and Green, G.W. (1958) *Geology of the Country Around Bridport and Yeovil*, Memoir of the Geological Survey of Great Britain, sheets 312 and 327 (England and Wales), HMSO, London, 239 pp.

References

Wimbledon, W.A., Benton, M.J., Bevins, R.E., Black, G.P., Bridgland, D.R., Cleal, C.J., Cooper, R.G. and May, V.J. (1995) The development of a methodology for the selection of British geological sites for conservation: Part I. *Modern Geology*, **20**, 159–202.

Wood, A. (1942) The development of hillside slopes. *Proceedings of the Geologists' Association*, **53**, 128–40.

Wood, J.N. (1949) The Sheffield Water Undertaking. *Journal of the Institution of Water*, **3**, 395–413.

Worssam, B.C. (1963) *Geology of the Country Around Maidstone*, Memoir of the Geological Survey of Great Britain, Sheet 288 (England and Wales), HMSO, London, 152 pp.

WP/WLI (1993) *The Multilingual Landslide Glossary*, The International Geotechnical Societies' UNESCO Working Party for World Landslide Inventory, Published by the Canadian Geotechnical Society.

Young, C.J. (1980) The pottery. In Excavations at the Roman Fort at Lympne, Kent 1976–78 (ed. B.W. Cunliffe). *Britannia*, **11**, pp. 274–283.

Younger, M. (1990) Will the sea always win? Coastal management in northeast Norfolk. *Geography Review*, **3**, 2–6.

Zaruba, Q. and Mencl, V. (1969) *Landslides and Their Control*, Academia, Prague, 205 pp.

Zischinsky, U. (1966) On the deformation of high slopes. In *Proceedings of the First Congress of the International Society of Rock Mechanics Congress, Lisbon, 25th September–1st October 1966*, Laboratorio Nacional de Engenharia Civil, Lisbon, Vol. **2**, pp. 179–85.

Index

Note: Page numbers in **bold** and *italic* type refer to **tables** and *figures* respectively

Index

Index

Index

Index

Index

Index